Multiple-Criteria Decision-Making (MCDM) Techniques for Business Processes Information Management

Multiple-Criteria Decision-Making (MCDM) Techniques for Business Processes Information Management

Special Issue Editors

Edmundas Kazimieras Zavadskas
Jurgita Antucheviciene
Prasenjit Chatterjee

MDPI • Basel • Beijing • Wuhan • Barcelona • Belgrade

Special Issue Editors
Edmundas Kazimieras Zavadskas
Vilnius Gediminas Technical University
Lithuania

Jurgita Antucheviciene
Vilnius Gediminas Technical University
Lithuania

Prasenjit Chatterjee
MCKV Institute of Engineering
India

Editorial Office
MDPI
St. Alban-Anlage 66
4052 Basel, Switzerland

This is a reprint of articles from the Special Issue published online in the open access journal *Information* (ISSN 2078-2489) from 2018 to 2019 (available at: https://www.mdpi.com/journal/information/special_issues/MCDM_Business_Processes)

For citation purposes, cite each article independently as indicated on the article page online and as indicated below:

LastName, A.A.; LastName, B.B.; LastName, C.C. Article Title. *Journal Name* **Year**, *Article Number*, Page Range.

ISBN 978-3-03897-642-4 (Pbk)
ISBN 978-3-03897-643-1 (PDF)

Contents

About the Special Issue Editors

Edmundas Kazimieras Zavadskas, Ph.D., DSc, Professor at the Department of Construction Management and Real Estate, Chief Research Fellow at the Laboratory of Operational Research, Research Institute of Sustainable Construction, Vilnius Gediminas Technical University, Lithuania. He received his PhD in Building Structures (1973), Dr Sc. (1987) in Building Technology and Management. He is a member of Lithuanian and several foreign Academies of Sciences, Doctor Honoris Causa from Poznan, Saint Petersburg and Kiev Universities, the Honorary International Chair Professor in the National Taipei University of Technology. Awarded by International Association of Grey Systems and Uncertainty Analysis (GSUA) for huge input in the Grey System field; has been elected to Honorary Fellowship of the International Association of Grey Systems and Uncertainty Analysis, which is a part of IEEE (2016); awarded by the "Neutrosophic Science International Association" for distinguished achievement in neutrosophics and has been conferred an honorary membership (2016); awarded the Thomson Reuters certificate for access to the list of the most highly cited scientists (2014). Highly Cited Researcher in the field of Cross-Field (2018), recognized for exceptional research performance demonstrated by the production of multiple highly cited papers that rank in the top 1% by citations for field and year in Web of Science. Main research interests: multi-criteria decision-making, operations research, decision support systems, multiple-criteria optimization in construction technology and management. Over 440 publications in Clarivate Analytic Web of Science, h = 53, a number of monographs in Lithuanian, English, German, and Russian. Editor-in-Chief of the journals *Technological and Economic Development of Economy* and the *Journal of Civil Engineering and Management*, Guest Editor of over ten Special Issues related to decision making in engineering and management.

Jurgita Antucheviciene, Professor at the Department of Construction Management and Real Estate at Vilnius Gediminas Technical University, Lithuania. She received her PhD in Civil Engineering in 2005. Her research interests include multiple-criteria decision-making theory and applications, sustainable development, and construction technology and management. Over 80 publications in Clarivate Analytic Web of Science, h = 23. A member of IEEE SMC, Systems Science and Engineering Technical Committee: Grey Systems and of two EURO Working Groups: Multicriteria Decision Aiding (EWG—MCDA) and Operations Research in Sustainable Development and Civil Engineering (EWG—ORSDCE). Deputy Editor-in-Chief of the *Journal of Civil Engineering and Management*, Editorial Board Member of *Applied Soft Computing* and *Sustainability* journals. Guest Editor of several Special Issues: "Decision Making Methods and Applications in Civil Engineering" (2015) and "Mathematical Models for Dealing with Risk in Engineering" (2016) in *Mathematical Problems in Engineering*, "Managing Information Uncertainty and Complexity in Decision-Making" (2017) in *Complexity*, "Civil Engineering and Symmetry" and "Solution Models based on Symmetric and Asymmetric Information" (2018) in *Symmetry*, "Sustainability in Construction Engineering" (2018) in *Sustainability*, as well as "Multiple-Criteria Decision-Making (MCDM) Techniques for Business Processes Information Management" (2018) in *Information*.

Prasenjit Chatterjee, Associate Professor of Mechanical Engineering Department at MCKV Institute of Engineering, India. He received his PhD in Engineering from the Department of Production Engineering, Jadavpur University in 2013. He has over 60 research papers in various international journals and peer-reviewed conferences. He has received numerous awards, including Best Track Paper Award, Outstanding Reviewer Award, Best Paper Award, Bright Researcher Award, Outstanding Researcher Award, and University Gold Medal, to name a few. He is the Guest Editor of Special Issues: "Algorithms for Multi-Criteria Decision-Making" (2019) in *Algorithms*, "Recent Trends in Intelligent Decision-Making Approaches for Sustainability Modeling" in the *International Journal of Decision Support System Technology* (2018), and "Multiple-Criteria Decision-Making (MCDM) Techniques for Business Processes Information Management" (2018) in *Information*. He is currently editing books on decision-making approaches and sustainability.

information

MDPI

Editorial

Multiple-Criteria Decision-Making (MCDM) Techniques for Business Processes Information Management

Edmundas Kazimieras Zavadskas [1,2], Jurgita Antucheviciene [1,*] and Prasenjit Chatterjee [3]

1 Department of Construction Management and Real Estate, Vilnius Gediminas Technical University, Sauletekio al. 11, Vilnius LT-10223, Lithuania; edmundas.zavadskas@vgtu.lt
2 Institute of Sustainable Construction, Vilnius Gediminas Technical University, Sauletekio al. 11, Vilnius LT-10223, Lithuania
3 Department of Mechanical Engineering, MCKV Institute of Engineering, Howrah-711204, India; prasenjit2007@gmail.com
* Correspondence: jurgita.antucheviciene@vgtu.lt; Tel.: +370-5-274-5233

Received: 21 December 2018; Accepted: 21 December 2018; Published: 23 December 2018

Abstract: Information management is a common paradigm in modern decision-making. A wide range of decision-making techniques have been proposed in the literature to model complex business processes. In this Special Issue, 16 selected and peer-reviewed original research articles contribute to business information management in various current real-world problems by proposing crisp or uncertain multiple-criteria decision-making (MCDM) models and techniques, mostly including multi-attribute decision-making (MADM) approaches in addition to a single paper proposing an interactive multi-objective decision-making (MODM) approach. The papers are mainly concentrated in three application areas: supplier selection and rational order allocation, the evaluation and selection of goods or facilities, and personnel selection/partner selection. A number of new approaches are proposed that are expected to attract great interest from the research community.

Keywords: multiple-criteria decision-making (MCDM); multi-attribute decision-making (MADM); fuzzy sets; neutrosophic sets; rough sets; aggregation operators; adaptive neuro-fuzzy inference system (ANFIS)

1. Introduction

Complex information management is an important part of activity in modern decision-making. Today's real-world problems involve multiple data sets, some precise or objective and some uncertain or subjective.

A wide range of statistical and non-statistical decision-making techniques have been proposed in the literature to model complex business or engineering processes. Multiple-criteria decision-making (MCDM) methods are among the techniques that have recently been gaining extraordinary popularity and wide applications [1].

Due to lack of precise data in real-word problems, statistical methods (i.e., probability theory) are useful in modeling processes with incomplete or inaccurate data. Meanwhile, non-statistical methods (i.e., fuzzy set theory, rough set theory, possibility theory, or fuzzy neural networks) are useful for modeling complex systems with imprecise, ambiguous, or vague data. Fuzzy MCDM techniques and their applications are constantly developing [2,3], starting from type-1 fuzzy sets and further extending to complex fuzzy sets [4]. A neuro-fuzzy approach recently has emerged as a popular technique for addressing problem-solving in the business environment [5]. A new emerging tool for uncertain data processing, known as neutrosophic set, has also been successfully applied for decision-making problems [6]. In addition, rough set theory is a powerful method for dealing within formation systems that demonstrate inconsistency, and fuzzy-rough models are able to analyze inconsistent and vague data [7]. On the basis of fuzzy set theory and aggregation operator theory, numerous decision-making theories have been developed and information aggregation methods under fuzzy aggregation operator have been suggested [8].

Discussions of the relationship and combination of fuzzy and probabilistic representations of uncertainties in multiple-attribute engineering and management problems have lasted for many years and do not seem to be finished to date [9]. Therefore, hybrid MCDM models are quickly emerging as alternative methods for information modeling [10–12].

One can notice that crisp, fuzzy, or hybrid decision-making techniques are extremely widely applied for transportation, logistics, and supplier selection problems, requiring the effective management of information when evaluating alternative solutions and making optimal decisions [13–18]. The next rather frequent application of MCDM techniques is for the assessment of service quality in different industries and various types of economic activities [19,20].

Therefore, based on the above-discussed items that highlight the topicality of the issue, we invited authors to submit their original research articles and disseminate their new ideas related to MCDM models and techniques to rationalize the complex process of business information management and optimal decision-making. Reviewers and editors approved 16 papers from all the received submissions. Next, we discuss the contribution of the published papers to the aim of the Special Issue in terms of proposed decision-making approaches and application areas.

2. Contributions

This Special Issue includes 16 original research articles. The papers contribute to decision-making techniques for business processes information management by offering optimal choice benefits through a variety of methodologies and tools, mainly including novel or extended decision-making models and methods in uncertain environments.

The topics of the Special Issue gained attention in Europe and Asia. A total of 48 authors from seven countries contributed to the Issue (Figure 1).

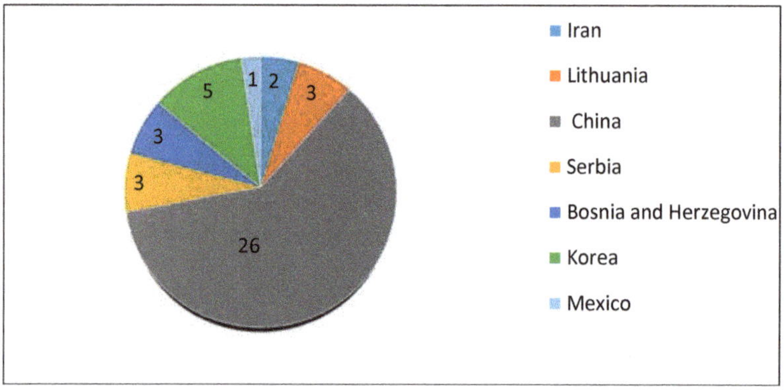

Figure 1. Distribution of authors by country.

The distribution of papers according to authors' affiliation is presented in Table 1. Authors from China contributed to 10 papers; authors and co-authors from Bosnia and Herzegovina, Serbia, and Lithuania contributed three papers. Authors from Korea prepared two papers, an author from Mexico contributed one paper, and researchers from Iran co-authored a single paper with Lithuanian researchers.

Table 1. Publications by country.

Countries	Number of Papers
China	10
Korea	2
Bosnia and Herzegovina–Serbia	1
Bosnia and Herzegovina–Serbia–Lithuania	1
Iran–Lithuania	1
Mexico	1

The papers are classified according to decision-making approaches into several groups, as presented in Figure 2. Mostly the papers propose uncertain multi-attribute decision-making (MADM) models and techniques, while a single paper proposes a multi-objective decision-making (MODM) approach. A significant part of papers are related to information aggregation operators (nine papers), including Pythagorean fuzzy and Dombioperators, probabilistic fuzzy and hesitant information aggregation (seven papers). Two papers analyze operations and aggregation methods of neutrosophic numbers, covering neutrosophic Bonferroni mean operators and linguistic neutrosophic aggregation. Two research works propose rough and fuzzy-rough MADM approaches and one paper presents a fuzzy MADM dynamic approach for the optimal choice of alternatives. Another research paper develops an adaptive neuro-fuzzy inference system (ANFIS) for the optimization of problem solutions, one paper proposes an interactive multi-objective optimization model, one paper uses an interval multiplicative preference relations (IMPRs) approach, and one paper develops a deterministic finite automata-based model.

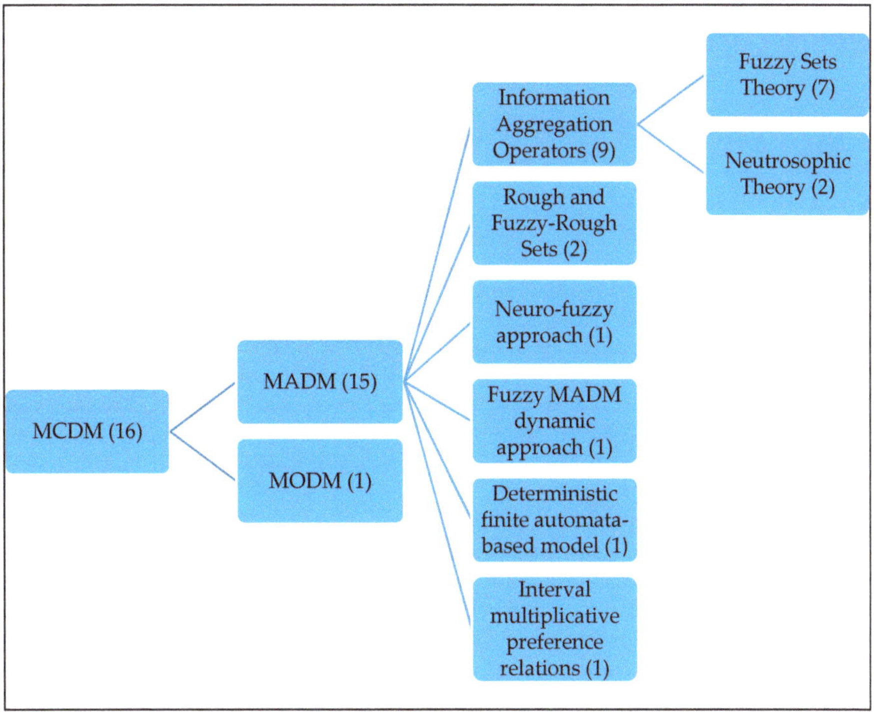

Figure 2. Decision-making approaches.

The case studies and application examples of the proposed approaches presented in the papers are mainly concentrated in three areas (Figure 3). One group of papers suggests different approaches for supplier assessment and selection, including the topical question of green supplier selection, as well as order allocation and the optimization of queuing systems in warehouses. The next group of applications is related to the evaluation of facilities or goods from the viewpoint of consumers such as brand selection, airline evaluation, or choosing a hospital, and from the view point of enterprise managers such as enterprise resource planning (ERP) system selection, technology selection, or the evaluation of investment alternatives. The last group of papers suggests multiple-criteria approaches for personnel or partner selection optimization as well as job options evaluation.

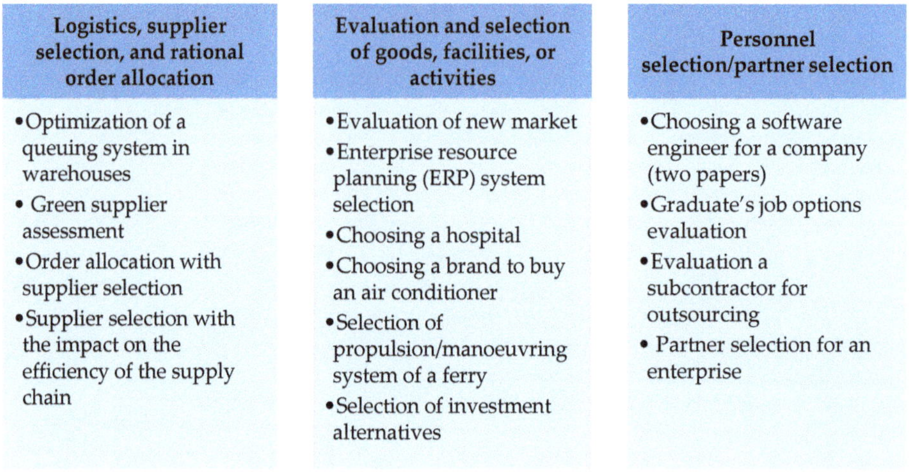
Logistics, supplier selection, and rational order allocation
•Optimization of a queuing system in warehouses
• Green supplier assessment
•Order allocation with supplier selection
•Supplier selection with the impact on the efficiency of the supply chain
Evaluation and selection of goods, facilities, or activities
•Evaluation of new market
•Enterprise resource planning (ERP) system selection
•Choosing a hospital
•Choosing a brand to buy an air conditioner
•Selection of propulsion/manoeuvring system of a ferry
•Selection of investment alternatives
Personnel selection/partner selection
•Choosing a software engineer for a company (two papers)
•Graduate's job options evaluation
•Evaluation a subcontractor for outsourcing
• Partner selection for an enterprise

Figure 3. Research areas of case studies.

The Special Issue received a paper related to subcontractor evaluation in a dynamic multi-attribute group decision-making (MAGDM) environment using a new fuzzy approach based on the EDAS (Evaluation based on Distance from Average Solution) method [21]. Another paper proposes an extended bi-directional projection method for solving MADM problems with Pythagorean uncertain linguistic variables [22]. A paper focusing on supplier selection in a polyvinyl chloride (PVC) carpentry company develops a new model using rough analytical hierarchical process (AHP) for criteria weight calculation, followed by the application of the rough weighted aggregated sum product assessment (WASPAS) method to determine the ranking preorders of the alternative suppliers [23]. One more paper related to supply chain management (SCM) offers an interactive multi-objective optimization approach for the supplier selection and order allocation problem. In this paper, the concept of desirability is also incorporated into the optimization model to take into account the principles of diminishing marginal utility [24]. For uncertainty and incompleteness in assessing green suppliers, an approach based on rough ANP and evidence theory is proposed [25]. A compound expression tool of interval-valued dual hesitant fuzzy unbalanced linguistic set (IVDHFUBLS) is proposed to help decision-makers by elucidating their assessments more comprehensively and completely. It also prioritizes weighted aggregation operator for IVDHFUBLS-based decision-making scenarios and then analyzes its properties and special cases [26]. A novel convex combination of multi-hesitant fuzzy numbers (MHFNs) is introduced along with some aggregation operators and their corresponding properties are discussed in detail [27]. Two papers focus on the analytical investigation of properties and some special cases related to the parameter vector for a Muirhead mean (MM) operator in a Pythagorean fuzzy context [28,29]. A MADM model is developed in a linguistic cubic variable (LCV) setting on the basis of Dombi-weighted geometric average (DWGA) [30]. One paper presents the concept of a single-valued linguistic neutrosophic interval linguistic number and some weighted arithmetic averaging and weighted geometric averaging operators are proposed and their properties are investigated [31]. Score function and accuracy function are proposed, satisfying the ranking principle and extending the operators under a neutrosophic environment [32]. Hesitant probabilistic fuzzy MAGDM is studied [33]. An adaptive neuro-fuzzy inference system (ANFIS) model is developed for a warehouse system with two servers to define queuing system optimization parameters [34]. A quick algorithm for a binary discernibility matrix simplification using deterministic finite automata is proposed along with definitions of row and column relations [35]. A methodology is proposed to provide reliable consistent and in consensus Interval Multiplicative Preference Relations (IMPRs) using Hadamard's operator [36].

3. Conclusions

The topics of this Special Issue piqued the interest of researchers both in Asia and in Europe; researchers from seven countries authored and co-authored papers published in the Issue.

Although the announced topics of the Special Issue covered MADM and MODM theories, mainly MADM approaches have been suggested, while a single research paper proposed a MODM model. Therefore, multi-attribute decision-making techniques proved to be highly applicable for business processes information management.

Most approaches suggested decision models under uncertainty, proposing extensions of decision-making methods in combination with fuzzy, rough, and neutrosophic sets theory. Particular attention was devoted to information aggregation operators; 65 percent of papers dealt with the item.

The application areas of proposed MCDM techniques mainly covered logistics and supply chains optimization, the selection of goods or facilities, and personnel selection.

Author Contributions: All authors contributed equally to this work.

Acknowledgments: The authors express their gratitude to the journal *Information* for offering an academic platform for researchers to contribute and exchange their recent findings in a sustainable environment.

Conflicts of Interest: The authors declare no conflict of interest.

References

1. Mardani, A.; Jusoh, A.; Nor, K.M.D.; Khalifah, Z.; Zakwan, N.; Valipour, A. Multiple criteria decision-making techniques and their applications- A review of the literature from 2000 to 2014. *Econ. Res.-Ekon. Istraz.* **2015**, *28*, 516–571. [CrossRef]
2. Mardani, A.; Jusoh, A.; Zavadskas, E.K. Fuzzy multiple criteria decision-making techniques and applications—Two decades review from 1994 to 2014. *Expert Syst. Appl.* **2015**, *42*, 4126–4148. [CrossRef]
3. Kahraman, C.; Onar, S.C.; Oztaysi, B. Fuzzy Multicriteria Decision-Making: A Literature Review. *Int. J. Comput. Intell. Syst.* **2015**, *8*, 637–666. [CrossRef]
4. Yazdanbakhsh, O.; Dick, S. A systematic review of complex fuzzy sets and logic. *Fuzzy Sets Syst.* **2018**, *338*, 1–22. [CrossRef]
5. Rajab, S.; Sharma, V. A review on the applications of neuro-fuzzy systems in business. *Artif. Intell. Rev.* **2018**, *49*, 481–510. [CrossRef]
6. Khan, M.; Son, L.H.; Ali, M.; Chau, H.T.M.; Na, N.T.N.; Smarandache, F. Systematic review of decision making algorithms in extended neutrosophic sets. *Symmetry* **2018**, *10*, 314. [CrossRef]
7. Mardani, A.; Nilashi, M.; Antucheviciene, J.; Tavana, M.; Bausys, R.; Ibrahim, O. Recent fuzzy generalisations of rough sets theory: A systematic review and methodological critique of the literature. *Complexity* **2017**, *2017*, 1608147. [CrossRef]
8. Mardani, A.; Nilashi, M.; Zavadskas, E.K.; Awang, S.R.; Zare, H.; Jamal, N.M. Decision making methods based on fuzzy aggregation operators: Three decades review from 1986 to 2017. *Int. J. Inf. Technol. Decis. Mak.* **2018**, *17*, 391–466. [CrossRef]

9. Antucheviciene, J.; Kala, Z.; Marzouk, M.; Vaidogas, E.R. Solving Civil Engineering Problems by Means of Fuzzy and Stochastic MCDM Methods: Current State and Future Research. *Math. Probl. Eng.* **2015**, *2015*, 362579. [CrossRef]
10. Zavadskas, E.K.; Antucheviciene, J.; Turskis, Z.; Adeli, H. Hybrid multiple-criteria decision-making methods: A review of applications in engineering. *Sci. Iran.* **2016**, *23*, 1–20.
11. Zavadskas, E.K.; Govindan, K.; Antucheviciene, J.; Turskis, Z. Hybrid multiple criteria decision-making methods: A review of applications for sustainability issues. *Econ. Res.-Ekon. Istraz.* **2016**, *29*, 857–887. [CrossRef]
12. Shen, K.Y.; Zavadskas, E.K.; Tzeng, G.H. Updated discussions on "Hybrid multiple criteria decision-making methods: A review of applications for sustainability issues". *Econ. Res.-Ekon. Istraz.* **2018**, *31*, 1437–1452. [CrossRef]
13. Mardani, A.; Zavadskas, E.K.; Khalifah, Z.; Jusoh, A.; Nor, K. Multiple criteria decision-making techniques in transportation systems: A systematic review of the state of the art literature. *Transport* **2016**, *31*, 359–385. [CrossRef]
14. Keshavarz Ghorabaee, M.; Amiri, M.; Zavadskas, E.K.; Antucheviciene, J. Supplier evaluation and selection in fuzzy environments: A review of MADM approaches. *Econ. Res.-Ekon. Istraz.* **2017**, *30*, 1073–1118. [CrossRef]
15. Govindan, K.; Hasanagic, M. A systematic review on drivers, barriers, and practices towards circular economy: a supply chain perspective. *Int. J. Prod. Res.* **2018**, *56*, 278–311. [CrossRef]
16. Govindan, K.; Soleimani, H. A review of reverse logistics and closed-loop supply chains: A Journal of Cleaner Production focus. *J. Clean. Prod.* **2017**, *142*, 371–384. [CrossRef]
17. Correia, E.; Carvalho, H.; Azevedo, S.G.; Govindan, K. Maturity models in supply chain sustainability: A systematic literature review. *Sustainability* **2017**, *9*, 64. [CrossRef]
18. Govindan, K.; Rajendran, S.; Sarkis, J.; Murugesan, P. Multi criteria decision making approaches for green supplier evaluation and selection: A literature review. *J. Clean. Prod.* **2015**, *98*, 66–83. [CrossRef]
19. Mardani, A.; Jusoh, A.; Zavadskas, E.K.; Kazemilari, M.; Ahmad, U.N.U.; Khalifah, Z. Application of multiple criteria decision making techniques in tourism and hospitality industry: A systematic review. *Transform. Bus. Econ.* **2016**, *15*, 192–213.
20. Mardani, A.; Jusoh, A.; Zavadskas, E.K.; Khalifah, Z.; Nor, K.M. Application of multiple-criteria decision-making techniques and approaches to evaluating of service quality: A systematic review of the literature. *J. Bus. Econ. Manag.* **2015**, *16*, 1034–1068. [CrossRef]
21. Keshavarz-Ghorabaee, M.; Amiri, M.; Zavadskas, E.K.; Turskis, Z.; Antucheviciene, J. A dynamic fuzzy approach based on the EDAS method for multi-criteria subcontractor evaluation. *Information* **2018**, *9*, 68. [CrossRef]
22. Wang, H.; He, S.; Pan, X. A new bi-directional projection model based on pythagorean uncertain linguistic variable. *Information* **2018**, *9*, 104. [CrossRef]
23. Stojić, G.; Stević, Ž.; Antuchevičienė, J.; Pamučar, D.; Vasiljević, M. A novel rough WASPAS approach for supplier selection in a company manufacturing PVC carpentry products. *Information* **2018**, *9*, 121. [CrossRef]
24. Lee, P.; Kang, S. An interactive multiobjective optimization approach to supplier selection and order allocation problems using the concept of desirability. *Information* **2018**, *9*, 130. [CrossRef]
25. Li, L.; Wang, H. A green supplier assessment method for manufacturing enterprises based on rough ANP and evidence theory. *Information* **2018**, *9*, 162. [CrossRef]
26. Qi, X.-W.; Zhang, J.-L.; Liang, C.-Y. Multiple attributes group decision making under interval-valued dual hesitant fuzzy unbalanced linguistic environment with prioritized attributes and unknown decision makers' weights. *Information* **2018**, *9*, 145. [CrossRef]
27. Mei, Y.; Peng, J.; Yang, J. Convex aggregation operators and their applications to multi-hesitant fuzzy multi-criteria decision making. *Information* **2018**, *9*, 207. [CrossRef]
28. Xu, Y.; Shang, X.; Wang, J. Pythagorean fuzzy interaction Muirhead means with their application to multi-attribute group decision making. *Information* **2018**, *9*, 157. [CrossRef]
29. Zhu, J.; Li, Y. Pythagorean fuzzy Muirhead mean operators and their application in multiple-criteria group decision-making. *Information* **2018**, *9*, 142. [CrossRef]
30. Lu, X.; Ye, J. Dombi Aggregation operators of linguistic cubic variables for multiple attribute decision making. *Information* **2018**, *9*, 188. [CrossRef]

31. Ye, J.; Cui, W. Operations and aggregation methods of single-valued linguistic neutrosophic interval linguistic numbers and their decision making method. *Information* **2018**, *9*, 196. [CrossRef]
32. Mo, J.; Huang, H.-L. Dual generalized nonnegative normal Neutrosophic Bonferroni mean operators and their application in multiple attribute decision making. *Information* **2018**, *9*, 201. [CrossRef]
33. Park, J.H.; Park, Y.K.; Son, M.J. Hesitant Probabilistic Fuzzy Information Aggregation Using Einstein Operations. *Information* **2018**, *9*, 226. [CrossRef]
34. Stojčić, M.; Pamučar, D.; Mahmutagić, E.; Stević, Ž. Development of an ANFIS model for the optimization of a queuing system in the warehouse. *Information* **2018**, *9*, 240. [CrossRef]
35. Zhang, N.; Li, B.; Zhang, Z.; Guo, Y. A quick algorithm for binary discernibility matrix simplification using deterministic finite automata. *Information* **2018**, *9*, 314. [CrossRef]
36. López-Morales, V. Multiplecriteria decision-making in heterogeneous groups of management experts. *Information* **2018**, *9*, 300. [CrossRef]

information

MDPI

Article

A Dynamic Fuzzy Approach Based on the EDAS Method for Multi-Criteria Subcontractor Evaluation

Mehdi Keshavarz-Ghorabaee [1], Maghsoud Amiri [1], Edmundas Kazimieras Zavadskas [2,*], Zenonas Turskis [2] and Jurgita Antucheviciene [2]

1 Department of Industrial Management, Faculty of Management and Accounting, Allameh Tabataba'i University, Tehran 1489684511, Iran; m.keshavarz_gh@ieee.org (M.K.-G.); amiri@atu.ac.ir (M.A.)

2 Department of Construction Management and Real Estate, Vilnius Gediminas Technical University, Sauletekio al. 11, LT-10223 Vilnius, Lithuania; zenonas.turskis@vgtu.lt (Z.T.); jurgita.antucheviciene@vgtu.lt (J.A.)

* Correspondence: edmundas.zavadskas@vgtu.lt; Tel.: +370-5-274-4910

Received: 2 March 2018; Accepted: 17 March 2018; Published: 19 March 2018

Abstract: Selection of appropriate subcontractors for outsourcing is very important for the success of construction projects. This can improve the overall quality of projects and promote the qualification and reputation of the main contractors. The evaluation of subcontractors can be made by some experts or decision-makers with respect to some criteria. If this process is done in different time periods, it can be defined as a dynamic multi-criteria group decision-making (MCGDM) problem. In this study, we propose a new fuzzy dynamic MCGDM approach based on the EDAS (Evaluation based on Distance from Average Solution) method for subcontractor evaluation. In the procedure of the proposed approach, the sets of alternatives, criteria and decision-makers can be changed at different time periods. Also, the proposed approach gives more weight to newer decision information for aggregating the overall performance of alternatives. A numerical example is used to illustrate the proposed approach and show the application of it in subcontractor evaluation. The results demonstrate that the proposed approach is efficient and useful in real-world decision-making problems.

Keywords: multi-criteria decision-making; group decision-making; subcontractor evaluation; MCDM; MADM; fuzzy sets; fuzzy EDAS

---

1. Introduction

Subcontracting is one of the most important characteristics of the construction industry. In many construction projects, the main contractor has usually the role of project coordinator, and a high percentage of work is done by subcontractors [1,2]. The completion time of a construction project (project delivery) and the reputation of the main contractor are heavily dependent on cooperation between a subcontractor and its main contractor [3]. Therefore, the performance of subcontractors could have a significant effect on the success of construction projects. Because of the increasing use of subcontracting in the construction industry, evaluation of subcontractors can be considered as an essential problem for the main contractors.

The subcontractor evaluation process (SEP) usually involves several alternatives (subcontractors), multiple criteria and a group of decision-makers (experts). Thus, we can consider this process as a multi-criteria group decision-making (MCGDM) problem [4]. Moreover, the main contractor generally needs to evaluate its subcontractors in multiple periods of time. This process makes the SEP into a dynamic MCGDM problem. In a dynamic MCGDM problem, the set of alternatives, criteria and decision-makers can be changed in different time periods [5]. Thus, we can make the evaluation process with a high degree of flexibility. In addition, the assessments of experts can be made under uncertainty in the SEP. The fuzzy sets theory is a useful tool to deal with the uncertainty of evaluation process [6–11].

There have been some studies on the problems related to the SEP and multi-criteria decision-making (MCDM) methods under certain and uncertain environments. Cheng, et al. [12] proposed a hierarchical structure for the target and factors for evaluation of subcontractors, and used the analytic hierarchy process (AHP) to select an appropriate subcontractor. Kargi and Öztürk [13] used the AHP method and the Expert Choice software for evaluation of subcontractors in a Turkish company. Yayla, et al. [14] presented a case study for selection of the optimal subcontractor in a Turkish textile firm. They used generalized Choquet integral methodology and a hierarchical decision model to solve the selection problem. Ng and Skitmore [15] proposed an approach based on the balanced scorecard methodology for evaluation of subcontractor and performed a questionnaire survey administered in Hong Kong. Abbasianjahromi, et al. [16] developed a model for subcontractor evaluation based on the fuzzy preference selection index. In their model, the weighting criteria phase is eliminated in the evaluation process of subcontractors. Shahvand, et al. [17] developed a multi-criteria fuzzy expert system for supplier and subcontractor evaluation in the construction industry and used it in three companies. Polat [2] presented an integrated MCDM approach based on AHP and preference ranking organization method for enrichment evaluations (PROMETHEE), and applied it to the subcontractor selection problem. Ulubeyli and Kazaz [18] proposed a fuzzy multi-criteria decision-making approach, called CoSMo (Construction Subcontractor selection Model), for evaluation of subcontractors in the construction projects. Abbasianjahromi, et al. [19] developed a new model to allocate the tasks of a construction project to some subcontractors for optimization of the portfolio of subcontractors and main contractor. Polat, et al. [20] proposed an integrated approach based on the AHP and Evidential Reasoning (ER) methods. They used AHP and ER to find the criteria weights for evaluation of subcontractors and rank the alternatives, respectively.

Dynamic MCDM approaches have been used by researchers in several fields. Campanella and Ribeiro [21] introduced a flexible framework for dynamic MCDM that can be used in many dynamic decision processes, and applied it to a small helicopter landing problem. Wei [22] utilized grey relational analysis (GRA) to develop a dynamic MCDM approach. Chen and Li [23] proposed a dynamic MCDM method based on triangular intuitionistic fuzzy numbers. Wang, et al. [24] presented a three-dimensional grey interval relational degree approach for dynamic multi-criteria decision-making problems. They applied the presented approach to the investment decision-making problems. Junhua, et al. [25] developed a dynamic stochastic MCDM approach based on conjoint analysis and prospect theory. Li, et al. [26] proposed a dynamic fuzzy MCDM method using a mathematical programming model and fuzzy technique for order preference by similarity to ideal solution (TOPSIS). Yan, et al. [27] presented a dynamic grey target MCDM method using interval numbers and based on the status of alternatives. Liu, et al. [28] proposed a dynamic fuzzy framework based on GRA and used it for evaluation of emergency treatment technology. Yan, et al. [29] developed a new dynamic MCDM approach with three-parameter grey numbers. In their approach, not only the attribute values of alternatives at all periods are aggregated, but also changes of these values between the adjacent periods are considered.

The EDAS (Evaluation based on Distance from Average Solution) method is a new and efficient method which introduced by Keshavarz Ghorabaee, et al. [30] and extended for using in the fuzzy environment [31]. The evaluation process in the EDAS method is made based on the distances of alternative from an average solution. Two types of distances (positive and negative) are defined for alternatives in this method, and the utility of alternatives is determined based on these distances. This method has been developed for using in different uncertain environments such as intuitionistic fuzzy sets [32], interval-valued neutrosophic sets [33], interval-valued fuzzy soft sets [34], neutrosophic soft sets [35], interval grey numbers [36] and interval type-2 fuzzy sets [37]. Also, the EDAS method has been applied to some real-world MCDM problems such as life cycle and sustainability assessment [38], supplier selection [39], architectural shape of the buildings [40], cultural heritage structures [41], quality assurance [42], evaluation in logistics [43,44] and stairs shape assessment [45].

In this study, we propose a new dynamic fuzzy MCGDM approach based on the EDAS method for evaluation of subcontractors. The main advantage of the proposed approach is its flexibility so that we can define different sets of alternatives, criteria and decision-makers in different time periods and make the evaluation in a fuzzy environment. Because of the importance of new information, we use a function that gives greater weights to newer time periods for aggregating the performance score of each alternative. A numerical example of subcontractor evaluation is presented to illustrate the proposed approach and show the efficiency of it.

The rest of this article is organized as follows. Section 2 describes the methodology. In this section, first, we present concepts and some definitions related to the fuzzy sets theory and the arithmetic operations of the fuzzy numbers, then the steps and flowchart of the proposed approach is depicted in detail. In Section 3, a numerical example is used to show the application of the proposed approach in subcontractor evaluation. Conclusions are briefly discussed in Section 4.

2. Methodology

In this section, we first present some concepts and definitions about the fuzzy sets theory, and then an extended dynamic fuzzy EDAS is described for multi-criteria group decision-making.

2.1. Concepts and Definitions of Fuzzy Sets

To deal with the uncertainty of information in real-world problems, the fuzzy sets theory was developed by Zadeh [46]. The membership of elements in a fuzzy set is described by means of a membership function with a range in [0, 1]. Therefore, fuzzy sets generalize classical sets in which the membership of elements has a two-valued condition (zero or one). The fuzzy set theory has been applied to many problems in different fields of science and engineering. To describe this theory, some definitions are presented as follows:

Definition 1. *Let denote by X a universal set. Then a fuzzy set $\widetilde{G}$ can be defined by a membership function $\mu_{\widetilde{G}}(x)$ as follows [47]:*

$$\widetilde{G} = \left\{ \left(x,\ \mu_{\widetilde{G}}(x)\right) \middle| x \in X \right\} \tag{1}$$

In the above equation, x denotes the elements belong to X, and $\mu_{\widetilde{G}}(x) : X \to [0,\ 1]$.

Definition 2. *A fuzzy number can be defined as a special case of a fuzzy set which is convex and normal [48].*

Definition 3. *If the membership function of a fuzzy number $\widetilde{G}$ is defined by the following equation then we can call it a triangular fuzzy number [49]:*

$$\mu_{\widetilde{G}}(x) = \begin{cases} (x - g_1)/(g_2 - g_1), & g_1 \le x \le g_2 \\ (g_3 - x)/(g_3 - g_2), & g_2 \le x \le g_3 \\ 0, & otherwise \end{cases} \tag{2}$$

A triplet $\widetilde{G} = (g_1, g_2, g_3)$ can also be used to define this fuzzy number. Figure 1 represents an example of triangular fuzzy numbers.

In this study, we use the triangular fuzzy sets due to their simplicity of presentation and computation. However, the other types of fuzzy numbers such as trapezoidal fuzzy number can also be used in the methodology proposed in the following sub-section.

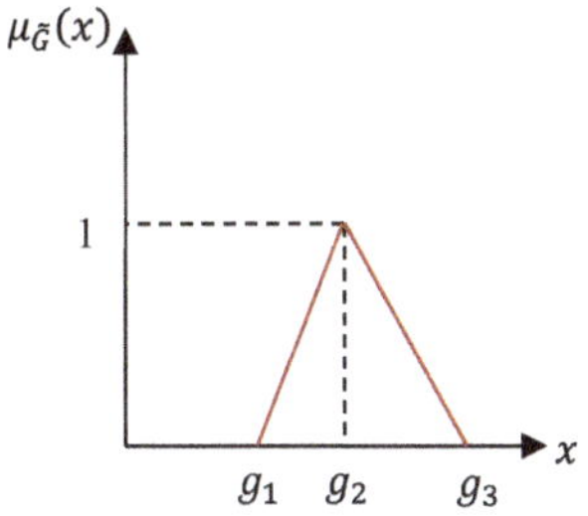

Figure 1. A triangular fuzzy number.

Definition 4. *Let us define $\widetilde{G} = (g_1, g_2, g_3)$ and $\widetilde{H} = (h_1, h_2, h_3)$ as two triangular fuzzy numbers which are also positive (i.e., $g_1 \geq 0$ and $h_1 \geq 0$), and suppose that q is a crisp number. In the following equations, the arithmetic operations of these fuzzy numbers are presented [49]:*

- Addition:

$$\widetilde{G} \oplus \widetilde{H} = (g_1 + h_1, g_2 + h_2, g_3 + h_3) \tag{3}$$

$$\widetilde{G} + q = (g_1 + q, g_2 + q, g_3 + q) \tag{4}$$

- Subtraction:

$$\widetilde{G} \ominus \widetilde{H} = (g_1 - h_3, g_2 - h_2, g_3 - h_1) \tag{5}$$

$$\widetilde{G} - q = (g_1 - q, g_2 - q, g_3 - q) \tag{6}$$

- Multiplication:

$$\widetilde{G} \otimes \widetilde{H} = (g_1 \times h_1, g_2 \times h_2, g_3 \times h_3) \tag{7}$$

$$\widetilde{G} \times q = \begin{cases} (g_1 \times q, g_2 \times q, g_3 \times q) & \text{if } q \geq 0 \\ (g_3 \times q, g_2 \times q, g_1 \times q) & \text{if } q < 0 \end{cases} \tag{8}$$

- Division:

$$\widetilde{G} \oslash \widetilde{H} = (g_1/h_3, g_2/h_2, g_3/h_1) \tag{9}$$

$$\widetilde{G}/q = \begin{cases} (g_1/q, g_2/q, g_3/q) & \text{if } q > 0 \\ (g_3/q, g_2/q, g_1/q) & \text{if } q < 0 \end{cases} \tag{10}$$

Definition 5. *The defuzzified or crisp value of a triangular fuzzy number $\widetilde{G} = (g_1, g_2, g_3)$ can be defined by the following equation [50]:*

$$\mathfrak{D}\left(\widetilde{G}\right) = \frac{1}{3}(g_1 + g_2 + g_3) \tag{11}$$

Definition 6. *To find the maximum between a triangular fuzzy number $\widetilde{G} = (g_1, g_2, g_3)$ and zero, the following function can be used [31].*

$$\mathcal{S}\left(\widetilde{A}\right) = \begin{cases} \widetilde{G} \;\; if \;\; \mathfrak{D}\left(\widetilde{G}\right) > 0 \\ \widetilde{0} \;\; if \;\; \mathfrak{D}\left(\widetilde{G}\right) \leq 0 \end{cases} \tag{12}$$

where $\widetilde{0} = (0, 0, 0)$.

2.2. Dynamic Fuzzy EDAS

The EDAS method is a new and efficient MCDM method introduced by Keshavarz Ghorabaee, Zavadskas, Olfat and Turskis [30], and has been extended to deal with fuzzy MCDM problems [31]. In this section, a new approach is proposed to handle dynamic fuzzy multi-criteria group decision-making based on the EDAS method, which is called dynamic fuzzy EDAS.

In a dynamic multi-criteria group decision-making, the multi-criteria evaluation process is made by multiple decision-makers in multiple periods. In each period, we have a set of alternatives that needs to be evaluated with respect to a set of criteria. Suppose that there are T periods and DM_t, CR_t and AL_t denote the sets of decision-makers, criteria, and alternatives at period t, respectively. The cardinality of these sets can be defined as $|DM_t| = k_t$, $|CR_t| = m_t$ and $|AL_t| = n_t$. In other words, we have k_t decision-makers, m_t criteria and n_t alternatives at period t.

Step 1: Start with the first period (t = 1).

Step 2: Define the sets of decision-makers, criteria, and alternatives (DM_t, CR_t and AL_t) at period t.

Step 3: Determine the union of the sets of alternatives at period t denoted by AL_t^T, where $AL_t^T = AL_{t-1}^T \cup AL_t$ and $AL_0^T = \varnothing$.

Step 4: Construct the decision-matrix and the matrix of criteria weights related to each decision-maker at period t as follows:

$$X_{pt} = \left[\widetilde{x}_{ijpt}\right]_{n_t \times m_t} \tag{13}$$

$$W_{pt} = \left[\widetilde{w}_{jpt}\right]_{1 \times m_t} \tag{14}$$

where $\widetilde{x}_{ijpt}$ denotes the rating of ith alternative (A_i) on jth criterion (C_j) given by pth decision-maker, and $\widetilde{w}_{jpt}$ shows the importance or weight of jth criterion given by pth decision-maker ($1 \leq i \leq n_t$, $1 \leq j \leq m_t$ and $1 \leq p \leq k_t$).

Step 5: Determine the average decision-matrix at period t using the following equations:

$$X_t = \left[\widetilde{x}_{ijt}\right]_{n_t \times m_t} \tag{15}$$

$$\widetilde{x}_{ijt} = \frac{1}{k_t} \bigoplus_{p=1}^{k_t} \widetilde{x}_{ijpt} \tag{16}$$

where $\widetilde{x}_{ijt}$ shows the average ratings at period t. If the decision-makers or experts, depending on their experience and knowledge, have different importance in the process of decision-making, we can use a weighted average instead of ordinary average of Equation (16).

Step 6: Compute the average matrix of criteria weights at period t presented as follows:

$$W_t = \left[\widetilde{w}_{jt}\right]_{1 \times m_t} \tag{17}$$

$$\widetilde{w}_{jt} = \frac{1}{k_t} \bigoplus_{p=1}^{k_t} \widetilde{w}_{jpt} \tag{18}$$

where $\widetilde{w}_{jt}$ denotes the average weights of criteria at period t. Like the previous step, we can also use a weighted average instead of ordinary average of Equation (18) if there are different weights for decision-makers.

It should be noted that if we have a problem with a hierarchical structure including some criteria and sub-criteria, we should calculate the average weights of criteria and sub-criteria first. Then the global weights of sub-criteria should be determined by multiplying the average calculated weights of them by the average weights of their upper level criterion.

Step 7: Calculate average solutions at period t using the following formula:

$$\tilde{g}_{jt} = \frac{1}{n_t} \bigoplus_{i=1}^{n_t} \tilde{x}_{ijt} \tag{19}$$

Step 8: Let denote by BC_t and NC_t the sets of beneficial and non-beneficial criteria at period t, respectively. The values of positive and negative distances from the average solutions at each period are calculated as follows:

$$\widetilde{pd}_{ijt} = \begin{cases} \frac{S(\tilde{x}_{ijt} \ominus \tilde{g}_{jt})}{\mathfrak{D}(\tilde{g}_{jt})} & if \;\; j \in BC_t \\ \frac{S(\tilde{g}_{jt} \ominus \tilde{x}_{ijt})}{\mathfrak{D}(\tilde{g}_{jt})} & if \;\; j \in NC_t \end{cases} \tag{20}$$

$$\widetilde{nd}_{ijt} = \begin{cases} \frac{S(\tilde{g}_{jt} \ominus \tilde{x}_{ijt})}{\mathfrak{D}(\tilde{g}_{jt})} & if \;\; j \in BC_t \\ \frac{S(\tilde{x}_{ijt} \ominus \tilde{g}_{jt})}{\mathfrak{D}(\tilde{g}_{jt})} & if \;\; j \in NC_t \end{cases} \tag{21}$$

where $\widetilde{pd}_{ijt}$ and $\widetilde{nd}_{ijt}$ denote the values of positive and negative distances from the average solutions at period t, respectively.

Step 9: Compute the weighted sum of the positive and negative distances for each alternative at period t using the following equations:

$$\tilde{sp}_{it} = \bigoplus_{j=1}^{m_t} \left(\tilde{w}_{jt} \otimes \widetilde{pd}_{ijt} \right) \tag{22}$$

$$\tilde{sn}_{it} = \bigoplus_{j=1}^{m_t} \left(\tilde{w}_{jt} \otimes \widetilde{nd}_{ijt} \right) \tag{23}$$

Step 10: Calculate the normalized values of $\tilde{sp}_{it}$ and $\tilde{sn}_{it}$ as follows:

$$\widetilde{np}_{it} = \frac{\tilde{sp}_{it}}{\max_l (\mathfrak{D}(\tilde{sp}_{lt}))} \tag{24}$$

$$\widetilde{nn}_{it} = 1 - \frac{\tilde{sn}_{it}}{\max_l (\mathfrak{D}(\tilde{sn}_{lt}))} \tag{25}$$

Step 11: Compute the overall performance score of ith alternative at period t ($\tilde{U}_{it}$) by the following formula:

$$\tilde{U}_{it} = \frac{1}{2}(\widetilde{np}_{it} \oplus \widetilde{nn}_{it}) \tag{26}$$

Step 12: Calculate the dynamic scores (S_{it}) for all alternatives which are the elements of the set AL_t^T ($A_i \in AL_t^T$) by the following equation:

$$S_{it} = \begin{cases} \mathfrak{D}\left(\tilde{U}_{it}\right) & if \;\; A_i \in AL_t \\ 0 & if \;\; A_i \notin AL_t \end{cases} \tag{27}$$

Step 13: Let ρ_t denotes the weight or importance of period t. Compute the aggregated dynamic scores (H_{it}) of the alternatives belong to the set AL_t^T as follows:

$$H_{it} = (1 - \rho_t) H_{i(t-1)} + \rho_t S_{it} \tag{28}$$

where $H_{i0} = 0$, and if $A_i \notin AL_{t-1}^T$ then $H_{i(t-1)} = \min_{l, l \in AL_{t-1}^T} H_{l(t-1)}$.

Because newer information is more important in decision-making, a weight function of periods that gives greater weight to the current period should be defined. We define the following function for setting the weights of periods:

$$\rho_t = \frac{t}{2t-1} \tag{29}$$

In Equation (29), the value of ρ_t is equal to 1 for the first period ($t = 1$), and it is always greater than 0.5.

Step 14: Increase the value of period by 1 ($t \leftarrow t+1$). If $t < T$ go to Step 2, otherwise continue.

Step 15: Evaluate the alternatives according to the values of aggregated dynamic scores (H_{it}). The higher values of H_{it} get the better alternatives.

To make the proposed approach clear, its procedure is depicted by a flowchart in Figure 2.

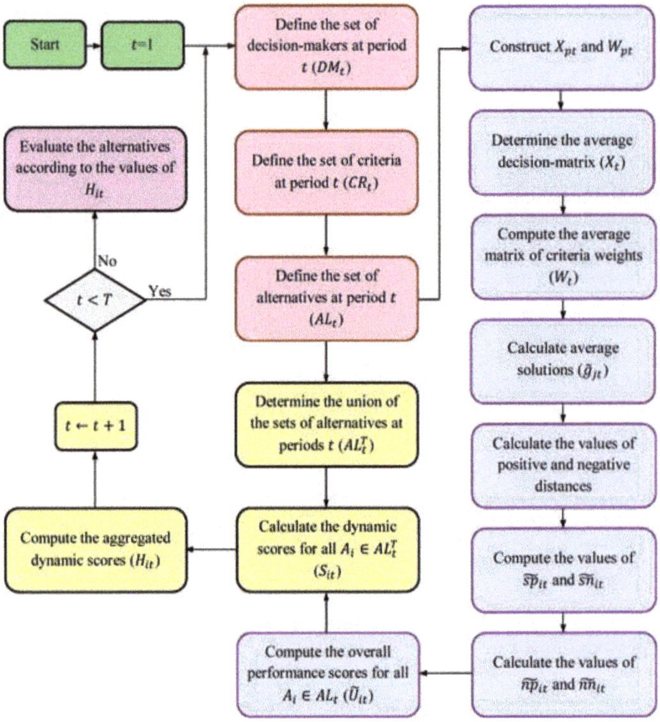

Figure 2. The flowchart of the proposed approach.

3. Illustrative Example (Subcontractor Evaluation)

In this section, the proposed approach is applied to a dynamic multi-criteria subcontractor evaluation problem in a construction project. The evaluation process is made by the main contractor of the project in four periods. According to the procedure of the proposed approach, we can define any number of decision-makers, criteria, and alternatives at each period. In this problem, four criteria are defined for evaluation of subcontractors based on the study of Lin, et al. [51]. These criteria are defined as follows:

- ***Reliability*** (C_1): This criterion is related to evaluation of subcontractors with respect to their records, reputation, and financial condition. It is clear that a subcontractor with good reputation and better financial condition is more favorable.
- ***Schedule-control ability*** (C_2): This criterion is related to the mobilization and efficiency of subcontractors. Activation of the subcontractor's physical and manpower resources for transfer to a construction site until the completion of the contract can be measured by this criterion.
- ***Management ability*** (C_3): The level of safety, quality and environmental management of subcontractors is very important in the overall performance of a subcontractor. This criterion can be used to assess these dimensions of subcontractors.
- ***Labor quality*** (C_4): This criterion can be used for assessment of the level of workers' skill and the coordination of managers and workers. The quality of the outcomes of a construction project is significantly affected by this criterion.

The criteria defined are used in all the periods. In other words, we can define the set of criteria as $CR_t = \{C_1, C_2, C_3, C_4\}$ where $t \in \{1, 2, 3, 4\}$. The evaluation process is made based on the assessments of some experts of the main contractor which are considered as decision-makers. In each period, some of the decision-makers may be available and some may be not available for the assessment. In this problem, the sets of decision-makers at each period are as follows:

$$DM_1 = \{D_1,\ D_2, D_3, D_4\},$$
$$DM_2 = \{D_1, D_3, D_4\},$$
$$DM_3 = \{D_1, D_2, D_3, D_4\},$$
$$DM_4 = \{D_1, D_2, D_3\}.$$

The number of subcontractors also varies from period to period. Here, we have four sets of alternatives (subcontractors):

$$AL_1 = \{A_1, A_2, A_3, A_4, A_5, A_6, A_7, A_8, A_9\},$$
$$AL_2 = \{A_1, A_2, A_3, A_4, A_6, A_7, A_{10}, A_{11}, A_{12}, A_{13}\},$$
$$AL_3 = \{A_1, A_2, A_5, A_6, A_7, A_8, A_9, A_{10}, A_{12}\},$$
$$AL_4 = \{A_2, A_3, A_6, A_7, A_8, A_{10}, A_{13}\}.$$

The decision-makers give the importance of criteria and rating of alternatives at each period using linguistic variables. The linguistic variables and their fuzzy equivalents are presented in Table 1 [52]. Because we use a spectrum from "Very poor" to "Very good" for rating of alternatives, all the criteria in the problem should be considered as beneficial criteria. Based on the linguistic variables defined in Table 1, the decision-matrix and the matrix of criteria weights related to each decision-maker can be constructed at each period. The decision-matrices of different periods are presented in Tables 2–6 presents the matrices of criteria weights in different periods.

Based on the steps of the proposed approach and Tables 1–6, we can determine the overall performance scores of alternatives at each period. According to the defuzzified values of overall performance scores, the rank of each alternative at each period can be obtained. The results of each period are shown in Table 7. Also, in this table, we present the ranking results which are obtained by using defuzzified decision-matrices and criteria weights and the TOPSIS method [53]. In addition, to show the validity of the ranking result of each period, the Spearman's rank correlation coefficients (r_s) between the results of the fuzzy EDAS and TOPSIS methods are calculated. As can be seen in Table 7, all the correlation values are greater than 0.9, and we can say that there is a strong relationship between the results in all the periods.

Table 1. The linguistic variables and their fuzzy equivalents.

	Linguistic Variables	Triangular Fuzzy Number
Importance of criteria	Very low (VL)	(0, 0, 0.1)
	Low (L)	(0, 0.1, 0.3)
	Medium low (ML)	(0.1, 0.3, 0.5)
	Medium (M)	(0.3, 0.5, 0.7)
	Medium high (MH)	(0.5, 0.7, 0.9)
	High (H)	(0.7, 0.9, 1)
	Very high (VH)	(0.9, 1,1)
Rating of alternatives	Very poor (VP)	(0, 0, 1)
	Poor (P)	(0, 1, 3)
	Medium poor (MP)	(1, 3, 5)
	Fair (F)	(3, 5, 7)
	Medium good (MG)	(5, 7, 9)
	Good (G)	(7, 9, 10)
	Very good (VG)	(9, 10, 10)

Table 2. The decision-matrix of each decision-maker at first period (t = 1).

	D_1				D_2				D_3				D_4			
	C_1	C_2	C_3	C_4	C_1	C_2	C_3	C_4	C_1	C_2	C_3	C_4	C_1	C_2	C_3	C_4
A_1	P	F	MP	MP	P	P	MP	P	VP	MP	MP	P	VP	F	F	VP
A_2	P	MP	P	P	MP	F	MP	VP	MP	MG	MP	VP	MP	F	P	MP
A_3	F	G	MP	G	F	F	F	G	P	G	MP	G	P	F	MG	P
A_4	VG	G	G	MG	VG	VG	MG	MG	G	VG	F	G	G	G	F	G
A_5	MG	G	VG	VG	F	F	VG	MG	F	G	MG	VG	G	F	MG	G
A_6	MP	F	MG	MG	F	F	MG	F	F	G	MP	F	MP	G	F	MP
A_7	MP	F	F	F	VP	P	MP	F	VP	F	MP	MG	P	P	F	P
A_8	F	VP	F	P	F	MP	MP	MP	MP	VP	P	P	MG	VP	MP	MG
A_9	F	MG	MG	G	G	MG	VG	G	G	G	G	F	G	MG	VG	G

Table 3. The decision-matrix of each decision-maker at second period (t = 2).

	D_1				D_2				D_3				D_4			
	C_1	C_2	C_3	C_4	C_1	C_2	C_3	C_4	C_1	C_2	C_3	C_4	C_1	C_2	C_3	C_4
A_1	P	MP	MP	VP	—	—	—	—	P	MP	MG	P	VP	MP	F	MP
A_2	P	MP	F	VP	—	—	—	—	P	MG	P	VP	VP	MG	MP	P
A_3	P	F	MP	MG	—	—	—	—	P	MG	MG	G	P	MG	F	F
A_4	MG	VG	MG	MG	—	—	—	—	MG	G	MG	G	G	G	MG	G
A_6	P	MG	F	MP	—	—	—	—	P	G	MP	MP	F	G	MG	F
A_7	P	F	P	MP	—	—	—	—	P	P	MP	F	P	P	F	MG
A_{10}	VG	VG	G	MG	—	—	—	—	G	VG	G	G	G	G	G	MG
A_{11}	P	MG	G	MG	—	—	—	—	MP	MG	G	MG	F	G	MG	F
A_{12}	P	VP	P	MP	—	—	—	—	VP	VP	MP	P	VP	VP	P	MP
A_{13}	MG	P	VP	MP	—	—	—	—	MP	P	P	P	F	MP	P	MP

Table 4. The decision-matrix of each decision-maker at third period (t = 3).

	D_1				D_2				D_3				D_4			
	C_1	C_2	C_3	C_4	C_1	C_2	C_3	C_4	C_1	C_2	C_3	C_4	C_1	C_2	C_3	C_4
A_1	P	P	MP	P	VP	F	MP	MP	P	MP	F	MP	VP	F	F	MP
A_2	MP	F	P	VP	VP	MP	P	P	P	MG	F	MP	MP	MP	F	MP
A_5	MG	G	MG	VG	F	G	MG	MG	MG	G	VG	G	G	F	G	MG
A_6	P	G	MG	MP	MP	MG	F	F	MP	F	F	MP	P	F	MG	F
A_7	VP	P	MP	MG	MP	MP	MP	F	VP	F	P	F	VP	MP	P	MP
A_8	F	MP	F	P	MG	MP	P	MP	MG	P	F	MP	MG	P	F	F
A_9	MG	G	VG	F	MG	G	VG	F	G	VG	G	F	MG	VG	VG	G
A_{10}	G	G	G	G	MG	G	VG	MG	G	G	G	F	MG	VG	VG	G
A_{12}	VP	P	P	P	P	P	P	F	MP	VP	MP	F	P	P	P	F

Table 5. The decision-matrix of each decision-maker at fourth period (t = 4).

	D_1				D_2				D_3				D_4			
	C_1	C_2	C_3	C_4	C_1	C_2	C_3	C_4	C_1	C_2	C_3	C_4	C_1	C_2	C_3	C_4
A_2	MP	MG	F	P	MP	MP	F	MP	MP	MG	MP	VP	—	—	—	—
A_3	F	MG	MP	F	F	G	MG	MG	MP	F	MP	F	—	—	—	—
A_6	MP	F	F	MP	MP	MG	MP	MG	P	F	F	MP	—	—	—	—
A_7	VP	P	P	MP	P	F	MP	F	MP	MP	P	MP	—	—	—	—
A_8	MP	P	P	MP	MP	VP	MP	MP	MP	VP	MP	F	—	—	—	—
A_{10}	MG	G	VG	G	VG	VG	G	MG	VG	G	G	G	—	—	—	—
A_{13}	MP	VP	VP	P	F	VP	VP	MP	MG	MP	P	VP	—	—	—	—

Table 6. The matrices of criteria weights in different periods.

		D_1	D_2	D_3	D_4
t = 1	C_1	ML	L	M	M
	C_2	M	ML	M	M
	C_3	MH	VH	MH	H
	C_4	H	MH	M	MH
t = 2	C_1	L	—	ML	L
	C_2	ML	—	M	M
	C_3	VH	—	VH	MH
	C_4	H	—	MH	M
t = 3	C_1	L	L	M	L
	C_2	ML	ML	MH	ML
	C_3	MH	MH	H	VH
	C_4	M	M	M	H
t = 4	C_1	ML	M	L	—
	C_2	ML	ML	MH	—
	C_3	MH	MH	H	—
	C_4	MH	MH	H	—

According to the results presented in Table 7 and Steps 12 and 13 of the proposed approach, the dynamic and aggregated dynamic scores of alternatives can be calculated.

It should be noted that we use Equation (29) to set the weights for aggregating the dynamic scores. However, this function can be replaced with any custom function which can consider the importance of newer decision information. Also, the user of the proposed approach can set the weights manually without defining a function.

The values of S_{it}, H_{it} and the rank of each alternative related to each period are represented in Table 8. We also show the changes in the members of AL_t and AL_t^T in this table. The members of these sets should be known for the calculations of Steps 12 and 13."

As it can be seen in Table 8, A_9 is the best alternative (subcontractor) in the first period (t = 1), but this alternative is not available in the second period. The unavailability of A_9, and availability of some better alternatives in the second period lead to a decrease in the value of the aggregated dynamic score for this alternative. Therefore, the rank of A_9 is changed from 1 to 6 at t = 2. On the other hand, the rank of A_4, which has the second rank at t = 1, is changed to 1 in the second period, and A_{10}, which is a new available subcontractor, has the second rank in the second period. We can say that the rank of alternatives is dynamic and changes in different periods according to the new information of decision-making process.

In this example, the changes in the rank of subcontractors at different time periods are depicted in Figure 3.

Table 7. The overall performance scores and ranking results at each period.

		$\tilde{U}_{it}$	$\mathfrak{D}(\tilde{U}_{it})$	Rank	
				Fuzzy EDAS	TOPSIS
$t = 1$	A_1	(−0.716, 0.05, 0.67)	0.0013	8	9
	A_2	(−0.712, 0.0290, 0.683)	0	9	8
	A_3	(−0.0580, 0.617, 1.305)	0.6216	4	4
	A_4	(0.349, 0.914, 1.614)	0.9591	2	3
	A_5	(0.366, 0.898, 1.562)	0.9421	3	2
	A_6	(−0.137, 0.550, 1.258)	0.5572	5	5
	A_7	(−0.524, 0.252, 0.903)	0.2105	6	6
	A_8	(−0.639, 0.101, 0.757)	0.0730	7	7
	A_9	(0.415, 0.953, 1.632)	1	1	1
				$r_s = 0.97$	
$t = 2$	A_1	(−0.402, 0.263, 0.762)	0.2075	7	7
	A_2	(−0.453, 0.166, 0.707)	0.1402	8	8
	A_3	(0.037, 0.59, 1.125)	0.5838	4	4
	A_4	(0.439, 0.874, 1.472)	0.9281	2	2
	A_6	(−0.089, 0.517, 1.111)	0.5130	5	5
	A_7	(−0.353, 0.309, 0.874)	0.2767	6	6
	A_{10}	(0.514, 0.965, 1.522)	1	1	1
	A_{11}	(0.249, 0.77, 1.334)	0.7843	3	3
	A_{12}	(−0.636, 0.041, 0.595)	0	10	10
	A_{13}	(−0.526, 0.091, 0.694)	0.0864	9	9
				$r_s = 1$	
$t = 3$	A_1	(−0.652, 0.194, 0.848)	0.1298	7	7
	A_2	(−0.65, 0.15, 0.801)	0.1005	8	8
	A_5	(0.436, 0.875, 1.501)	0.9376	3	2
	A_6	(−0.215, 0.489, 1.144)	0.4725	4	4
	A_7	(−0.569, 0.184, 0.782)	0.1324	6	6
	A_8	(−0.473, 0.278, 0.991)	0.2652	5	5
	A_9	(0.448, 0.9, 1.482)	0.9434	2	3
	A_{10}	(0.503, 0.947, 1.55)	1	1	1
	A_{12}	(−0.745, 0.05, 0.695)	0	9	9
				$r_s = 0.98$	
$t = 4$	A_2	(−0.26, 0.357, 0.93)	0.3424	4	4
	A_3	(0.179, 0.614, 1.102)	0.6318	2	2
	A_6	(−0.026, 0.507, 1.016)	0.4992	3	3
	A_7	(−0.586, 0.242, 0.946)	0.2007	6	6
	A_8	(−0.592, 0.244, 0.969)	0.2070	5	5
	A_{10}	(0.59, 0.971, 1.438)	1	1	1
	A_{13}	(−0.599, 0.043, 0.611)	0.0183	7	7
				$r_s = 1$	

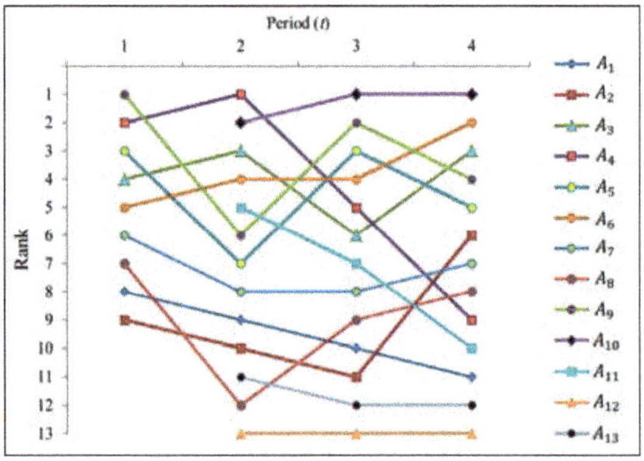

Figure 3. The changes in the rank of alternatives at different time periods.

Table 8. The final scores and ranks of alternatives at each period.

		A_1	A_2	A_3	A_4	A_5	A_6	A_7	A_8	A_9	A_{10}	A_{11}	A_{12}	A_{13}
	AL_t	√	√	√	√	√	√	√	√	√	×	×	×	×
	AL_t^T	√	√	√	√	√	√	√	√	√	×	×	×	×
$t = 1$	S_{it}	0.0013	0	0.6216	0.9591	0.9421	0.5572	0.2105	0.0730	1	—	—	—	—
	H_{it}	0.0013	0	0.6216	0.9591	0.9421	0.5572	0.2105	0.0730	1	—	—	—	—
	Rank	8	9	4	2	3	5	6	7	1	—	—	—	—
	AL_t	√	√	√	√	×	√	√	×	×	√	√	√	√
	AL_t^T	√	√	√	√	√	√	√	√	√	√	√	√	√
$t = 2$	S_{it}	0.2075	0.1402	0.5838	0.9281	0	0.5130	0.2767	0	0	1	0.7843	0	0.0864
	H_{it}	0.1388	0.0935	0.5964	0.9384	0.314	0.5277	0.2546	0.0243	0.3333	0.6667	0.5229	0	0.0576
	Rank	9	10	3	1	7	4	8	12	6	2	5	13	11
	AL_t	√	√	×	×	√	√	√	√	√	√	×	√	×
	AL_t^T	√	√	√	√	√	√	√	√	√	√	√	√	√
$t = 3$	S_{it}	0.1298	0.1005	0	0	0.9376	0.4725	0.1324	0.2652	0.9434	1	0	0	0
	H_{it}	0.1334	0.0977	0.2386	0.3754	0.6882	0.4946	0.1813	0.1689	0.6994	0.8667	0.2091	0	0.023
	Rank	10	11	6	5	3	4	8	9	2	1	7	13	12
	AL_t	×	√	√	×	×	√	√	√	×	√	×	×	√
	AL_t^T	√	√	√	√	√	√	√	√	√	√	√	√	√
$t = 4$	S_{it}	0	0.3424	0.6318	0	0	0.4992	0.2007	0.2070	0	1	0	0	0.0183
	H_{it}	0.0572	0.2375	0.4633	0.1609	0.2949	0.4972	0.1924	0.1907	0.2997	0.9429	0.0896	0	0.0203
	Rank	11	6	3	9	5	2	7	8	4	1	10	13	12

According to the evaluation of the last period (t = 4), A_{10} is the best alternative, and the final ranking is as follows:

$$A_{10} \succ A_6 \succ A_3 \succ A_9 \succ A_5 \succ A_2 \succ A_7 \succ A_8 \succ A_4 \succ A_{11} \succ A_{13} \succ A_{12}$$

Although the final evaluation can be made based on the above-mentioned ranking, the main contractor should be cautious about the subcontractors which have higher degree of fluctuation in their ranks at different periods. The fluctuation in the rank of subcontractors could be occurred due to the unavailability of them or their low performance in some periods. Both reasons lead to unreliability of a subcontractor. As we can see in Figure 3, the ranks of A_{10}, A_6, A_7, A_{12} and A_{13} have lower fluctuation than the other alternatives. Therefore, the main contractor can select A_{10} as a reliable subcontractor and consider A_6 as a backup alternative.

4. Conclusions

In the management of a contract, it has become usual to outsource specialized tasks by the main contractor. This can be done to ensure the quality of construction projects. Although outsourcing most tasks to a subcontractor is convenient and safe for the main contractor, the failure of the selected subcontractor can lead to the failure of the entire project. Hence the process of evaluation and selection of subcontractors can be considered as one of the important actions that should be carried out by the main contractor.

In this study, we have defined the subcontractor evaluation process as a dynamic multi-criteria group decision-making problem. Due to the uncertainty of information in the process of evaluation, a fuzzy dynamic MCGDM approach has been proposed to deal with SEP. The proposed approach has been designed based on the EDAS method which is a new and efficient MCDM approach. In the procedure of the proposed approach, we can define different sets of alternatives, criteria, and decision-makers in different time periods. The performance of each alternative is updated in each period by an aggregation function which gives greater weights to newer information. Thus, we can ensure that the final evaluation involves the importance of up-to-date decision information.

We have used an example of subcontractor evaluation problem to illustrate the process of the proposed approach and show the utility of it in real-world decision-making problems. Because the weights of criteria as well as the set of criteria can be changed at each period in the process of using the proposed dynamic approach, the sensitivity analysis on the weights of criteria has not been made in this study. Lack of this analysis can be considered as a limitation of this study. To make the sensitivity analysis in the proposed approach, we need to devise a new research methodology, and this can be addressed in future research. Also, future research can examine the effect of different weight functions for aggregation of the dynamic scores of alternatives and apply the proposed approach to the other MCDM problems such as supplier evaluation, service quality assessment and risk evaluation. Moreover, other types of fuzzy sets such as interval type-2 fuzzy sets, intuitionistic fuzzy sets and hesitant fuzzy sets can be used to extend the propose approach.

Author Contributions: Mehdi Keshavarz-Ghorabaee, Maghsoud Amiri and Edmundas Kazimieras Zavadskas designed the research, analyzed the data and the obtained results, and performed the development of the paper. Zenonas Turskis and Jurgita Antucheviciene provided extensive advice throughout the study, regarding the research design, methodology, findings and revised the manuscript. All the authors have read and approved the final manuscript.

Conflicts of Interest: The authors declare no conflict of interest.

References

1. Hinze, J.; Tracey, A. The contractor-subcontractor relationship: The subcontractor's view. *J. Constr. Eng. Manag.* **1994**, *120*, 274–287. [CrossRef]
2. Polat, G. Subcontractor selection using the integration of the ahp and promethee methods. *J. Civ. Eng. Manag.* **2016**, *22*, 1042–1054. [CrossRef]
3. Hartmann, A.; Ling, F.Y.Y.; Tan, J.S. Relative importance of subcontractor selection criteria: Evidence from singapore. *J. Constr. Eng. Manag.* **2009**, *135*, 826–832. [CrossRef]
4. Arslan, G.; Kivrak, S.; Birgonul, M.T.; Dikmen, I. Improving sub-contractor selection process in construction projects: Web-based sub-contractor evaluation system (WEBSES). *Autom. Constr.* **2008**, *17*, 480–488. [CrossRef]
5. Zhao, J.; Yu, L.; Li, L. Application of GRA method, dynamic analysis and fuzzy set theory in evaluation and selection of emergency treatment technology for large scale phenol spill incidents. *AIP Conf. Proc.* **2017**, *1839*, 1–6.
6. Keshavarz Ghorabaee, M.; Amiri, M.; Zavadskas, E.K.; Antucheviciene, J. Supplier evaluation and selection in fuzzy environments: A review of madm approaches. *Econ. Res. Ekon. Istraž.* **2017**, *30*, 1073–1118. [CrossRef]
7. Sun, Y.; Zhang, G.; Hong, Z.; Dong, K. How uncertain information on service capacity influences the intermodal routing decision: A fuzzy programming perspective. *Information* **2018**, *9*, 24. [CrossRef]
8. Wu, Y.; Liu, L.; Gao, J.; Chu, H.; Xu, C. An extended vikor-based approach for pumped hydro energy storage plant site selection with heterogeneous information. *Information* **2017**, *8*, 106. [CrossRef]
9. Wang, R.; Li, Y. Generalized single-valued neutrosophic hesitant fuzzy prioritized aggregation operators and their applications to multiple criteria decision-making. *Information* **2018**, *9*, 10. [CrossRef]
10. Lu, Z.; Ye, J. Exponential operations and an aggregation method for single-valued neutrosophic numbers in decision making. *Information* **2017**, *8*, 62. [CrossRef]
11. Gao, J.; Liu, H. A new prospect projection multi-criteria decision-making method for interval-valued intuitionistic fuzzy numbers. *Information* **2016**, *7*, 64. [CrossRef]
12. Cheng, M.; Ma, G.X.; Sun, J.K. AHP-based research on the selection of construction project subcontractor. *Adv. Mater. Res.* **2012**, *594–597*, 3035–3039. [CrossRef]
13. Kargi, V.S.A.; Öztürk, A. Subcontractor selection using analytic hierarchy process. *Bus. Econ. Res. J.* **2012**, *3*, 121–143.
14. Yayla, A.Y.; Yildiz, A.; Yildiz, K. Generalised Choquet integral algorithm for subcontractor selection in the textile industry—A case study for turkey. *Fibres Text. East. Eur.* **2013**, *21*, 16–21.
15. Ng, S.T.; Skitmore, M. Developing a framework for subcontractor appraisal using a balanced scorecard. *J. Civ. Eng. Manag.* **2014**, *20*, 149–158. [CrossRef]
16. Abbasianjahromi, H.; Rajaie, H.; Shakeri, E. A framework for subcontractor selection in the construction industry. *J. Civ. Eng. Manag.* **2013**, *19*, 158–168. [CrossRef]
17. Shahvand, E.; Sebt, M.; Banki, M. Developing fuzzy expert system for supplier and subcontractor evaluation in construction industry. *Sci. Iran. Trans. A Civ. Eng.* **2016**, *23*, 842–855. [CrossRef]
18. Ulubeyli, S.; Kazaz, A. Fuzzy multi-criteria decision making model for subcontractor selection in international construction projects. *Technol. Econ. Dev. Econ.* **2016**, *22*, 210–234. [CrossRef]

19. Abbasianjahromi, H.; Rajaie, H.; Shakeri, E.; Kazemi, O. A new approach for subcontractor selection in the construction industry based on portfolio theory. *J. Civ. Eng. Manag.* **2016**, *22*, 346–356. [CrossRef]
20. Polat, G.; Cetindere, F.; Damci, A.; Bingol, B.N. Smart home subcontractor selection using the integration of ahp and evidential reasoning approaches. *Procedia Eng.* **2016**, *164*, 347–353. [CrossRef]
21. Campanella, G.; Ribeiro, R.A. A framework for dynamic multiple-criteria decision making. *Decis. Support Syst.* **2011**, *52*, 52–60. [CrossRef]
22. Wei, G. Grey relational analysis model for dynamic hybrid multiple attribute decision making. *Knowl. Based Syst.* **2011**, *24*, 672–679. [CrossRef]
23. Chen, Y.; Li, B. Dynamic multi-attribute decision making model based on triangular intuitionistic fuzzy numbers. *Sci. Iran.* **2011**, *18*, 268–274. [CrossRef]
24. Wang, Y.; Shi, X.; Sun, J.; Qian, W. A grey interval relational degree-based dynamic multiattribute decision making method and its application in investment decision making. *Math. Probl. Eng.* **2014**, *2014*, 6. [CrossRef]
25. Junhua, H.; Peng, C.; Liu, Y. Dynamic stochastic multi-criteria decision making method based on prospect theory and conjoint analysis. *Manag. Sci. Eng.* **2014**, *8*, 65–71.
26. Li, G.; Kou, G.; Peng, Y. Dynamic fuzzy multiple criteria decision making for performance evaluation. *Technol. Econ. Dev. Econ.* **2015**, *21*, 705–719. [CrossRef]
27. Yan, S.; Liu, S.; Liu, J.; Wu, L. Dynamic grey target decision making method with grey numbers based on existing state and future development trend of alternatives. *J. Intell. Fuzzy Syst.* **2015**, *28*, 2159–2168. [CrossRef]
28. Liu, J.; Guo, L.; Jiang, J.; Hao, L.; Liu, R.; Wang, P. Evaluation and selection of emergency treatment technology based on dynamic fuzzy GRA method for chemical contingency spills. *J. Hazard. Mater.* **2015**, *299*, 306–315. [CrossRef] [PubMed]
29. Yan, S.; Liu, S.; Liu, X. Dynamic grey target decision making method with three-parameter grey numbers. *Grey Syst. Theory Appl.* **2016**, *6*, 169–179. [CrossRef]
30. Keshavarz Ghorabaee, M.; Zavadskas, E.K.; Olfat, L.; Turskis, Z. Multi-criteria inventory classification using a new method of evaluation based on distance from average solution (EDAS). *Informatica* **2015**, *26*, 435–451. [CrossRef]
31. Keshavarz Ghorabaee, M.; Zavadskas, E.K.; Amiri, M.; Turskis, Z. Extended EDAS method for fuzzy multi-criteria decision-making: An application to supplier selection. *Int. J. Comput. Commun. Control* **2016**, *11*, 358–371. [CrossRef]
32. Kahraman, C.; Keshavarz Ghorabaee, M.; Zavadskas, E.K.; Cevik Onar, S.; Yazdani, M.; Oztaysi, B. Intuitionistic fuzzy EDAS method: An application to solid waste disposal site selection. *J. Environ. Eng. Landsc. Manag.* **2017**, *25*, 1–12. [CrossRef]
33. Karaşan, A.; Kahraman, C. Interval-valued neutrosophic extension of EDAS method. In *Advances in Fuzzy Logic and Technology 2017, Proceedings of the 10th Conference of the European Society for Fuzzy Logic and Technology (EUSFLAT 2017), Warsaw, Poland, 11–15 September 2017*; Kacprzyk, J., Szmidt, E., Zadrożny, S., Atanassov, K.T., Krawczak, M., Eds.; Springer: Cham, Switzerland, 2018; Volume 2, pp. 343–357.
34. Peng, X.; Dai, J.; Yuan, H. Interval-valued fuzzy soft decision making methods based on MABAC, similarity measure and EDAS. *Fundam. Inform.* **2017**, *152*, 373–396. [CrossRef]
35. Peng, X.; Liu, C. Algorithms for neutrosophic soft decision making based on EDAS, new similarity measure and level soft set. *J. Intell. Fuzzy Syst.* **2017**, *32*, 955–968. [CrossRef]
36. Stanujkic, D.; Zavadskas, E.K.; Keshavarz Ghorabaee, M.; Turskis, Z. An extension of the EDAS method based on the use of interval grey numbers. *Stud. Inform. Control* **2017**, *26*, 5–12. [CrossRef]
37. Keshavarz Ghorabaee, M.; Amiri, M.; Zavadskas, E.K.; Turskis, Z. Multi-criteria group decision-making using an extended EDAS method with interval type-2 fuzzy sets. *E M Ekon. Manag.* **2017**, *20*, 48–68.
38. Ren, J.; Toniolo, S. Life cycle sustainability decision-support framework for ranking of hydrogen production pathways under uncertainties: An interval multi-criteria decision making approach. *J. Clean. Prod.* **2018**, *175*, 222–236. [CrossRef]
39. Stević, Ž.; Pamučar, D.; Vasiljević, M.; Stojić, G.; Korica, S. Novel integrated multi-criteria model for supplier selection: Case study construction company. *Symmetry* **2017**, *9*, 279. [CrossRef]
40. Juodagalvienė, B.; Turskis, Z.; Šaparauskas, J.; Endriukaitytė, A. Integrated multi-criteria evaluation of house's plan shape based on the EDAS and SWARA methods. *Eng. Struct. Technol.* **2017**, *9*, 117–125. [CrossRef]

41. Turskis, Z.; Morkunaite, Z.; Kutut, V. A hybrid multiple criteria evaluation method of ranking of cultural heritage structures for renovation projects. *Int. J. Strateg. Prop. Manag.* **2017**, *21*, 318–329. [CrossRef]
42. Trinkūnienė, E.; Podvezko, V.; Zavadskas, E.K.; Jokšienė, I.; Vinogradova, I.; Trinkūnas, V. Evaluation of quality assurance in contractor contracts by multi-attribute decision-making methods. *Econ. Res. Ekon. Istraž.* **2017**, *30*, 1152–1180. [CrossRef]
43. Ecer, F. Third-party logistics (3pls) provider selection via fuzzy AHP and EDAS integrated model. *Technol. Econ. Dev. Econ.* **2018**, *24*, 615–634. [CrossRef]
44. Stević, Ž.; Vasiljević, M.; Vesković, S. Evaluation in logistics using combined AHP and EDAS method. In Proceedings of the XLIII International Symposium on Operational Research, Belgrade, Serbia, 20–23 September 2016; pp. 309–313.
45. Turskis, Z.; Juodagalvienė, B. A novel hybrid multi-criteria decision-making model to assess a stairs shape for dwelling houses. *J. Civ. Eng. Manag.* **2016**, *22*, 1078–1087. [CrossRef]
46. Zadeh, L.A. Fuzzy sets. *Inf. Control* **1965**, *8*, 338–353. [CrossRef]
47. Zimmermann, H.J. Fuzzy set theory. *Wiley Interdiscip. Rev. Comput. Stat.* **2010**, *2*, 317–332. [CrossRef]
48. Wang, Y.-J.; Lee, H.-S. Generalizing topsis for fuzzy multiple-criteria group decision-making. *Comput. Math. Appl.* **2007**, *53*, 1762–1772. [CrossRef]
49. Chen, S.-J.; Hwang, C.-L. *Fuzzy Multiple Attribute Decision Making: Methods and Applications*; Springer: Berlin/Heidelberg, Germany, 1992.
50. Wang, Y.-M.; Yang, J.-B.; Xu, D.-L.; Chin, K.-S. On the centroids of fuzzy numbers. *Fuzzy Sets Syst.* **2006**, *157*, 919–926. [CrossRef]
51. Lin, Y.-H.; Lee, P.-C.; Ting, H.-I. Dynamic multi-attribute decision making model with grey number evaluations. *Expert Syst. Appl.* **2008**, *35*, 1638–1644. [CrossRef]
52. Chen, C.-T. Extensions of the TOPSIS for group decision-making under fuzzy environment. *Fuzzy Sets Syst.* **2000**, *114*, 1–9. [CrossRef]
53. Opricovic, S.; Tzeng, G.-H. Compromise solution by mcdm methods: A comparative analysis of VIKOR and TOPSIS. *Eur. J. Oper. Res.* **2004**, *156*, 445–455. [CrossRef]

information

Article

A New Bi-Directional Projection Model Based on Pythagorean Uncertain Linguistic Variable

Huidong Wang *, Shifan He and Xiaohong Pan

School of Management Science and Engineering, Shandong University of Finance and Economics, Jinan 250014, China; shifanhe0828@163.com (S.H.); xiaohongpan0303@126.com (X.P.)
* Correspondence: huidong.wang@ia.ac.cn; Tel.: +86-531-8852-5933

Received: 9 April 2018; Accepted: 21 April 2018; Published: 26 April 2018

Abstract: To solve the multi-attribute decision making (MADM) problems with Pythagorean uncertain linguistic variable, an extended bi-directional projection method is proposed. First, we utilize the linguistic scale function to convert uncertain linguistic variable and provide a new projection model, subsequently. Then, to depict the bi-directional projection method, the formative vectors of alternatives and ideal alternatives are defined. Furthermore, a comparative analysis with projection model is conducted to show the superiority of bi-directional projection method. Finally, an example of graduate's job option is given to demonstrate the effectiveness and feasibility of the proposed method.

Keywords: multi-attribute decision making; projection model; bi-directional projection model; Pythagorean uncertain linguistic variable

1. Introduction

Multi-attribute decision making (MADM) problem is to select the optimal alternative(s) or get the ranking order of all alternatives with multiple attributes. For the complexity of the decision making environment and the limitation of decision makers' knowledge, vagueness and uncertainty are typical factors we must take into account. To describe the vague and uncertain information accurately, Zadeh [1] proposed the concept of fuzzy set. However, applying fuzzy set to solve decision making problems is confined to the lacking of information. Intuitionistic fuzzy set (IFS) as the extension of fuzzy set, can capture uncertain information more appropriately. Recently, IFS have been extensively applied to MADM area, because the superiority in dealing with vague and uncertain information [2–5]. Whereas, IFS is difficult to depict vague and uncertain information when the sum of membership degree and non-membership degree is bigger than 1.

To express fuzzy information more effectively, Yager [6] proposed the Pythagorean fuzzy set (PFS) to capture the vague and uncertain information. Different from IFS, the sum of membership degree and non-membership degree of PFS may be bigger than one, but the square sum of them is less than one. As a useful extension of IFS, the PFS can depict the problem which the IFS cannot. For example, if the membership degree and non-membership degree are 0.8 and 0.6, respectively. It is easily to see, the IFS cannot describe this situation because of $0.8 + 0.6 > 1$, but the PFS can effectively solve the issue due to $0.8^2 + 0.6^2 \leq 1$. Since PFS appeared, multi-attribute decision making problems under PFS environment have got a lot of attention, and some research results have been obtained. Du et al. [7] proposed a new score function and a new accurate function of PFS. Liang et al. [8], Liang and Xu [9], and Zhang and Xu [10] extended the TOPSIS (Technique for Order Performance by Similarity to Ideal Solution) method under PFS and hesitant Pythagorean fuzzy set circumstances, respectively. A new closeness index of Pythagorean fuzzy set was proposed by Zhang [11] and the QUALIFLEX (QUALItative FLEXible multiple criteria method) method was extended based on the closeness index subsequently. Ren et al. [12] presented an extended TODIM (An Acronym in

Portuguese of Interactive and Multiple Attribute Decision Making) method based on PFS, and made an emulational analysis for the result. Chen [13] proposed a new distance formula for PFS and an extended VIKOR (the Serbian name: Vlsekriterijumska Optimizacijia I Kompromisno Resenje) method was presented based on the distance formula. Following the pioneering work of Yager, Garg [14,15] developed a new relevant coefficient of PFS and an extended accurate function of interval Pythagorean fuzzy set (IPFS), respectively. A new MADM method was proposed by Peng and Dai [16] based on prospect theory and regret theory. Furthermore, Xue et al. [17] defined the concept of entropy of PFS and extended the LINMAP (The Linear Programming Technique for Multidimensional Analysis of Preference) method based on the concept. Liang et al. [18] developed a weighted Pythagorean fuzzy geometric mean operator and extended the projection method based on the geometric mean operator. Peng and Yang [19] proposed an extended ELECTRE (ELimination Et Choice Translating REality) method based on IPFS. The new accurate function and similarity measure of PFS were developed by Zhang [20], respectively.

The projection model can simultaneously consider the angle and distance between two evaluative values [18]. Therefore, the projection model has been widely applied to replace the single distance measure in multi-attribute decision making domain. Tsao and Chen [21] developed a projection model of interval intuitionistic fuzzy set (IIFS) and an extended VIKOR method was proposed based on the projection model. Sun et al. [22] proposed a projection model of hesitant linguistic variable and extended the multi-attributive border approximation area comparison (MABAC) method to hesitant linguistic circumstance. To overcome the drawback of the extant TODIM method, Ji et al. [23] developed a projection-based TODIM method with multi-valued neutrosophic sets (MVNSs). Wu et al. [24] proposed an extended projection model based on hesitant linguistic variable to handle the hospital management problem. Inspired by the advantage of projection model, Liang et al. [25] proposed an extended PROMETHEE (Preference Ranking Organization Method for Enrichment Evaluations) method based on the projection model.

Recently, projection model has been extensively applied to solve the MADM problems due to the advantage of capturing vague and uncertain information. However, projection model cannot effectively get the ranking order when alternatives distribute on the perpendicular bisector of ideal alternatives [26]. Motivated by the drawback of the projection model and the advantage of linguistic variables, we developed an extended bi-directional projection model of Pythagorean uncertain linguistic variables [27]. Our model can not only utilize the advantage of both Pythagorean uncertain linguistic variable and projection models but it can also effectively overcome the defects of the projection model.

This paper is organized as follows. Section 2 presents some basic definitions of IFS, linguistic variables, and the Pythagorean uncertain linguistic variables. In Section 3, we propose a new bi-directional projection model. Comparative analysis of the proposed model and projection model is provided in Section 4 and the MADM Procedures are listed in Section 5. In Section 6, the effectiveness of the proposed method is demonstrated by a practical MADM problem. Finally, Section 7 comes to some conclusions.

2. Preliminaries

Definition 1 [2]. *Let X be a crisp set, an intuitionistic fuzzy set on X can be defined as*

$$A = \langle x, u_A(x), v_A(x) | x \in X \rangle.$$

where, $u_A(x)$*:* $X \to [0,1]$ *and* $v_A(x)$*:* $X \to [0,1]$ *denote membership function and non-membership function of* $x \in X$*, respectively, with* $0 \le u_A(x) + v_A(x) \le 1$*.* $\pi(x) = 1 - u_A(x) - v_A(x)$ *denote the hesitation function of* $x \in X$*.*

Definition 2 [28]. *Let $S = \{s_i | i = 0, 1, \cdots, 2z\}$ be linguistic term set, where z is a positive integer and s_i denotes an evaluation value of linguistic variable. We call $\widetilde{s} = [s_\alpha, s_\beta]$ as the uncertain linguistic variable, where $s_\alpha, s_\beta \in S$ and $0 \le \alpha \le \beta \le z$, besides, α and β are positive integers. s_α, s_β denote the upper bound and the lower bound, respectively.*

Definition 3 [27]. *Let X be a fixed set. $\widetilde{\alpha} = \left\{\left\langle x_i | \left([s_\alpha, s_\beta], \widetilde{P}\left(u_{\widetilde{p}}(x_i), v_{\widetilde{p}}(x_i)\right)\right)\right\rangle | x_i \in X\right\}$ denote the Pythagorean uncertain linguistic variable on X, where function $u_{\widetilde{p}}(x)$: $X \to [0,1]$ and $v_{\widetilde{p}}(x)$: $X \to [0,1]$ denote membership function and non-membership function of $x \in X$, respectively, with $u^2{}_{\widetilde{p}}(x) + v^2{}_{\widetilde{p}}(x) \le 1$. To expediently depict the evaluation value, we call $\alpha = \left\langle [s_\alpha, s_\beta], \widetilde{P}\left(u_{\widetilde{p}}(x_i), v_{\widetilde{p}}(x_i)\right)\right\rangle$ as the Pythagorean uncertain linguistic number.*

Definition 4 [25]. *If $\eta_i \in [0,1]$ is a numerical value, then the linguistic scale function f can defined as $f : s_i \to \eta_i (i = 0, 1, \cdots, 2z)$, where $0 \le \eta_0 < \eta_1 < \cdots < \eta_{2z}$. η_i represent the preference of decision maker on the chosen linguistic term s_i.*

$$f(s_i) = \eta_i = \begin{cases} \frac{z^\delta - (z-i)^\delta}{2z^\delta}, 0 \le i \le z \\ \frac{z^\gamma + (i-z)^\gamma}{2z^\gamma}, z < i \le 2z \end{cases} \tag{1}$$

where δ, γ denote the sensibility coefficient, $\delta, \gamma \in [0,1]$ and f is a monotone increasing function.

Definition 5 *Let $X_p = \left(\left\langle \left[\underline{s}_{pj}, \overline{s}_{pj}\right], \widetilde{P}(u_{pj}, v_{pj})\right\rangle\right)$ and $X_q = \left(\left\langle \left[\underline{s}_{qj}, \overline{s}_{qj}\right], \widetilde{P}(u_{qj}, v_{qj})\right\rangle\right)$ be two Pythagorean uncertain linguistic variable on X. If we convert X_p and X_q to $X_p = \left(\left\langle \left[f\left(\underline{s}_{pj}\right), f(\overline{s}_{pj})\right], \widetilde{P}(u_{pj}, v_{pj})\right\rangle\right)$ and $X_q = \left(\left\langle \left[f\left(\underline{s}_{qj}\right), f(\overline{s}_{qj})\right], \widetilde{P}(u_{qj}, v_{qj})\right\rangle\right)$ via linguistic scale function, then the formative vector of X_p and X_q is computed as*

$$X_pX_q = \left(\left\langle [\min f(s_{pj}), \max f(s_{pj})], \widetilde{P}(|u_{qj} - u_{pj}|, |v_{qj} - v_{pj}|)\right\rangle\right) \tag{2}$$

where

$$\min f(s_{pj}) = \min\left(\left|f\left(\underline{s}_{qj}\right) - f\left(\underline{s}_{pj}\right)\right|, |f(\overline{s}_{qj}) - f(\overline{s}_{pj})|\right)$$
$$\max f(s_{pj}) = \max\left(\left|f\left(\underline{s}_{qj}\right) - f\left(\underline{s}_{pj}\right)\right|, |f(\overline{s}_{qj}) - f(\overline{s}_{pj})|\right)$$

Example 1. *Let $X_p = \left(\left\langle [s_3, s_5], \widetilde{P}(0.6, 0.5)\right\rangle\right)$ and $X_q = \left(\left\langle [s_4, s_5], \widetilde{P}(0.7, 0.4)\right\rangle\right)$ be two Pythagorean uncertain linguistic numbers, where $z = 4$, $\alpha = 0.6$, $\gamma = 0.8$.*

According to Definition 4, we can obtain

$$X_p = \left(\left\langle [0.28, 0.66], \widetilde{P}(0.6, 0.5)\right\rangle\right)$$
$$X_q = \left(\left\langle [0.5, 0.66], \widetilde{P}(0.7, 0.4)\right\rangle\right)$$

Then, the formative vector of X_p and X_q is obtained via (2).

$$X_pX_q = \left(\left\langle [0, 0.22], \widetilde{P}(0.1, 0.1)\right\rangle\right)$$

3. Bi-Directional Projection Model

3.1. Projection Model

Let $\alpha = \left\langle \left[\underline{s}_{\alpha_j}, \overline{s}_{\alpha_j}\right], \widetilde{P}\left(u_{\alpha_j}, v_{\alpha_j}\right)\right\rangle$ and $\beta = \left\langle \left[\underline{s}_{\beta_j}, \overline{s}_{\beta_j}\right], \widetilde{P}\left(u_{\beta_j}, v_{\beta_j}\right)\right\rangle$, which are two Pythagorean uncertain linguistic variables, then the cosine of α and β is defined as

$$\cos(\alpha,\beta) = \frac{\sum_{j=1}^{n}\left(f\left(\underline{s}_{\alpha_j}\right)\cdot f\left(\underline{s}_{\beta_j}\right)+f\left(\overline{s}_{\alpha_j}\right)\cdot f\left(\overline{s}_{\beta_j}\right)+u_{\alpha_j}^2\cdot u_{\beta_j}^2+v_{\alpha_j}^2\cdot v_{\beta_j}^2\right)}{\sqrt{\sum_{j=1}^{n}\left(f\left(\underline{s}_{\alpha_j}\right)\right)^2+\left(f\left(\overline{s}_{\alpha_j}\right)\right)^2+\left(u_{\alpha_j}\right)^4+\left(v_{\alpha_j}\right)^4}\cdot\sqrt{\sum_{j=1}^{n}\left(f\left(\underline{s}_{\beta_j}\right)\right)^2+\left(f\left(\overline{s}_{\beta_j}\right)\right)^2+\left(u_{\beta_j}\right)^4+\left(v_{\beta_j}\right)^4}} \tag{3}$$

$|\alpha| = \sqrt{\sum_{j=1}^{n}\left(f\left(\underline{s}_{\alpha_j}\right)\right)^2+\left(f\left(\overline{s}_{\alpha_j}\right)\right)^2+\left(u_{\alpha_j}\right)^4+\left(v_{\alpha_j}\right)^4}$ and $|\beta| = \sqrt{\sum_{j=1}^{n}\left(f\left(\underline{s}_{\beta_j}\right)\right)^2+\left(f\left(\overline{s}_{\beta_j}\right)\right)^2+\left(u_{\beta_j}\right)^4+\left(v_{\beta_j}\right)^4}$ denote the modules of α and β. f is linguistic scale function.

Therefore, the projection of α and β is defined as

$$\begin{aligned} prj_\beta(\alpha) &= |\alpha|\cdot\cos(\alpha,\beta) \\ &= \frac{\sum_{j=1}^{n}\left(f\left(\underline{s}_{\alpha_j}\right)\cdot f\left(\underline{s}_{\beta_j}\right)+f\left(\overline{s}_{\alpha_j}\right)\cdot f\left(\overline{s}_{\beta_j}\right)+u_{\alpha_j}^2\cdot u_{\beta_j}^2+v_{\alpha_j}^2\cdot v_{\beta_j}^2\right)}{\sqrt{\sum_{j=1}^{n}\left(f\left(\underline{s}_{\beta_j}\right)\right)^2+\left(f\left(\overline{s}_{\beta_j}\right)\right)^2+\left(u_{\beta_j}\right)^4+\left(v_{\beta_j}\right)^4}} \end{aligned} \tag{4}$$

Theorem 1 [25]. *The cosine of α and β meets the following several properties*

(1) $\cos(\alpha,\beta) = \cos(\beta,\alpha)$
(2) $0 \le \cos(\alpha,\beta) \le 1$
(3) $\alpha = \beta \Leftrightarrow \cos(\alpha,\beta) = 1$

3.2. Bi-Directional Projection Model

Let $X_i = \left\langle \left[f\left(\underline{s}_{ij}\right), f\left(\overline{s}_{ij}\right)\right], \widetilde{P}\left(u_{ij}, v_{ij}\right)\right\rangle$ be an alternative with Pythagorean uncertain linguistic variable information. The positive and negative ideal alternatives are denoted as: $X^+ = \left\langle \left[\max_{1\le i\le m} f\left(\underline{s}_{ij}\right), \max_{1\le i\le m} f\left(\overline{s}_{ij}\right)\right], \widetilde{P}\left(\max_{1\le i\le m} u_{ij}, \min_{1\le i\le m} v_{ij}\right)\right\rangle$ and $X^- = \left\langle \left[\min_{1\le i\le m} f\left(\underline{s}_{ij}\right), \min_{1\le i\le m} f\left(\overline{s}_{ij}\right)\right], \widetilde{P}\left(\min_{1\le i\le m} u_{ij}, \max_{1\le i\le m} v_{ij}\right)\right\rangle$, respectively, m represents the number of alternatives. Then, the formative vectors of X_i and ideal alternatives are denoted as

$$X^-X^+ = \left\langle \left[f\left(\underline{s}_{ij}^t\right), f\left(\overline{s}_{ij}^t\right)\right], \widetilde{P}\left(u_{ij}^t, v_{ij}^t\right)\right\rangle \tag{5}$$

$$X^-X_i = \left\langle \left[f\left(\underline{s}_{ij}^-\right), f\left(\overline{s}_{ij}^-\right)\right], \widetilde{P}\left(u_{ij}^-, v_{ij}^-\right)\right\rangle \tag{6}$$

$$X_iX^+ = \left\langle \left[f\left(\underline{s}_{ij}^+\right), f\left(\overline{s}_{ij}^+\right)\right], \widetilde{P}\left(u_{ij}^+, v_{ij}^+\right)\right\rangle \tag{7}$$

where

$$f\left(\underline{s}_{ij}^t\right) = \min\left(\left(\max_{1\le i\le m} f\left(\underline{s}_{ij}\right) - \min_{1\le i\le m} f\left(\underline{s}_{ij}\right)\right), \left(\max_{1\le i\le m} f\left(\overline{s}_{ij}\right) - \min_{1\le i\le m} f\left(\overline{s}_{ij}\right)\right)\right)$$

$$f\left(\overline{s}_{ij}^t\right) = \max\left(\left(\max_{1\le i\le m} f\left(\underline{s}_{ij}\right) - \min_{1\le i\le m} f\left(\underline{s}_{ij}\right)\right), \left(\max_{1\le i\le m} f\left(\overline{s}_{ij}\right) - \min_{1\le i\le m} f\left(\overline{s}_{ij}\right)\right)\right)$$

$$u_{ij}^t = \max_{1\le i\le m} u_{ij}^2 - \min_{1\le i\le m} u_{ij}^2, v_{ij}^t = \max_{1\le i\le m} v_{ij}^2 - \min_{1\le i\le m} v_{ij}^2$$

$$f\left(\underline{s}_{ij}^-\right) = \min\left(\left(f\left(\underline{s}_{ij}\right) - \min_{1\le i\le m} f\left(\underline{s}_{ij}\right)\right), \left(f\left(\overline{s}_{ij}\right) - \min_{1\le i\le m} f\left(\overline{s}_{ij}\right)\right)\right)$$

$$
\begin{aligned}
f\left(\bar{s}_{ij}^{-}\right) &= \max\left(\left(f\left(\underline{s}_{ij}\right) - \min_{1\le i\le m} f\left(\underline{s}_{ij}\right)\right), \left(f(\bar{s}_{ij}) - \min_{1\le i\le m} f(\bar{s}_{ij})\right)\right) \\
u_{ij}^{-} &= u_{ij}^{2} - \min_{1\le i\le m} u_{ij}^{2}, v_{ij}^{-} = v_{ij}^{2} - \min_{1\le i\le m} v_{ij}^{2} \\
f\left(\underline{s}_{ij}^{+}\right) &= \min\left(\left(\max_{1\le i\le m} f\left(\underline{s}_{ij}\right) - f\left(\underline{s}_{ij}\right)\right), \left(\max_{1\le i\le m} f(\bar{s}_{ij}) - f(\bar{s}_{ij})\right)\right) \\
f\left(\bar{s}_{ij}^{+}\right) &= \max\left(\left(\max_{1\le i\le m} f\left(\underline{s}_{ij}\right) - f\left(\underline{s}_{ij}\right)\right), \left(\max_{1\le i\le m} f(\bar{s}_{ij}) - f(\bar{s}_{ij})\right)\right) \\
u_{ij}^{+} &= \max_{1\le i\le m} u_{ij}^{2} - u_{ij}^{2}, v_{ij}^{+} = \max_{1\le i\le m} v_{ij}^{2} - v_{ij}^{2}
\end{aligned} \quad (1)
$$

The modules are computed as

$$
|X^{-}X^{+}| = \sqrt{\sum_{j=1}^{n}\left(\left(f\left(\underline{s}_{ij}^{t}\right)\right)^{2} + \left(f\left(\bar{s}_{ij}^{t}\right)\right)^{2} + \left(u_{ij}^{t}\right)^{4} + \left(v_{ij}^{t}\right)^{4}\right)} \quad (8)
$$

$$
|X^{-}X_{i}| = \sqrt{\sum_{j=1}^{n}\left(\left(f\left(\underline{s}_{ij}^{-}\right)\right)^{2} + \left(f\left(\bar{s}_{ij}^{-}\right)\right)^{2} + \left(u_{ij}^{-}\right)^{4} + \left(v_{ij}^{-}\right)^{4}\right)} \quad (9)
$$

The cosine of $X^{-}X^{+}$ and $X^{-}X_{i}$ is expressed as

$$
\cos\left(X^{-}X_{i}, X^{-}X^{+}\right) = \frac{\sum_{j=1}^{n}\left(f\left(\underline{s}_{ij}^{t}\right)\cdot f\left(\underline{s}_{ij}^{-}\right) + f\left(\bar{s}_{ij}^{t}\right)\cdot f\left(\bar{s}_{ij}^{-}\right) + \left(u_{ij}^{t}\cdot u_{ij}^{-}\right)^{2} + \left(v_{ij}^{t}\cdot v_{\alpha_{ij}}^{-}\right)^{2}\right)}{|X^{-}X^{+}|\cdot|X^{-}X_{i}|} \quad (10)
$$

The projection value of $X^{-}X_{i}$ on $X^{-}X^{+}$ and $X^{-}X^{+}$ on $X_{i}X^{+}$ are calculated as

$$
\begin{aligned}
prj_{X^{-}X^{+}}(X^{-}X_{i}) &= |X^{-}X_{i}|\cdot\cos(X^{-}X_{i}, X^{-}X^{+}) \\
&= \frac{\sum_{j=1}^{n}\left(f\left(\underline{s}_{ij}^{t}\right)\cdot f\left(\underline{s}_{ij}^{-}\right) + f\left(\bar{s}_{ij}^{t}\right)\cdot f\left(\bar{s}_{ij}^{-}\right) + \left(u_{ij}^{t}\cdot u_{ij}^{-}\right)^{2} + \left(v_{ij}^{t}\cdot v_{ij}^{-}\right)^{2}\right)}{|X^{-}X^{+}|}
\end{aligned} \quad (11)
$$

$$
\begin{aligned}
prj_{X_{i}X^{+}}(X^{-}X^{+}) &= |X^{-}X^{+}|\cdot\cos(X_{i}X^{+}, X^{-}X^{+}) \\
&= \frac{\sum_{j=1}^{n}\left(f\left(\underline{s}_{ij}^{t}\right)\cdot f\left(\underline{s}_{ij}^{+}\right) + f\left(\bar{s}_{ij}^{t}\right)\cdot f\left(\bar{s}_{ij}^{+}\right) + \left(u_{ij}^{t}\cdot u_{ij}^{+}\right)^{2} + \left(v_{ij}^{t}\cdot v_{ij}^{+}\right)^{2}\right)}{|X_{i}X^{+}|}
\end{aligned} \quad (12)
$$

Theorem 2. *The bigger the value of $prj_{X^{-}X^{+}}(X^{-}X_{i})$, the closer the alternative X_i to positive ideal alternative X^{+}. Analogously, the bigger the value of $prj_{X_{i}X^{+}}(X^{-}X^{+})$, the closer the alternative X_i will be to negative ideal alternative X^{-} (as shown in Figure 1 [29]).*

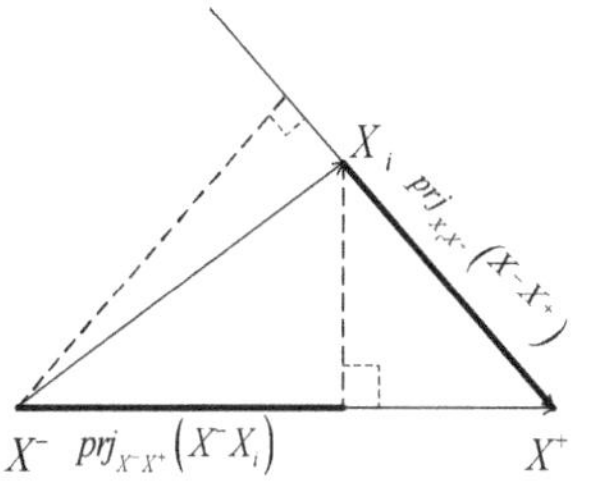

Figure 1. The graphical representation of $prj_{X^{-}X^{+}}(X^{-}X_{i})$ and $prj_{X_{i}X^{+}}(X^{-}X^{+})$.

4. Comparative Analysis of Projection Model and Bi-Directional Projection Model

Let $X_i = \left\langle \left[f\left(\underline{s}_{ij}\right), f\left(\overline{s}_{ij}\right)\right], \widetilde{P}\left(u_{ij}, v_{ij}\right)\right\rangle$ and $X_l = \left\langle \left[f\left(\underline{s}_{lj}\right), f\left(\overline{s}_{lj}\right)\right], \widetilde{P}\left(u_{lj}, v_{lj}\right)\right\rangle$, which are two alternatives with Pythagorean uncertain linguistic variable information, the positive and negative ideal alternatives are defined as $X^+ = \left\langle \left[\max\limits_{1\le i\le m} f\left(\underline{s}_{ij}\right), \max\limits_{1\le i\le m} f\left(\overline{s}_{ij}\right)\right], \widetilde{P}\left(\max\limits_{1\le i\le m} u_{ij}, \min\limits_{1\le i\le m} v_{ij}\right)\right\rangle$ and $X^- = \left\langle \left[\min\limits_{1\le i\le m} f\left(\underline{s}_{ij}\right), \min\limits_{1\le i\le m} f\left(\overline{s}_{ij}\right)\right], \widetilde{P}\left(\min\limits_{1\le i\le m} u_{ij}, \max\limits_{1\le i\le m} v_{ij}\right)\right\rangle$, respectively. The X_i and X_l distribute on the perpendicular bisector of X^- and X^+ (shown as Figure 2). We compare the X_i and X_l via projection and bi-directional projection models, respectively.

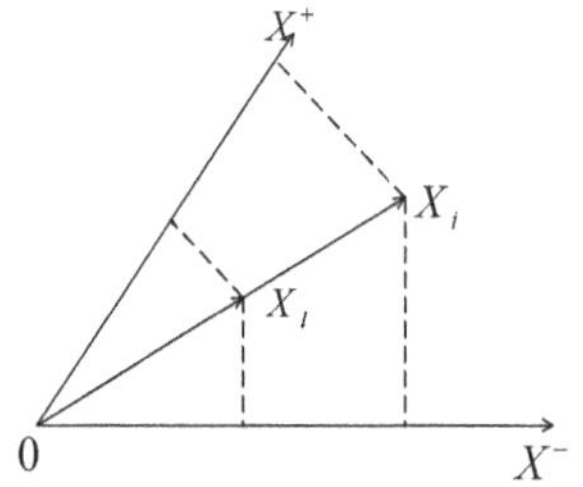

Figure 2. The closeness degree of projection model.

4.1. Projection Model

Step 1: Compute the projection of X_i and X_l on the positive and negative ideal alternatives via (4).

$$prj_{X^+}(X_i) = |X_i| \cdot \cos(X_i, X^+) = \frac{\sum\limits_{j=1}^{n}\left(f\left(\underline{s}_{ij}\right)\cdot \max\limits_{1\le i\le m} f\left(\underline{s}_{ij}\right) + f\left(\overline{s}_{ij}\right)\cdot \max\limits_{1\le i\le m} f\left(\overline{s}_{ij}\right) + u_{ij}^2 \cdot \max\limits_{1\le i\le m} u_{ij}^2 + v_{ij}^2 \cdot \min\limits_{1\le i\le m} v_{ij}^2\right)}{\sqrt{\sum\limits_{j=1}^{n}\left(\max\limits_{1\le i\le m} f\left(\underline{s}_{ij}\right)\right)^2 + \left(\max\limits_{1\le i\le m} f\left(\overline{s}_{ij}\right)\right)^2 + \left(\max\limits_{1\le i\le m} u_{ij}\right)^4 + \left(\min\limits_{1\le i\le m} v_{ij}\right)^4}} \tag{13}$$

$$prj_{X^-}(X_i) = |X_i| \cdot \cos(X_i, X^-) = \frac{\sum\limits_{j=1}^{n}\left(f\left(\underline{s}_{ij}\right)\cdot \min\limits_{1\le i\le m} f\left(\underline{s}_{ij}\right) + f\left(\overline{s}_{ij}\right)\cdot \min\limits_{1\le i\le m} f\left(\overline{s}_{ij}\right) + u_{ij}^2 \cdot \min\limits_{1\le i\le m} u_{ij}^2 + v_{ij}^2 \cdot \max\limits_{1\le i\le m} v_{ij}^2\right)}{\sqrt{\sum\limits_{j=1}^{n}\left(\min\limits_{1\le i\le m} f\left(\underline{s}_{ij}\right)\right)^2 + \left(\min\limits_{1\le i\le m} f\left(\overline{s}_{ij}\right)\right)^2 + \left(\min\limits_{1\le i\le m} u_{ij}\right)^4 + \left(\max\limits_{1\le i\le m} v_{ij}\right)^4}} \tag{14}$$

$$prj_{X^+}(X_l) = |X_l| \cdot \cos(X_l, X^+) = \frac{\sum\limits_{j=1}^{n}\left(f\left(\underline{s}_{lj}\right)\cdot \max\limits_{1\le l\le m} f\left(\underline{s}_{lj}\right) + f\left(\overline{s}_{lj}\right)\cdot \max\limits_{1\le l\le m} f\left(\overline{s}_{lj}\right) + u_{lj}^2 \cdot \max\limits_{1\le l\le m} u_{lj}^2 + v_{lj}^2 \cdot \min\limits_{1\le l\le m} v_{lj}^2\right)}{\sqrt{\sum\limits_{j=1}^{n}\left(\max\limits_{1\le l\le m} f\left(\underline{s}_{lj}\right)\right)^2 + \left(\max\limits_{1\le l\le m} f\left(\overline{s}_{lj}\right)\right)^2 + \left(\max\limits_{1\le l\le m} u_{lj}\right)^4 + \left(\min\limits_{1\le l\le m} v_{lj}\right)^4}} \tag{15}$$

$$prj_{X^-}(X_l) = |X_l| \cdot \cos(X_l, X^-) = \frac{\sum\limits_{j=1}^{n}\left(f\left(\underline{s}_{lj}\right)\cdot \min\limits_{1\le l\le m} f\left(\underline{s}_{lj}\right) + f\left(\overline{s}_{lj}\right)\cdot \min\limits_{1\le l\le m} f\left(\overline{s}_{lj}\right) + u_{lj}^2 \cdot \min\limits_{1\le l\le m} u_{lj}^2 + v_{lj}^2 \cdot \max\limits_{1\le l\le m} v_{lj}^2\right)}{\sqrt{\sum\limits_{j=1}^{n}\left(\min\limits_{1\le l\le m} f\left(\underline{s}_{lj}\right)\right)^2 + \left(\min\limits_{1\le l\le m} f\left(\overline{s}_{lj}\right)\right)^2 + \left(\min\limits_{1\le l\le m} u_{lj}\right)^4 + \left(\max\limits_{1\le l\le m} v_{lj}\right)^4}} \tag{16}$$

Step 2: Calculate the closeness degree of X_i and X_l to ideal alternatives, respectively.

$$C(X_i) = \frac{prj_{X^+}(X_i)}{prj_{X^-}(X_i) + prj_{X^+}(X_i)} \tag{17}$$

$$C(X_l) = \frac{prj_{X^+}(X_l)}{prj_{X^-}(X_l) + prj_{X^+}(X_l)} \tag{18}$$

we can see $prj_{X^+}(X_i) = prj_{X^-}(X_i)$ and $prj_{X^+}(X_l) = prj_{X^-}(X_l)$ because the X_i and X_l distribute on the perpendicular bisector of X^- and X^+. Therefore, $C(X_i) = C(X_l) = \frac{1}{2}$, $X_i \sim X_l$.

Where the closeness degree is ranking indicators. The bigger value of the closeness degree, the better the preference order of the alternatives.

4.2. Bi-Directional Projection Model

Step 1: Compute the formative vector: X_iX^+, X^-X_i, X_lX^+, X^-X_l, respectively.

$$X^-X^+ = \left\langle \left[f\left(\underline{s}_{ij}^t\right), f\left(\overline{s}_{ij}^t\right) \right], \widetilde{P}\left(u_{ij}^t, v_{ij}^t\right) \right\rangle \tag{19}$$

$$X^-X_i = \left\langle \left[f\left(\underline{s}_{ij}^-\right), f\left(\overline{s}_{ij}^-\right) \right], \widetilde{P}\left(u_{ij}^-, v_{ij}^-\right) \right\rangle \tag{20}$$

$$X_iX^+ = \left\langle \left[f\left(\underline{s}_{ij}^+\right), f\left(\overline{s}_{ij}^+\right) \right], \widetilde{P}\left(u_{ij}^+, v_{ij}^+\right) \right\rangle \tag{21}$$

$$X_lX^+ = \left\langle \left[f\left(\underline{s}_{lj}^+\right), f\left(\overline{s}_{lj}^+\right) \right], \widetilde{P}\left(u_{lj}^+, v_{lj}^+\right) \right\rangle \tag{22}$$

$$X^-X_l = \left\langle \left[f\left(\underline{s}_{lj}^-\right), f\left(\overline{s}_{lj}^-\right) \right], \widetilde{P}\left(u_{lj}^-, v_{lj}^-\right) \right\rangle \tag{23}$$

Step 2: Calculate the projection value of formative vector X^-X_i and X^-X_l to X^-X^+, denoted as $prj_{X^-X^+}(X^-X_i)$, $prj_{X^-X^+}(X^-X_l)$, and the projection value of formative vector X^-X^+ to X_iX^+ and X_lX^+, denoted as $prj_{X_iX^+}(X^-X^+)$, $prj_{X_lX^+}(X^-X^+)$, respectively (shown as Figure 3).

$$\begin{aligned} prj_{X^-X^+}(X^-X_i) &= |X^-X_i| \cdot \cos(X^-X_i, X^-X^+) \\ &= \frac{\sum_{j=1}^{n}\left(f\left(\underline{s}_{ij}^t\right)\cdot f\left(\underline{s}_{ij}^-\right) + f\left(\overline{s}_{ij}^t\right)\cdot f\left(\overline{s}_{ij}^-\right) + \left(u_{ij}^t \cdot u_{ij}^-\right)^2 + \left(v_{ij}^t \cdot v_{ij}^-\right)^2 \right)}{|X^-X^+|} \end{aligned} \tag{24}$$

$$\begin{aligned} prj_{X_iX^+}(X^-X^+) &= |X^-X^+| \cdot \cos(X_iX^+, X^-X^+) \\ &= \frac{\sum_{j=1}^{n}\left(f\left(\underline{s}_{ij}^t\right)\cdot f\left(\underline{s}_{ij}^+\right) + f\left(\overline{s}_{ij}^t\right)\cdot f\left(\overline{s}_{ij}^+\right) + \left(u_{ij}^t \cdot u_{ij}^+\right)^2 + \left(v_{ij}^t \cdot v_{ij}^+\right)^2 \right)}{|X_iX^+|} \end{aligned} \tag{25}$$

$$\begin{aligned} prj_{X^-X^+}(X^-X_l) &= |X^-X_l| \cdot \cos(X^-X_l, X^-X^+) \\ &= \frac{\sum_{j=1}^{n}\left(f\left(\underline{s}_{lj}^t\right)\cdot f\left(\underline{s}_{lj}^-\right) + f\left(\overline{s}_{lj}^t\right)\cdot f\left(\overline{s}_{lj}^-\right) + \left(u_{lj}^t \cdot u_{lj}^-\right)^2 + \left(v_{lj}^t \cdot v_{lj}^-\right)^2 \right)}{|X^-X^+|} \end{aligned} \tag{26}$$

$$\begin{aligned} prj_{X_lX^+}(X^-X^+) &= |X^-X^+| \cdot \cos(X_lX^+, X^-X^+) \\ &= \frac{\sum_{j=1}^{n}\left(f\left(\underline{s}_{lj}^t\right)\cdot f\left(\underline{s}_{lj}^+\right) + f\left(\overline{s}_{lj}^t\right)\cdot f\left(\overline{s}_{lj}^+\right) + \left(u_{lj}^t \cdot u_{lj}^+\right)^2 + \left(v_{lj}^t \cdot v_{lj}^+\right)^2 \right)}{|X_iX^+|} \end{aligned} \tag{27}$$

Step 3: Compute the closeness degree of X_i and X_l to ideal alternatives.

$$C(X_i) = \frac{prj_{X^-X^+}(X^-X_i)}{prj_{X^-X^+}(X^-X_i) + prj_{X_iX^+}(X^-X^+)} \tag{28}$$

$$C(X_l) = \frac{prj_{X^-X^+}(X^-X_l)}{prj_{X^-X^+}(X^-X_l) + prj_{X_lX^+}(X^-X^+)} \tag{29}$$

We can see $prj_{X^-X^+}(X^-X_i) = prj_{X^-X^+}(X^-X_l)$ because X_i and X_l distribute on the perpendicular bisector of ideal alternatives, as shown in Figure 3 $prj_{X_lX^+}(X^-X^+) > prj_{X_iX^+}(X^-X^+)$. Therefore, $C(X_i) > C(X_l)$, $X_i \succ X_l$.

From the foregoing analysis, we can know the projection model is difficult to obtain the ranking order of X_i and X_l, when they distribute on the perpendicular bisector of ideal alternatives. Whereas, the bi-directional projection model can remarkably overcome the drawback and get the rational ranking order of X_i and X_l.

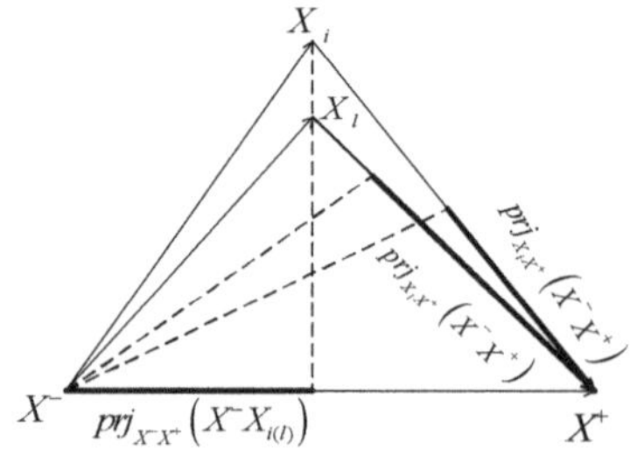

Figure 3. The closeness degree of bi-directional projection model.

5. Decision Making Steps of Bi-Directional Projection Model

To solve certain decision making problems, we propose a new bi-directional projection model based on Pythagorean uncertain linguistic variables. $X = \{X_1, X_2, \cdots, X_n\}$ denotes the set of alternatives, the set of attributes are denoted by $C = \{C_1, C_2, \cdots, C_n\}$ and the weights are represented by $W = \{w_1, w_2, \cdots, w_n\}$, where $w_j \in [0,1]$, $\sum_{j=1}^{n} w_j = 1$. $\alpha = \left\langle \left[\underline{s}_{ij}, \overline{s}_{ij}\right], \widetilde{P}(u_{ij}, v_{ij}) \right\rangle$ is the evaluation value of X_i under C_j with Pythagorean uncertain linguistic variable information. The linguistic term set is $S = \{s_0, s_1, \cdots, s_{2z}\}$, I_b and I_c denote the benefit attribute and cost attribute, respectively. In general, the proposed method involves the following steps:

Step 1: Construct the decision making matrix $\alpha = (\alpha_{ij})$ with Pythagorean uncertain linguistic variable information, and normalize the decision matrix.

$$\alpha_{ij} = \begin{cases} \alpha_{ij}\, , benefit\ attribute\ I_b \\ (\alpha_{ij})^c, cost\ attribute\ I_c \end{cases}$$

where, $(\alpha_{ij})^c$ is the complement of α_{ij}, and the form of $(\alpha_{ij})^c$ is defined as $(\alpha_{ij})^c = \left\langle \left[f^-(f(s_t) - f(\overline{s}_{ij})), f^-\left(f(s_t) - f\left(\underline{s}_{ij}\right)\right)\right], P(v_{ij}, u_{ij}) \right\rangle$.

Step 2: Convert the Pythagorean uncertain linguistic variable to Pythagorean uncertain linguistic function via the linguistic scale function.

Step 3: Determine the ideal alternatives $\beta^+ = \{\beta_1^+, \beta_2^+, \cdots, \beta_n^+\}$ and $\beta^- = \{\beta_1^-, \beta_2^-, \cdots, \beta_n^-\}$.

Where,

$$\beta_j^+ = \left\langle \left[\max_{1 \le i \le m} f\left(\underline{s}_{ij}\right), \max_{1 \le i \le m} f(\overline{s}_{ij})\right], \widetilde{P}\left(\max_{1 \le i \le m} u_{ij}, \min_{1 \le i \le m} v_{ij}\right) \right\rangle$$

$$\beta_j^- = \left\langle \left[\min_{1 \le i \le m} f\left(\underline{s}_{ij}\right), \min_{1 \le i \le m} f(\overline{s}_{ij})\right], \widetilde{P}\left(\min_{1 \le i \le m} u_{ij}, \max_{1 \le i \le m} v_{ij}\right) \right\rangle$$

Step 4: Compute the formative vector of ideal alternative and X_i via (2).

Step 5: Calculate the projection value of formative vector X^-X_i to X^-X^+, denoted as $prj_{X^-X^+}(X^-X_i)$, and the projection value of formative vector X^-X^+ to X_iX^+, denoted as $prj_{X_iX^+}(X^-X^+)$, respectively.

Step 6: Develop a closeness degree formula based on TOPSIS method, and obtain the ranking order of all alternatives via closeness degree.

$$C(X_i) = \frac{prj_{X^-X^+}(X^-X_i)}{prj_{X^-X^+}(X^-X_i) + prj_{X_iX^+}(X^-X^+)}$$

6. Numerical Example

It is essential to choose the right enterprise for graduates' future development. In order to provide reasonable employment guidance for graduates, we get the influence factors by questionnaires of 260 college graduates of Shandong province. After eliminating the invalid and incomplete questionnaires, seven main attributes are selected to evaluate the alternative companies according to 210 valid questionnaires. The set of attributes is: {*prospects of company, working strength, wage level, personal prospects, social insurance and house funding, professional relevance, geographical*}. For the sake of convenience, the set of attributes is denoted by $\{C_1, C_2, C_3, C_4, C_5, C_6, C_7\}$. C_2 is cost attribute and the rest are benefit attributes. The detailed guidance process of a graduate is shown as follows: The set of companies are denoted as: $\{A_1, A_2, \cdots, A_{10}\}$ and the weights are $\{w_1, w_2, \cdots, w_7\}$, where, $w_1 = 0.12, w_2 = 0.06, w_3 = 0.2, w_4 = 0.24, w_5 = 0.1, w_6 = 0.21, w_7 = 0.07$, which are given by experts, $\alpha = \left\langle \left[\underline{s}_{ij}, \bar{s}_{ij}\right], \widetilde{P}(u_{ij}, v_{ij}) \right\rangle$ is the evaluation value of X_i under C_j with Pythagorean uncertain linguistic variable information. The linguistic term set is: S = {s_0 = *extremely bad*, s_1 = *very bad*, s_2 = *bad*, s_3 = *slightly bad*, s_4 = *fair*, s_5 = *good*, s_6 = *slightly good*, s_7 = *very good*, s_8 = *extremely good*}. Determine the ranking order of the 10 companies based on bi-directional projection model.

Step 1: Construct the decision making matrix $\alpha = (\alpha_{ij})$ with Pythagorean uncertain linguistic variable information, and normalize the decision matrix.

	c_1	c_2	c_3	c_4	c_5
A_1	$\langle[s_4,s_6],(0.8,0.6)\rangle$	$\langle[s_5,s_6],(0.6,0.8)\rangle$	$\langle[s_5,s_7],(0.8,0.6)\rangle$	$\langle[s_4,s_6],(0.7,0.6)\rangle$	$\langle[s_4,s_5],(0.8,0.6)\rangle$
A_2	$\langle[s_5,s_7],(0.8,0.4)\rangle$	$\langle[s_4,s_7],(0.6,0.7)\rangle$	$\langle[s_6,s_7],(0.9,0.3)\rangle$	$\langle[s_7,s_8],(0.8,0.3)\rangle$	$\langle[s_6,s_7],(0.9,0.2)\rangle$
A_3	$\langle[s_4,s_7],(0.7,0.6)\rangle$	$\langle[s_4,s_6],(0.6,0.7)\rangle$	$\langle[s_4,s_6],(0.8,0.5)\rangle$	$\langle[s_4,s_5],(0.6,0.5)\rangle$	$\langle[s_6,s_8],(0.7,0.4)\rangle$
A_4	$\langle[s_4,s_5],(0.7,0.5)\rangle$	$\langle[s_5,s_8],(0.5,0.7)\rangle$	$\langle[s_3,s_5],(0.6,0.5)\rangle$	$\langle[s_5,s_6],(0.7,0.4)\rangle$	$\langle[s_4,s_5],(0.6,0.3)\rangle$
A_5	$\langle[s_5,s_7],(0.8,0.4)\rangle$	$\langle[s_5,s_7],(0.5,0.8)\rangle$	$\langle[s_5,s_6],(0.6,0.4)\rangle$	$\langle[s_6,s_7],(0.6,0.5)\rangle$	$\langle[s_3,s_6],(0.6,0.5)\rangle$
A_6	$\langle[s_6,s_8],(0.7,0.5)\rangle$	$\langle[s_6,s_7],(0.4,0.5)\rangle$	$\langle[s_3,s_5],(0.8,0.5)\rangle$	$\langle[s_4,s_7],(0.7,0.5)\rangle$	$\langle[s_5,s_7],(0.7,0.4)\rangle$
A_7	$\langle[s_5,s_6],(0.6,0.5)\rangle$	$\langle[s_6,s_8],(0.6,0.7)\rangle$	$\langle[s_4,s_5],(0.6,0.5)\rangle$	$\langle[s_6,s_8],(0.7,0.3)\rangle$	$\langle[s_6,s_7],(0.7,0.5)\rangle$
A_8	$\langle[s_4,s_6],(0.7,0.6)\rangle$	$\langle[s_5,s_6],(0.3,0.9)\rangle$	$\langle[s_5,s_7],(0.6,0.4)\rangle$	$\langle[s_4,s_5],(0.6,0.4)\rangle$	$\langle[s_4,s_5],(0.8,0.5)\rangle$
A_9	$\langle[s_6,s_8],(0.7,0.6)\rangle$	$\langle[s_4,s_7],(0.4,0.8)\rangle$	$\langle[s_4,s_5],(0.7,0.5)\rangle$	$\langle[s_5,s_6],(0.6,0.5)\rangle$	$\langle[s_4,s_7],(0.6,0.5)\rangle$
A_{10}	$\langle[s_4,s_7],(0.8,0.5)\rangle$	$\langle[s_3,s_5],(0.5,0.7)\rangle$	$\langle[s_4,s_6],(0.6,0.5)\rangle$	$\langle[s_4,s_6],(0.7,0.5)\rangle$	$\langle[s_4,s_5],(0.7,0.4)\rangle$

	c_6	c_7
A_1	$\langle[s_6,s_7],(0.7,0.6)\rangle$	$\langle[s_6,s_8],(0.8,0.5)\rangle$
A_2	$\langle[s_6,s_8],(0.8,0.6)\rangle$	$\langle[s_6,s_7],(0.8,0.4)\rangle$
A_3	$\langle[s_4,s_5],(0.7,0.5)\rangle$	$\langle[s_4,s_7],(0.6,0.4)\rangle$
A_4	$\langle[s_5,s_6],(0.6,0.4)\rangle$	$\langle[s_5,s_7],(0.6,0.5)\rangle$
A_5	$\langle[s_5,s_7],(0.7,0.4)\rangle$	$\langle[s_4,s_7],(0.7,0.4)\rangle$
A_6	$\langle[s_5,s_8],(0.7,0.3)\rangle$	$\langle[s_5,s_7],(0.8,0.6)\rangle$
A_7	$\langle[s_6,s_7],(0.8,0.4)\rangle$	$\langle[s_4,s_6],(0.7,0.5)\rangle$
A_8	$\langle[s_4,s_5],(0.9,0.3)\rangle$	$\langle[s_6,s_7],(0.7,0.6)\rangle$
A_9	$\langle[s_4,s_6],(0.6,0.4)\rangle$	$\langle[s_7,s_8],(0.6,0.3)\rangle$
A_{10}	$\langle[s_5,s_7],(0.7,0.5)\rangle$	$\langle[s_4,s_5],(0.7,0.4)\rangle$

Step 2: Convert the Pythagorean uncertain linguistic variable to Pythagorean uncertain linguistic function via the linguistic scale function, where $\alpha = 0.6$, $\gamma = 0.8$.

	c_1	c_2	c_3	c_4	c_5
A_1	$\langle[0.5,0.79],(0.8,0.6)\rangle$	$\langle[0.66,0.79],(0.6,0.8)\rangle$	$\langle[0.66,0.9],(0.8,0.6)\rangle$	$\langle[0.5,0.79],(0.7,0.6)\rangle$	$\langle[0.5,0.66],(0.8,0.6)\rangle$
A_2	$\langle[0.66,0.9],(0.8,0.4)\rangle$	$\langle[0.5,0.9],(0.6,0.7)\rangle$	$\langle[0.79,0.9],(0.9,0.3)\rangle$	$\langle[0.9,1],(0.8,0.3)\rangle$	$\langle[0.79,0.9],(0.9,0.2)\rangle$
A_3	$\langle[0.5,0.9],(0.7,0.6)\rangle$	$\langle[0.5,0.79],(0.6,0.7)\rangle$	$\langle[0.5,0.79],(0.8,0.5)\rangle$	$\langle[0.5,0.66],(0.6,0.5)\rangle$	$\langle[0.79,1],(0.7,0.4)\rangle$
A_4	$\langle[0.5,0.66],(0.7,0.5)\rangle$	$\langle[0.66,1],(0.5,0.7)\rangle$	$\langle[0.28,0.66],(0.6,0.5)\rangle$	$\langle[0.66,0.79],(0.7,0.4)\rangle$	$\langle[0.5,0.66],(0.6,0.3)\rangle$
A_5	$\langle[0.66,0.9],(0.8,0.4)\rangle$	$\langle[0.66,0.9],(0.5,0.8)\rangle$	$\langle[0.66,0.79],(0.6,0.4)\rangle$	$\langle[0.79,0.9],(0.6,0.5)\rangle$	$\langle[0.28,0.79],(0.6,0.5)\rangle$
A_6	$\langle[0.79,1],(0.7,0.5)\rangle$	$\langle[0.79,0.9],(0.4,0.5)\rangle$	$\langle[0.28,0.66],(0.8,0.5)\rangle$	$\langle[0.5,0.9],(0.7,0.5)\rangle$	$\langle[0.66,0.89],(0.7,0.4)\rangle$
A_7	$\langle[0.66,0.79],(0.6,0.5)\rangle$	$\langle[0.79,1],(0.6,0.7)\rangle$	$\langle[0.5,0.66],(0.6,0.5)\rangle$	$\langle[0.79,1],(0.7,0.3)\rangle$	$\langle[0.79,0.9],(0.7,0.5)\rangle$
A_8	$\langle[0.5,0.79],(0.7,0.6)\rangle$	$\langle[0.66,0.79],(0.3,0.9)\rangle$	$\langle[0.66,0.9],(0.6,0.4)\rangle$	$\langle[0.5,0.66],(0.6,0.4)\rangle$	$\langle[0.5,0.66],(0.8,0.5)\rangle$
A_9	$\langle[0.79,1],(0.7,0.6)\rangle$	$\langle[0.5,0.9],(0.4,0.8)\rangle$	$\langle[0.5,0.66],(0.7,0.5)\rangle$	$\langle[0.66,0.79],(0.6,0.5)\rangle$	$\langle[0.5,0.9],(0.6,0.5)\rangle$
A_{10}	$\langle[0.5,1],(0.8,0.5)\rangle$	$\langle[0.28,0.66],(0.5,0.7)\rangle$	$\langle[0.5,0.79],(0.6,0.5)\rangle$	$\langle[0.5,0.79],(0.7,0.5)\rangle$	$\langle[0.5,0.66],(0.7,0.4)\rangle$

	c_6	c_7
A_1	$\langle[0.79,0.9],(0.7,0.6)\rangle$	$\langle[0.79,1],(0.8,0.5)\rangle$
A_2	$\langle[0.79,1],(0.8,0.6)\rangle$	$\langle[0.79,0.9],(0.8,0.4)\rangle$
A_3	$\langle[0.5,0.66],(0.7,0.5)\rangle$	$\langle[0.5,0.9],(0.6,0.4)\rangle$
A_4	$\langle[0.66,0.79],(0.6,0.4)\rangle$	$\langle[0.66,0.9],(0.6,0.5)\rangle$
A_5	$\langle[0.66,0.9],(0.7,0.4)\rangle$	$\langle[0.5,0.9],(0.7,0.4)\rangle$
A_6	$\langle[0.66,1],(0.7,0.3)\rangle$	$\langle[0.66,0.9],(0.8,0.6)\rangle$
A_7	$\langle[0.79,0.9],(0.8,0.4)\rangle$	$\langle[0.5,0.79],(0.7,0.5)\rangle$
A_8	$\langle[0.5,0.66],(0.9,0.3)\rangle$	$\langle[0.79,0.9],(0.7,0.6)\rangle$
A_9	$\langle[0.5,0.79],(0.6,0.4)\rangle$	$\langle[0.9,1],(0.6,0.3)\rangle$
A_{10}	$\langle[0.66,0.9],(0.7,0.5)\rangle$	$\langle[0.5,0.66],(0.7,0.4)\rangle$

Step 3: Determine the positive and negative ideal alternatives X^+ and X^-.

$$X^+ = \{\langle[0.79,1],(0.8,0.4)\rangle, \langle[0.79,1],(0.6,0.5)\rangle, \langle[0.79,1],(0.9,0.3)\rangle, \langle[0.9,1],(0.8,0.3)\rangle, \langle[0.79,1],(0.9,0.3)\rangle, \langle[0.79,1],(0.9,0.3)\rangle, \langle[0.9,1],(0.8,0.3)\rangle\}$$

$$X^- = \{\langle[0.5,0.66],(0.6,0.6)\rangle, \langle[0.28,0.66],(0.3,0.9)\rangle, \langle[0.28,0.66],(0.6,0.6)\rangle, \langle[0.5,0.66],(0.6,0.6)\rangle, \langle[0.28,0.66],(0.6,0.6)\rangle, \langle[0.5,0.66],(0.6,0.6)\rangle, \langle[0.5,0.66],(0.6,0.6)\rangle\}$$

Step 4: Compute the formative vector of ideal alternative and X_i via (2).

$$X^-X^+ = \{\langle[0.29,0.34],(0.2,0.2)\rangle, \langle[0.34,0.51],(0.3,0.4)\rangle, \langle[0.34,0.51],(0.3,0.3)\rangle, \langle[0.34,0.4],(0.2,0.3)\rangle, \langle[0.24,0.51],(0.3,0.3)\rangle, \langle[0.29,0.34],(0.3,0.3)\rangle, \langle[0.34,0.4],(0.2,0.3)\rangle\}$$

Similarly, we can get the formative vector of ideal alternative and .

Step 5: Calculate the projection value of formative vector X^-X_i to X^-X^+, denoted as $prj_{X^-X^+}(X^-X_i)$, and the projection value of formative vector X^-X^+ to X_iX^+, denoted as $prj_{X_iX^+}(X^-X^+)$, respectively.

$prj_{X^-X^+}(X^-X_i)$	0.487	0.822	0.429	0.357	0.508	0.59	0.607	0.412	0.467	0.341
$prj_{X_iX^+}(X^-X^+)$	1.39	2.057	1.322	1.288	1.488	1.395	1.512	1.261	1.403	1.269

Step 6: Compute the closeness degree and obtain the ranking order of all alternatives via closeness degree.

$C(X_i)$	0.259	0.285	0.245	0.217	0.254	0.297	0.287	0.246	0.25	0.212

where,

$$C(X_6) > C(X_7) > C(X_2) > C(X_1) > C(X_5) > C(X_9) > C(X_8) > C(X_3) > C(X_4) > C(X_{10})$$

Therefore, $A_6 \succ A_7 \succ A_2 \succ A_1 \succ A_5 \succ A_9 \succ A_8 \succ A_3 \succ A_4 \succ A_{10}$, and A_6 is the best company for this graduate.

7. Conclusions

To solve multi-attribute decision making (MADM) problems with Pythagorean uncertain linguistic variables, we proposed an extended bi-directional projection model. The extended model can take the advantages of the Pythagorean uncertain linguistic variable and projection models, and effectively overcome the drawbacks of the single distance measure. The feasibility of the proposed method is demonstrated by the graduates' job-hunting problem.

The superiority of our bi-directional projection model is that it can consider the angle and distance between two evaluation values simultaneously. Compared with projection model, the proposed model can handle the real-life case of alternatives distribution on the perpendicular bisector of positive and negative ideal alternatives, which made it widely suitable in MADM. However, the proposed method does not consider the psychological risk factors of decision makers in this paper, which will be explored in the future research.

Author Contributions: Huidong Wang conceived the idea; Shifan He and Xiaohong Pan wrote the paper and revised the paper. The authors have read and approved the final manuscript.

Funding: This work was supported by NSFC Foundation under Grant No. 61402260.

Conflicts of Interest: The authors declare no conflict of interest.

References

1. Zadeh, L.A. Fuzzy sets. *Inf. Control* **1965**, *8*, 338–353. [CrossRef]
2. Shen, F.; Ma, X.S.; Li, Z.Y.; Cai, D.L. An extended intuitionistic fuzzy TOPSIS method based on a new distance measure with an application to credit risk evaluation. *Inf. Sci.* **2018**, *428*, 105–119. [CrossRef]
3. Qiu, J.D.; Li, L. A new approach for multiple attribute group decision making with interval-valued intuitionistic fuzzy information. *Appl. Soft Comput.* **2017**, *61*, 111–121. [CrossRef]
4. Li, C.D.; Gao, J.L.; Yi, J.Q.; Zhang, G.Q. Analysis and design of functionally weighted single-input-rule-modules connected fuzzy inference systems. *IEEE Trans. Fuzzy Syst.* **2018**, *26*, 56–71. [CrossRef]
5. Cheng, S.H. Autocratic multiattribute group decision making for hotel location selection based on interval-valued intuitionistic fuzzy sets. *Inf. Sci.* **2018**, *427*, 77–87. [CrossRef]
6. Yager, R.R. Pythagorean membership grades in multicriteria decision making. *IEEE Trans. Fuzzy Syst.* **2014**, *22*, 958–965. [CrossRef]
7. Du, Y.Q.; Hu, F.J.; Zafar, W.; Yu, Q.; Zhai, Y. A Novel Method for Multi-attribute Decision Making with Interval-Valued Pythagorean Fuzzy Linguistic Information. *Int. J. Intell. Syst.* **2017**, *32*, 1085–1112. [CrossRef]
8. Liang, D.C.; Xu, Z.S.; Liu, D.; Wu, Y. Method for Three-Way Decisions using Ideal TOPSIS Solutions at Pythagorean Fuzzy Information. *Inf. Sci.* **2018**, *435*, 282–295. [CrossRef]
9. Liang, D.C.; Xu, Z.S. The new extension of TOPSIS method for multiple criteria decision making with hesitant Pythagorean fuzzy sets. *Appl. Soft Comput.* **2017**, *60*, 167–179. [CrossRef]
10. Zhang, X.L.; Xu, Z.S. Extension of TOPSIS to Multiple Criteria Decision Making with Pythagorean Fuzzy Sets. *Int. J. Intell. Syst.* **2014**, *29*, 1061–1078. [CrossRef]
11. Zhang, X.L. Multicriteria Pythagorean fuzzy decision analysis: A hierarchical QUALIFLEX approach with the closeness index-based ranking methods. *Inf. Sci.* **2016**, *330*, 104–124. [CrossRef]
12. Ren, P.J.; Xu, Z.S.; Gou, X.J. Pythagorean fuzzy TODIM approach to multi-criteria decision making. *Appl. Soft Comput.* **2016**, *42*, 246–259. [CrossRef]
13. Chen, T.Y. Remoteness index-based Pythagorean fuzzy VIKOR methods with a generalized distance measure for multiple criteria decision analysis. *Inf. Fusion* **2018**, *41*, 129–150. [CrossRef]

14. Garg, H. A Novel Correlation Coefficients between Pythagorean Fuzzy Sets and Its Applications to Decision-Making Processes. *Int. J. Intell. Syst.* **2016**, *12*, 1234–1252. [CrossRef]
15. Garg, H. A Novel Improved Accuracy Function for Interval Valued Pythagorean Fuzzy Sets and Its Applications in the Decision-Making Process. *Int. J. Intell. Syst.* **2017**, *32*, 1247–1260. [CrossRef]
16. Peng, X.D.; Dai, J.G. Approaches to Pythagorean Fuzzy Stochastic Multi-criteria Decision Making Based on Prospect Theory and Regret Theory with New Distance Measure and Score Function. *Int. J. Intell. Syst.* **2017**, *32*, 1187–1214. [CrossRef]
17. Xue, W.T.; Xu, Z.S.; Zhang, X.L.; Tian, X.L. Pythagorean Fuzzy LINMAP Method Based on the Entropy Theory for Railway Project Investment Decision Making. *Int. J. Intell. Syst.* **2018**, *33*, 93–125. [CrossRef]
18. Liang, D.C.; Xu, Z.S.; Peter, A. Projection Model for Fusing the Information of Pythagorean Fuzzy Multicriteria Group Decision Making Based on Geometric Bonferroni Mean. *Int. J. Intell. Syst.* **2017**, *32*, 966–987. [CrossRef]
19. Peng, X.D.; Yang, Y. Fundamental Properties of Interval-Valued Pythagorean Fuzzy Aggregation Operators. *Int. J. Intell. Syst.* **2015**, *31*, 444–487. [CrossRef]
20. Zhang, X.L. A Novel Approach Based on Similarity Measure for Pythagorean Fuzzy Multiple Criteria Group Decision Making. *Int. J. Intell. Syst.* **2016**, *31*, 593–611. [CrossRef]
21. Tsao, C.Y.; Chen, T.Y. A projection-based compromising method for multiple criteria decision analysis with interval-valued intuitionistic fuzzy information. *Appl. Soft Comput.* **2016**, *45*, 207–223. [CrossRef]
22. Sun, R.X.; Hu, J.H.; Zhou, J.D.; Chen, X. A Hesitant Fuzzy Linguistic Projection-Based MABAC Method for Patients' Prioritization. *Int. J. Fuzzy Syst.* **2017**, *1*, 1–17. [CrossRef]
23. Ji, P.; Zhang, H.Y.; Wang, J.Q. A projection-based TODIM method under multi-valued neutrosophic environments and its application in personnel selection. *Neural Comput. Appl.* **2018**, *29*, 221–234. [CrossRef]
24. Wu, H.Y.; Xu, Z.S.; Ren, P.J.; Liao, H.C. Hesitant fuzzy linguistic projection model to multi-criteria decision making for hospital decision support systems. *Comput. Ind. Eng.* **2018**, *115*, 449–458. [CrossRef]
25. Liang, R.X.; Wang, J.Q.; Zhang, H.Y. Projection-based PROMETHEE methods based on hesitant fuzzy linguistic term sets. *Int. J. Fuzzy Syst.* **2017**, 1–14. [CrossRef]
26. Ye, J. Bidirectional projection method for multiple attribute group decision making with neutrosophic numbers. *Neural Comput. Appl.* **2017**, *28*, 1021–1029. [CrossRef]
27. Liu, Z.M.; Liu, P.D.; Liu, W.L. An extended VIKOR method based on Pythagorean uncertain linguistic variable. *Control Decis.* **2017**, *32*, 2145–2152. (In Chinese)
28. Li, D.Q.; Zeng, W.Y.; Li, J.H. Note on uncertain linguistic Bonferroni mean operators and their application to multiple attribute decision making. *Appl. Math. Model.* **2015**, *39*, 894–900. [CrossRef]
29. Liu, X.D.; Zhu, J.J.; Liu, S.F. An bi-directional projection method based on hesitant fuzzy information. *Syst. Eng. Theory Pract.* **2014**, *34*, 2637–2644. (In Chinese)

information

MDPI

Article

A Novel Rough WASPAS Approach for Supplier Selection in a Company Manufacturing PVC Carpentry Products

Gordan Stojić [1], Željko Stević [2,*], Jurgita Antuchevičienė [3], Dragan Pamučar [4] and Marko Vasiljević [2]

1. Faculty of Technical Sciences, University of Novi Sad, Trg Dositeja Obradovića 6, 21000 Novi Sad, Serbia; gordan@uns.ac.rs
2. Faculty of Transport and Traffic Engineering, University of East Sarajevo, Vojvode Mišića 52, 74000 Doboj, Bosnia and Herzegovina; drmarkovasiljevic@gmail.com
3. Department of Construction Management and Real Estate, Vilnius Gediminas Technical University, LT-10223 Vilnius, Lithuania; jurgita.antucheviciene@vgtu.lt
4. Department of Logistics, University of Defence in Belgrade, Pavla Jurisica Sturma 33, 11000 Belgrade, Serbia; dpamucar@gmail.com

* Correspondence: zeljkostevic88@yahoo.com; Tel.: +387-66-795-413

Received: 23 April 2018; Accepted: 13 May 2018; Published: 16 May 2018

Abstract: The decision-making process requires the prior definition and fulfillment of certain factors, especially when it comes to complex areas such as supply chain management. One of the most important items in the initial phase of the supply chain, which strongly influences its further flow, is to decide on the most favorable supplier. In this paper a selection of suppliers in a company producing polyvinyl chloride (PVC) carpentry was made based on the new approach developed in the field of multi-criteria decision making (MCDM). The relative values of the weight coefficients of the criteria are calculated using the rough analytical hierarchical process (AHP) method. The evaluation and ranking of suppliers is carried out using the new rough weighted aggregated sum product assessment (WASPAS) method. In order to determine the stability of the model and the ability to apply the developed rough WASPAS approach, the paper analyzes its sensitivity, which involves changing the value of the coefficient λ in the first part. The second part of the sensitivity analysis relates to the application of different multi-criteria decision-making methods in combination with rough numbers that have been developed in the very recent past. The model presented in the paper is solved by using the following methods: rough Simple Additive Weighting (SAW), rough Evaluation based on Distancefrom Average Solution (EDAS), rough MultiAttributive Border Approximation area Comparison (MABAC), rough Višekriterijumsko kompromisno rangiranje (VIKOR), rough MultiAttributiveIdeal-Real Comparative Analysis (MAIRCA) and rough Multi-objective optimization by ratio analysis plus the full multiplicative form (MULTIMOORA). In addition, in the third part of the sensitivity analysis, the Spearman correlation coefficient (SCC) of the ranks obtained was calculated which confirms the applicability of all the proposed approaches. The proposed rough model allows the evaluation of alternatives despite the imprecision and lack of quantitative information in the information-management process.

Keywords: rough number; rough weighted aggregated sum product assessment (WASPAS); rough analytical hierarchical process (AHP); multiple criteria decision making (MCDM); supplier

1. Introduction

The concept of the supply chain changes over time, but essentially retains its original form; it is growing in importance and, according to Petrović et al. [1], information management and control of the supply chain are the strategic focus of the leading manufacturing companies. This is caused by very rapid changes in the environment in which companies operate, with the globalization of the market and the very high demands of users for whom the high quality of products and services becomes a priority. In today's supply chains, supply as a subsystem and the choice of an adequate supplier as the most important process in the procurement subsystem are issues of strategic importance for the functioning of production and other companies, and the goal is to model the supply chain in such a way as to provide profitable outputs for all parts of the supply chain and its participants. The basic participants and elements in the supply chain in relation to the time when this concept emerged are still almost the same, with increasing attention paid to the end-user of services and the satisfaction of their requirements and needs. In order for this to be fulfilled and, on the other hand, to generate profit and efficiently carry out a set of activities in the supply chain with as little cost as possible, it is necessary to take into account the method of suppliers' selection. Since the 1990s, organizational skills were further enhanced according to Monczka et al. [2], and managers began to realize that material inputs from suppliers have a major impact on their ability to respond to the fulfilment of user needs. For this reason, a growing focus has been placed on suppliers as an important factor for the formation of the final product price.

A reliable supplier who performs all his contractual obligations in an adequate way and a smooth flow of goods can be distinguished as the most important goals, on which can largely depend the complete flow of the supply chain and the achievement of the goals of its participants. The choice of suppliers is one of the more important items for supply chain management [3], while managing and developing relationships with suppliers is a critical issue for achieving a competitive advantage [4]. In addition to the aforementioned supply chain processes, information flows and the additional value of material flows are important.

This work has two primary goals, whereby the first objective relates to the possibility of improving the methodology for the treatment of imprecision when it comes to the field of group multiple criteria decision making (MCDM) through the development of the new rough weighted aggregated sum product assessment (WASPAS) approach. The second goal of this paper is to enrich the evaluation methodology and selection of suppliers through a new approach to the treatment of imprecision that is based on rough numbers.

The paper is structured in six sections. In the first section, introductory considerations about the importance and effects of selection the most appropriate suppliers are given. In the second section, a literature review is carried out presenting the application of the WASPAS method in different areas and demonstrating rough sets theory applications. The third section presents the novel rough WASPAS approach with a detailed explanation of each step. The fourth section presents a practical example of selection of a supplier in a polyvinyl chloride (PVC) manufacturing company using a pre-developed approach. The fifth section presents a sensitivity analysis consisting of three parts on the basis of which the stability of the proposed approach is determined. In the same section, in addition to sensitivity analysis, the results obtained are discussed. The sixth section contains concluding remarks.

2. Literature Review

A short literature review is carried out presenting successful applications of the WASPAS method in different precise or uncertain decision-making situations, as well as demonstrating cases of rough sets theory applications in MCDM problems.

2.1. Applications of Weighted Aggregated Sum Product Assessment (WASPAS) Method

The WASPAS method falls within a group of recent MCDM methods. It was developed by Zavadskas et al. in 2012 [5] and so far has been successfully applied in various areas for solving problems of a different nature. Ighravwe and Oke [6] use the WASPAS method for evaluating maintenance performance systems, while Mathew et al. [7] make a selection of an industrial robot. It is also applied to the determination of the location areas of wind farms in [8], while in [9] in combination with factor relationship (FARE) it is applied in hard magnetic material selection. Solving the location problem for the construction of a shopping center was discussed in [10], where this method is also applied. Zavadskas et al. [11] evaluated apartments in residential buildings using the WASPAS method. The following studies in different areas use the WASPAS method [12,13]. The combination of the analytical hierarchical process (AHP) and WASPAS methods is not rare, so a number of publications using AHP for determining the weight values of the criteria and WASPAS for the choice of alternatives can be found in the literature [14,15]. Madić et al. [16] evaluate the machining process with combination of AHP and WASPAS method, while Turskis et al. [17] use the fuzzy form of these methods for construction site selection. A combination of the classic form of these methods is applied to laser cutting in [18]. The combination of Step-wise Weight Assessment Ratio Analysis (SWARA) and WASPAS is used for solar power plant site selection in [19], and in [20] the combination of these two methods is applied in the nanotechnology industry. The integration of the SWARA, Quality function deployment (QFD) and WASPAS methods is proposed in [21] to resolve the selection of suppliers. SWARA was also used to determine the significance of the criteria. This combination is also integrated into [22] where it is used for the selection of staff in tourism. The combination of methods has been also applied in many decision-making problems and environments [23].

The WASPAS method has a number of extensions. Zavadskas et al. in 2015 [24] developed a new WASPAS-G which is a combination of the classic WASPAS method with grey values, while Keshavarz Ghorabaee et al. in 2016 [25] developed a WASPAS method with interval type-2 fuzzy sets to evaluate and select suppliers in the green supply chain. The same approach is combined with the CRiteria Importance Through Inter-criteria Correlation (CRITIC) method used for a third-party logistics (3PL) provider in [26]. A combination of the WASPAS and single-valued neutrosophic set is applied in [27,28]. WASPAS combined with interval-valued intuitionistic fuzzy numbers (IVIF) was developed in [29]. Solving solar-wind power station location problem using the WASPAS method with interval neutrosophic sets was considered in [30].

2.2. Applications of Rough Sets in Multiple Criteria Decision Making (MCDM)

The popularization of rough sets is lately evident and is increasingly used to make decisions in different areas. Song et al. in his paper [31] used a rough Technique for Ordering Preference by Similarity to Ideal Solution (TOPSIS) approach in uncertain environments. Integration of rough AHP and MABAC are proposed in [32], while integration interval rough AHP and interval rough MABAC is proposed in [33] for evaluation of university websites. A rough AHP and rough TOPSIS approach is also used in [34]. Rough numbers in integration with MCDM methods, according Stević et al. [35], give good results, so lately we can notice the popularization of the use of rough numbers [36–38]. Besides the AHP method and its rough form, it is also possible to apply rough best worst method (BWM). So far, rough BWM has been applied in several publications. A rough BWM model was applied for determining the importance of the criteria for selecting a wagon for a logistics company in [39]. The evaluation of suppliers can be executed using a new believable rough set approach which was developed in [40] and, according to the authors, it provided good results.

In comparison with other concepts, a novel rough WASPAS approach has some advantages that can be described as follows. The first reason is its advantage in comparison with grey theory. Grey relation analysis provides a well-structured analytical framework for a multi-criteria decision-making process, but it lacks the capability to characterize the subjective perceptions of designers in the evaluation process. Rough set theory may help here, because rough sets can facilitate effective representation of vague information or imprecise data [41]. According to Khoo et al. [42], a very important advantage of using rough set theory to handle vagueness and uncertainty is that it expresses vagueness by means of the boundary region of a set instead of membership function. In addition, the integration of rough numbers in MCDM methods gives the possibility to explore subjective and unclear evaluation of the experts and to avoid assumptions, which is not the case when applying fuzzy theory [35]. According to Hashemkhani Zolfani et al. [10], the main advantage of the WASPAS method is its high degree of reliability. Integration of rough numbers and the WASPAS method with advantages of both concepts presents a very important support in decision-making in everyday conflicting situations.

The purpose of the fuzzy tehnique in the decision making process is to enable the transformation of crisp numbers into fuzzy numbers that show uncertainties in real world systems using the membership function. As opposed to fuzzy sets theory, which requires a subjective approach in determining partial functions and fuzzy set boundaries, rough set theory determines set boundaries based on real values and depends on the degree of certainty of the decision maker. Since rough set theory deals solely with internal knowledge, i.e., operational data, there is no need to rely on assumption models. In other words, when applying rough sets, only the structure of the given data is used instead of various additional/external parameters [43]. Duntsch and Gediga [44] believe that the logic of rough set theory is based solely on data that speak for themselves. When dealing with rough sets, the measurement of uncertainty is based on the vagueness already contained in the data [45]. In this way, the objective indicators contained in the data can be determined. In addition, rough set theory is suitable for application on sets characterized by irrelevant data where the use of statistical methods does not seem appropriate [46].

3. Methods

3.1. Rough Set Theory

In rough set theory, any vague idea can be represented as a couple of exact concepts based on the lower and upper approximations.

Suppose U is the universe which contains all the objects, Y is an arbitrary object of U, R is a set of t classes $\{G_1, G_2, \ldots, G_t\}$ that cover all the objects in U, $R = \{G_1, G_2, \ldots, G_t\}$. If these classes are ordered as $G_1 < G_2 < \ldots < G_t\}$, then $\forall Y \in U, G_q \in R, 1 \leq q \leq t$, by $R\ (Y)$ we mean the class to which the object belongs, the lower approximation $(\underline{Apr}(G_q))$, upper approximation $(\overline{Apr}(G_q))$ and boundary region $(\overline{Bnd}(G_q))$ of class G_q are, according to [47], defined as:

$$\underline{Apr}(G_q) = \{Y \in U/R(Y) \leq G_q\} \quad (1)$$

$$\overline{Apr}(G_q) = \{Y \in U/R(Y) \geq G_q\} \quad (2)$$

$$Bnd(G_q) = \{Y \in U/R(Y) \neq G_q\} = \{Y \in U/R(Y) > G_q\} \cup \{Y \in U/R(Y) < G_q\} \quad (3)$$

Then G_q can be shown as rough number $(RN(G_q))$, which is determined by its corresponding lower limit $(\underline{Lim}(G_q))$ and upper limit $(\overline{Lim}(G_q))$ where:

$$\underline{Lim}(G_q) = \frac{1}{M_L} \sum \left\{Y \in \underline{Apr}(G_q)\right\} R(Y) \quad (4)$$

$$\overline{Lim}(G_q) = \frac{1}{M_U} \sum \{Y \in \overline{Apr}(G_q)\} R(Y) \quad (5)$$

$$RN(G_q) = \left[\underline{Lim}(G_q), \overline{Lim}(G_q)\right] \quad (6)$$

where M_L, M_U are the numbers of objects that are contained in $\underline{Apr}(G_q)$ and $\overline{Apr}(G_q)$, respectively.

3.2. A Novel Rough WASPAS Approach

The WASPAS method [5] represents a relatively new MCDM method, that has been proved to be robust in a number of publications. Bearing in mind all the advantages of using rough theory [48,49] in the MCDM to represent ambiguity, vagueness and uncertainty, the authors have decided in this paper to modify the WASPAS algorithm using rough numbers, which is an original contribution.

The proposed rough WASPAS method consists of the following steps:

Step 1: Formulation of the model, which consists of m alternatives and n criteria.

Step 2: Formation a team of k experts for the evaluation of alternatives according all criteria using the linguistic scale (Table 1).

Table 1. Linguistic scale for evaluating alternatives depending on the type of criteria [39].

Linguistic Scale	For Criteria of Type Max (Benefit Criteria)	For Criteria of Type Min (Cost Criteria)
Very Poor—VP	1	9
Poor—P	3	7
Medium—M	5	5
Good—G	7	3
Very Good—VG	9	1

Step 3: Formation of initial individual matrices based on evaluations made by experts. It is necessary to form as many individual matrices as there are experts. If the model includes e.g., 5 experts it is necessary to form 5 individual matrices.

Step 4: Converting an individual matrix into a group rough matrix. Each individual matrix of experts $k_1, k_2, \ldots, k_n$ needs to be converted into a rough group matrix (RGM) (1) using Equations (1)–(6):

$$RGM = \begin{bmatrix} [x_{11}^L, x_{11}^U] & [x_{12}^L, x_{12}^U] & \cdots & [x_{1n}^L, x_{1n}^U] \\ [x_{21}^L, x_{21}^U] & [x_{22}^L, x_{22}^U] & \cdots & [x_{2n}^L, x_{2n}^U] \\ \vdots & \vdots & \ddots & \vdots \\ [x_{m1}^L, x_{m1}^U] & [x_{m2}^L, x_{m2}^U] & \cdots & [x_{mn}^L, x_{mn}^U] \end{bmatrix} \tag{7}$$

Step 5: In this step it is necessary to normalize the previous matrix using Equations (8) and (9):

$$n_{ij} = \frac{\left[x_{ij}^L; x_{ij}^U\right]}{max\left[x_{ij}^{+L}; x_{ij}^{+U}\right]} \; for\ C_1, C_2, \ldots, C_n \in B \tag{8}$$

$$n_{ij} = \frac{min\left[x_{ij}^{-L}; x_{ij}^{-U}\right]}{\left[x_{ij}^L; x_{ij}^U\right]} \; for\ C_1, C_2, \ldots, C_n \in C \tag{9}$$

where $\left[x_{ij}{}^L; x_{ij}{}^U\right]$ denotes the values from the initial rough group matrix, $\max\left[x_{ij}{}^{+L}; x_{ij}{}^{+U}\right]$ represent the maximum value of a criterion if the same belongs a set of benefit criteria and $\min\left[x_{ij}{}^{-L}; x_{ij}{}^{-U}\right]$ represent minimal value of a criterion if the same belongs a set of cost criteria.

With "+" and "−" values are marked in terms of easier recognition of the same criteria that belong to a different type of criteria.

Equations (8) and (9) can simpler be written as:

$$n_{ij} = \left[\frac{x_{ij}^L}{x_{ij}^{+U}}; \frac{x_{ij}^U}{x_{ij}^{+L}}\right] for\ C_1, C_2, \ldots, C_n \in B \tag{10}$$

$$n_{ij} = \left[\frac{x_{ij}^{-L}}{x_{ij}^U}; \frac{x_{ij}^{-U}}{x_{ij}^L}\right] for\ C_1, C_2, \ldots, C_n \in C \tag{11}$$

and get a normalized matrix that looks like (12):

$$NM = \begin{bmatrix} [n_{11}^L, n_{11}^U] & [n_{12}^L, n_{12}^U] & \cdots & [n_{1n}^L, n_{1n}^U] \\ [n_{21}^L, n_{21}^U] & [n_{22}^L, n_{22}^U] & \cdots & [n_{2n}^L, n_{2n}^U] \\ \vdots & \vdots & \ddots & \vdots \\ [n_{m1}^L, n_{m1}^U] & [n_{m2}^L, n_{m2}^U] & \cdots & [n_{mn}^L, n_{mn}^U] \end{bmatrix} \tag{12}$$

Step 6: Weighting of the normalized matrix by multiplying the previously obtained matrix with weighted values of the criteria (13):

$$\begin{aligned} Vn &= \left[v_{ij}^L; v_{ij}^U\right]_{m\times n} \\ v_{ij}^L &= w_J^L \times n_{ij}^L,\ i = 1,2,\ldots m, j \\ v_{ij}^U &= w_J^U \times n_{ij}^U,\ i = 1,2,\ldots m, j \end{aligned} \tag{13}$$

where $w_J{}^L$ is lower limit, and $w_J{}^U$ is the upper limit of the weight value of the criterion obtained by applying one of the MCDM methods to determine the significance of the criteria.

Step 7: Summing all the values of the alternatives obtained (14):

$$\begin{aligned} Q_i &= \left[q_{ij}^L; q_{ij}^U\right]_{1\times m} \\ q_{ij}^L &= \sum_{j=1}^{n} v_{ij}^L;\ q_{ij}^U = \sum_{j=1}^{n} v_{ij}^U \end{aligned} \tag{14}$$

Step 8: Determination of the weighted product model using Equation (15):

$$\begin{aligned} P_i &= \left[p_{ij}^L; p_{ij}^U\right]_{1\times m} \\ p_{ij}^L &= \prod_{j=1}^{n} \left(v_{ij}^L\right)^{w_J^L} \\ p_{ij}^U &= \prod_{j=1}^{n} \left(v_{ij}^U\right)^{w_J^U} \end{aligned} \tag{15}$$

Step 9: Determination of the relative values of the alternative A_i (16):

$$\begin{aligned} A_i &= \left[a_{ij}^L; a_{ij}^U\right]_{1xm} \\ A_i &= \lambda \times Q_i + (1-\lambda) \times P_i \end{aligned} \tag{16}$$

Coefficient λ can be crisp values in range 0, 0.1, 0.2, ...1.0, but it is recommended to apply the Equation (17) for its calculation:

$$\lambda = 0.5 + \frac{\sum P_i}{\sum Q_i + \sum P_i} = 0.5 + \frac{\sum\left[p_{ij}^L; p_{ij}^U\right]}{\sum\left[q_{ij}^L; q_{ij}^U\right] + \sum\left[p_{ij}^L; p_{ij}^U\right]} \tag{17}$$

Step 10: Ranking the alternatives. The highest value of the alternative marks the best ranked, while the smallest value reflects the worst alternative.

4. Supplier Selection in a Company Manufacturing Polyvinyl Chloride (PVC) Carpentry

Supplier selection in the company manufacturing PVC carpentry was carried out on the basis of the combination of nine quantitative and qualitative criteria: quality of the material, price of the material, certification of the products, delivery time, reputation, volume discounts, warranty period, reliability, and the method of payments. The second and the fourth criteria (the price of the material and delivery time) are the Expenses criteria (type min), while the others are the Benefit criteria (type max). Criteria used in this paper were selected and verified through two-year research related to the evaluation of suppliers in the manufacturing companies of the supply chain presented in [50]. The market is filled with a large number of manufacturers of PVC carpentry products which need an adequate supplier for ensuring the low cost of a product and a good position in the market. In the research, six suppliers (alternatives) were selected from different countries which were evaluated using a developed rough model. In this study, a group of five experts took part in the assessment process. After the interview with the experts, the collected data were processed, and the aggregation of the expert opinion was obtained.

The rough AHP method [51] was used to determine the weight values of the criteria with the following calculation procedure:

Step 1: After the experts' evaluation of criteria, by applying Saaty's scale, five matrices of comparison were constructed in criteria sets (Table 2).

Table 2. Expert evaluation of the criteria.

	E_1									E_2								
	C_1	C_2	C_3	C_4	C_5	C_6	C_7	C_8	C_9	C_1	C_2	C_3	C_4	C_5	C_6	C_7	C_8	C_9
C_1	1.00	7.00	2.00	5.00	6.00	4.00	3.00	3.00	8.00	1.00	8.00	2.00	6.00	6.00	4.00	5.00	3.00	9.00
C_2	0.14	1.00	0.17	0.33	0.50	0.33	0.25	0.17	2.00	0.13	1.00	0.14	0.25	0.50	0.33	0.25	0.17	2.00
C_3	0.50	6.00	1.00	4.00	5.00	6.00	2.00	2.00	7.00	0.50	7.00	1.00	4.00	5.00	6.00	2.00	2.00	7.00
C_4	0.20	3.00	0.25	1.00	2.00	0.50	0.33	0.25	4.00	0.20	4.00	0.25	1.00	2.00	0.50	0.33	0.25	4.00
C_5	0.17	2.00	0.20	0.50	1.00	0.33	0.25	0.20	3.00	0.17	2.00	0.20	0.50	1.00	0.33	0.25	0.20	3.00
C_6	0.25	3.00	0.17	2.00	3.00	1.00	0.50	0.33	5.00	0.17	3.00	0.17	2.00	3.00	1.00	0.50	0.33	5.00
C_7	0.33	4.00	0.50	3.00	4.00	2.00	1.00	0.50	5.00	0.20	4.00	0.50	3.00	4.00	2.00	1.00	0.50	5.00
C_8	0.33	6.00	0.50	4.00	5.00	3.00	2.00	1.00	6.00	0.33	6.00	0.50	4.00	5.00	3.00	2.00	1.00	6.00
C_9	0.13	0.50	0.14	0.25	0.33	0.20	0.20	0.17	1.00	0.11	0.50	0.14	0.25	0.33	0.20	0.20	0.17	1.00
	E_3									E_4								
	C_1	C_2	C_3	C_4	C_5	C_6	C_7	C_8	C_9	C_1	C_2	C_3	C_4	C_5	C_6	C_7	C_8	C_9
C_1	1.00	8.00	1.00	6.00	6.00	4.00	5.00	3.00	8.00	1.00	6.00	0.50	4.00	5.00	4.00	2.00	2.00	7.00
C_2	0.13	1.00	0.14	0.25	0.50	0.33	0.25	0.17	1.00	0.17	1.00	0.14	0.33	0.50	0.33	0.25	0.17	2.00
C_3	1.00	7.00	1.00	6.00	6.00	4.00	5.00	3.00	8.00	2.00	7.00	1.00	5.00	6.00	6.00	3.00	3.00	8.00
C_4	0.20	4.00	0.20	1.00	2.00	0.50	0.33	0.25	4.00	0.25	3.00	0.20	1.00	2.00	0.50	0.33	0.25	4.00
C_5	0.17	2.00	0.17	0.50	1.00	0.33	0.25	0.20	3.00	0.20	2.00	0.17	0.50	1.00	0.33	0.25	0.20	3.00
C_6	0.17	3.00	0.17	2.00	3.00	1.00	0.50	0.33	5.00	0.25	3.00	0.17	2.00	3.00	1.00	0.50	0.33	5.00
C_7	0.20	4.00	0.20	3.00	4.00	2.00	1.00	0.50	5.00	0.50	4.00	0.33	3.00	4.00	2.00	1.00	0.50	5.00
C_8	0.33	6.00	0.33	4.00	5.00	3.00	2.00	1.00	6.00	0.50	6.00	0.33	4.00	5.00	3.00	2.00	1.00	6.00
C_9	0.13	1.00	0.13	0.25	0.33	0.20	0.20	0.17	1.00	0.14	0.50	0.13	0.25	0.33	0.20	0.20	0.17	1.00
	E_5																	
	C_1	C_2	C_3	C_4	C_5	C_6	C_7	C_8	C_9									
C_1	1.00	6.00	0.50	4.00	5.00	4.00	2.00	2.00	7.00									
C_2	0.17	1.00	0.14	0.33	0.50	0.33	0.25	0.17	2.00									
C_3	2.00	7.00	1.00	5.00	6.00	6.00	3.00	3.00	8.00									
C_4	0.25	3.00	0.20	1.00	2.00	0.50	0.33	0.25	4.00									
C_5	0.20	2.00	0.17	0.50	1.00	0.33	0.25	0.25	3.00									
C_6	0.25	3.00	0.17	2.00	3.00	1.00	0.50	0.50	5.00									
C_7	0.50	4.00	0.33	3.00	4.00	2.00	1.00	1.00	5.00									
C_8	0.50	6.00	0.33	4.00	4.00	2.00	1.00	1.00	5.00									
C_9	0.14	0.50	0.13	0.25	0.33	0.20	0.20	0.20	1.00									

By applying Expressions (1)–(6), each of presented sequences is transformed in rough sequence. So, for the sequence $\tilde{x}_{12} = \{7, 8, 8, 6, 6\}$ we get:

$$\underline{Lim}(6) = 6.00, \overline{Lim}(6) = \tfrac{1}{5}(7+8+8+6+6) = 7.00$$

$$\underline{Lim}(7) = \tfrac{1}{3}(7+6+6) = 6.33, \overline{Lim}(7) = \tfrac{1}{3}(7+8+8) = 7.67$$

$$\underline{Lim}(8) = \tfrac{1}{5}(7+8+8+6+6) = 7.00, \overline{Lim}(8) = 8.00$$

$$RN(x_{12}^1) = [6.33; 7.67]; RN(x_{12}^2) = RN(x_{12}^3) = [7.00; 8.00];\ RN(x_{12}^4) = RN(x_{12}^5) = [6.00; 7.00]$$

$$x_{12}^L = \frac{x_{12}^1 + x_{12}^2 + x_{12}^s + x_{12}^4 + x_{12}^5}{S} = \frac{6.33+7.00+7.00+6.00+6.00}{5} = 6.47$$

$$x_{12}^U = \frac{x_{12}^1 + x_{12}^2 + x_{12}^s + x_{12}^4 + x_{12}^5}{S} = \frac{7.67+8.00+8.00+7.00+7.00}{5} = 7.53$$

After formation of the group rough matrix shown in Table 3, it is necessary to calculate the geometric middle of the upper and lower limits of the group matrix of the criteria. From the obtained matrix maximum value, the upper limit is chosen, and all other values are divided by that. In that way, we obtain the final values of the criteria weight:

$$w_j \begin{bmatrix} [0.803, 0.992] \\ [0.090,\ 0.095] \\ [0.819,\ 1.000] \\ [0.187,\ 0.194] \\ [0.127,\ 0.132] \\ [0.245,\ 0.254] \\ [0,367,\ 0,416] \\ [0.505,\ 0.552] \\ [0.067,\ 0.071] \end{bmatrix} \tag{18}$$

Table 3. Group rough matrix.

	C_1	C_2	C_3	C_4	C_5	C_6	C_7	C_8	C_9
C_1	[1, 1]	[6.47, 7.53]	[0.81, 1.61]	[4.47, 5.53]	[5.36, 5.84]	[4, 4]	[2.63, 4.23]	[2.36, 2.84]	[7.36, 8.25]
C_2	[0.14, 0.16]	[1, 1]	[0.14, 0.15]	[0.28, 0.32]	[0.5, 0.5]	[0.33, 0.33]	[0.25, 0.25]	[0.17, 0.17]	[1.64, 1.96]
C_3	[0.81, 1.61]	[6.64, 6.96]	[1, 1]	[4.36, 5.25]	[5.36, 5.84]	[5.28, 5.92]	[2.4, 3.67]	[2.36, 2.84]	[7.36, 7.84]
C_4	[0.21, 0.23]	[3.16, 3.64]	[0.21, 0.23]	[1, 1]	[2, 2]	[0.5, 0.5]	[0.33, 0.33]	[0.25, 0.25]	[4, 4]
C_5	[0.17, 0.19]	[2, 2]	[0.17, 0.19]	[0.5, 0.5]	[1, 1]	[0.33, 0.33]	[0.25, 0.25]	[0.2, 0.22]	[3, 3]
C_6	[0.2, 0.24]	[3, 3]	[0.17, 0.17]	[2, 2]	[3, 3]	[1, 1]	[0.5, 0.5]	[0.34, 0.4]	[5, 5]
C_7	[0.27, 0.43]	[4, 4]	[0.3, 0.44]	[3, 3]	[4, 4]	[2, 2]	[1, 1]	[0.52, 0.68]	[5, 5]
C_8	[0.36, 0.44]	[6, 6]	[0.36, 0.44]	[4, 4]	[4.64, 4.96]	[2.64, 2,96]	[1.64, 1,96]	[1, 1]	[5.64, 5.96]
C_9	[0.12, 0.14]	[0.52, 0.68]	[0.13, 0.14]	[0.25, 0.25]	[0.33, 0.33]	[0.2, 0.2]	[0.2, 0.2]	[0.17, 0.18]	[1, 1]

After obtaining the weight of the criteria, the expert team performed an evaluation of the alternatives (Table 4).

After the first two steps of the rough WASPAS method which imply the setting of a model by choosing the criteria and alternatives in the first step and determining the expert assessment in the second step (Table 2), it is necessary to convert all the individual matrices from the third step into a group rough matrix, which is the fourth step.

Converting individual matrices into rough matrices is executed in the same way as was the case in determining the weight values of the criteria. The example of the group matrix elements evaluation is presented in Table 5.

$$
\begin{aligned}
\widetilde{x}_{11} &= \{7, 9, 5, 5, 7\} \\
\underline{Lim}(5) = 5.00, \overline{Lim}(5) &= \tfrac{1}{5}(7+9+5+5+7) = 6.60 \\
\underline{Lim}(7) = \tfrac{1}{4}(7+5+5+7) = 6.00, \overline{Lim}(7) &= \tfrac{1}{3}(7+9+7) = 7.67 \\
\underline{Lim}(9) = \tfrac{1}{5}(7+9+5+5+7) = 6.60, \overline{Lim}(9) &= 9.00 \\
RN(x_{11}^{1}) = RN(x_{11}^{5}) = [6.00; 7.67]; RN(x_{11}^{2}) = [6.60; 9.00];\ & RN(x_{11}^{3}) = RN(x_{11}^{4}) = [5.00; 6.60] \\
x_{11}^{L} = \tfrac{x_{11}^{1}+x_{11}^{2}+x_{11}^{s}+x_{11}^{4}+x_{11}^{5}}{S} &= \tfrac{6.00+6.60+5.00+5.00+6.00}{5} = 5.72 \\
x_{11}^{U} = \tfrac{x_{11}^{1}+x_{11}^{2}+x_{11}^{s}+x_{11}^{4}+x_{11}^{5}}{S} &= \tfrac{7.67+9.00+6.60+6.60+7.67}{5} = 7.51
\end{aligned}
\tag{19}
$$

Table 4. Evaluation of the alternatives based on the criteria of five experts.

	A_1					A_2					A_3				
	E_1	E_2	E_3	E_4	E_5	E_1	E_2	E_3	E_4	E_5	E_1	E_2	E_3	E_4	E_5
C_1	7	9	5	5	7	7	7	3	5	7	5	3	5	7	5
C_2	1	1	1	3	1	3	3	5	1	3	7	9	3	5	7
C_3	3	3	1	3	1	7	9	5	5	7	7	7	3	5	5
C_4	9	7	7	9	9	9	5	5	7	9	5	1	7	7	5
C_5	1	9	1	3	3	3	7	3	5	5	7	5	3	5	9
C_6	3	7	3	3	7	5	7	5	3	7	7	5	3	5	9
C_7	5	5	3	3	5	5	3	5	1	5	7	7	5	3	7
C_8	3	5	1	1	7	5	7	3	3	9	5	5	5	3	9
C_9	3	7	3	1	7	3	5	1	3	5	7	5	3	5	7
	A_4					A_5					A_6				
C_1	5	3	7	7	5	5	3	5	7	5	3	5	3	5	5
C_2	3	7	5	3	5	9	9	5	5	9	7	7	7	5	7
C_3	3	3	1	3	1	9	9	5	7	7	7	9	3	5	7
C_4	5	3	7	5	5	3	1	5	5	1	3	3	3	5	3
C_5	7	5	5	5	9	7	5	9	7	9	5	5	3	3	5
C_6	3	3	5	3	9	5	5	3	7	5	3	5	3	5	9
C_7	9	7	5	3	9	9	7	5	3	9	3	3	1	3	3
C_8	5	5	3	3	7	5	5	5	5	7	3	5	3	5	5
C_9	5	5	5	3	9	7	5	5	5	9	3	1	5	5	3

Table 5. Group rough matrix.

	C_1	C_2	C_3	C_4	C_5	C_6	C_7	C_8	C_9
A_1	[5.72, 7.51]	[1.08, 1.72]	[1.72, 2.68]	[7.72, 8.68]	[1.88, 5.16]	[3.64, 5.56]	[3.72, 4.68]	[1.91, 4.96]	[2.81, 5.64]
A_2	[4.88, 6.66]	[2.3, 3.7]	[5.72, 7.51]	[5.93, 8.07]	[3.72, 5.51]	[4.49, 6.28]	[2.88, 4.66]	[3.91, 6.96]	[2.49, 4.28]
A_3	[4.3, 5.7]	[4.84, 7.51]	[4.49, 6.28]	[3.67, 6.2]	[4.49, 7.16]	[4.49, 7.16]	[4.88, 6.66]	[4.38, 6.48]	[4.49, 6.28]
A_4	[4.49, 6.28]	[3.72, 5.51]	[1.72, 2.68]	[4.3, 5.7]	[5.34, 7.12]	[3.42, 5.96]	[5.04, 8.09]	[3.72, 5.51]	[4.38, 6.48]
A_5	[4.3, 5.7]	[6.44, 8.36]	[6.49, 8.28]	[1.93, 4.07]	[6.49, 8.28]	[4.3, 5.7]	[5.04, 8.09]	[5.08, 5.72]	[5.34, 7.12]
A_6	[3.72, 4.68]	[6.28, 6.92]	[4.84, 7.51]	[3.08, 3.72]	[3.72, 4.68]	[3.8, 6.33]	[2.28, 2.92]	[3.72, 4.68]	[2.49, 4.28]

In the fifth step, it is necessary to normalize a rough group matrix using Equations (8) and (9) (Table 6) in the following way:

Normalization of the group matrix elements for benefit criteria was carried out in the following way:

$$
n_{11} = \left[\frac{x_{ij}^{L}}{x_{ij}^{+U}};\ \frac{x_{ij}^{U}}{x_{ij}^{+L}}\right] = \left[\frac{4.88}{7.51};\ \frac{6.66}{5.72}\right] \rightarrow n_{11} = [0.65;\ 1.16] \tag{20}
$$

and for the cost criteria:

$$n_{22} = \left[\frac{x_{ij}^{-L}}{x_{ij}^{U}}; \frac{x_{ij}^{-U}}{x_{ij}^{L}} \right] = \left[\frac{1.08}{3.70}; \frac{1.72}{2.30} \right] \rightarrow n_{22} = [0.29;\ 0.75]$$

Step 6: Weighting of the normalized matrix multiplying the previously obtained matrix by the weighted values of the criteria using Equation (13) (Table 7):

$$v_{13}^{L} = \left[w_3^L \times n_{13}^L; w_3^U \times n_{13}^U \right] = [0.82 \times 0.21;\ 1.00 \times 0.41] \rightarrow v_{13}^{L} = [0, 17; 0.41]$$

Table 6. Normalized matrix.

	C_1	C_2	C_3	C_4	C_5	C_6	C_7	C_8	C_9
A_1	[0.76, 1.31]	[0.63, 1.59]	[0.21, 0.41]	[0.22, 0.48]	[0.23, 0.8]	[0.51, 1.24]	[0.46, 0.93]	[0.27, 0.98]	[0.39, 1.06]
A_2	[0.65, 1.16]	[0.29, 0.75]	[0.69, 1.16]	[0.24, 0.63]	[0.45, 0.85]	[0.63, 1.4]	[0.36, 0.92]	[0.56, 1.37]	[0.35, 0.8]
A_3	[0.57, 1]	[0.14, 0.36]	[0.54, 0.97]	[0.31, 1.01]	[0.54, 1.1]	[0.63, 1.59]	[0.6, 1.32]	[0.63, 1.28]	[0.63, 1.18]
A_4	[0.6, 1.1]	[0.2, 0.46]	[0.21, 0.41]	[0.34, 0.87]	[0.64, 1.1]	[0.48, 1.33]	[0.62, 1.61]	[0.53, 1.08]	[0.62, 1.21]
A_5	[0.57, 1]	[0.13, 0.27]	[0.78, 1.28]	[0.47, 1.93]	[0.78, 1.28]	[0.6, 1.27]	[0.62, 1.61]	[0.73, 1.13]	[0.75, 1.33]
A_6	[0.5, 0.82]	[0.16, 0.27]	[0.58, 1.16]	[0.52, 1.21]	[0.45, 0.72]	[0.53, 1.41]	[0.28, 0.58]	[0.53, 0.92]	[0.35, 0.8]

Table 7. Weighted normalized matrix.

	C_1	C_2	C_3	C_4	C_5	C_6	C_7	C_8	C_9
A_1	[0.61, 1.3]	[0.06, 0.15]	[0.17, 0.41]	[0.04, 0.09]	[0.03, 0.1]	[0.12, 0.31]	[0.17, 0.39]	[0.14, 0.54]	[0.03, 0.08]
A_2	[0.52, 1.16]	[0.03, 0.07]	[0.57, 1.16]	[0.04, 0.12]	[0.06, 0.11]	[0.15, 0.36]	[0.13, 0.38]	[0.28, 0.76]	[0.02, 0.06]
A_3	[0.46, 0.99]	[0.01, 0.03]	[0.44, 0.97]	[0.06, 0.2]	[0.07, 0.15]	[0.15, 0.41]	[0.22, 0.55]	[0.32, 0.7]	[0.04, 0.08]
A_4	[0.48, 1.09]	[0.02, 0.04]	[0.17, 0.41]	[0.06, 0.17]	[0.08, 0.14]	[0.12, 0.34]	[0.23, 0.67]	[0.27, 0.6]	[0.04, 0.09]
A_5	[0.46, 0.99]	[0.01, 0.03]	[0.64, 1.28]	[0.09, 0.37]	[0.1, 0.17]	[0.15, 0.32]	[0.23, 0.67]	[0.37, 0.62]	[0.05, 0.1]
A_6	[0.4, 0.81]	[0.01, 0.03]	[0.48, 1.16]	[0.1, 0.23]	[0.06, 0.09]	[0.13, 0.36]	[0.1, 0.24]	[0.27, 0.51]	[0.02, 0.06]

Step 7: Summing all the values of the alternatives obtained (summing by rows) (14):

$$\begin{aligned} Q_i &= \left[q_{ij}^L; q_{ij}^U \right]_{1 \times m} \\ q_{ij}^L &= \sum_{j=1}^{n} v_{ij}^L;\ q_{ij}^U = \sum_{j=1}^{n} v_{ij}^U \\ q_{11}^L &= [0.61 + 0.06 + 0.17 + 0.04 + 0.03 + 0.12 + 0.17 + 0.14 + 0.03] = 1.37 \\ q_{11}^U &= [1.3 + 0.15 + 0.41 + 0.09 + 0.1 + 0.31 + 0.39 + 0.54 + 0.08] = 3.38 \\ WSM &= \begin{bmatrix} [1.37, 3.38] \\ [1.81, 4.17] \\ [1.78,\ 4.08] \\ [1.47,\ 3.55] \\ [2.10,\ 4.54] \\ [1.57,\ 3.49] \end{bmatrix} \end{aligned} \tag{21}$$

Step 8: Determination of the weighted product model using Equation (15):

$$\begin{aligned} P_i &= \left[p_{ij}^L; p_{ij}^U \right]_{1 \times m} \\ p_{ij}^L &= \prod_{j=1}^{n} \left(v_{ij}^L \right)^{w_j^L} \\ p_{ij}^U &= \prod_{j=1}^{n} \left(v_{ij}^U \right)^{w_j^U} \end{aligned}$$

$$p_{11}^L = \left[(0.61^{0.80}) \times (0.06^{0.09}) \times (0.17^{0.82}) \times (0.04^{0.19}) \times (0.03^{0.13}) \times (0.12^{0.24}) \times (0.17^{0.37}) \times (0.14^{0.50}) \times (0.03^{0.07}) \right] = 0.004$$

$$p_{11}^u = \left[(1.30^{0.99}) \times (0.15^{0.09}) \times (0.41^{1.00}) \times (0.09^{0.19}) \times (0.1^{0.13}) \times (0.31^{0.25}) \times (0.39^{0.42}) \times (0.54^{0.55}) \times (0.08^{0.07}) \right] = 0.063$$

$$WPM = \begin{bmatrix} [0.004, 0.063] \\ [0.013, 0.187] \\ [0.013,\ 0.168] \\ [0.006,\ 0.074] \\ [0.021,\ 0.240] \\ [0.008,\ 0.088] \end{bmatrix}$$

Step 9: Determination of the relative values of the alternative A_i (Equation (16)):

$$A_i = \left[a_{ij}^L; a_{ij}^U\right]_{1xm}$$
$$A_i = \lambda \times Q_i + (1 - \lambda) \times P_i \quad (22)$$

Coeficient λ can be in the range of 0, 0.1, 0.2, ... , 1.0.
However, it is recommended to apply Equation (17) for its calculation:

$$\lambda = 0.5 + \frac{\sum P_i}{\sum Q_i + \sum P_i} = 0.5 + \frac{[0.065, 0.820]}{[10.087, 23.203] + [0.065, 0.820]} = [0.503, 0.581]$$

Table 8 shows the calculations of Equation (16). At first it is necessary to calculate the product of the coefficient λ with the values of the Qi matrix from the 7th step. After that, it is necessary to detract a rough number of the coefficient λ from one (1) and multiply it with the values of the P_i matrix from the 8th step. In addition, Table 8 shows rough values for each alternative, their crisp number and ranking.

Table 8. Determining the relative values of the alternatives and their ranking.

	$\lambda \times Q_i$	$(1-\lambda) \times P_i$	A_i	Crisp A_i	Rank
A$_1$	[0.687, 1.963]	[0.002, 0.031]	[0.689, 1.995]	1.342	**6**
A$_2$	[0.908, 2.422]	[0.005, 0.093]	[0.914, 2.515	1.714	**2**
A$_3$	[0.894, 2.367]	[0.005, 0.083]	[0.899, 2.450]	1.675	**3**
A$_4$	[0.739, 2.061]	[0.002, 0.037]	[0.741, 2.098]	1.419	**5**
A$_5$	[1.053, 2.636]	[0.009, 0.119]	[1.063, 2.755	1.909	**1**
A$_6$	[0.790, 2.026]	[0.004, 0.044]	[0.793, 2.070]	1.432	**4**

After applying the previous 1–9 steps of the rough WASPAS method, in the last, 10th step it is necessary to perform the ranking of alternatives. The highest value represents the best alternative, which in this case is alternative **A$_5$**, while the worst alternative is alternative **A$_1$**.

5. Sensitivity Analysis

The sensitivity analysis performed in this paper consists of three parts. In step 9, it is indicated that besides Equation (17) which is recommended for the calculation of the coefficient λ, this can also have a crisp value in the range of 0.1, 0.2, 0.3, ..., 1.0. Therefore, in the first part of the sensitivity analysis, a change in the coefficient λ was made, which is shown in Table 9.

Table 9. Relative values of the alternatives depending on the value of the coefficient λ.

	λ = 0	λ = 0.1	λ = 0.2	λ = 0.3	λ = 0.4	λ = 0.5	λ = 0.6	λ = 0.7	λ = 0.8	λ = 0.9	λ = 1.0
A$_1$	0.033	0.267	0.501	0.735	0.969	1.203	1.437	1.671	1.905	2.139	2.374
A$_2$	0.100	0.389	0.678	0.967	1.255	1.544	1.833	2.122	2.411	2.699	2.989
A$_3$	0.090	0.374	0.658	0.942	1.225	1.509	1.792	2.076	2.359	2.643	2.927
A$_4$	0.040	0.287	0.534	0.781	1.028	1.275	1.521	1.768	2.015	2.262	2.509
A$_5$	0.131	0.449	0.768	1.086	1.405	1.724	2.042	2.361	2.680	2.998	3.317
A$_6$	0.048	0.296	0.545	0.793	1.041	1.289	1.537	1.785	2.034	2.282	2.530

Table 9 and Figure 1 show the relative values of the alternatives depending on the value of the coefficient λ. It can be noted that the values of the coefficient λ do not affect the change in the rank of the alternative, but actually retain their starting rank, as shown in Table 8. As the value of λ increases, the relative values of the alternatives are also increased.

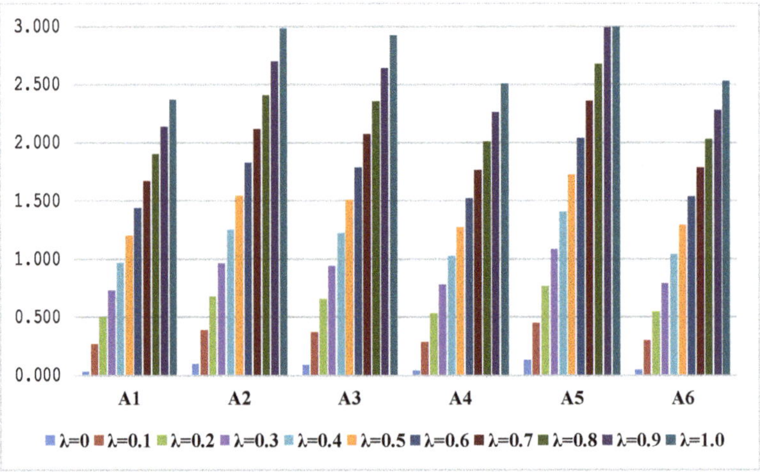

Figure 1. Results of sensitivity analysis dependent on coefficient λ.

The relative value of the alternative A_1 increases by 0.234 with increasing values of λ, while the second alternative (A_2) increases by 0.289. A slightly smaller increase compared to the previous alternative is in alternative A_3 (by 0.284), while in alternative A_4 it is 0.247. The best alternative is A_5 and it is logical that its value is increased depending on λ at most by 0.319. The last alternative, A_6, is closest to the fourth alternative (A_4) and has an increase of 0.248.

The second part of the sensitivity analysis relates to the application of different MCDM methods in combination with rough numbers that are very recently developed. The model presented in the paper is solved by using the following methods: rough SAW [39], rough EDAS [35], rough MABAC [32], rough VIKOR [47], rough MAIRCA [43] and rough MULTIMOORA [35]. Their results and comparison with the rough WASPAS approach are shown in Figure 2.

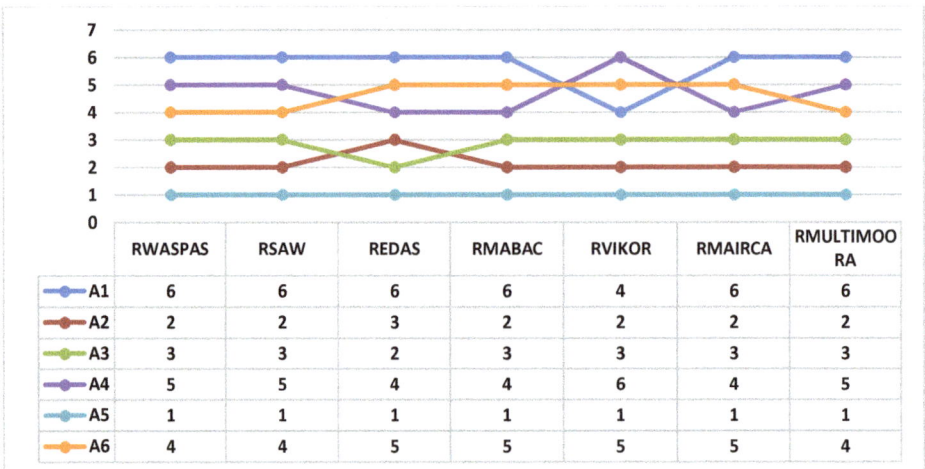

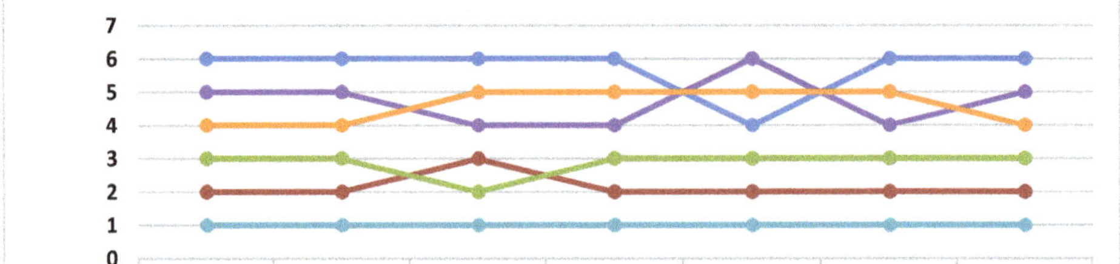

	RWASPAS	RSAW	REDAS	RMABAC	RVIKOR	RMAIRCA	RMULTIMOORA
A1	6	6	6	6	4	6	6
A2	2	2	3	2	2	2	2
A3	3	3	2	3	3	3	3
A4	5	5	4	4	6	4	5
A5	1	1	1	1	1	1	1
A6	4	4	5	5	5	5	4

Figure 2. Results of sensitivity analysis in comparison with other rough methods.

Figure 2 shows a comparison of the initial rank obtained by applying a newly developed rough WASPAS approach with other similar approaches to determine the validity of the developed approach. The best alternative does not change its ranking, it is always ranked the first, regardless of the applied approach. The alternative A_1 is at the last place in the application of all other methods, except when using rough VIKOR when it is in fourth place. An alternative A_2 is in the second position, in all cases, except in the rough EDAS method, when it is in third place. The alternative number three (A_3) also retains its position in the ranks of the other approaches, except in the rough EDAS approach where a rotation of position appears, so the alternative three occupy the second place. The alternative four (A_4) is three times in fifth place, three times in fourth place and at the last, sixth place, when the rough VIKOR method is applied. Alternative six (A_6) occupies the fourth place at rough SAW and rough MULTIMOORA, while in the other methods it is in fifth position. Based on all the results and rankings in all the approaches, stability in the rank correlation can be seen, but in order to validate them Spearman's coefficient of correlation (rk) for statistical comparison of ranks was applied. A comparison of the ranks was done through a comparison of all 7 hybrid models, as shown in Table 10.

Table 10. Statistical comparison of ranks for tested models.

Methods	RWASPAS	RSAW	REDAS	RMABAC	RVIKOR	RMAIRCA	RMULTI-MOORA	Average
RWASPAS	1.000	1.000	0.886	0.943	0.829	0.943	1.000	0.943
RSAW	-	1.000	0.886	0.943	0.829	0.943	1.000	0.933
REDAS	-	-	1.000	0.943	0.714	0.943	0.886	0.897
RMABAC	-	-	-	1.000	0.771	1.000	0.943	0.929
RVIKOR	-	-	-	-	1.000	0.771	0.829	0.867
RMAIRCA	-	-	-	-	-	1.000	0.943	0.971
RMULTI-MOORA	-	-	-	-	-	-	1.000	1.000
			Overall average					0.934

From Table 10 it is possible to notice that there is a strong correlation of the ranges between the considered models, since the total mean value of the correlation coefficient is rk = 0.934. The smallest correlation of ranks is when comparing the Rough EDAS approach with Rough VIKOR, Rough MABAC with Rough VIKOR, and Rough VIKOR with Rough MAIRCA, where the values 0.714, 0.717 and 0.717 were obtained, respectively. These are unique situations where rk < 0.80. In other ranking comparison situations, the coefficient of correlation ranges from 0.829 to1.00. Rough WASPAS and Rough SAW have identical ranks, and the correlation coefficient is equal to 1.00, so these two approaches have rk = 0.829 in comparison with the Rough VIKOR approach. The same correlation coefficient value has Rough VIKOR with Rough MULTIMOORA, while the values of 0.886 have Rough WASPAS and Rough SAW with Rough MABAC, and Rough EDAS with Rough MULTIMOORA. The value of the correlation coefficient of 0.943 has Rough WASPAS, Rough SAW and Rough EDAS with Rough MABAC and with Rough MAIRCA, Rough MABAC and Rough MAIRCA with Rough MULTIMOORA. A complete correlation of ranks, besides already mentioned Rough WASPAS and Rough SAW interdependence, the latter methods have with Rough MULTIMOORA, and Rough MABAC with Rough MAIRCA. It can be concluded that there is an extremely strong correlation of ranks and that the ranks obtained by the proposed new approach are confirmed and credible.

6. Conclusions

The developed approach presented in this research refers to the integration of the rough AHP and rough WASPAS methods, where rough AHP is used to calculate the weight values of the criteria, and rough WASPAS is applied for the evaluation and ranking of suppliers. The model is verified through the process of selecting suppliers in the company for the production of PVC furniture based on nine criteria. The results obtained using the rough WASPAS approach show that the fifth alternative is the best solution, in both parts of the sensitivity analysis, that involves changing the value of the coefficient λ and solving the set model with various approaches developed in recent times. Analysis of the results obtained through the calculation of Spearman's correlation coefficient found that the rough WASPAS approach is in complete correlation with the ranks of other approaches.

Through the research carried out in this paper, two contributions can be distinguished, and one of them is the development of a new rough AHP–rough WASPAS approach which provides an objective aggregation of expert decisions with full observance of inaccuracies and subjectivity that prevails in group decision making. The development of a new approach contributes to the improvement of literature that considers the theoretical and practical application of MCDM methods. The approach that has been developed allows evaluation of alternatives, regardless of imprecision and lack of quantitative information in decision-making. Another contribution of this paper is to improve the methodology of evaluation and selection of suppliers in the production of PVC furniture through a new approach in the treatment of inaccuracies, because the application of this or a similar approach in the selection of suppliers in the field of PVC furniture production has not been identified in the literature. Applying the developed approach, it is possible in a very simple way to solve the problem of MCDM and perform an evaluation and selection of suppliers that has a significant impact on the efficiency of the complete supply chain. The approach developed, besides the considered problem, can also be used for decision making in other areas. Its flexibility is reflected in the fact that verification can be carried out by the integration of any of the multi-criteria decision-making methods for determining the weight values of the criteria. Future research relates to the use of rough numbers in integration with other methods and an attempt to develop a new method in this area.

Author Contributions: All authors designed the research, analyzed the data and the obtained results, and performed the development of the approach in the paper. All the authors have read and approved the final version of manuscript.

Conflicts of Interest: The authors declare no conflict of interest.

References

1. Petrovic, D.; Xie, Y.; Burnham, K.; Petrovic, R. Coordinated control of distribution supply chains in the presence of fuzzy customer demand. *Eur. J. Oper. Res.* **2008**, *185*, 146–158. [CrossRef]
2. Monczka, R.M.; Handfield, R.B.; Giunipero, L.C.; Patterson, J.L. *Purchasing and and Supply Chain Management*; Cengage Learning: Boston, MA, USA, 2015; ISBN 978-1-285-86968-1.
3. Zhong, L.; Yao, L. An ELECTRE I-based multi-criteria group decision making method with interval type-2 fuzzy numbers and its application to supplier selection. *Appl. Soft Comput.* **2017**, *57*, 556–576. [CrossRef]
4. Bai, C.; Sarkis, J. Evaluating supplier development programs with a grey based rough set methodology. *Expert Syst. Appl.* **2011**, *38*, 13505–13517. [CrossRef]
5. Zavadskas, E.K.; Turskis, Z.; Antucheviciene, J.; Zakarevicius, A. Optimization of weighted aggregated sum product assessment. *Elektron. Elektrotech.* **2012**, *122*, 3–6. [CrossRef]
6. Ighravwe, D.E.; Oke, S.A. A fuzzy-grey-weighted aggregate sum product assessment methodical approach for multi-criteria analysis of maintenance performance systems. *Int. J. Syst. Assur. Eng. Manag.* **2017**, *8*, 961–973. [CrossRef]
7. Mathew, M.; Sahu, S.; Upadhyay, A.K. Effect of normalization techniques in robot selection using weighted aggregated sum product assessment. *Int. J. Innov. Res. Adv. Stud.* **2017**, *4*, 59–63.
8. Bagočius, V.; Zavadskas, E.K.; Turskis, Z. Multi-person selection of the best wind turbine based on the multi-criteria integrated additive-multiplicative utility function. *J. Civ. Eng. Manag.* **2014**, *20*, 590–599. [CrossRef]
9. Yazdani, M. New approach to select materials using MADM tools. *Int. J. Bus. Syst. Res.* **2018**, *12*, 25–42. [CrossRef]
10. Hashemkhani Zolfani, S.; Aghdaie, M.H.; Derakhti, A.; Zavadskas, E.K.; Varzandeh, M.H.M. Decision making on business issues with foresight perspective; an application of new hybrid MCDM model in shopping mall locating. *Expert Syst. Appl.* **2013**, *40*, 7111–7121. [CrossRef]
11. Zavadskas, E.K.; Kalibatas, D.; Kalibatiene, D. A multi-attribute assessment using WASPAS for choosing an optimal indoor environment. *Arch. Civ. Mech. Eng.* **2016**, *16*, 76–85. [CrossRef]
12. Suresh, R.K.; Krishnaiah, G.; Venkataramaiah, P. An experimental investigation towards multi objective optimization during hard turning of tool steel using a novel MCDM technique. *Int. J. Appl. Eng. Res.* **2017**, *12*, 1899–1907.

13. Džiugaitė-Tumėnienė, R.; Lapinskienė, V. The multicriteria assessment model for an energy supply system of a low energy house. *Eng. Struct. Technol.* **2014**, *6*, 33–41. [CrossRef]
14. Turskis, Z.; Daniūnas, A.; Zavadskas, E.K.; Medzvieckas, J. Multicriteria evaluation of building foundation alternatives. *Comput. Aided Civ. Infrastruct. Eng.* **2016**, *31*, 717–729. [CrossRef]
15. Emovon, I. A Model for determining appropriate speed breaker mechanism for power generation. *J. Appl. Sci. Process Eng.* **2018**, *5*, 256–265.
16. Madić, M.; Gecevska, V.; Radovanović, M.; Petković, D. Multicriteria economic analysis of machining processes using the WASPAS method. *J. Prod. Eng.* **2014**, *17*, 1–6.
17. Turskis, Z.; Zavadskas, E.K.; Antucheviciene, J.; Kosareva, N. A hybrid model based on fuzzy AHP and fuzzy WASPAS for construction site selection. *Int. J. Comput. Commun. Control* **2015**, *10*, 113–128. [CrossRef]
18. Madic, M.; Antucheviciene, J.; Radovanovic, M.; Petkovic, D. Determination of manufacturing process conditions by using MCDM methods: Application in laser cutting. *Eng. Econ.* **2016**, *27*, 144–150. [CrossRef]
19. Vafaeipour, M.; Zolfani, S.H.; Varzandeh, M.H.M.; Derakhti, A.; Eshkalag, M.K. Assessment of regions priority for implementation of solar projects in Iran: New application of a hybrid multi-criteria decision making approach. *Energy Convers. Manag.* **2014**, *86*, 653–663. [CrossRef]
20. Ghorshi Nezhad, M.R.; Zolfani, S.H.; Moztarzadeh, F.; Zavadskas, E.K.; Bahrami, M. Planning the priority of high tech industries based on SWARA-WASPAS methodology: The case of the nanotechnology industry in Iran. *Econ. Res.-Ekon. Istraz.* **2015**, *28*, 1111–1137. [CrossRef]
21. Yazdani, M.; Hashemkhani Zolfani, S.; Zavadskas, E.K. New integration of MCDM methods and QFD in the selection of green suppliers. *J. Bus. Econ. Manag.* **2016**, *17*, 1097–1113. [CrossRef]
22. Urosevic, S.; Karabasevic, D.; Stanujkic, D.; Maksimovic, M. An approach to personnel selection in the tourism industry based on the SWARA and the WASPAS methods. *Econ. Comput. Econ. Cybern. Stud. Res.* **2017**, *51*, 75–88.
23. Mardani, A.; Nilashi, M.; Zakuan, N.; Loganathan, N.; Soheilirad, S.; Saman, M.Z.M.; Ibrahim, O. A systematic review and meta-Analysis of SWARA and WASPAS methods: Theory and applications with recent fuzzy developments. *Appl. Soft Comput.* **2017**, *57*, 265–292. [CrossRef]
24. Zavadskas, E.K.; Turskis, Z.; Antucheviciene, J. Selecting a contractor by using a novel method for multiple attribute analysis: Weighted Aggregated Sum Product Assessment with grey values (WASPAS-G). *Stud. Inform. Control* **2015**, *24*, 141–150. [CrossRef]
25. Keshavarz Ghorabaee, M.; Zavadskas, E.K.; Amiri, M.; Esmaeili, A. Multi-criteria evaluation of green suppliers using an extended WASPAS method with interval type-2 fuzzy sets. *J. Clean. Prod.* **2016**, *137*, 213–229. [CrossRef]
26. Keshavarz Ghorabaee, M.; Amiri, M.; Zavadskas, E.K.; Antuchevičienė, J. Assessment of third-party logistics providers using a CRITIC–WASPAS approach with interval type-2 fuzzy sets. *Transport* **2017**, *32*, 66–78. [CrossRef]
27. Zavadskas, E.K.; Baušys, R.; Stanujkic, D.; Magdalinovic-Kalinovic, M. Selection of lead-zinc flotation circuit design by applying WASPAS method with single-valued neutrosophic set. *Acta Montan. Slov.* **2016**, *21*, 85–92.
28. Baušys, R.; Juodagalvienė, B. Garage location selection for residential house by WASPAS-SVNS method. *J. Civ. Eng. Manag.* **2017**, *23*, 421–429. [CrossRef]
29. Zavadskas, E.K.; Antucheviciene, J.; Hajiagha, S.H.R.; Hashemi, S.S. Extension of weighted aggregated sum product assessment with interval-valued intuitionistic fuzzy numbers (WASPAS-IVIF). *Appl. Soft Comput.* **2014**, *24*, 1013–1021. [CrossRef]
30. Nie, R.X.; Wang, J.Q.; Zhang, H.Y. Solving solar-wind power station location problem using an extended weighted aggregated sum product assessment (WASPAS) technique with interval neutrosophic sets. *Symmetry* **2017**, *9*, 106. [CrossRef]
31. Song, W.; Ming, X.; Wu, Z.; Zhu, B. A rough TOPSIS approach for failure mode and effects analysis in uncertain environments. *Qual. Reliab. Eng. Int.* **2014**, *30*, 473–486. [CrossRef]
32. Roy, J.; Chatterjee, K.; Bandhopadhyay, A.; Kar, S. Evaluation and selection of Medical Tourism sites: A rough AHP based MABAC approach. *arXiv*, 2016; arXiv:1606.08962.
33. Pamučar, D.; Stević, Ž.; Zavadskas, E.K. Integration of interval rough AHP and interval rough MABAC methods for evaluating university web pages. *Appl. Soft Comput.* **2018**, *67*, 141–163. [CrossRef]

34. Song, W.; Ming, X.; Wu, Z. An integrated rough number-based approach to design concept evaluation under subjective environments. *J. Eng. Des.* **2013**, *24*, 320–341. [CrossRef]
35. Stević, Ž.; Pamučar, D.; Vasiljević, M.; Stojić, G.; Korica, S. Novel integrated multi-criteria model for supplier selection: Case study construction company. *Symmetry* **2017**, *9*, 279. [CrossRef]
36. Cao, J.; Song, W. Risk assessment of co-creating value with customers: A rough group analytic network process approach. *Expert Syst. Appl.* **2016**, *55*, 145–156. [CrossRef]
37. Vasiljevic, M.; Fazlollahtabar, H.; Stevic, Z.; Veskovic, S. A rough multicriteria approach for evaluation of the supplier criteria in automotive industry. *Decis. Mak. Appl. Manag. Eng.* **2018**, *1*, 82–96. [CrossRef]
38. Karavidic, Z.; Projovic, D. A multi-criteria decision-making (MCDM) model in the security forces operations based on rough sets. *Decis. Mak. Appl. Manag. Eng.* **2018**, *1*, 97–120. [CrossRef]
39. Stević, Ž.; Pamučar, D.; Zavadskas, E.K.; Ćirović, G.; Prentkovskis, O. The selection of wagons for the internal transport of a logistics company: A novel approach based on Rough BWM and Rough SAW Methods. *Symmetry* **2017**, *9*, 264. [CrossRef]
40. Chai, J.; Liu, J.N. A novel believable rough set approach for supplier selection. *Expert Syst. Appl.* **2014**, *41*, 92–104. [CrossRef]
41. Zhai, L.Y.; Khoo, L.P.; Zhong, Z.W. Design concept evaluation in product development using rough sets and grey relation analysis. *Expert Syst. Appl.* **2009**, *36*, 7072–7079. [CrossRef]
42. Khoo, L.P.; Tor, S.B.; Zhai, L.Y. A rough-set based approach for classification and rule induction. *Int. J. Adv. Manuf. Technol.* **1999**, *15*, 438–444. [CrossRef]
43. Pamučar, D.; Mihajlović, M.; Obradović, R.; Atanasković, P. Novel approach to group multi-criteria decision making based on interval rough numbers: Hybrid DEMATEL-ANP-MAIRCA model. *Expert Syst. Appl.* **2017**, *88*, 58–80. [CrossRef]
44. Duntsch, I.; Gediga, G. The rough set engine GROBIAN. In Proceedings of the 15th IMACS World Congress; Sydow, A., Ed.; Berlin Wissenschaft und Technik Verlag: Berlin, Germany, 1997; Volume 4, pp. 613–618.
45. Khoo, L.-P.; Zhai, L.-Y. A prototype genetic algorithm enhanced rough set-based rule induction system. *Comput. Ind.* **2001**, *46*, 95–106. [CrossRef]
46. Pawlak, Z. Rough sets. *Int. J. Comput. Inf. Sci.* **1982**, *11*, 341–356. [CrossRef]
47. Zhu, G.N.; Hu, J.; Qi, J.; Gu, C.C.; Peng, J.H. An integrated AHP and VIKOR for design concept evaluation based on rough number. *Adv. Eng. Inform.* **2015**, *29*, 408–418. [CrossRef]
48. Pawlak, Z. *Rough Sets: Theoretical Aspects of Reasoning about Data*; Springer: Berlin, Germany, 1991.
49. Pawlak, Z. Anatomy of conflicts. *Bull. Eur. Assoc. Theor. Comput. Sci.* **1993**, *50*, 234–247.
50. Stević, Ž. Integrisani Model Vrednovanja Dobavljača u Lancima Snabdevanja. Ph.D. Thesis, Univerzitet u Novom Sadu, Fakultet Tehničkih Nauka, Novi Sad, Serbia, 2018. (In Bosnian)
51. Zhai, L.Y.; Khoo, L.P.; Zhong, Z.W. A rough set based QFD approach to the management of imprecise design information in product development. *Adv. Eng. Inform.* **2009**, *23*, 222–228. [CrossRef]

information

Article

An Interactive Multiobjective Optimization Approach to Supplier Selection and Order Allocation Problems Using the Concept of Desirability

Pyoungsoo Lee [1,*] **and Sungmin Kang** [2]

[1] Korea E-Trade Research Institute, Chung-Ang University, 84 Heukseok, Dongjak, Seoul 06974, Korea
[2] College of Business and Economics, Chung-Ang University, 84 Heukseok-ro, Dongjak-gu, Seoul 06974, Korea; smkang@cau.ac.kr
* Correspondence: pyoungsoo@cau.ac.kr; Tel.: +82-2-820-6504

Received: 9 May 2018; Accepted: 22 May 2018; Published: 23 May 2018

Abstract: In supply chain management, selecting the right supplier is one of the most important decision-making processes for improving corporate competitiveness. In particular, when a buyer considers selecting multiple suppliers, one should consider the issue of order allocation with supplier selection. In this article, an interactive multiobjective optimization approach is proposed for the supplier selection and order allocation problem. Also, the concept of desirability is incorporated into the optimization model to take into account the principles of diminishing marginal utility. The results are presented by comparing them with the solutions from the weighting methods. This study shows the advantage of the proposed method in that the decision-maker directly checks the degree of desirability and learns his/her preference structure through improved solutions.

Keywords: supplier selection; order allocation; multiobjective optimization; interactive approach; desirability function

1. Introduction

With the development of the Internet and network technologies, the digital divide between businesses is being addressed. In this situation, companies are using strategies to enhance outsourcing to focus on core competences along with information. In particular, such rapid developments in information technology are pushing many companies into a race that transcends time and space. Furthermore, the global business environment surrounding corporations is changing from competition between individual companies to competition between supply chains. In order to ensure the competitive edge of the supply chain, companies have tried to find the right suppliers offering higher quality, reduced costs, and shorter lead times. Therefore, in supply chain management, selecting the right suppliers is one of the most important decision-making processes for improving corporate competitiveness [1–3].

In many cases, a single supplier may not be able to meet the buyer's requirements. In such cases, selecting multiple suppliers—multiple sourcing—would be a reasonable alternative [4]. Whereas single sourcing significantly increases the disruption risk in the supply chain, multiple sourcing increases the fixed cost in terms of administrative and negotiating costs [5]. However, multiple sourcing is preferred over a single sourcing, ensuring order flexibility [6]. Therefore, multiple sourcing inevitably includes the problem of order allocation. Furthermore, the relationship between the buyer and supplier is influenced by order allocation decisions based on strategic purchase decisions [7]. Therefore, the overall supplier selection problem should not only cover the selection of the right supplier but also the determination of the orders assigned to the selected supplier based on the given objectives and constraints [8]. To date, however, only a few mathematical programming models to analyze such decisions have been published [9–12]. In the field of supply chain management, it is necessary to develop a precise decision support model which simultaneously considers supplier selection and order allocation. In particular, researchers need to advance research in this direction to guide the decision-maker (DM)'s rational choice.

Supplier selection is inherently a multi-criteria decision-making (MCDM) problem since some conflicting performance criteria have an influence on the selection of suppliers [13,14]. It is also considered one of the most familiar problems in MCDM [15]. This problem has been studied from a variety of perspectives, such as green supplier selection [6,16] and global supplier selection [17]. Generally, the performance criteria involves the factors such as cost, quality, and lead time. Each supplier has its own strengths and weaknesses, so it is very difficult to select a superior supplier at all dimensions of criteria. Thus, supplier selection problem has made MCDM a very challenging task. Ho et al. [18] provided comprehensive reviews for MCDM approaches to vendor selection problems between 2000 and 2008. In addition, a more recent review from Chai et al. [19] provides guidelines for a MCDM-based supplier selection model.

For solving a multiobjective optimization (MOO) problem in MCDM, the DM seeks a compromise solution which provides the greatest satisfaction in the presence of conflicting objectives. Thus, the DM's preference information plays a critical role in finding the solution. In the literature, the DM's preference information can enter the solving process of MOO problems in three different ways: (1) a priori; (2) a posteriori; and (3) progressive (interactive) articulation [20,21]. For an a priori setting, multiple objective functions combined with preference convert into one single objective. For an a posteriori setting, the DM's preference information is articulated after optimization process by selecting the most preferred one from a set of non-dominated solutions, usually called a Pareto optimal set. For progressive optimization, the preference of the DM is incorporated into the solution search process. Iterative dialogues between the DM and optimization model contribute to find the most satisfactory solution in this optimization approach. For this reason, the progressive articulation approach is also referred to as the interactive approach.

Most studies in supplier selection and order allocation are categorized into the prior approach by the MOO categorization scheme. However, the prior approach may decrease the reliability of solutions because of the unrealistic assumption that the DM can specify the preference information in advance. Although the interactive approach has been broadly applied to various fields as an alternative to overcome this limitation on the DM's preference, it is rarely used to solve supplier selection and order allocation problems. To the best of the author's knowledge, Demirtas and Üstün [4]'s research is the only material related to using the interactive MOO method to solve supplier selection and order allocation problem. Demirtas and Üstün [4] proposed an interactive MOO method, and a reservation level driven Tchebycheff procedure. In this model, the solution process makes the DM express adjusting some or all of reservation values of objective functions, by generating candidate solutions with sampling weights. The interactive method that we propose in this paper has something in common with Demirtas and Üstün [4]'s research in terms of a kind of objective space reduction. Also, two methods share the Tchebycheff framework to find a solution by utilizing the concept of distance to the ideal vector. However, the proposed approach provides an integrated use of the desirability function approach in response to surface methodology and the step method (STEM). Our model has advantages in that it reflects the satisfaction of the DM more realistically by using the concept of desirability function, and it reduces the burden on the DM to express preference information.

Based on the above-mentioned background, the purpose of the research is summarized as follows. We aim to solve the multiple sourcing problem which deals with order allocation at the same time as supplier selection by using an interactive MOO method. The devised method, which progressively articulates the DM's preference information, is applied to a problem with three important criteria: cost, quality, and delivery. Also, to intuitively utilize the level of satisfaction, we borrow the concept of desirability from the research field of product and process design. We show that our method can be utilized effectively in the supplier selection and order allocation problem.

The organization of this paper is as follows. Section 2 presents previous works to be addressed in this study. The proposed interactive desirability function approach to supplier selection is presented in Section 3. Section 4 analyzes results and provides discussions. Conclusions are provided in Section 5.

2. Literature Review

This section briefly describes the concept of desirability function that is the basis for the proposed methodology. Then we deal with the consideration of the DM's preference information in MOO problems. Also, the STEM is introduced.

2.1. Desirability Function Approach

The desirability function approach is in fact designed to solve the problem of product and process design (also called multiple response optimization). The desirability function approach, proposed by Harrington [22] and Derringer and Suich [23], is one of the most widely used methods in product and process design. This approach transforms an estimated response function into a scale-free function, called the desirability function, which ranges from zero to one. Thus, the value of the desirability function presents the degree of desirability or satisfaction for the corresponding response. If the larger-the-better (LTB) response is the objective that has to be maximized, the individual desirability function is defined as

$$d_i = \begin{cases} 0, & \text{for } \hat{y}_i(\mathbf{x}) \leq \hat{y}_i^{\min} \\ \left(\frac{\hat{y}_i(\mathbf{x}) - \hat{y}_i^{\min}}{\hat{y}_i^{\max} - \hat{y}_i^{\min}} \right)^r, & \text{for } \hat{y}_i^{\min} < \hat{y}_i(\mathbf{x}) < \hat{y}_i^{\max} \\ 1, & \text{for } \hat{y}_i(\mathbf{x}) \geq \hat{y}_i^{\max}, \end{cases} \tag{1}$$

where $\hat{y}_i^{\min}$ is the minimum and $\hat{y}_i^{\max}$ is the maximum acceptable value of response i and r is the parameter ($r > 0$) that determines the shape of desirability functions. The function is convex if $r > 1$, and is concave if $0 < r < 1$. If a response is of the smaller-the-better (STB) type, it is defined in a similar way to (1). Also, for a certain response, if the specified target value has to be attained, the response is referred to as the nominal-the-best (NTB) type. In this case, the individual desirability function is derived in a slightly different form from STB and LTB type. For more information, readers may refer to Derringer and Suich [23].

There are several ways to draw the optimal solution considering multiple individual desirability functions [22–26] but the most widely used method is to optimize by converting multiple desirability functions into single desirability. In most cases the concept of geometric means is often used for this aggregation.

2.2. Interactive Approach and Decision Maker's Preference Information

The supplier selection is inherently a MCDM problem, since some conflicting criteria have influence on the selection of suppliers [13,14]. MOO is a mathematical optimization technique of MCDM, which involves more than one objective function to be optimized simultaneously. The ultimate goal of MOO is to find the best compromise solution from a set of the non-dominated solutions that satisfies multiple objectives simultaneously. Thus, it necessarily requires the involvement of the DM who provides the preference information among conflicting objectives. According to the timing of articulating the DM's preference information, the MOO is classified into three categories: the prior approach, the posterior approach, and the interactive approach [20,21]. More detailed descriptions of the theories and applications of MOO studies can be found in the classical MOO textbooks: [20,27,28].

The interactive approach (also referred as the progressive approach) allows the DM to articulate his/her preference information progressively while solving the problem. The optimization process is repeated until the DM has found the most preferred solution to extracting preference information in an interactive manner. Specifically, at each iteration in an algorithm, the DM is asked to express some preference information given one or more solutions generated from the previous iteration. This information reflects the judgment based on the DM's implicit value function, and the process updates the DM's preference through adding the new information to an optimization model. Typically, a general interactive method has the following three steps [27].

- Step 1: Find an initial solution.
- Step 2: Interact with the DM.
- Step 3: Generate one or some new solution(s). If the DM is satisfied with the current solution, stop. Otherwise, go to step 2.

Over the years, a number of various interactive methods have been developed. The most well-known interactive methods are the step method (STEM) [29], the Geoffrion-Dyer-Feinberg (GDF) method [30], the Zionts-Wallenius (ZW) method [31], the reference direction approach (RDA) [32], and the Nondifferentiable Interactive Multiobjective BUndle-based optimization System (NIMBUS) [27]. Each of interactive methods has a different scheme from others in terms of which type of preference information is asked from the DM. No matter what kind of preference information is asked, an iterative process is valuable for the DM in that he/she can learn about the structure of the problem and expect the result of the interaction. Furthermore, each employs different computational algorithms. Therefore, a particular interactive method may be suitable for a specified MOO problem given mathematical properties or assumptions. In other words, there is no the sole interactive method that is decidedly superior to the others, in a universal sense. Even though a lot of extensions and variants of the interactive methods can be applied to the supplier selection problem, in this paper we propose an interactive approach based on the STEM with some modifications.

2.3. STEM

The STEM is proposed by Benayoun et al. [29] and it is known as the first interactive MOO method. Originally STEM is designed for the linear MOO problems, and it also can be applied to integer and nonlinear MOO problems [28]. The interactive process in the STEM begins with generating an initial feasible solution and identifying the ideal objective function values. The ideal objective function values are obtained by optimizing each individual objective function over the initial feasible region. At each iteration, the DM examines the current solution with the ideal objective function value, and presents his/her preference depending on the concept of level of satisfaction. In particular, the DM provides an amount of relaxation by which the objective function value can be sacrificed, in order to improve some other unsatisfied objectives. From this numerical information presented from the DM, it generates a new solution. If the DM satisfied with this new solution, this solution becomes a final solution as a compromise and the interactive process terminates. Otherwise this process is repeated until no further relaxation is accepted. The mathematical procedure for the STEM has been given in the following steps.

Step 0: Construct the pay-off table. In this step, the ideal objective function vector is obtained by maximizing each objective function individually. Let $\mathbf{f}^*$ be the ideal solution vector of the following K problems, $\mathbf{f}^* = (f_1^*, f_2^*, \ldots, f_K^*)$.

$$\begin{aligned} & f_i^* = \max f_i(\mathbf{x}) \\ & \text{subject to} \\ & \mathbf{x} \in S \end{aligned} \tag{2}$$

Step 1: Let h be iteration counter and set be zero ($h = 0$). Calculate π_i values for use in weighting the objectives:

$$\pi_i = \begin{cases} \frac{f_i^* - n_i}{f_i^*} \left[\sum_{j=1}^{n} c_{ij}^2 \right]^{-1/2}, \text{ if } f_i^* > 0, \\ \frac{n_i - f_i^*}{n_i} \left[\sum_{j=1}^{n} c_{ij}^2 \right]^{-1/2}, \text{ if } f_i^* \leq 0, \end{cases} \tag{3}$$

where c_{ij} are the coefficients of the ith objective. The first term is to place the most weight on the objectives with the greatest relative ranges. The second term normalizes the gradients of the objective functions.

Step 2: Let $h = h + 1$, and $S^{h+1} = S$. $S^1 = S$ means the solution process begin with the original feasible region. At this point, the relaxation index set be null ($J = \varnothing$). Compute relative weights:

$$\lambda_i^h = \begin{cases} 0, & i \in J \\ \frac{\pi_i}{\sum_j \pi_j}, & i \notin J \end{cases} \tag{4}$$

Step 3: Solve the weighted minimax program:

$$\begin{aligned} & \min \alpha \\ & \text{subject to} \\ & \alpha \geq \lambda_i^h \left[f_i^* - f_i(\mathbf{x}) \right], \quad i = 1, \ldots, k \\ & \mathbf{x} \in S^h \\ & \alpha \geq 0 \end{aligned} \tag{5}$$

If all objective function values are satisfactory, then stop; the current solution may become the final solution.

Step 4: Determine satisfactory objective and then specify the amount of relaxation, Δ_i ($i \in J$).

Step 5: Form reduced feasible region:

$$S^{h+1} = \begin{cases} \mathbf{x} \in S \\ f_i(\mathbf{x}) \geq f_i(\mathbf{x}^h), i \notin J \\ f_i(\mathbf{x}) \geq f_i(\mathbf{x}^h) - \Delta_i, i \in J \end{cases} \tag{6}$$

where $h + 1$ is increased iteration counter. Thus, S^{h+1} is the feasible region in the $(h + 1)$th iteration. $f_i(\mathbf{x})^h$ is the objective function value in the hth iteration and Δ_i is the amount of relaxation of particular objective function. Go to **Step 2** with the reduced feasible region.

The STEM is widely used to solve not only linear but also nonlinear MOO problems for the reason that it is simple and straightforward to generate the solution by reducing feasible regions in the objective space. Despite these advantages, the STEM shows two major drawbacks. First, the STEM does not take into account different degrees of satisfaction within the acceptable interval of a relaxed objective. For example, suppose that one objective f_e is determined by the DM as a satisfactory function, and that he/she decides to relax as much as the Δ_e. In this case, the weight of objective function e becomes zero since $e \notin J$, so this function does not serve as an objective function in the next iteration. Therefore, the solution of the next iteration depends on the assumption that the satisfaction of the DM is indifferent within the interval of $[f_e - \Delta_e, f_e^*]$. This disadvantage has been pointed out by Jeong and Kim [33]. The second drawback is in the dialogue scheme between the DM and the optimization model. The dialogue is made mainly by asking the DM to acceptable amount of relaxation of a satisfactory objective. However, it increases the burden on the DM because he/she must provide specific numerical information at each iteration. In this paper, we intend to utilize the STEM to supplier selection problem by modifying it in such a way that we enjoy its merits and complement the shortcomings mentioned above. More specifically, we try to overcome the first limitation by applying the concept of desirability function and the second one by proposing a modification to the DM's preference information representation.

3. Proposed Model

Across fifty years of evaluating suppliers, many researchers have proposed different sets of criteria.

The first set of criteria was proposed by Dickson [14], who identified 23 different criteria evaluated in supplier selection. Evans [34] and Shipley [35] agreed that price, quality and delivery are the most important criteria for evaluating suppliers. Ellram [36] proposed that the quality dimension should be divided into product quality and service quality, and it is suggested to use them with price and delivery time to select suppliers. Weber et al. [37] surveyed based on Dickson's 23 criteria and concluded that price, delivery, quality, production capacity, and localization are the most important criteria. Pi and Low [38] proposed quality, delivery, price and service for supplier evaluation. Amid et al. [39] uses price, quality, and service, and Jadidi et al. [11] utilized price, quality, and lead time to the supplier selection and order allocation model. As shown by the literature, the most important criteria for supplier selection problems are cost, quality, and delivery. We also use those criteria by using the measures: total purchasing costs, the number of rejects, and the number of late delivery for cost, quality, and delivery, respectively. We assume the buyer considers a single item that should be purchased under known total demand. Also, information about criteria and production capacity is already known for a set of potential suppliers. The notations of the proposed model are presented in Table 1.

Table 1. The notations.

k	index for objectives, $k = 1, 2, \ldots, K$
n	number of suppliers
x_i	number of units ordered to supplier i (decision variable)
$\mathbf{x}$	vector of decision variables
V_i	capacity of supplier i
c_i	unit purchasing price from supplier i
q_i	expected defect rate of supplier i
l_i	percentage of items delivered late by supplier i
D	demand

There are three original objective functions: f_1 (minimizing total purchasing cost), f_2 (minimizing total number of defects), f_3 (minimizing the number of late delivery). Each objective function can be shown as follows:

$$\begin{array}{l} f_1(\mathbf{x}) = \sum_i^n c_i x_i \\ f_2(\mathbf{x}) = \sum_i^n q_i x_i \\ f_3(\mathbf{x}) = \sum_i^n l_i x_i \end{array} \tag{7}$$

Using the functions defined above, MOO problem considering the supplier selection with order allocation can be formulated as follows [12]:

$$\begin{array}{l} \min f_1(\mathbf{x}), f_2(\mathbf{x}), f_3(\mathbf{x}) \\ \text{subject to} \\ \sum_i^n x_i = D \\ x_i \leq V_i \\ x_i \geq 0 \end{array} \tag{8}$$

The first and second constraint present demand satisfaction and capacity restriction, respectively, and the last constraint ensures non-negativity of the decision variables. Now we transform the original objective functions into the desirability functions. Because all original objective functions are to be minimized (STB type), the individual desirability is defined as

$$d_i = \begin{cases} 1, & \text{for } f_i(\mathbf{x}) \leq f_i^* \\ \left(\frac{f_i(\mathbf{x}) - f_i^{\max}}{f_i^* - f_i^{\max}} \right)^r, & \text{for } f_i^* < f_i(\mathbf{x}) \leq f_i^{\max} \\ 0, & \text{for } f_i(\mathbf{x}) \geq f_i^{\max}, \end{cases} \tag{9}$$

where $f_i^{\max}$ are the maximum values of the objective functions as obtained from the payoff table and $f_i^*(\mathbf{x})$ are the ideal points from maximizing the objective functions individually. Table 2 presents the payoff table that includes $f_i^{\max}$ and $f_i^*(\mathbf{x})$. For a STB type function, $f_i^{\max}$ are recognized as the *nadir points* which are maximum values from each column in the payoff table, and the ideal values are on the diagonal of the table.

Table 2. Payoff table.

Payoffs	f_1	f_2	f_3
f_1	58.75	5.325	3.675
f_2	82.25	3.225	5.05
f_3	61.25	5.07	3.425

We represent the desirability-based supplier selection problem in the form of a MOO problem:

$$\begin{aligned}
&\max d_1(f_1(\mathbf{x})), d_2(f_2(\mathbf{x})), d_3(f_3(\mathbf{x})) \\
&\text{subject to} \\
&\textstyle\sum_i^n x_i = D \\
&x_i \leq V_i \\
&x_i \geq 0
\end{aligned} \tag{10}$$

The individual desirability functions d_i ($i = 1, \ldots, k$) are objective functions to be maximized simultaneously.

The optimization process consists of five major steps. The overall procedure is presented in Figure 1. In addition, a pseudo code is described in Algorithm 1 to aid readers understanding.

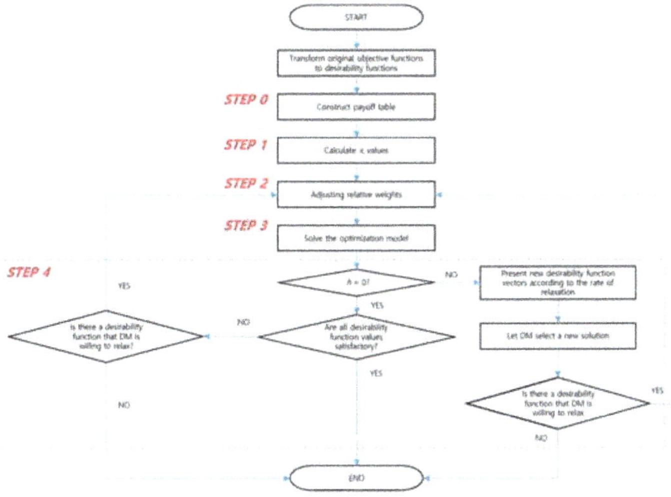

Figure 1. The overall procedure of the proposed method.

Algorithm 1. Pseudo code of the proposed method.

```
begin
        initialize: h ← 0
        initialize: J = ∅
        calculate d* and n*
        calculate π_i values
        compute relative weights λ^h
        compute an initial solution d^h
        ask if there is a desirability function value want to relax (j)
        while the preferred solution has not been found do
            J = {j}
            compute relative weights λ^h
            compute potential solutions based on the preferences of the DM
            present the solutions to the DM
            ask the most preferred solution d^h* from potential solutions
            ask if there is a desirability function value want to relax (j)
            h ← h + 1
        endwhile
end
```

3.1. *Case r = 1*

We apply a modified STEM to supplier selection problem. In order to present how the method works, we deal with the linear type desirability function transformation.

3.1.1. Initialization

Step 0: Construct the payoff table (Table 3) based on the desirability functions with $r = 1$.

Table 3. Payoff table of $r = 1$.

Payoffs	d_1	d_2	d_3
d_1	1	0	0.846
d_2	0	1	0
d_3	0.894	0.119	1

Therefore, the ideal vector and nadir vector are $\mathbf{d}^* = (1, 1, 1)$ and $\mathbf{n}^* = (0, 0, 0)$, respectively.

Step 1: Calculate π_i values for use in weighting the objectives: The coefficients c_{ij}, which reflect the gradients of the objective functions in the weight calculation is the coefficient of desirability function, not the original objective function. If the desirability functions are nonlinear, i.e., $r \neq 1$, the weights calculation is somewhat different from that of the linear case. As mentioned in Section 2.2, originally the STEM is developed for multiobjective linear problems, but nonlinear extensions have been proposed, for example, in Vanderpooten and Vincke [40], Eschenauer et al. [41], and Sawaragi et al. [42]. A weight calculation scheme for the nonlinear case ($r = 2$) is presented in following Section 3.2. At this point, calculated π_i values are shown as

$$\begin{aligned}\pi_1 &= \tfrac{d_1^* - n_1}{d_1^*}\left[\sum_j c_{1j}^2\right]^{-1/2} = 2.1588\\ \pi_2 &= 2.6158\\ \pi_3 &= 2.3039\end{aligned}$$

Step 2: Let h be iteration counter (set h as zero for the first iteration).
For $h = 0$, the relaxation index set be null ($J = \varnothing$).

$S^0 = S$, the original feasible region. Compute relative weights:

$$\begin{aligned}
&\lambda_1^{h=0} = \pi_1/(\pi_1 + \pi_2 + \pi_3) = 0.3050 \\
&\lambda_2^0 = 0.3695 \\
&\lambda_3^0 = 0.3255
\end{aligned}$$

Step 3: Formulate the weighted minimax program:

$$\begin{aligned}
&\min \alpha \\
&\text{subject to} \\
&\alpha \geq 0.3050(1 - d_1) \\
&\alpha \geq 0.3695(1 - d_2) \\
&\alpha \geq 0.3255(1 - d_3) \\
&\textstyle\sum_i^n x_i = D \\
&x_i \leq V_i \\
&x_i \geq 0 \\
&\alpha \geq 0
\end{aligned} \tag{11}$$

The obtained solution is $\mathbf{d}^{0^*} = (d_1{}^{0^*}, d_2{}^{0^*}, d_3{}^{0^*}) = (0.4867, 0.5755, 0.5181)$. We assume the DM does not satisfied with the obtained solution. Go to **Step 4**.

Step 4: Determine satisfactory objective by asking the DM to express satisfactory objectives. If $h = 0$, go to **Step 2** by setting $h = h + 1$. We assume the DM satisfied with d_2, so relax d_2.

3.1.2. First Iteration

Step 2: Compute relative weights ($h = 1$, $S^1 = S$): For $h = 1$, the objective function to be relaxed has been defined as d_2, thus $\{2\} = J$. Accordingly, the computed weights are $\lambda_1^1 = \pi_1/(\pi_1 + \pi_3) = 0.4847$, $\lambda_2^1 = 0$, and $\lambda_3^1 = 0.5163$.

Step 3: Formulate the weighted minimax program:

$$\begin{aligned}
&\min \alpha \\
&\text{subject to} \\
&\alpha \geq 0.4837(1 - d_1(\mathbf{x})) \\
&\alpha \geq 0.5163(1 - d_3(\mathbf{x})) \\
&d_1(\mathbf{x}) \geq d_1^{0*} \\
&d_2(\mathbf{x}) \geq (1 - \delta)d_2^{0*} \\
&d_3(\mathbf{x}) \geq d_3^{0*} \\
&\textstyle\sum_i^n x_i = D \\
&x_i \leq V_i \\
&x_i \geq 0 \\
&\alpha \geq 0
\end{aligned} \tag{12}$$

where δ is rate of relaxation.

Step 4: Present new desirability function vectors to the DM as shown in Table 4, who is asked to select an acceptable rate of relaxation, δ. Unlike original STEM, the proposed approach does not compel the DM to ask specified amount of relaxation from satisfactory level. Then, the optimal solution corresponding selected rate of relaxation becomes the new vector of desirability function values.

Table 4. Changes of desirability function values according to the rate of relaxation (h = 1).

	δ = 0.05	δ = 0.10	δ = 0.15	δ = 0.20	δ = 0.25	δ = 0.30
d_1	0.511	0.537	0.563	0.589	0.614	0.640
d_2	0.547	0.518	0.489	0.460	0.432	0.403
d_3	0.542	0.566	0.590	0.615	0.639	0.663

If the DM is satisfied with the new solution with selected rate of relaxation, δ^*, then stop. Otherwise, set $h = h + 1$ and set the new feasible reason S^{h+1}. Go to **Step 2**. Assume the DM choose 20% relaxation of d_2. Therefore, a new solution is $\mathbf{d}^{1^*}$ = (0.589, 0.460, 0.615) with **f** = (61.167, 5.083, 3.582). Also, we assume the DM satisfied with d_1, and relax d_1.

3.1.3. Second Iteration

Step 2: Let h = 2. Compute relative weights: $\lambda_1^2 = 0, \lambda_2^2 = 0.5317, \lambda_3^2 = 0.4683$.
Step 3: Formulate the weighted minimax program:

$$\begin{aligned}
&\min \alpha \\
&\text{subject to} \\
&\alpha \geq 0.5317(1 - d_2(\mathbf{x})) \\
&\alpha \geq 0.4683(1 - d_3(\mathbf{x})) \\
&d_1(\mathbf{x}) \geq (1 - \delta)d_1^{1*} \\
&d_2(\mathbf{x}) \geq d_2^{1*} \\
&d_3(\mathbf{x}) \geq d_3^{1*} \\
&\textstyle\sum_i^n x_i = D \\
&x_i \leq V_i \\
&x_i \geq 0 \\
&\alpha \geq 0
\end{aligned} \tag{13}$$

Step 4: As shown in Table 5, present new desirability function vectors as δ increases, then ask to the DM an acceptable rate of relaxation, δ.

Table 5. Changes of desirability function values according to the rate of relaxation (h = 2).

	δ = 0.05	δ = 0.10	δ = 0.15	δ = 0.20	δ = 0.25	δ = 0.30
d_1	0.559	0.530	0.500	0.471	0.441	0.412
d_2	0.493	0.526	0.550	0.573	0.596	0.619
d_3	0.626	0.614	0.614	0.614	0.614	0.614

Assume the DM choose 15% relaxation of d_1. Accordingly, a new solution is $\mathbf{d}^{2^*}$ = (0.500, 0.550, 0.614) with **f** = (70.494, 4.170, 4.052). This interactive process can be repeated until the final solution satisfies the DM. At this stage, we assume that the DM is satisfied with the current solution and terminate the iterative procedure.

3.2. *Case r* = 2

All three original objective functions considered in the supplier selection model are linear, and the defined desirability functions in Section 3.1 are also linear. This sub-section discusses the need for nonlinear desirability functions and presents a problem-solving process in r = 2 case.

The weighted sum method assumes that the DM's overall desirability (utility) is determined by the weighted sum of multiple objective functions. Thus, the marginal contribution to utility of each linear objective function is constant (see Figure 2a). Therefore, the optimization result is highly dependent upon the predetermined weights, although the objective functions are normalized by using the ideal points and the nadir points. On the other hand, the STEM modifies the concept of satisfaction by accepting the DM's preference information in the problem-solving process. In particular, the STEM assumes that the desirability of the DM is the same up to the acceptable level, i.e., $f_i(\mathbf{x}) + \Delta_i$ for a STB type objective function (see Figure 2b). Namely, the STEM considers that the DM's satisfaction is indifferent within the interval of $[f_i^*, f_i(\mathbf{x}) + \Delta_i]$. With this concept the STEM improves the other objective functions at a level that does not impair the satisfaction of the satisfactory objective. However, we need to consider the principles of diminishing marginal utility, overlooked in the STEM. The principles of diminishing marginal utility explain that the closer the objective function value is to the optimum value, the less contribution it will make. In this study, the desirability function enables the proposed model to consider the principles of diminishing marginal utility. If the parameter r of desirability functions is not equal to 1, the functions become nonlinear. The parameter r exhibits a diminishing marginal contribution to the maximum cumulative desirability. The nonlinearity of desirability function provides a clue to successfully deal with the principles of diminishing marginal utility in the proposed model. This characteristic is also highlighted in Jeong and Kim [33]'s STEM based on the desirability function. Figure 2c shows an example of a nonlinear desirability function.

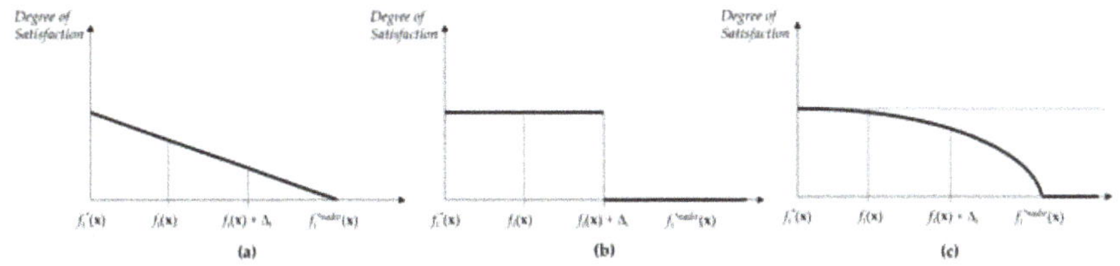

Figure 2. Examples of perceived satisfaction: (**a**) Constant marginal contribution; (**b**) Perceived satisfaction of STEM; (**c**) Nonlinear desirability function.

3.2.1. Initialization

Step 0: Construct the payoff table (Table 6) based on the desirability functions with $r = 2$.

Table 6. Payoff table of $r = 2$.

Payoffs	d_1	d_2	d_3
d_1	1	0	0.799
d_2	0	1	0.014
d_3	0.716	0	1

Therefore, the ideal vector and nadir vector are $\mathbf{d}^* = (1, 1, 1)$ and $\mathbf{n}^* = (0, 0, 0.014)$, respectively.

Step 1: Calculate π_i values for use in weighting the objectives. The several weight schemes used to extend the STEM to nonlinear MOO problems were suggested by various researchers (Eschenauer et al. [41]; Sawaragi et al. [42]; Vanderpooten and Vincke [40]). In this study, we utilize Vanderpooten and Vincke [40]'s calculation (assume that the denominators are not equal to zero) as follows:

$$\pi_i = \frac{d_i^* - n_i}{\max\{d_i^*, n_i\}}, \tag{14}$$

By (14), the determined weights are $\pi_1 = \pi_2 = 1$ and $\pi_3 = 0.986$.

Step 2: Compute relative weights: $\lambda_1^0 = \lambda_2^0 = 0.3349$, $\lambda_3^0 = 0.3302$

Step 3: Solve the weighted minimax program: $\mathbf{d}^{0^*} = (0.2789, 0.2789, 0.3622)$.

Step 4: We assume the DM satisfied with d_2, so relax d_2. Go to **Step 2**.

3.2.2. First Iteration

Step 2: Compute relative weights ($h = 1$, $S^1 = S$): For $h = 1$, the objective function to be relaxed has been defined as a d_2, $\{2\} = J$. Thus, the computed weights are $\lambda_1^1 = 0.5035$, $\lambda_2^1 = 0$, and $\lambda_3^1 = 0.4965$.

Step 3 and **Step 4**: Solve the weighted minimax program and present new desirability function vectors as δ increases, then ask to the DM an acceptable rate of relaxation, δ. Use the computation results as shown in Table 7.

Table 7. Changes of desirability function values according to the rate of relaxation (h = 1).

	δ = 0.05	δ = 0.10	δ = 0.15	δ = 0.20	δ = 0.25	δ = 0.30
d_1	0.292	0.305	0.319	0.334	0.350	0.366
d_2	0.265	0.251	0.237	0.223	0.209	0.195
d_3	0.362	0.362	0.362	0.362	0. 362	0.362

Assume the DM choose 10% relaxation of d_2. Therefore, a new solution is $\mathbf{d}^{1^*}$ = (0.305, 0.251, 0.362) with **f** = (69.271, 4.273, 4.072). We assume the DM satisfied with the current solution, and terminate the procedure.

4. Summary of Results and Discussion

This study successfully applied the concept of desirability function in response surface methodology, which is used in product and process design, to supplier selection problems. The desirability function diverts the DM's recognition system for each objective function from linear coupling and realistically reflects the degree of satisfaction for each objective function. Since the proposed method assumes that the DM's preference information is not completely known, it is difficult to discuss the superiority of solutions by directly comparing the results with other methods. However, we explain the advantages of the proposed method by comparing them with the solutions from two weighting methods: the weighted sum method and weighted geometric method. Several weighting vectors are assumed for the purpose of comparing solutions. We adopted three sets of weighting parameter **w**, proposed be Jadidi et al. [11] for the same supplier selection problem. The results are shown in Table 8.

The results of the weighted sum method show that the more weights are assigned to the first objective function (minimizing total purchasing costs), the larger the contribution is made to the value of the first desirability function. However, when w_1 is larger than 0.6, the individual desirability value for f_2 equals to 0. Namely, the second objective function, the number of late delivery, has the worst value. Furthermore, there is no differences between **w** = (0.6, 0.2, 0.2) and **w** = (0.8, 0.1, 0.1), although the overall objective function values $\sum w_i d_i$ differ. These extreme results show that the DM may not satisfied with the results.

Table 8. Comparison with some weighting methods.

	w	d	x	f
Weighted sum	0.33, 0.33, 0.33	0.894, 0.119, 1.000	5, 1.5, 3.5, 6, 0, 0	61.250, 5.075, 3.425
	0.60, 0.20, 0.20	1.000, 0.000, 0.846	5, 4, 3.5, 3.5, 0, 0	58.750, 5.325, 3.675
	0.80, 0.10, 0.10	1.000, 0.000, 0.846	5, 4, 3.5, 3.5, 0, 0	58.750, 5.325, 3.675
Weighted geometric mean	0.33, 0.33, 0.33	0.577, 0.409, 0.878	3.8, 0, 3.5, 6, 0, 2.7	68.695, 4.467, 3.623
	0.60, 0.20, 0.20	0.798, 0.226, 0.908	5, 0, 3.5, 6, 1.5, 0	63.500, 4.850, 3.575
	0.80, 0.10, 0.10	0.894, 0.119, 1.000	5, 1.5, 3.5, 6, 0, 0	61.250, 5.075, 3.425
Overall desirability function [1]	r = 1	0.577, 0.409, 0.878	3.8, 0, 3.5, 6, 0, 2.7	68.695, 4.467, 3.623
	r = 2	0.333, 0.167, 0.771	3.8, 0, 3.5, 6, 0, 2.7	68.695, 4.467, 3.623
Proposed method	r = 1, h = 0	0.487, 0.576, 0.518	1.4, 0, 3.5, 5.6, 5.5, 0	70.836, 4.116, 4.197
	r = 1, h = 1	0.589, 0.460, 0.615	2.5, 0, 3.5, 6, 4, 0	68.419, 4.358, 3.943
	r = 1, h = 2	0.500, 0.550, 0.614	1.7, 0, 3.5, 6, 4.4, 0.4	70.494, 4.170, 4.052
	r = 2, h = 0	0.279, 0.279, 0.363	0.5, 1.7, 3.5, 6, 4.2, 0	69.840, 4.216, 4.072
	r = 2, h = 1	0.305, 0.251, 0.362	2.2, 0, 3.4, 5.8, 4.6, 0	69.271, 4.273, 4.072

[1] A special case of weighted geometric mean with equal weights.

Next, we also tested the weighted geometric model using the same set of **w**. In fact, the weighted geometric mean is a popular method for unifying individual desirability functions to a single function in desirability function approach. The weighted geometric method results in a somewhat balanced solution avoiding extreme values in one objective function, even if the weight is to one side. This method, however, also may result in a controversial solution. In Table 8, we found the third desirability value for late delivery increases as w_3 decreases. Thus, the results do not necessarily guarantee that the DM can find the most satisfactory solution even if he/she can decide preference information in advance at one time. The proposed method can complement the shortcomings of these methods because the DM directly checks the degree of desirability and learns a preference structure through improved solutions. In other words, the advantage of the proposed method is that the solution changes in the object function space can be detected through the DM's preference information.

The last row section in Table 8 summarizes the solutions from the proposed method. The results show that the proposed method prevents the emergence of extreme values in two ways. First, the initial solution $\mathbf{d}^0$ describes this characteristic. In each case, for r, the initial solution is a neutral compromise solution because it is calculated by using the Tchebycheff metric without preference information. This setting eliminates the unnecessary iteration that causes a desirability function value to deviate from an extreme value. The results in the 12th and 15th row show reasonable levels of desirability functions, for both cases, $r = 1$ and $r = 2$. Second, the proposed method also prevents the extreme value in the problem-solving process. The original STEM allows the DM to relax one objective function, which may lead to significantly improved values than expected. In such a case, the DM might want to relax an excessively improved value for the purpose of adjustment. However, the proposed method does not ask the DM directly for the amount of relaxation, so it induces the satisfactory solution to find the change of desirability functions according to the rate of relaxation. Accordingly, the DM can choose the most satisfactory solution among the various possible solutions. Therefore, it helps to avoid occurrence of an extreme value by showing the alternatives that the DM can choose.

The proposed method presents the changes of the desirability function values to the DM according to the rate of relaxation of a particular desirability function in the form of a table. If a MOO method requires preference information that is difficult for the DM to express—for example, specific numerical information—it may be difficult to find the satisfactory solution. In this regard, the information presentation of the proposed method provides an additional advantage in that it can ease the burden of the DM by showing a set of expected candidate solutions and selecting the most satisfactory solution rather than requiring specific values. If the number of desirability functions increases, it is recommended to use the graph form. The graph form may also be more useful because the scales of the desirability are all the same. An alternative interactive MOO approach to supplier selection and order allocation is Demirtas and Üstün [4]. In this interactive method, the DM can control objective values directly in a similar way to our research. The authors uses the reservation level that represents an objective function value which must be equaled or exceeded to be considered acceptable, in the maximization context. The method repeatedly reduces the objective space by adjusting reservations level from the DM's preference information. Although there are differences in, for example, how local weights are used and how the DM expresses preference information, both studies show that interactive MOO approaches can help supplier selection and order allocation problems. We expect that a variety of interactive MOO methods can be used for supplier selection and order allocation problems.

5. Conclusions

In this study, a new interactive MOO approach is proposed for supplier selection and order allocation problems with the concept of desirability. The multiple objectives for this problem are defined as purchasing costs, quality, and lead time. Then, we applied an interactive MOO method based on STEM to this problem. We presented how to solve the problem step by step to deliver the problem-solving procedure. It is also shown that the principles of diminishing marginal utility can be used to determine the level of the satisfaction through the use of desirability function. Two cases involving linear and nonlinear desirability function are described to explain the detailed procedure of the proposed method. The obtained results are presented along with results from a prioiri methods and have been discussed accordingly.

This study has following salient features that contribute to the research stream on supplier selection and order allocation.

1. The use of desirability concept was shown to be an excellent tool for reflecting the level of satisfaction and has advantages in that the sensitivity of satisfaction can be adjusted by using desirability parameter.
2. The proposed method alleviates the appearance of extreme values that can be derived when the model uses pre-determined weights.
3. The interactive MOO method, which progressively articulates the DM's preference information to deal with supplier selection problem, may be an alternative.
4. Modification of the original STEM allowed us to relieve the DM's burden in terms of presenting preference information and reduce unnecessary iteration in the optimization process.

This study tried to deal with the concept of desirability more intuitively by repeatedly articulating the preference information of the DM. However, our model will also require the parameter value of the desirability function, in advance, to determine its shape. Therefore, it is necessary to develop a method that progressively reflects the information of this parameter, since the shape of the desirability function plays a critical role in deriving the solution. It is possible to complement this shortcoming if a method is developed that can estimate this parameter for each objective function utilized for the supplier selection and order allocation problems. Also, as stated in Section 4, we could not directly compare the solution quality of the developed method with solutions of other methods. Nonetheless, in order to compare the solution quality, we suggest that setting the proxy on the premise that the DM's preference information is known and comparing the posteriorly derived results is an alternative that can also improve the method's validation. Finally, developing an interactive method under incomplete information and applying it to supplier selection problems would also be a challenging work for practical application. Recent research considering incomplete weights presented by Liao and Xu [43] is expected to be useful in these respects.

Author Contributions: P.L. designed the research, analyzed the data and the obtained results, and performed the development of the paper. S.K. provided advice throughout the study, regarding the research design, methodology and findings. All the authors have read and approved the final manuscript.

Funding: This research was funded by the Ministry of Education of the Republic of Korea and the National Research Foundation of Korea (NRF-2015S1A5B8046893).

Conflicts of Interest: The authors declare no conflicts of interest.

References

1. Willis, T.H.; Huston, C.R.; Pohlkamp, F. Evaluation measures of just-in-time supplier performance. *Prod. Inventory Manag. J.* **1993**, *34*, 1–5.
2. Dobler, D.W.; Burt, D.N.; Lee, L. *Purchasing and Materials Management*; McGraw-Hill: New York, NY, USA, 1990.
3. Xia, W.; Wu, Z. Supplier selection with multiple criteria in volume discount environments. *Omega* **2007**, *35*, 494–504. [CrossRef]
4. Demirtas, E.A.; Üstün, Ö. An integrated multiobjective decision making process for supplier selection and order allocation. *Omega* **2008**, *36*, 76–90. [CrossRef]
5. Zhang, J.-L.; Zhang, M.-Y. Supplier selection and purchase problem with fixed cost and constrained order quantities under stochastic demand. *Int. J. Prod. Econ.* **2011**, *129*, 1–7. [CrossRef]
6. Kannan, D.; Khodaverdi, R.; Olfat, L.; Jafarian, A.; Diabat, A. Integrated fuzzy multi criteria decision making method and multi-objective programming approach for supplier selection and order allocation in a green supply chain. *J. Clean. Prod.* **2013**, *47*, 355–367. [CrossRef]

7. Nazari-Shirkouhi, S.; Shakouri, H.; Javadi, B.; Keramati, A. Supplier selection and order allocation problem using a two-phase fuzzy multi-objective linear programming. *Appl. Math. Model.* **2013**, *37*, 9308–9323. [CrossRef]
8. Ting, S.-C.; Cho, D.I. An integrated approach for supplier selection and purchasing decisions. *Supply Chain Manag. J.* **2008**, *13*, 116–127. [CrossRef]
9. Ghodsypour, S.H.; O'Brien, C. A decision support system for supplier selection using an integrated analytic hierarchy process and linear programming. *Int. J. Prod. Econ.* **1998**, *56*, 199–212. [CrossRef]
10. Gao, Z.; Tang, L. A multi-objective model for purchasing of bulk raw materials of a large-scale integrated steel plant. *Int. J. Prod. Econ.* **2003**, *83*, 325–334. [CrossRef]
11. Jadidi, O.; Cavalieri, S.; Zolfaghari, S. An improved multi-choice goal programming approach for supplier selection problems. *Appl. Math. Model.* **2015**, *39*, 4213–4222. [CrossRef]
12. Jadidi, O.; Zolfaghari, S.; Cavalieri, S. A new normalized goal programming model for multi-objective problems: A case of supplier selection and order allocation. *Int. J. Prod. Econ.* **2014**, *148*, 158–165. [CrossRef]
13. Aissaoui, N.; Haouari, M.; Hassini, E. Supplier selection and order lot sizing modeling: A review. *Comput. Oper. Res.* **2007**, *34*, 3516–3540. [CrossRef]
14. Dickson, G.W. An analysis of vendor selection systems and decisions. *J. Purch.* **1966**, *2*, 5–17. [CrossRef]
15. Timmerman, E. An approach to vendor performance evaluation. *J. Supply Chain Manag.* **1986**, *22*, 2–8. [CrossRef]
16. Liao, H.; Wu, D.; Huang, Y.; Ren, P.; Xu, Z.; Verma, M. Green logistic provider selection with a hesitant fuzzy linguistic thermodynamic method integrating cumulative prospect theory and PROMETHEE. *Sustainability* **2018**, *10*, 1291. [CrossRef]
17. Xu, Z.; Liao, H. Intuitionistic fuzzy analytic hierarchy process. *IEEE Trans. Fuzzy Syst.* **2014**, *22*, 749–761. [CrossRef]
18. Ho, W.; Xu, X.; Dey, P.K. Multi-criteria decision making approaches for supplier evaluation and selection: A literature review. *Eur. J. Oper. Res.* **2010**, *202*, 16–24. [CrossRef]
19. Chai, J.; Liu, J.N.; Ngai, E.W. Application of decision-making techniques in supplier selection: A systematic review of literature. *Expert Syst. Appl.* **2013**, *40*, 3872–3885. [CrossRef]
20. Hwang, C.; Yoon, K. *Multiple Attribute Decision Making*; Springer: Berlin/Heidelberg, Germany, 1981; pp. 58–191. ISBN 978-3-540-10558-9.
21. Korhonen, P.; Moskowitz, H.; Wallenius, J. Multiple criteria decision support—A review. *Eur. J. Oper. Res.* **1992**, *63*, 361–375. [CrossRef]
22. Harrington, E.C. The desirability function. *Ind. Qual. Control* **1965**, *21*, 494–498.
23. Derringer, G.; Suich, R. Simultaneous optimization of several response variables. *J. Qual. Technol.* **1980**, *12*, 214–219. [CrossRef]
24. Del Castillo, E.; Montgomery, D.C. A nonlinear programming solution to the dual response problem. *J. Qual. Technol.* **1993**, *25*, 199–204. [CrossRef]
25. Del Castillo, E.; Montgomery, D.C.; McCarville, D.R. Modified desirability functions for multiple response optimization. *J. Qual. Technol.* **1996**, *28*, 337–345. [CrossRef]
26. Kim, K.J.; Lin, D.K. Simultaneous optimization of mechanical properties of steel by maximizing exponential desirability functions. *J. R. Stat. Soc. Ser. C (Appl. Stat.)* **2000**, *49*, 311–325. [CrossRef]
27. Miettinen, K. *Nonlinear Multiobjective Optimization, Volume 12 of International Series in Operations Research and Management Science*; Kluwer Academic Publishers: Dordrecht, The Netherlands, 1999.
28. Steuer, R.E. *Multiple Criteria Optimization: Theory, Computation, and Applications*; Wiley: New York, NY, USA, 1986.
29. Benayoun, R.; De Montgolfier, J.; Tergny, J.; Laritchev, O. Linear programming with multiple objective functions: Step method (STEM). *Math. Program* **1971**, *1*, 366–375. [CrossRef]
30. Geoffrion, A.M.; Dyer, J.S.; Feinberg, A. An interactive approach for multi-criterion optimization, with an application to the operation of an academic department. *Manag. Sci.* **1972**, *19*, 357–368. [CrossRef]
31. Zionts, S.; Wallenius, J. An interactive programming method for solving the multiple criteria problem. *Manag. Sci.* **1976**, *22*, 652–663. [CrossRef]
32. Korhonen, P.J.; Laakso, J. A visual interactive method for solving the multiple criteria problem. *Eur. J. Oper. Res.* **1986**, *24*, 277–287. [CrossRef]

33. Jeong, I.-J.; Kim, K.-J. D-STEM: A modified step method with desirability function concept. *Comput. Oper. Res.* **2005**, *32*, 3175–3190. [CrossRef]
34. Evans, R.H. Choice criteria revisited. *J. Mark.* **1980**, *44*, 55–56. [CrossRef]
35. Shipley, D.D. Resellers' supplier selection criteria for different consumer products. *Eur. J. Mark.* **1985**, *19*, 26–36. [CrossRef]
36. Ellram, L.M. The supplier selection decision in strategic partnerships. *J. Supply Chain Manag.* **1990**, *26*, 8–14. [CrossRef]
37. Weber, C.A.; Current, J.R.; Benton, W. Vendor selection criteria and methods. *Eur. J. Oper. Res.* **1991**, *50*, 2–18. [CrossRef]
38. Pi, W.-N.; Low, C. Supplier evaluation and selection using Taguchi loss functions. *Int. J. Adv. Manuf. Technol.* **2005**, *26*, 155–160. [CrossRef]
39. Amid, A.; Ghodsypour, S.; O'Brien, C. A weighted max–min model for fuzzy multi-objective supplier selection in a supply chain. *Int. J. Prod. Econ.* **2011**, *131*, 139–145. [CrossRef]
40. Vanderpooten, D.; Vincke, P. Description and analysis of some representative interactive multicriteria procedures. *Math. Comput. Model.* **1989**, *12*, 1221–1238. [CrossRef]
41. Eschenauer, H.; Koski, J.; Osyczka, A. Multicriteria optimization—Fundamentals and motivation. In *Multicriteria Design Optimization*; Springer: Berlin/Heidelberg, Germany, 1990; pp. 1–32.
42. Sawaragi, Y.; Nakayama, H.; Tanino, T. *Theory of Multiobjective Optimization*; Elsevier: Orlando, FL, USA, 1985; Volume 176.
43. Liao, H.; Xu, Z. Satisfaction degree based interactive decision making under hesitant fuzzy environment with incomplete weights. *Int. J. Uncertain. Fuzziness Knowl.-Based Syst.* **2014**, *22*, 553–572. [CrossRef]

information

MDPI

Article

Pythagorean Fuzzy Muirhead Mean Operators and Their Application in Multiple-Criteria Group Decision-Making

Jianghong Zhu [1,*] and Yanlai Li [1,2]

1 School of Transportation and Logistics, Southwest Jiaotong University, Chengdu 611756, China; yanlaili@home.swjtu.edu.cn or yanlaili2010@163.com
2 National Laboratory of Railway Transportation, Southwest Jiaotong University, Chengdu 611756, China
* Correspondence: zhujianghong007@163.com

Received: 25 May 2018; Accepted: 7 June 2018; Published: 11 June 2018

Abstract: As a generalization of the intuitionistic fuzzy set (IFS), a Pythagorean fuzzy set has more flexibility than IFS in expressing uncertainty and fuzziness in the process of multiple criteria group decision-making (MCGDM). Meanwhile, the prominent advantage of the Muirhead mean (MM) operator is that it can reflect the relationships among the various input arguments through changing a parameter vector. Motivated by these primary characters, in this study, we introduced the MM operator into the Pythagorean fuzzy context to expand its applied fields. To do so, we presented the Pythagorean fuzzy MM (PFMM) operators and Pythagorean fuzzy dual MM (PFDMM) operator to fuse the Pythagorean fuzzy information. Then, we investigated their some properties and gave some special cases related to the parameter vector. In addition, based on the developed operators, two MCGDM methods under the Pythagorean fuzzy environment are proposed. An example is given to verify the validity and feasibility of our proposed methods, and a comparative analysis is provided to show their advantages.

Keywords: Pythagorean fuzzy set; Muirhead mean; multiple criteria group decision-making

1. Introduction

Multi-criteria group decision-making (MCGDM), a sub-field of decision-making, is a common and important activity in the real world, and is especially useful in the fields of engineering, economic, management, and the military. In practical applications, a critical problem is how to express the valuation information provided by decision makers. Due to the complexity and fuzziness of MCGDM problems, it is difficult for decision makers to give precise valuation information through employing crisp numbers. Fuzzy set (FS) theory, originally developed by Zadeh [1], is a particularly effective tool to capture uncertain and fuzzy information. However, due to the FS having only one membership degree, it cannot deal effectively with some complicated fuzzy information. Therefore, Atanassov and Rangasamy [2] developed intuitionistic fuzzy set (IFS) through introducing the non-membership degree into the FS. In IFS, the sum of the membership degree and non-membership degree needs to be equal to or less than 1. However, in some practical applications, IFS cannot solve the problem that the sum of the membership and non-membership is bigger than 1, but the square sum is equal to or less than 1. To overcome this drawback of IFS, Pythagorean fuzzy set (PFS), as a generalization of IFS, was introduced by Yager [3,4], of which the square sum of the membership degree and non-membership degree is less than or equal to 1. In other words, when we treat uncertainty and fuzziness in practical MCGDM problems, PFS is a more effective and flexible tool compared with IFS.

Based on some existing aggregation operators, various aggregation operators of Pythagorean fuzzy set have been developed by a number of researchers to solve multi-criteria decision-making

(MCDM) problems with Pythagorean fuzzy information. Depending on whether the input argument is independent, these operators can be divided into two categories: (1) the input argument is independent; (2) any two input arguments are correlated. Many operators fall into the former category. For example, Yager [3,4] developed the Pythagorean fuzzy weighted averaging (PFWA) and Pythagorean fuzzy weighted geometric (PFWG) operators, and used these to solve Pythagorean fuzzy MCDM problems. Based on the operational laws proposed by Zhang and Xu [5], Ma and Xu [6] presented two new PFWA and PFWG operators, symmetric Pythagorean fuzzy weighted geometric/averaging operators, and examined the relationships between these operators and the operators proposed by Yager. Rahman et al. [7] proposed the Pythagorean fuzzy Einstein weighted geometric operator and discussed its desirable properties and special cases. Garg [8] introduced the Einstein operational laws into the Pythagorean fuzzy environment to develop two generalized averaging aggregation operators, and utilized these operators to solve MCDM problems. Through incorporating the confidence level into each Pythagorean fuzzy number, Garg [9] presented a series of novel averaging and geometric operators. Zeng et al. [10] proposed the Pythagorean fuzzy ordered weighted averaging weighted averaging distance operator. On the other hand, Peng and Yang [11] extended the Choquet integral into the Pythagorean fuzzy environment to propose a Pythagorean fuzzy Choquet integral operator. Wei and Lu [12] presented some Pythagorean fuzzy power aggregation operators based on the power aggregation operator, and investigated the main characteristics of these operators. Liang et al. [13] developed the Pythagorean fuzzy Bonferroni mean operator and their weighted form. Moreover, some properties and cases of the proposed operators are explored and an accelerative calculating algorithm is designed to simplify the computation process of the presented operators. Liang et al. [14] proposed the Pythagorean fuzzy weighted geometric Bonferroni mean operator and applied it to handle MCGDM problems with Pythagorean fuzzy information. In real decision-making, however, a relationship may exist among more than two input arguments due to the complexity of decision-making problems. Thus it can be seen that it is difficult for the above operators to capture the relationships between three or more Pythagorean fuzzy input arguments.

The Muirhead mean (MM) operator, originally presented by Muirhead [15], is a well-known information fusion operator and provides us with a new fusion method for the correlation information. The primary characteristic of the MM operator is that it can reflect the relationship among any number of input arguments. In addition, some existing operators including the arithmetic and geometric averaging, Bonferroni mean [16] and Maclaurin symmetric mean [17] are special cases of it. Consequently, some researchers have extended the MM operator into various fuzzy environments. For instance, Qin and Liu [18] presented some 2-tuple linguistic MM operators by introducing the MM operator into the 2-tuple linguistic context, and utilized them to solve the supplier selection problems. Liu and You [19] developed some interval neutrosophic MM operators based on the MM operator, and presented two novel approaches to handle multiple attribute group decision-making problems in light of the proposed operators. Liu and Li [20] explored the MM operator under the intuitionistic fuzzy environment, and proposed some intuitionistic fuzzy MM operators. Liu et al. [21] introduced the MM operator into a hesitant fuzzy linguistic environment, and developed a hesitant fuzzy linguistic MM operator and its weighted form. Wang et al. [22] extended the MM operator to a hesitant fuzzy linguistic set, and proposed the hesitant fuzzy linguistic MM operator and hesitant fuzzy linguistic dual MM operator and their weighted forms. Based on the Archimedean t-norm and t-conorm, Liu and Teng [23] put forward some probabilistic linguistic Archimedean MM operators and further explored some special cases. Liu et al. [24] proposed an interval 2-tuple weighted MM operator by enlarging the scope of MM operator to the interval 2-tuple linguistic environment, and applied the proposed operator to present a large group dependence evaluation model for human reliability analysis. When we consider the relationship among any number of input arguments, however, the above operators fail to deal with the Pythagorean fuzzy information.

According to the above analysis, we know that the existing aggregation operators of Pythagorean fuzzy cannot capture the relationships between any number of input arguments in the information fusion process. At the same time, the MM operator can reflect the relationships between input arguments, so it is necessary to extend it to handle Pythagorean fuzzy information. Hence, inspired by the ideal characteristics of the MM operator, the present paper aims at developing some new aggregation operators of Pythagorean fuzzy to solve MCGDM problems in which we consider the interrelationship among any number of input arguments.

In order to accomplish this goal, the remainder of this paper is arranged as follows. In Section 2, we describe some basic concepts and operational laws of PFS. Based on the MM operator, we develop the Pythagorean fuzzy MM operator and Pythagorean fuzzy weighted MM operator, and the Pythagorean fuzzy dual MM operator and Pythagorean fuzzy dual weighted MM operator in Sections 3 and 4, respectively. In Section 5, we utilize these operators to present two MCGDM methods for the MCGDM problem with Pythagorean fuzzy information. In Section 6, an example is provided to demonstrate the effectiveness and feasibility of the developed approaches, and the advantages of the proposed operators are illustrated by comparing them with the existing operators. Finally, a brief conclusion and future work directions are given in Section 7.

2. Preliminaries

In this section, some fundamental concepts related to the Pythagorean fuzzy number (PFN) are briefly introduced below, which will be used in the following sections.

Definition 1 [3,4]. *Let $X = \{x_1, x_2, \cdots, x_n\}$ be a finite nonempty set, and a PFS P in X is defined as follows*

$$X = \{< x, \mu_P(x), \nu_P(x) > | x \in X\} \tag{1}$$

where $\mu_P(x) \in [0,1]$ and $\nu_P(x) \in [0,1]$ are defined as the degree of membership and non-membership of the element $x \in X$ to P, respectively, and satisfy $\mu_P^2(x) + \nu_P^2(x) \leq 1$. For every $x \in X$, we designate $\pi_P(x)$ as the degree of indeterminacy of the PFS, where $\pi_P(x) = \sqrt{1 - \mu_P^2(x) - \nu_P^2(x)}$. For convenience, $\alpha = (\mu_P, \nu_P)$ is called as a PFN, and $\mu_P^2 + \nu_P^2 \leq 1$ and $\pi_P = \sqrt{1 - \mu_P^2 - \nu_P^2}$.

Definition 2 [5]. *Let $\alpha_1 = (\mu_{P_1}, \nu_{P_1})$, $\alpha_2 = (\mu_{P_2}, \nu_{P_2})$ and $\alpha = (\mu_P, \nu_P)$ be three PFNs, and $\lambda > 0$. Then the basic operational laws of PFN can be defined as follows:*

(1) $\alpha_1 \oplus \alpha_2 = \left(\sqrt{\mu_{P_1}^2 + \mu_{P_2}^2 - \mu_{P_1}^2\mu_{P_2}^2}, \nu_{P_1}\nu_{P_2}\right)$;

(2) $\alpha_1 \otimes \alpha_2 = \left(\mu_{P_1}\mu_{P_2}, \sqrt{\nu_{P_1}^2 + \nu_{P_2}^2 - \nu_{P_1}^2\nu_{P_2}^2}\right)$;

(3) $\lambda\alpha = \left(\sqrt{1 - (1 - \mu_P^2)^\lambda}, (\nu_P)^\lambda\right)$;

(4) $\alpha^\lambda = \left((\mu_P)^\lambda, \sqrt{1 - (1 - \mu_P^2)^\lambda}\right)$.

Definition 3 [25]. *Let $\alpha = (\mu_P, \nu_P)$ be a PFN, then the score and accuracy function of α is defined respectively as follows*

$$S(\alpha) = \frac{1}{2}(1 + \mu_P^2 - \nu_P^2), \tag{2}$$

$$H(\alpha) = \mu_P^2 + \nu_P^2. \tag{3}$$

Definition 4 [25]. *Let $\alpha = (\mu_{P_1}, \nu_{P_1})$ and $\beta = (\mu_{P_2}, \nu_{P_2})$ be any two PFNs, $S(\alpha)$ and $H(\alpha)$ be the score and accuracy function of α, and $S(\beta)$ and $H(\beta)$ be the score and accuracy function of β, then*

(1) If $S(\alpha) > S(\beta)$, then α is superior to β, $\alpha > \beta$;

(2) If $S(\alpha) = S(\beta)$, then

(a) If $H(\alpha) > H(\beta)$, then α is superior to β, $\alpha > \beta$;

(b) If $H(\alpha) = H(\beta)$, then α is equivalent to β $\alpha = \beta$.

Definition 5 [4,5]. *Let* $\alpha_1 = (\mu_{P_1}, \nu_{P_1})$ *and* $\alpha_2 = (\mu_{P_2}, \nu_{P_2})$ *be two PFNs, the ordering relationship on the PFNs is defined as follows:* $\alpha_1 \geq \alpha_2$ *if and only if* $\mu_{P_1} \geq \mu_{P_2}$ *and* $\nu_{P_1} \leq \nu_{P_2}$.

3. Some Pythagorean Fuzzy Muirhead Operators

The Pythagorean fuzzy MM (PFMM) operator and Pythagorean fuzzy weighted MM (PFWMM) operator are defined in Sections 3.1 and 3.2, respectively.

3.1. The PFMM Operator

Definition 6 [15]. *Let* $Q = (q_1, q_2, \cdots, q_n) \in R^n$ *be a parameter vector, and* $\alpha_i (i = 1, 2, \cdots, n)$ *be a collection of nonnegative real numbers. If*

$$MM^Q(\alpha_1, \alpha_2, \cdots, \alpha_n) = \left(\frac{1}{n!}\sum_{\theta \in S_n}\prod_{j=1}^{n}\alpha_{\theta(j)}^{q_j}\right)^{\frac{1}{\sum_{j=1}^{n} q_j}}, \tag{4}$$

where MM^Q *is called the Muirhead mean (MM) operator and* $\theta(j)$ $(j = 1, 2, \cdots, n)$ *is any a permutation of* $(1, 2, \cdots, n)$*, and* S_n *is the collection of all permutation of* $(1, 2, \cdots, n)$.

Definition 7. *Let* $Q = (q_1, q_2, \cdots, q_n) \in R^n$ *be a parameter vector, and* $\alpha_i = (\mu_{P_i}, \nu_{P_i})$, $(i = 1, 2, \cdots, n)$ *be a collection of PFNs. If*

$$PFMM^Q(\alpha_1, \alpha_2, \cdots, \alpha_n) = \left(\frac{1}{n!}\sum_{\theta \in S_n}\prod_{j=1}^{n}\alpha_{\theta(j)}^{q_j}\right)^{\frac{1}{\sum_{j=1}^{n} q_j}}. \tag{5}$$

where $PFMM^Q$ *is called the PFMM operator and* $\theta(j)$ $(j = 1, 2, \cdots, n)$ *is any a permutation of* $(1, 2, \cdots, n)$*, and* S_n *is the collection of all permutation of* $(1, 2, \cdots, n)$.

Theorem 1. *Let* $\alpha_i = (\mu_{P_i}, \nu_{P_i})$, $(i = 1, 2, \cdots, n)$ *be a collection of PFNs, then the aggregated value by using the PFMM operator is also a PFN, and*

$$PFMM^Q(\alpha_1, \alpha_2, \cdots, \alpha_n) = \left(\left(\sqrt{1-\left(\prod_{\theta \in S_n}(1-\prod_{j=1}^{n}\mu_{\theta(j)}^{2q_j})\right)^{\frac{1}{n!}}}\right)^{\frac{1}{\sum_{j=1}^{n} q_j}}, \sqrt{1-\left(1-\left(\prod_{\theta \in S_n}\left(1-\prod_{j=1}^{n}(1-\nu_{\theta(j)}^2)^{q_j}\right)\right)^{\frac{1}{n!}}\right)^{\frac{1}{\sum_{j=1}^{n} q_j}}}\right) \tag{6}$$

Proof. We need to prove that Equation (6) holds and is a PFN.

(1) Firstly, we prove that Equation (6) holds.
According to the operational laws (4) and (2) of Definition 2,

$$\alpha_{\theta(j)}^{q_j} = \left(\mu_{\theta(j)}^{q_j}, \sqrt{1-(1-\nu_{\theta(j)}^2)^{q_j}}\right), \text{ and, } \prod_{j=1}^{n}\alpha_{\theta(j)}^{q_j} = \left(\prod_{j=1}^{n}\mu_{\theta(j)}^{q_j}, \sqrt{1-\prod_{j=1}^{n}(1-\nu_{\theta(j)}^2)^{q_j}}\right),$$

then

$$\sum_{\theta \in S_n}\prod_{j=1}^{n}\alpha_{\theta(j)}^{q_j} = \left(\sqrt{1-\prod_{\theta \in S_n}(1-\prod_{j=1}^{n}\mu_{\theta(j)}^{2q_j})}, \prod_{\theta \in S_n}\sqrt{1-\prod_{j=1}^{n}(1-\nu_{\theta(j)}^2)^{q_j}}\right),$$

further, based on operational law (3), we can get

$$\frac{1}{n!}\sum_{\theta\in S_n}\prod_{j=1}^{n}\alpha_{\theta(j)}^{q_j}=\left(\sqrt{1-\left(\prod_{\theta\in S_n}(1-\prod_{j=1}^{n}\mu_{\theta(j)}^{2q_j})\right)^{\frac{1}{n!}}},\left(\prod_{\theta\in S_n}\sqrt{1-\prod_{j=1}^{n}(1-v_{\theta(j)}^{2})^{q_j}}\right)^{\frac{1}{n!}}\right).$$

Consequently, we have

$$\left(\frac{1}{n!}\sum_{\theta\in S_n}\prod_{j=1}^{n}\alpha_{\theta(j)}^{q_j}\right)^{\frac{1}{\sum_{j=1}^{n}q_j}}=\left(\left(\sqrt{1-\left(\prod_{\theta\in S_n}(1-\prod_{j=1}^{n}\mu_{\theta(j)}^{2q_j})\right)^{\frac{1}{n!}}}\right)^{\frac{1}{\sum_{j=1}^{n}q_j}},\sqrt{1-\left(1-\left(\prod_{\theta\in S_n}\left(1-\prod_{j=1}^{n}(1-v_{\theta(j)}^{2})^{q_j}\right)\right)^{\frac{1}{n!}}\right)^{\frac{1}{\sum_{j=1}^{n}q_j}}}\right),$$

which illustrates that Equation (6) holds.

(2) In what follows, we will prove that Equation (6) is a PFN.

$$\text{Let } \mu_P=\left(\sqrt{1-\left(\prod_{\theta\in S_n}(1-\prod_{j=1}^{n}\mu_{\theta(j)}^{2q_j})\right)^{\frac{1}{n!}}}\right)^{\frac{1}{\sum_{j=1}^{n}q_j}},\ v_P=\sqrt{1-\left(1-\left(\prod_{\theta\in S_n}\left(1-\prod_{j=1}^{n}(1-v_{\theta(j)}^{2})^{q_j}\right)\right)^{\frac{1}{n!}}\right)^{\frac{1}{\sum_{j=1}^{n}q_j}}}.$$

Then we need to prove that Equation (6) satisfies the following two conditions.

(a) $0\le\mu_P\le1$, and $0\le v_P\le1$;

(b) $\mu_P^2+v_P^2\le1$.

(a) According to Definition 1,

$$\mu_{\theta(j)}^{2q_j}\in[0,1]\text{ and }\prod_{j=1}^{n}\mu_{\theta(j)}^{2q_j}\in[0,1],$$

then we have

$$\prod_{\theta\in S_n}\left(1-\prod_{j=1}^{n}\mu_{\theta(j)}^{2q_j}\right)\in[0,1],\left(\prod_{\theta\in S_n}\left(1-\prod_{j=1}^{n}\mu_{\theta(j)}^{2q_j}\right)\right)^{\frac{1}{n!}}\in[0,1],\text{ and }\sqrt{1-\left(\prod_{\theta\in S_n}\left(1-\prod_{j=1}^{n}\mu_{\theta(j)}^{2q_j}\right)\right)^{\frac{1}{n!}}}\in[0,1],$$

further

$$\left(\sqrt{1-\left(\prod_{\theta\in S_n}\left(1-\prod_{j=1}^{n}\mu_{\theta(j)}^{2q_j}\right)\right)^{\frac{1}{n!}}}\right)^{\frac{1}{\sum_{j=1}^{n}q_j}}\in[0,1],\text{ i.e., }0\le\mu_P\le1.$$

Similarly, we can get $0\le v_P\le1$. So condition (a) is satisfied.

(b) Based on $\mu_{\theta(j)}^2+v_{\theta(j)}^2\le1$, then $\mu_{\theta(j)}^2\le1-v_{\theta(j)}^2$, we yield the inequality as follows:

$$\mu_P^2+v_P^2=\left(1-\left(\prod_{\theta\in S_n}(1-\prod_{j=1}^{n}\mu_{\theta(j)}^{2q_j})\right)^{\frac{1}{n!}}\right)^{\frac{1}{\sum_{j=1}^{n}q_j}}+1-\left(1-\left(\prod_{\theta\in S_n}\left(1-\prod_{j=1}^{n}(1-v_{\theta(j)}^{2})^{q_j}\right)\right)^{\frac{1}{n!}}\right)^{\frac{1}{\sum_{j=1}^{n}q_j}}$$

$$\le\left(1-\left(\prod_{\theta\in S_n}\left(1-\prod_{j=1}^{n}(1-v_{\theta(j)}^{2})^{q_j}\right)\right)^{\frac{1}{n!}}\right)^{\frac{1}{\sum_{j=1}^{n}q_j}}+1-\left(1-\left(\prod_{\theta\in S_n}\left(1-\prod_{j=1}^{n}(1-v_{\theta(j)}^{2})^{q_j}\right)\right)^{\frac{1}{n!}}\right)^{\frac{1}{\sum_{j=1}^{n}q_j}}=1,$$

i.e., $\mu_P^2+v_P^2\le1$. Consequently, condition (b) is satisfied.

Based on the proof above, we know that theorem 1 holds. □

In what follows, we will explore some properties of the PFMM operator.

Property 1 (Idempotency). *Let $\alpha_i = (\mu_{P_i}, \nu_{P_i}), (i = 1, 2, \cdots, n)$ are equal, i.e., $\alpha_i = \alpha = (\mu_P, \nu_P)$ for all i, then*

$$PFMM^Q(\alpha_1, \alpha_2, \cdots, \alpha_n) = \alpha = (\mu_P, \nu_P).$$

Proof. Since $\alpha_i = \alpha = (\mu_P, \nu_P)$, according to the Theorem 1 yields

$$PFMM^Q(\alpha_1, \alpha_2, \cdots, \alpha_n) = \left(\left(\sqrt{1 - \left(\prod_{\theta \in S_n} (1 - \prod_{j=1}^{n} \mu_P^{2q_j})\right)^{\frac{1}{n!}}}\right)^{\frac{1}{\sum_{j=1}^{n} q_j}}, \sqrt{1 - \left(1 - \left(\prod_{\theta \in S_n}\left(1 - \prod_{j=1}^{n} (1 - \nu_P^2)^{q_j}\right)\right)^{\frac{1}{n!}}\right)^{\frac{1}{\sum_{j=1}^{n} q_j}}}\right),$$

$$= \left(\left(\sqrt{1 - \left(\prod_{\theta \in S_n} (1 - \mu_P^{2\sum_{j=1}^{n} q_j})\right)^{\frac{1}{n!}}}\right)^{\frac{1}{\sum_{j=1}^{n} q_j}}, \sqrt{1 - \left(1 - \left(\prod_{\theta \in S_n}\left(1 - (1 - \nu_P^2)^{\sum_{j=1}^{n} q_j}\right)\right)^{\frac{1}{n!}}\right)^{\frac{1}{\sum_{j=1}^{n} q_j}}}\right)$$

$$= \left(\left(\sqrt{1 - \left((1 - \mu_P^{2\sum_{j=1}^{n} q_j})^{n!}\right)^{\frac{1}{n!}}}\right)^{\frac{1}{\sum_{j=1}^{n} q_j}}, \sqrt{1 - \left(1 - \left(\left(1 - (1 - \nu_P^2)^{\sum_{j=1}^{n} q_j}\right)^{n!}\right)^{\frac{1}{n!}}\right)^{\frac{1}{\sum_{j=1}^{n} q_j}}}\right)$$

$$= \left(\left(\sqrt{\mu_P^{2\sum_{j=1}^{n} q_j}}\right)^{\frac{1}{\sum_{j=1}^{n} q_j}}, \sqrt{1 - \left((1 - \nu_P^2)^{\sum_{j=1}^{n} q_j}\right)^{\frac{1}{\sum_{j=1}^{n} q_j}}}\right) = (\mu_P, \nu_P).$$

□

Property 2 (Monotonicity). *Let $\alpha_i = (\mu_{P_i}, \nu_{P_i})$ and $\hat{\alpha}_i = (\hat{\mu}_{P_i}, \hat{\nu}_{P_i}), (i = 1, 2, \cdots, n)$ be two collections of PFNs. Through using the PFMM operator, if $\mu_{P_i} \geq \hat{\mu}_{P_i}$ and $\nu_{P_i} \leq \hat{\nu}_{P_i}$,*

$$PFMM^Q(\alpha_1, \alpha_2, \cdots, \alpha_n) \geq PFMM^Q(\hat{\alpha}_1, \hat{\alpha}_2, \cdots, \hat{\alpha}_n).$$

Proof. Let

$$PFMM^Q(\alpha_1, \alpha_2, \cdots, \alpha_n)$$

$$= \left(\left(\sqrt{1 - \left(\prod_{\theta \in S_n}\left(1 - \prod_{j=1}^{n} \mu_{\theta(j)}^{2q_j}\right)\right)^{\frac{1}{n!}}}\right)^{\frac{1}{\sum_{j=1}^{n} q_j}}, \sqrt{1 - \left(1 - \left(\prod_{\theta \in S_n}\left(1 - \prod_{j=1}^{n} (1 - \nu_{\theta(j)}^2)^{q_j}\right)\right)^{\frac{1}{n!}}\right)^{\frac{1}{\sum_{j=1}^{n} q_j}}}\right) = (\mu_P, \nu_P), \text{ and}$$

$$PFMM^Q(\hat{\alpha}_1, \hat{\alpha}_2, \cdots, \hat{\alpha}_n)$$

$$= \left(\left(\sqrt{1 - \left(\prod_{\theta \in S_n}\left(1 - \prod_{j=1}^{n} \hat{\mu}_{\theta(j)}^{2q_j}\right)\right)^{\frac{1}{n!}}}\right)^{\frac{1}{\sum_{j=1}^{n} q_j}}, \sqrt{1 - \left(1 - \left(\prod_{\theta \in S_n}\left(1 - \prod_{j=1}^{n} (1 - \hat{\nu}_{\theta(j)}^2)^{q_j}\right)\right)^{\frac{1}{n!}}\right)^{\frac{1}{\sum_{j=1}^{n} q_j}}}\right) = (\hat{\mu}_P, \hat{\nu}_P).$$

Since $\mu_{P_i} \geq \hat{\mu}_{P_i}$, based on the operational laws of Definition 2, we have

$$\mu_{\theta(j)}^{2q_j} \geq \hat{\mu}_{\theta(j)}^{2q_j} \text{ and } \prod_{j=1}^{n} \mu_{\theta(j)}^{2q_j} \geq \prod_{j=1}^{n} \hat{\mu}_{\theta(j)}^{2q_j} \text{ then } \prod_{\theta \in S_n}\left(1 - \prod_{j=1}^{n} \mu_{\theta(j)}^{2q_j}\right) \leq \prod_{\theta \in S_n}\left(1 - \prod_{j=1}^{n} \hat{\mu}_{\theta(j)}^{2q_j}\right),$$

and

$$\left(\prod_{\theta \in S_n}\left(1 - \prod_{j=1}^{n} \mu_{\theta(j)}^{2q_j}\right)\right)^{\frac{1}{n!}} \leq \left(\prod_{\theta \in S_n}\left(1 - \prod_{j=1}^{n} \hat{\mu}_{\theta(j)}^{2q_j}\right)\right)^{\frac{1}{n!}}.$$

Further,

$$1-\left(\prod_{\theta\in S_n}\left(1-\prod_{j=1}^{n}\mu_{\theta(j)}^{2q_j}\right)\right)^{\frac{1}{n!}}\geq 1-\left(\prod_{\theta\in S_n}\left(1-\prod_{j=1}^{n}\hat{\mu}_{\theta(j)}^{2q_j}\right)\right)^{\frac{1}{n!}},\sqrt{1-\left(\prod_{\theta\in S_n}\left(1-\prod_{j=1}^{n}\mu_{\theta(j)}^{2q_j}\right)\right)^{\frac{1}{n!}}}\geq\sqrt{1-\left(\prod_{\theta\in S_n}\left(1-\prod_{j=1}^{n}\hat{\mu}_{\theta(j)}^{2q_j}\right)\right)^{\frac{1}{n!}}},$$

and

$$\left(\sqrt{1-\left(\prod_{\theta\in S_n}\left(1-\prod_{j=1}^{n}\mu_{\theta(j)}^{2q_j}\right)\right)^{\frac{1}{n!}}}\right)^{\frac{1}{\sum_{j=1}^{n}q_j}}\geq\left(\sqrt{1-\left(\prod_{\theta\in S_n}\left(1-\prod_{j=1}^{n}\hat{\mu}_{\theta(j)}^{2q_j}\right)\right)^{\frac{1}{n!}}}\right)^{\frac{1}{\sum_{j=1}^{n}q_j}}.$$

i.e., $\mu_P \geq \hat{\mu}_P$. Similarly, we also yield $\nu_P \leq \hat{\nu}_P$.

Consequently, $PFMM^Q(\alpha_1, \alpha_2, \cdots, \alpha_n) \geq PFMM^Q(\hat{\alpha}_1, \hat{\alpha}_2, \cdots, \hat{\alpha}_n)$ holds. □

Property 3 (Boundedness). *Let* $\alpha_i = (\mu_{P_i}, \nu_{P_i}), (i = 1, 2, \cdots, n)$ *be a collections of PFNs,* $\alpha^+ = (max(\mu_{P_i}), min(\nu_{P_i}))$ *and* $\alpha^- = (min(\mu_{P_i}), max(\nu_{P_i}))$, *then*

$$\alpha^+ \geq PFMM^Q(\alpha_1, \alpha_2, \cdots, \alpha_n) \geq \alpha^-.$$

Proof. Based on Properties 1 and 2,

$PFMM^Q(\alpha_1, \alpha_2, \cdots, \alpha_n) \leq PFMM^Q(\alpha^+, \alpha^+, \cdots, \alpha^+) = \alpha^+$,

$PFMM^Q(\alpha_1, \alpha_2, \cdots, \alpha_n) \geq PFMM^Q(\alpha^-, \alpha^-, \cdots, \alpha^-) = \alpha^-$.

So, we can get $\alpha^+ \geq PFMM^Q(\alpha_1, \alpha_2, \cdots, \alpha_n) \geq \alpha^-$.□

In what follows, we will discuss some special cases of the PFMM operator through changing the values of parameter vector Q.

(1) When $Q = (1, 0, \cdots, 0)$, Equation (6) is transformed into a Pythagorean fuzzy arithmetic averaging operator.

$$PFMM^{(1,0,\cdots,0)}(\alpha_1, \alpha_2, \cdots, \alpha_n) = \frac{1}{n}\sum_{i=1}^{n}\alpha_i = \left(\sqrt{1-\prod_{i=1}^{n}(1-\mu_{P_i}^2)^{\frac{1}{n}}}, \prod_{i=1}^{n}(\nu_{P_i})^{\frac{1}{n}}\right). \tag{7}$$

(2) When $Q = (\lambda, 0, \cdots, 0)$, Equation (6) is transformed into a Pythagorean fuzzy generalized arithmetic averaging operator:

$$PFMM^{(\lambda,0,\cdots,0)}(\alpha_1, \alpha_2, \cdots, \alpha_n) = \left(\left(\sqrt{1-\prod_{i=1}^{n}(1-\mu_{P_i}^{2\lambda})^{\frac{1}{n}}}\right)^{\frac{1}{\lambda}}, \sqrt{1-\left(1-\prod_{i=1}^{n}\left(1-(1-\nu_{P_i}^2)^{\lambda}\right)^{\frac{1}{n}}\right)^{\frac{1}{\lambda}}}\right). \tag{8}$$

(3) When $Q = (1, 1, 0, 0 \cdots, 0)$, Equation (6) is transformed into a Pythagorean fuzzy BM operator:

$$PFMM^{(1,1,0,0\cdots,0)}(\alpha_1, \alpha_2, \cdots, \alpha_n) = \left(\left(\sqrt{1-\left(\prod_{\substack{i,j=1\\ i\neq j}}^{n}(1-\mu_{P_i}^2\mu_{P_j}^2)\right)^{\frac{1}{n(n-1)}}}\right)^{\frac{1}{2}}, \sqrt{1-\left(1-\prod_{\substack{i,j=1\\ i\neq j}}^{n}\left(1-(1-\nu_{P_i}^2)(1-\nu_{P_j}^2)\right)^{\frac{1}{n(n-1)}}\right)^{\frac{1}{2}}}\right). \tag{9}$$

(4) When $Q = (\underbrace{1,1,\cdots,1}_{k},\underbrace{0,0,\cdots,0}_{n-k})$, Equation (6) is transformed into a Pythagorean fuzzy MSM operator [25]:

$$PFMM^{(\underbrace{1,1,\cdots,1}_{k},\underbrace{0,0,\cdots,0}_{n-k})}(\alpha_1,\alpha_2,\cdots,\alpha_n) = \left(\left(\sqrt{1-\left(\prod_{1\le i1\prec\cdots\prec ik\le n}(1-\prod_{j=1}^{k}\mu_{P_{ij}}^2)\right)^{\frac{1}{C_n^k}}}\right)^{\frac{1}{k}}, \sqrt{1-\left(1-\left(\prod_{1\le i1\prec\cdots\prec ik\le n}\left(1-\prod_{j=1}^{k}(1-v_{P_{ij}}^2)\right)\right)^{\frac{1}{C_n^k}}\right)^{\frac{1}{k}}}\right). \quad (10)$$

(5) When $Q = (1,1,\cdots,1)$, Equation (6) is transformed into a Pythagorean fuzzy geometric averaging operator:

$$PFMM^{(1,1,\cdots,1)}(\alpha_1,\alpha_2,\cdots,\alpha_n) = \left(\prod_{i=1}^{n}\alpha_i\right)^{\frac{1}{n}} = \left(\left(\prod_{i=1}^{n}\mu_{P_i}\right)^{\frac{1}{n}}, \sqrt{1-\prod_{i=1}^{n}(1-v_{P_i}^2)^{\frac{1}{n}}}\right). \quad (11)$$

(6) When $Q = (1/n,1/n,\cdots,1/n)$, Equation (6) is transformed into a Pythagorean fuzzy geometric averaging operator:

$$PFMM^{(1/n,1/n,\cdots,1/n)}(\alpha_1,\alpha_2,\cdots,\alpha_n) = \prod_{i=1}^{n}\alpha_i^{\frac{1}{n}} = \left(\left(\prod_{i=1}^{n}\mu_{P_i}\right)^{\frac{1}{n}}, \sqrt{1-\prod_{i=1}^{n}(1-v_{P_i}^2)^{\frac{1}{n}}}\right). \quad (12)$$

3.2. The PFWMM Operator

Definition 8. *Let $Q = (q_1,q_2,\cdots,q_n) \in R^n$ be a parameter vector, $\alpha_i = (\mu_{P_i},v_{P_i}),(i = 1,2,\cdots,n)$ be a collection of PFNs, and $w = (w_1,w_2,\cdots,w_n)^T$ be the weight vector of α_i, where w_i indicates the importance degree of α_i, satisfying $w_i \in [0,1]$ and $\sum_{i=1}^{n} w_i = 1$. If*

$$PFWMM^Q(\alpha_1,\alpha_2,\cdots,\alpha_n) = \left(\frac{1}{n!}\sum_{\theta\in S_n}\prod_{j=1}^{n}(nw_{\theta(j)}\alpha_{\theta(j)})^{q_j}\right)^{\frac{1}{\sum_{j=1}^{n}q_j}}, \quad (13)$$

where $PFWMM^Q$ is called the PFWMM operator and $\theta(j)(j = 1,2,\cdots,n)$ is any a permutation of $(1,2,\cdots,n)$, and S_n is the collection of all permutation of $(1,2,\cdots,n)$.

Theorem 2. *Let $Q = (q_1,q_2,\cdots,q_n) \in R^n$ be a parameter vector, $\alpha_i = (\mu_{P_i},v_{P_i}),(i = 1,2,\cdots,n)$ be a collection of PFNs, and $w = (w_1,w_2,\cdots,w_n)^T$ be the weight vector of α_i, where w_i indicates the importance degree of α_i, satisfying $w_i \in [0,1]$ and $\sum_{i=1}^{n} w_i = 1$. Then, the aggregated value by using the PFWMM operator is also a PFN, and*

$$PFWMM^Q(\alpha_1,\alpha_2,\cdots,\alpha_n)$$
$$= \left(\left(\sqrt{1-\left(\prod_{\theta\in S_n}\left(1-\prod_{j=1}^{n}(1-(1-\mu_{\theta(j)}^2)^{nw_{\theta(j)}})^{q_j}\right)\right)^{\frac{1}{n!}}}\right)^{\frac{1}{\sum_{j=1}^{n}q_j}}, \sqrt{1-\left(1-\left(\prod_{\theta\in S_n}\left(1-\prod_{j=1}^{n}(1-v_{\theta(j)}^{2nw_{\theta(j)}})^{q_j}\right)\right)^{\frac{1}{n!}}\right)^{\frac{1}{\sum_{j=1}^{n}q_j}}}\right). \quad (14)$$

Proof. Based on the operational law (3) in Definition 2, $nw_{\theta(j)}\alpha_{\theta(j)} = \left(\sqrt{1-(1-\mu_{\theta(j)}^2)^{nw_{\theta(j)}}}, v_{\theta(j)}^{nw_{\theta(j)}}\right)$, and we can replace $\mu_{\theta(j)}$ and $v_{\theta(j)}$ with $\sqrt{1-(1-\mu_{\theta(j)}^2)^{nw_{\theta(j)}}}$ and $v_{\theta(j)}^{nw_{\theta(j)}}$, respectively, in Equation (6), thus obtaining Equation (14). Since $\alpha_{\theta(j)}$ is a PFN, then $nw_{\theta(j)}\alpha_{\theta(j)}$ is also a PFN. Similar to the proof of Theorem 1, we know Equation (14) is also a PFN. □

In the following, we will discuss some desirable properties of the PFWMM operator.

Property 4 (Monotonicity). *Let* $\alpha_i = (\mu_{P_i}, \nu_{P_i})$ *and* $\hat{\alpha}_i = (\hat{\mu}_{P_i}, \hat{\nu}_{P_i})$, $(i = 1, 2, \cdots, n)$ *be two collections of PFNs. Through using the PFMM operator, if* $\mu_{P_i} \geq \hat{\mu}_{P_i}$ *and* $\nu_{P_i} \leq \hat{\nu}_{P_i}$, *then*

$$PFWMM^Q(\alpha_1, \alpha_2, \cdots, \alpha_n) \geq PFWMM^Q(\hat{\alpha}_1, \hat{\alpha}_2, \cdots, \hat{\alpha}_n).$$

The Proof of Property 4 is similar to that of Property 2, so is omitted here.

Property 5 (Boundedness). *Let* $\alpha_i = (\mu_{P_i}, \nu_{P_i}), (i = 1, 2, \cdots, n)$ *be a collections of PFNs,* $\alpha^+ = (max(\mu_{P_i}), min(\nu_{P_i}))$ *and* $\alpha^- = (min(\mu_{P_i}), max(\nu_{P_i}))$, *then*

$$\alpha^+ \geq PFWMM^Q(\alpha_1, \alpha_2, \cdots, \alpha_n) \geq \alpha^-.$$

The Proof of Property 5 is similar to that of Property 3, so is omitted here.

Theorem 3. *The PFMM operator is a special case of the PFWMM operator.*

Proof. When $w = (1/n, 1/n, \cdots, 1/n)^T$

$$PFWMM^Q(\alpha_1, \alpha_2, \cdots, \alpha_n)$$

$$= \left(\left(\sqrt{1 - \left(\prod_{\theta \in S_n}\left(1 - \prod_{j=1}^{n}(1 - (1 - \mu_{\theta(j)}^2)^{n \times \frac{1}{n}})^{q_j}\right)\right)^{\frac{1}{n!}}}\right)^{\frac{1}{\sum_{j=1}^{n} q_j}}, \sqrt{1 - \left(1 - \left(\prod_{\theta \in S_n}\left(1 - \prod_{j=1}^{n}(1 - \nu_{\theta(j)}^{2n \times \frac{1}{n}})^{q_j}\right)\right)^{\frac{1}{n!}}\right)^{\frac{1}{\sum_{j=1}^{n} q_j}}}\right)$$

$$= \left(\left(\sqrt{1 - \left(\prod_{\theta \in S_n}(1 - \prod_{j=1}^{n} \mu_{\theta(j)}^{2q_j})\right)^{\frac{1}{n!}}}\right)^{\frac{1}{\sum_{j=1}^{n} q_j}}, \sqrt{1 - \left(1 - \left(\prod_{\theta \in S_n}\left(1 - \prod_{j=1}^{n}(1 - \nu_{\theta(j)}^2)^{q_j}\right)\right)^{\frac{1}{n!}}\right)^{\frac{1}{\sum_{j=1}^{n} q_j}}}\right).$$

□

Theorem 4. The Pythagorean fuzzy weighted averaging operator [6] is a special case of the PFWMM operator.

Proof. When $Q = (1, 0, \cdots, 0)$,

$$PFWMM^{(1,0,\cdots,0)}(\alpha_1, \alpha_2, \cdots, \alpha_n)$$

$$= \left(\left(\sqrt{1 - \left(\prod_{\theta \in S_n}\left(1 - \prod_{j=1}^{n}(1 - (1 - \mu_{\theta(j)}^2)^{nw_{\theta(j)}})^{q_j}\right)\right)^{\frac{1}{n!}}}\right)^{\frac{1}{\sum_{j=1}^{n} q_j}}, \sqrt{1 - \left(1 - \left(\prod_{\theta \in S_n}\left(1 - \prod_{j=1}^{n}(1 - \nu_{\theta(j)}^{2nw_{\theta(j)}})^{q_j}\right)\right)^{\frac{1}{n!}}\right)^{\frac{1}{\sum_{j=1}^{n} q_j}}}\right)$$

$$= \left(\sqrt{1 - \left(\prod_{j=1}^{n}(1 - \mu_{P_j}^2)^{nw_j}\right)^{\frac{1}{n}}}, \sqrt{1 - \left(1 - \left(\prod_{j=1}^{n} \nu_{P_j}^{2nw_j}\right)^{\frac{1}{n}}\right)}\right)$$

$$= \left(\sqrt{1 - \prod_{j=1}^{n}(1 - \mu_{P_j}^2)^{w_j}}, \prod_{j=1}^{n} \nu_{P_j}^{w_j}\right).$$

□

4. Some Pythagorean Fuzzy Dual MM Operators

In this section, we will define the Pythagorean fuzzy dual MM (PFDMM) operator and Pythagorean fuzzy dual weighted MM (PFDWMM) operator.

4.1. The PFDMM Operator

Definition 9. *Let* $Q = (q_1, q_2, \cdots, q_n) \in R^n$ *be a parameter vector, and* $\alpha_i = (\mu_{P_i}, \nu_{P_i}), (i = 1, 2, \cdots, n)$ *be a collection of PFNs. If*

$$PFDMM^Q(\alpha_1, \alpha_2, \cdots, \alpha_n) = \frac{1}{\sum_{j=1}^{n} q_j} \left(\prod_{\theta \in S_n} \sum_{j=1}^{n} (q_j \alpha_{\theta(j)}) \right)^{\frac{1}{n!}}, \tag{15}$$

where $PFDMM^Q$ *is called the PFDMM operator and* $\theta(j)(j = 1, 2, \cdots, n)$ *is any a permutation of* $(1, 2, \cdots, n)$*, and* S_n *is the collection of all permutations of* $(1, 2, \cdots, n)$*.*

Theorem 5. *Let* $\alpha_i = (\mu_{P_i}, \nu_{P_i}), (i = 1, 2, \cdots, n)$ *be a collection of PFNs, then the aggregated value by using the PFDMM operator is also a PFN, and*

$$PFDMM^Q(\alpha_1, \alpha_2, \cdots, \alpha_n) = \left(\sqrt{1 - \left(1 - \left(\prod_{\theta \in S_n} \left(1 - \prod_{j=1}^{n} (1 - \mu_{\theta(j)}^2)^{q_j}\right)\right)^{\frac{1}{n!}}\right)^{\frac{1}{\sum_{j=1}^{n} q_j}}}, \left(\sqrt{1 - \left(\prod_{\theta \in S_n} (1 - \prod_{j=1}^{n} \nu_{\theta(j)}^{2q_j})\right)^{\frac{1}{n!}}}\right)^{\frac{1}{\sum_{j=1}^{n} q_j}} \right). \tag{16}$$

Proof. We need to prove that Equation (16) holds and is a PFN.

(1) Firstly, we will prove that Equation (16) holds.

According to laws (3) and (1) in Definition 2,

$$q_j \alpha_{\theta(j)} = \left(\sqrt{1 - (1 - \mu_{\theta(j)}^2)^{q_j}}, \nu_{\theta(j)}^{q_j}\right) \text{ and } \sum_{j=1}^{n} (q_j \alpha_{\theta(j)}) = \left(\sqrt{1 - \prod_{j=1}^{n} (1 - \mu_{\theta(j)}^2)^{q_j}}, \prod_{j=1}^{n} \nu_{\theta(j)}^{q_j} \right).$$

then, based on laws (1) and (3) in Definition 2, we can obtain

$$\prod_{\theta \in S_n} \sum_{j=1}^{n} (q_j \alpha_{\theta(j)}) = \left(\prod_{\theta \in S_n} \sqrt{1 - \prod_{j=1}^{n} (1 - \mu_{\theta(j)}^2)^{q_j}}, \sqrt{1 - \prod_{\theta \in S_n} (1 - \prod_{j=1}^{n} \nu_{\theta(j)}^{2q_j})} \right)$$

and

$$\left(\prod_{\theta \in S_n} \sum_{j=1}^{n} (q_j \alpha_{\theta(j)}) \right)^{\frac{1}{n!}} = \left(\left(\prod_{\theta \in S_n} \sqrt{1 - \prod_{j=1}^{n} (1 - \mu_{\theta(j)}^2)^{q_j}} \right)^{\frac{1}{n!}}, \sqrt{1 - \left(\prod_{\theta \in S_n} (1 - \prod_{j=1}^{n} \nu_{\theta(j)}^{2q_j}) \right)^{\frac{1}{n!}}} \right).$$

Further,

$$\frac{1}{\sum_{j=1}^{n} q_j} \left(\prod_{\theta \in S_n} \sum_{j=1}^{n} (q_j \alpha_{\theta(j)}) \right)^{\frac{1}{n!}} = \left(\sqrt{1 - \left(1 - \left(\prod_{\theta \in S_n} \left(1 - \prod_{j=1}^{n} (1 - \mu_{\theta(j)}^2)^{q_j}\right)\right)^{\frac{1}{n!}}\right)^{\frac{1}{\sum_{j=1}^{n} q_j}}}, \left(\sqrt{1 - \left(\prod_{\theta \in S_n} (1 - \prod_{j=1}^{n} \nu_{\theta(j)}^{2q_j})\right)^{\frac{1}{n!}}}\right)^{\frac{1}{\sum_{j=1}^{n} q_j}} \right),$$

which illustrates that Equation (16) holds.

(2) In the following, we will prove that Equation (16) is a PFN.

Let $\mu_P = \sqrt{1 - \left(1 - \left(\prod_{\theta \in S_n} \left(1 - \prod_{j=1}^{n} (1 - \mu_{\theta(j)}^2)^{q_j}\right)\right)^{\frac{1}{n!}}\right)^{\frac{1}{\sum_{j=1}^{n} q_j}}}$,

$$\nu_P = \left(\sqrt{1 - \left(\prod_{\theta \in S_n} (1 - \prod_{j=1}^{n} \nu_{\theta(j)}^{2q_j})\right)^{\frac{1}{n!}}}\right)^{\frac{1}{\sum_{j=1}^{n} q_j}}.$$

Then we also need to prove that Equation (16) satisfies the following two conditions.

(a) $0 \le \mu_P \le 1$, and $0 \le \nu_P \le 1$;

(b) $\mu_P^2 + \nu_P^2 \le 1$.

(a) Based on Definition 1,

$$1 - \mu_{\theta(j)}^2 \in [0,1] \text{ and } (1 - \mu_{\theta(j)}^2)^{q_j} \in [0,1],$$

we get

$$\prod_{j=1}^{n} (1 - \mu_{\theta(j)}^2)^{q_j} \in [0,1] \text{ and } \prod_{\theta \in S_n} \left(1 - \prod_{j=1}^{n} (1 - \mu_{\theta(j)}^2)^{q_j}\right) \in [0,1].$$

Further,

$$\left(\prod_{\theta \in S_n}\left(1 - \prod_{j=1}^{n} (1 - \mu_{\theta(j)}^2)^{q_j}\right)\right)^{\frac{1}{n!}} \in [0,1] \text{ and } \left(1 - \left(\prod_{\theta \in S_n}\left(1 - \prod_{j=1}^{n} (1 - \mu_{\theta(j)}^2)^{q_j}\right)\right)^{\frac{1}{n!}}\right)^{\frac{1}{\sum_{j=1}^{n} q_j}} \in [0,1],$$

then

$$\sqrt{1 - \left(1 - \left(\prod_{\theta \in S_n}\left(1 - \prod_{j=1}^{n} (1 - \mu_{\theta(j)}^2)^{q_j}\right)\right)^{\frac{1}{n!}}\right)^{\frac{1}{\sum_{j=1}^{n} q_j}}} \in [0,1],$$

i.e., $0 \le \mu_P \le 1$. Similarly, we can yield $0 \le \nu_P \le 1$. Therefore, condition (a) is satisfied.

(b) Because $\mu_{\theta(j)}^2 + \nu_{\theta(j)}^2 \le 1$, then $\nu_{\theta(j)}^2 \le 1 - \mu_{\theta(j)}^2$, we can obtain the inequality as follows:

$$\begin{aligned}
\mu_{P_i}^2 + \nu_{P_i}^2 &= 1 - \left(1 - \left(\prod_{\theta \in S_n}\left(1 - \prod_{j=1}^{n} (1 - \mu_{\theta(j)}^2)^{q_j}\right)\right)^{\frac{1}{n!}}\right)^{\frac{1}{\sum_{j=1}^{n} q_j}} + \left(1 - \left(\prod_{\theta \in S_n} (1 - \prod_{j=1}^{n} \nu_{\theta(j)}^{2q_j})\right)^{\frac{1}{n!}}\right)^{\frac{1}{\sum_{j=1}^{n} q_j}} \\
&\le 1 - \left(1 - \left(\prod_{\theta \in S_n}\left(1 - \prod_{j=1}^{n} (1 - \mu_{\theta(j)}^2)^{q_j}\right)\right)^{\frac{1}{n!}}\right)^{\frac{1}{\sum_{j=1}^{n} q_j}} + \left(1 - \left(\prod_{\theta \in S_n}\left(1 - \prod_{j=1}^{n} (1 - \mu_{\theta(j)}^2)^{q_j}\right)\right)^{\frac{1}{n!}}\right)^{\frac{1}{\sum_{j=1}^{n} q_j}} = 1,
\end{aligned}$$

i.e., $\mu_P^2 + \nu_P^2 \le 1$. Consequently, condition (b) is satisfied.

Based on the proof above, we know that Theorem 5 holds. □

Similar to the properties of the PFMM operator, we can easily obtain some properties of the PFDMM operator as follows.

Property 6 (Idempotency). *Let* $\alpha_i = (\mu_{P_i}, \nu_{P_i}), (i = 1, 2, \cdots, n)$ *are equal, i.e.,* $\alpha_i = \alpha = (\mu_P, \nu_P)$ *for all* i*, then*

$$PFDMM^Q(\alpha_1, \alpha_2, \cdots, \alpha_n) = \alpha = (\mu_P, \nu_P).$$

The Proof of Property 6 is similar to that of Property 1, so is omitted here.

Property 7 (Monotonicity). *Let* $\alpha_i = (\mu_{P_i}, \nu_{P_i})$ *and* $\hat{\alpha}_i = (\hat{\mu}_{P_i}, \hat{\nu}_{P_i})$, $(i = 1, 2, \cdots, n)$ *be two collections of PFNs. If* $\mu_{P_i} \ge \hat{\mu}_{P_i}$ *and* $\nu_{P_i} \le \hat{\nu}_{P_i}$ *for all* i*, then*

$$PFDMM^Q(\alpha_1, \alpha_2, \cdots, \alpha_n) \ge PFDMM^Q(\hat{\alpha}_1, \hat{\alpha}_2, \cdots, \hat{\alpha}_n).$$

The Proof of Property 7 is similar to that of Property 2, so is omitted here.

Property 8 (Boundedness). *Let* $\alpha_i = (\mu_{P_i}, \nu_{P_i}), (i = 1, 2, \cdots, n)$ *be a collections of PFNs,* $\alpha^+ = (max(\mu_{P_i}), min(\nu_{P_i}))$ *and* $\alpha^- = (min(\mu_{P_i}), max(\nu_{P_i}))$, *then*

$$\alpha^+ \geq PFDMM^Q(\alpha_1, \alpha_2, \cdots, \alpha_n) \geq \alpha^-.$$

The Proof of Property 8 is similar to that of Property 3, so is omitted here.

In what follows, we will discuss some special cases of the PFDMM operator through changing the values of parameter vector Q.

(1). When $Q = (1, 0, \cdots, 0)$, Equation (16) is transformed into a Pythagorean fuzzy geometric averaging operator:

$$PFDMM^{(1,0,\cdots,0)}(\alpha_1, \alpha_2, \cdots, \alpha_n) = \left(\prod_{i=1}^{n} (\mu_{P_i})^{\frac{1}{n}}, \sqrt{1 - \prod_{i=1}^{n} (1 - \nu_{P_i}^2)^{\frac{1}{n}}} \right). \tag{17}$$

(2). When $Q = (\lambda, 0, \cdots, 0)$, Equation (16) is transformed into a Pythagorean fuzzy generalized geometric averaging operator:

$$PFDMM^{(\lambda,0,\cdots,0)}(\alpha_1, \alpha_2, \cdots, \alpha_n) = \left(\sqrt{1 - \left(1 - \prod_{i=1}^{n} \left(1 - (1 - \mu_{P_i}^2)^{\lambda}\right)^{\frac{1}{n}}\right)^{\frac{1}{\lambda}}}, \left(\sqrt{1 - \prod_{i=1}^{n} (1 - \nu_{P_i}^{2\lambda})^{\frac{1}{n}}}\right)^{\frac{1}{\lambda}} \right). \tag{18}$$

(3). When $Q = (1, 1, 0, 0 \cdots, 0)$, Equation (16) is transformed into a Pythagorean fuzzy geometric BM operator:

$$PFDMM^{(1,1,0,0,\cdots,0)}(\alpha_1, \alpha_2, \cdots, \alpha_n) = \left(\sqrt{1 - \left(1 - \prod_{\substack{i,j=1 \\ i \neq j}}^{n} \left(1 - (1 - \mu_{P_i}^2)(1 - \mu_{P_j}^2)\right)^{\frac{1}{n(n-1)}}\right)^{\frac{1}{2}}}, \left(\sqrt{1 - \left(\prod_{\substack{i,j=1 \\ i \neq j}}^{n} (1 - \nu_{P_i}^2 \nu_{P_j}^2)\right)^{\frac{1}{n(n-1)}}}\right)^{\frac{1}{2}} \right). \tag{19}$$

(4). When $Q = (\underbrace{1, 1, \cdots, 1}_{k}, \underbrace{0, 0, \cdots, 0}_{n-k})$, Equation (16) is transformed into a Pythagorean fuzzy geometric MSM operator:

$$PFDMM^{(\underbrace{1,1,\cdots,1}_{k}\underbrace{0,0,\cdots,0}_{n-k})}(\alpha_1, \alpha_2, \cdots, \alpha_n) = \left(\sqrt{1 - \left(1 - \left(\prod_{1 \leq i1 \prec \cdots \prec ik \leq n} \left(1 - \prod_{j=1}^{k} (1 - \mu_{P_{ij}}^2)\right)\right)^{\frac{1}{C_n^k}}\right)^{\frac{1}{k}}}, \left(\sqrt{1 - \left(\prod_{1 \leq i1 \prec \cdots \prec ik \leq n} (1 - \prod_{j=1}^{k} \nu_{P_{ij}}^2)\right)^{\frac{1}{C_n^k}}}\right)^{\frac{1}{k}} \right). \tag{20}$$

(5). When $Q = (1, 1, \cdots, 1)$, Equation (16) is transformed into a Pythagorean fuzzy arithmetic averaging operator:

$$PFDMM^{(1,1,\cdots,1)}(\alpha_1, \alpha_2, \cdots, \alpha_n) = \left(\sqrt{1 - \prod_{i=1}^{n} (1 - \mu_{P_i}^2)^{\frac{1}{n}}}, \left(\prod_{i=1}^{n} \nu_{P_i}\right)^{\frac{1}{n}} \right). \tag{21}$$

(6). When $Q = (1/n, 1/n, \cdots, 1/n)$, Equation (16) is transformed into a Pythagorean fuzzy arithmetic averaging operator:

$$PFDMM^{(1/n,1/n,\cdots,1/n)}(\alpha_1, \alpha_2, \cdots, \alpha_n) = \left(\sqrt{1 - \prod_{i=1}^{n} (1 - \mu_{P_i}^2)^{\frac{1}{n}}}, \left(\prod_{i=1}^{n} \nu_{P_i}\right)^{\frac{1}{n}} \right). \tag{22}$$

4.2. The PFDWMM Operator

Definition 10. *Let $Q = (q_1, q_2, \cdots, q_n) \in R^n$ is parameter vector, $\alpha_i = (\mu_{P_i}, \nu_{P_i}), (i = 1, 2, \cdots, n)$ be a collection of PFNs, and $w = (w_1, w_2, \cdots, w_n)^T$ be the weighted vector of α_i, where w_i indicates the importance degree of α_i, satisfying $w_i \in [0, 1]$ and $\sum\limits_{i=1}^{n} w_i = 1$, If*

$$PFDWMM^Q(\alpha_1, \alpha_2, \cdots, \alpha_n) = \frac{1}{\sum\limits_{j=1}^{n} q_j} \left(\prod_{\theta \in S_n} \sum_{j=1}^{n} (q_j \alpha_{\theta(j)}^{n w_{\theta(j)}}) \right)^{\frac{1}{n!}}. \tag{23}$$

then $PFDWMM^Q$ is called the PFDWMM operator.

Theorem 6. *Let $Q = (q_1, q_2, \cdots, q_n) \in R^n$ be a parameter vector, $\alpha_i = (\mu_{P_i}, \nu_{P_i}), (i = 1, 2, \cdots, n)$ be a collection of PFNs, and $w = (w_1, w_2, \cdots, w_n)^T$ be the weight vector of α_i, where w_i indicates the importance degree of α_i, satisfying $w_i \in [0, 1]$ and $\sum\limits_{i=1}^{n} w_i = 1$. Then, the aggregated value by using the PFDWMM operator is also a PFN, and*

$$PFDWMM^Q(\alpha_1, \alpha_2, \cdots, \alpha_n)$$
$$= \left(\sqrt{1 - \left(1 - \left(\prod_{\theta \in S_n} \left(1 - \prod_{j=1}^{n} (1 - \mu_{\theta(j)}^{2 n w_{\theta(j)}})^{q_j} \right) \right)^{\frac{1}{n!}} \right)^{\frac{1}{\sum_{j=1}^{n} q_j}}}, \left(\sqrt{1 - \left(\prod_{\theta \in S_n} \left(1 - \prod_{j=1}^{n} (1 - (1 - v_{\theta(j)}^2)^{n w_{\theta(j)}})^{q_j} \right) \right)^{\frac{1}{n!}}} \right)^{\frac{1}{\sum_{j=1}^{n} q_j}} \right). \tag{24}$$

Proof. According to operational law (4) in Definition 2, then $\alpha_{\theta(j)}^{n w_{\theta(j)}} = \left(\mu_{\theta(j)}^{n w_{\theta(j)}}, \sqrt{1 - (1 - \nu_{\theta(j)}^2)^{n w_{\theta(j)}}} \right)$, and we can replace $\mu_{\theta(j)}$ and $\nu_{\theta(j)}$ with $\mu_{\theta(j)}^{n w_{\theta(j)}}$ and $\sqrt{1 - (1 - \nu_{\theta(j)}^2)^{n w_{\theta(j)}}}$, respectively, in Equation (16), to obtain Equation (24). Since $\alpha_{\theta(j)}$ is a PFN, $\alpha_{\theta(j)}^{n w_{\theta(j)}}$ is also a PFN. Similar to the proof of Theorem 5, we know Equation (24) is also a PFN. □

In the following, we will discuss some desirable properties of the PFDWMM operator.

Property 9 (Monotonicicty). *Let $\alpha_i = (\mu_{P_i}, \nu_{P_i})$ and $\hat{\alpha}_i = (\hat{\mu}_{P_i}, \hat{\nu}_{P_i})$, $(i = 1, 2, \cdots, n)$ be two collections of PFNs. Through using the PFDWMM operator, if $\mu_{P_i} \geq \hat{\mu}_{P_i}$ and $\nu_{P_i} \leq \hat{\nu}_{P_i}$, then*

$$PFDWMM^Q(\alpha_1, \alpha_2, \cdots, \alpha_n) \geq PFDWMM^Q(\hat{\alpha}_1, \hat{\alpha}_2, \cdots, \hat{\alpha}_n).$$

The Proof of Property 9 is similar to that of Property 2, so is omitted here.

Property 10 (Boundedness). *Let $\alpha_i = (\mu_{P_i}, \nu_{P_i}), (i = 1, 2, \cdots, n)$ be a collections of PFNs, $\alpha^+ = (max(\mu_{P_i}), min(\nu_{P_i}))$ and $\alpha^- = (min(\mu_{P_i}), max(\nu_{P_i}))$, then*

$$\alpha^+ \geq PFDWMM^Q(\alpha_1, \alpha_2, \cdots, \alpha_n) \geq \alpha^-.$$

The Proof Property 10 is similar to that of Property 3, so is omitted here.

Theorem 7. *The PFDMM operator is a special case of the PFDWMM operator.*

Proof. When $w = (1/n, 1/n, \cdots, 1/n)^T$,

$$
\begin{aligned}
&PFDWMM^{Q}(\alpha_1, \alpha_2, \cdots, \alpha_n) \\
&= \left(\sqrt{1 - \left(1 - \left(\prod_{\theta \in S_n}\left(1 - \prod_{j=1}^{n}\left(1 - \mu_{\theta(j)}^{2n \times \frac{1}{n}}\right)^{q_j}\right)\right)^{\frac{1}{n!}}\right)^{\frac{1}{\sum_{j=1}^{n} q_j}}}, \left(\sqrt{1 - \left(\prod_{\theta \in S_n}\left(1 - \prod_{j=1}^{n}\left(1 - \left(1 - v_{\theta(j)}^2\right)^{n \times \frac{1}{n}}\right)^{q_j}\right)\right)^{\frac{1}{n!}}}\right)^{\frac{1}{\sum_{j=1}^{n} q_j}} \right) \\
&= \left(\sqrt{1 - \left(1 - \left(\prod_{\theta \in S_n}\left(1 - \prod_{j=1}^{n}\left(1 - \mu_{\theta(j)}^{2}\right)^{q_j}\right)\right)^{\frac{1}{n!}}\right)^{\frac{1}{\sum_{j=1}^{n} q_j}}}, \left(\sqrt{1 - \left(\prod_{\theta \in S_n}\left(1 - \prod_{j=1}^{n} v_{\theta(j)}^{2q_j}\right)\right)^{\frac{1}{n!}}}\right)^{\frac{1}{\sum_{j=1}^{n} q_j}} \right).
\end{aligned}
$$

□

Theorem 8. *The Pythagorean fuzzy weighted geometric averaging operator [6] is a special case of the PFDWMM operator.*

Proof. When $Q = (1, 0, \cdots, 0)$,

$$
\begin{aligned}
&PFDWMM^{(1,0,\cdots,0)}(\alpha_1, \alpha_2, \cdots, \alpha_n) \\
&= \left(\sqrt{1 - \left(1 - \left(\prod_{\theta \in S_n}\left(1 - \prod_{j=1}^{n}\left(1 - \mu_{\theta(j)}^{2nw_{\theta(j)}}\right)^{q_j}\right)\right)^{\frac{1}{n!}}\right)^{\frac{1}{\sum_{j=1}^{n} q_j}}}, \left(\sqrt{1 - \left(\prod_{\theta \in S_n}\left(1 - \prod_{j=1}^{n}\left(1 - \left(1 - v_{\theta(j)}^2\right)^{nw_{\theta(j)}}\right)^{q_j}\right)\right)^{\frac{1}{n!}}}\right)^{\frac{1}{\sum_{j=1}^{n} q_j}} \right) \\
&= \left(\sqrt{\left(\prod_{j=1}^{n} \mu_j^{2nw_j}\right)^{\frac{1}{n}}}, \sqrt{1 - \left(\prod_{j=1}^{n}\left(1 - v_j^2\right)^{nw_j}\right)^{\frac{1}{n}}} \right) \\
&= \left(\prod_{j=1}^{n} \mu_{P_j}^{w_j}, \sqrt{1 - \prod_{j=1}^{n}\left(1 - v_{P_j}^2\right)^{w_j}} \right).
\end{aligned}
$$

□

5. New Approach to MCGDM with Pythagorean Fuzzy Information

In this section, we propose a new MCGDM method under the Pythagorean fuzzy environment based on the PFWMM operator or PFDWMM operator. A typical MCGDM problem with Pythagorean fuzzy information can be described as follows. Let $A = \{A_1, A_2, \cdots, A_m\}$ be a discrete set of alternatives, and $C = \{C_1, C_2, \cdots, C_n\}$ be a finite set of criteria with the weight vector is $w = \{w_1, w_2, \cdots, w_n\}$, satisfying $w_j \in [0,1]$ $(j = 1, 2, \cdots, n)$ and $\sum_{j=1}^{n} w_j = 1$. Assume that $E = \{E_1, E_2, \cdots, E_p\}$ be a finite set of experts with the weight vector is $\eta = \{\eta_1, \eta_2, \cdots, \eta_p\}$, satisfying $\eta_k \in [0,1] (k = 1, 2, \cdots, p)$ and $\sum_{k=1}^{p} \eta_k = 1$. The evaluation information of alternative $A_i (i = 1, 2, \cdots, m)$ with respect to criteria $C_j (j = 1, 2, \cdots, n)$ provided by the expert $E_k (k = 1, 2, \cdots, p)$ can be denoted as $\alpha_{ij}^k = (\mu_{P_{ij}}^k, v_{P_{ij}}^k)$, where α_{ij}^k is a PFN. Therefore, the Pythagorean fuzzy evaluation matrix $R^k = \left(\alpha_{ij}^k\right)_{m \times n}$ provided by the expert E_k is obtained.

In what follows, a novel approach based on the PFWMM operator or PFDWMM operator is proposed to solve the MCGDM problem with Pythagorean fuzzy information, and the detailed steps are depicted as follows.

Step 1: Generally, there are two types of criteria, i.e., benefit criterion and cost criterion. Therefore, the Pythagorean fuzzy evaluation matrix should be normalized by

$$
\alpha_{ij}^k = \begin{cases} (\mu_{P_{ij}}^k, v_{P_{ij}}^k), & \text{for benefit criterion.} \\ (v_{P_{ij}}^k, \mu_{P_{ij}}^k), & \text{for cos t criterion.} \end{cases}
$$

Step 2: Construct the group decision matrix $R = (\alpha_{ij})_{m\times n}$ by applying the PFWMM operator or PFDWMM operators to aggregate all individual evaluation matrix R^k.

$$\alpha_{ij} = PFWMM(\alpha_{ij}^1, \alpha_{ij}^2, \cdots, \alpha_{ij}^p), \alpha_{ij} = PFDWMM(\alpha_{ij}^1, \alpha_{ij}^2, \cdots, \alpha_{ij}^p).$$

Step 3: Calculate the comprehensive evaluation value α_i by using the PFWMM or PFDWMM operators to aggregate all the performance values of alternative with regard to each criterion.

$$\alpha_i = PFWMM(\alpha_{i1}, \alpha_{i2}, \cdots, \alpha_{in}), \alpha_i = PFDWMM(\alpha_{i1}, \alpha_{i2}, \cdots, \alpha_{in}).$$

Step 4: Determine the priority of alternatives according to the score value $S(\alpha_i)$.

6. An Example

To validate the effectiveness and feasibility of the proposed method, we adopt a numerical example that is about the selection decision of enterprise resource planning (ERP) system. An enterprise wants to select a suitable ERP system to improve the competitive capability of the company. In order to make a scientific decision, three experts are selected to form the expert team, denoted as $E = \{E_1, E_2, E_3\}$, where E_1 is a CIO, E_2 and E_3 are two senior representatives from the user department. Suppose that the importance of experts is equal, namely, $\eta = (1/3,\ 1/3,\ 1/3)^T$. Through analyzing the ERP system, the expert team determines the assessment criteria including function and technology (C_1), strategic fitness (C_2), vendor ability (C_3), and vendor reputation (C_4). According to the existing experience and knowledge, the weight vector of the criteria is assigned by experts as $w = (0.2,\ 0.1,\ 0.3,\ 0.4)^T$. Five potential ERP systems $A = \{A_1, A_2, A_3, A_4, A_5\}$ are chosen by the expert team as candidates. Subsequently, the experts adopt PFNs to provide the assessment information of the alternatives with regard to each criterion. The Pythagorean fuzzy decision matrices are provided by three experts, shown in Tables 1–3, respectively. In what follows, we apply the proposed method to obtain the best ERP system for the enterprise.

Table 1. The Pythagorean fuzzy decision matrix provided by the E_1.

Alternatives	C_1	C_2	C_3	C_4
A_1	(0.4, 0.8)	(0.8, 0.6)	(0.6, 0.7)	(0.3, 0.8)
A_2	(0.7, 0.5)	(0.8, 0.4)	(0.8, 0.5)	(0.3, 0.6)
A_3	(0.3, 0.4)	(0.3, 0.7)	(0.7, 0.4)	(0.6, 0.4)
A_4	(0.6, 0.6)	(0.7, 0.5)	(0.7, 0.2)	(0.4, 0.6)
A_5	(0.5, 0.7)	(0.6, 0.4)	(0.9, 0.3)	(0.6, 0.7)

Table 2. The Pythagorean fuzzy decision matrix provided by the E_2.

Alternatives	C_1	C_2	C_3	C_4
A_1	(0.3, 0.9)	(0.7, 0.6)	(0.5, 0.8)	(0.3, 0.6)
A_2	(0.7, 0.4)	(0.9, 0.2)	(0.8, 0.1)	(0.3, 0.5)
A_3	(0.3, 0.6)	(0.7, 0.7)	(0.7, 0.6)	(0.4, 0.4)
A_4	(0.4, 0.8)	(0.7, 0.5)	(0.6, 0.2)	(0.4, 0.7)
A_5	(0.2, 0.7)	(0.8, 0.2)	(0.8, 0.4)	(0.6, 0.6)

Table 3. The Pythagorean fuzzy decision matrix provided by the E_3.

Alternatives	C_1	C_2	C_3	C_4
A_1	(0.6, 0.8)	(0.7, 0.6)	(0.5, 0.8)	(0.5, 0.5)
A_2	(0.6, 0.5)	(0.9, 0.2)	(0.8, 0.1)	(0.3, 0.5)
A_3	(0.4, 0.7)	(0.7, 0.5)	(0.6, 0.1)	(0.2, 0.9)
A_4	(0.2, 0.9)	(0.5, 0.6)	(0.6, 0.2)	(0.1, 0.6)
A_5	(0.1, 0.6)	(0.8, 0.2)	(0.9, 0.2)	(0.6, 0.5)

6.1. Implementation of the Proposed Method

To obtain the best ERP system, the computation steps are shown in the following:

Step 1: The criterion value of the ERP system does not require normalization because all the criteria are benefit type.

Step 2: Based on the individual evaluation matrix $R^k(k = 1,2,3)$, we employ the PFWMM operator or PFDWMM operator (Suppose $Q = (1,1,1)$) to obtain the group decision matrix R, and the results are shown in Tables 4 and 5, respectively.

Table 4. The collective decision matrix obtained by the PFWMM operator.

Alternatives	C_1	C_2	C_3	C_4
A_1	(0.4160, 0.8421)	(0.7319, 0.6000)	(0.5313, 0.7718)	(0.3557, 0.6656)
A_2	(0.6649, 0.4702)	(0.8653, 0.2860)	(0.8000, 0.3123)	(0.3000, 0.5372)
A_3	(0.3302, 0.5919)	(0.5278, 0.6481)	(0.6649, 0.4354)	(0.3634, 0.6987)
A_4	(0.3634, 0.8047)	(0.6257, 0.5372)	(0.6316, 0.2000)	(0.2520, 0.6377)
A_5	(0.2154, 0.6707)	(0.7268, 0.2860)	(0.8653, 0.3131)	(0.6000, 0.6119)

Table 5. The collective decision matrix obtained by the PFDWMM operator.

Alternatives	C_1	C_2	C_3	C_4
A_1	(0.4605, 0.8320)	(0.7389, 0.6000)	(0.5372, 0.7652)	(0.3832, 0.6214)
A_2	(0.6707, 0.4642)	(0.8746, 0.2520)	(0.8000, 0.1710)	(0.3000, 0.5313)
A_3	(0.3376, 0.5518)	(0.6176, 0.6257)	(0.6707, 0.2884)	(0.4448, 0.5241)
A_4	(0.4448, 0.7560)	(0.6481, 0.5313)	(0.6377, 0.2000)	(0.3357, 0.6316)
A_5	(0.3267, 0.6649)	(0.7509, 0.2520)	(0.8746, 0.2884)	(0.6000, 0.5944)

Step 3: With the aid of the PFWMM operator or PFDWMM operator (suppose $Q = (1,1,1,1)$), the comprehensive evaluation value α_i of the ERP system can be obtained. The calculated results are shown in Table 6.

Table 6. The comprehensive evaluation value by PFWMM and PFDWMM operators.

Operator	A_1	A_2	A_3	A_4	A_5
PFWMM	(0.4700, 0.7707)	(0.5948, 0.4751)	(0.4150, 0.6672)	(0.5015, 0.5599)	(0.4269, 0.6635)
PFDWMM	(0.6483, 0.6583)	(0.7915, 0.3001)	(0.6264, 0.4482)	(0.7478, 0.3847)	(0.6142, 0.4285)

Step 4: Calculate the scores of the comprehensive evaluation value α_i of each ERP system (results shown in Table 7). The ranking of the ERP systems can be obtained according to the scores in descending order. The obtained rankings are shown in Table 7.

Table 7. The score values and the ranking results of five alternatives by two operators.

Operator	$S(\alpha_1)$	$S(\alpha_2)$	$S(\alpha_3)$	$S(\alpha_4)$	$S(\alpha_5)$	Ranking Order
PFWMM	0.3135	0.5640	0.3710	0.3568	0.4690	$A_2 \succ A_5 \succ A_3 \succ A_4 \succ A_1$
PFDWMM	0.4935	0.7682	0.5968	0.5957	0.7056	$A_2 \succ A_5 \succ A_3 \succ A_4 \succ A_1$

Based on the ranking results in Table 7, we know that the ranking order obtained by the PFWMM operator and PFDWMM operator are the same, and the best ERP system is A_2.

To further demonstrate the effectiveness and applicability of the presented approach, we employ the proposed method to solve two practical MCDM problems concerned with the investment decision respecting Internet stocks [11] and investment decision respecting R&D projects [11]. This paper refers to individual decision-making, while these two investment decision problems are group decision-making problems. Hence, the implementation of the presented approach is performed based on the comprehensive evaluation matrix that is obtained by employing the PFWA operator to aggregate the individual evaluation matrix.

The investment decision problem of Internet stocks involves four Internet stocks (A_1, A_2, A_3, A_4) and three benefit criteria (C_1, C_2, C_3), and the weight vector of the criterion is $w = (0.5,\ 0.2,\ 0.3)$. The comprehensive evaluation matrix of Internet stocks with respect to each criterion is shown in Table 8. The obtained results and ranking orders are shown in Table 9. From Table 9, we can see that the ranking order obtained by the proposed method and algorithm 1 is totally identical. The best and worst Internet stocks are A_1 and A_4, respectively.

Table 8. The collective evaluation matrix of Internet stocks regarding each criterion [11].

Alternatives	C_1	C_2	C_3
A_1	(0.77, 0.19)	(0.88, 0.18)	(0.77, 0.17)
A_2	(0.61, 0.67)	(0.51, 0.56)	(0.67, 0.18)
A_3	(0.68, 0.27)	(0.70, 0.51)	(0.67, 0.48)
A_4	(0.66, 0.62)	(0.56, 0.67)	(0.56, 0.36)
weight	0.5	0.2	0.3

Table 9. The score values and ranking results of four Internet stocks.

Operator	$S(A_1)$	$S(A_2)$	$S(A_3)$	$S(A_4)$	Ranking Order
Algorithm 1 [11]	0.5944	0.1183	0.3195	0.0807	$A_1 \succ A_3 \succ A_2 \succ A_4$
PFWMM	0.7818	0.5098	0.5865	0.4737	$A_1 \succ A_3 \succ A_2 \succ A_4$
PFDWMM	0.8313	0.6223	0.6741	0.5614	$A_1 \succ A_3 \succ A_2 \succ A_4$

The investment decision problem of the R&D project includes three potential R&D projects (A_1, A_2, A_3) and five benefit criteria (C_1, C_2, C_3, C_4, C_5); the weight vector of the criterion is $w = (0.2,\ 0.1,\ 0.3,\ 0.15,\ 0.25)$. The comprehensive evaluation matrix of R&D projects based on each criterion is shown in Table 10, and the ranking index and ranking results of three R&D projects are presented in Table 11. From Table 11, we know that the ranking order of R&D projects obtained in the developed approach is the same as that determined by algorithm 2. The preferred R&D project is A_1, and the worst one is A_3.

Table 10. The collective evaluation matrix of R&D project regarding each criterion [11].

Alternatives	C_1	C_2	C_3	C_4	C_5
A_1	(0.77, 0.21)	(0.71, 0.18)	(0.77, 0.17)	(0.75, 0.10)	(0.76, 0.20)
A_2	(0.59, 0.64)	(0.53, 0.46)	(0.67, 0.29)	(0.45, 0.65)	(0.80, 0.24)
A_3	(0.68, 0.34)	(0.68, 0.51)	(0.67, 0.51)	(0.80, 0.45)	(0.45, 0.77)
weight	0.2	0.1	0.3	0.15	0.25

Table 11. The ranking index and ranking results of three R&D projects.

Operator	$S(A_1)$	$S(A_2)$	$S(A_3)$	Ranking Order
Algorithm 2 [11]	0.9083	−1.1927	−2.2731	$A_1 \succ A_2 \succ A_3$
PFWMM	0.7331	0.5008	0.5279	$A_1 \succ A_2 \succ A_3$
PFDWMM	0.5931	0.4230	0.3449	$A_1 \succ A_2 \succ A_3$

Based on the above analysis on two investment decision problems, we can conclude that the method presented in this paper is effective and feasible.

6.2. Sensitivity Analysis

To illustrate the influence of different values of parameter vector Q, we change the values of parameter vector Q in our proposed method to rank the alternatives. The results are shown in Table 12.

Table 12. Ranking order of alternatives determined by different parameter vector Q.

Parameter Vector Q	Ranking Order by PFWMM	Ranking Order by PFDWMM
$Q = (3, 0, 0, 0)$	$A_5 \succ A_2 \succ A_4 \succ A_3 \succ A_1$	$A_5 \succ A_3 \succ A_2 \succ A_4 \succ A_1$
$Q = (2, 0, 0, 0)$	$A_5 \succ A_2 \succ A_4 \succ A_3 \succ A_1$	$A_5 \succ A_2 \succ A_3 \succ A_4 \succ A_1$
$Q = (1, 0, 0, 0)$	$A_2 \succ A_5 \succ A_4 \succ A_3 \succ A_1$	$A_2 \succ A_5 \succ A_3 \succ A_4 \succ A_1$
$Q = (1, 1, 0, 0)$	$A_2 \succ A_5 \succ A_3 \succ A_4 \succ A_1$	$A_2 \succ A_5 \succ A_3 \succ A_4 \succ A_1$
$Q = (1, 1, 1, 0)$	$A_2 \succ A_5 \succ A_3 \succ A_4 \succ A_1$	$A_2 \succ A_5 \succ A_3 \succ A_4 \succ A_1$
$Q = (1, 1, 1, 1)$	$A_2 \succ A_5 \succ A_3 \succ A_4 \succ A_1$	$A_2 \succ A_5 \succ A_3 \succ A_4 \succ A_1$
$Q = (0.25, 0.25, 0.25, 0.25)$	$A_2 \succ A_5 \succ A_3 \succ A_4 \succ A_1$	$A_2 \succ A_5 \succ A_3 \succ A_4 \succ A_1$

From Table 12, we know the ERP systems obtained with different parameter vectors Q are slightly different. The main reason is that the PFWMM operator highlights the impact of overall arguments, but the PFDWMM operator emphasizes the role of individual arguments. When $Q = (1, 0, 0, 0)$, it is worth noting that the PFWMM operator and PFDWMM operator will reduce to Pythagorean fuzzy weighted averaging operator and Pythagorean fuzzy weighted geometric operator, respectively. In addition, from Figures 1 and 2 we can make the following conclusions. For the PFWMM operator, when the parameter vector Q has only one real number and the rest are 0, we discover that the larger the real number of parameter vector Q, the greater the value of the score function will become. The more interdependent relationships of criteria we consider, the smaller the score function will become. Nevertheless, for the PFDWMM operator, the conclusion is just the opposite: that is, the greater the real number of parameter vector Q, the smaller the value of the score function will become. The more relationships between attributes we consider, the larger the value of the score function will become. Therefore, the experts can select different values of parameter vector Q based on different risk preferences.

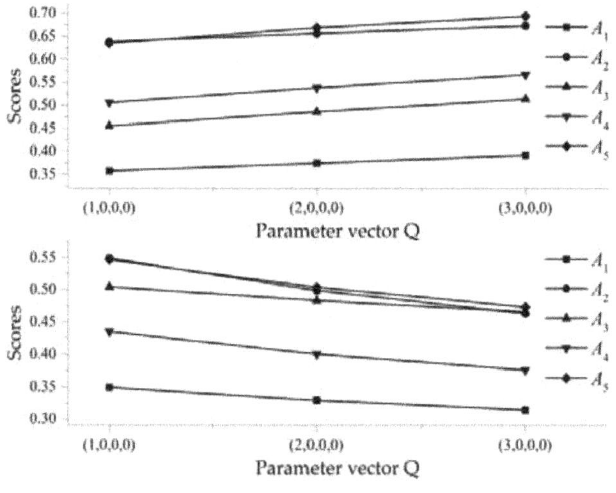

Figure 1. Scores of alternatives for different parameter vector Q.

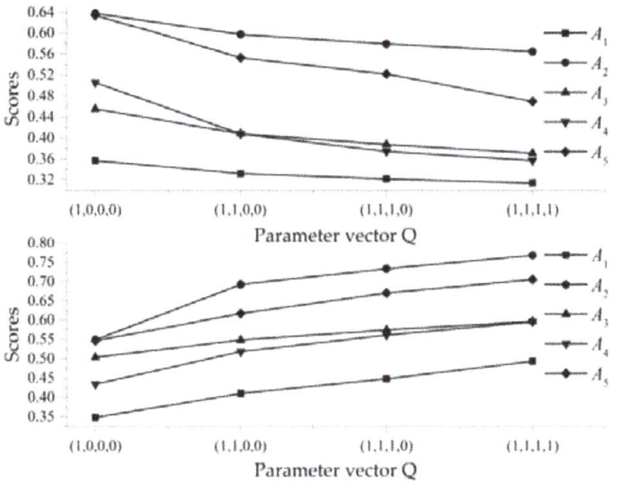

Figure 2. Scores of alternatives for interrelationship of different arguments.

6.3. *Comparative Analysis*

In order to further verify the validity and illustrate the advantage of the proposed approach, we compare our developed method with other existing MCGDM methods including the Pythagorean fuzzy weighted averaging (PFWA) operator and Pythagorean fuzzy weighted geometric (PFWG) operator [6], the symmetric Pythagorean fuzzy weighted averaging (SPFWA) operator and symmetric Pythagorean fuzzy weighted geometric (SPFWG) operator [6], the Pythagorean fuzzy weighted geometric Bonferroni mean (PFWGBM) operator [14], and the Pythagorean fuzzy weighted Maclaurin symmetric mean (PFWMSM) operator [25]. The ranking results are shown in Table 13.

Table 13. Ranking order of alternatives obtained by different methods.

Operator	Parameter	Ranking Order
PFWA	No	$A_2 \succ A_5 \succ A_3 \succ A_4 \succ A_1$
PFWG	No	$A_2 \succ A_5 \succ A_4 \succ A_3 \succ A_1$
SPFWA	No	$A_2 \succ A_5 \succ A_3 \succ A_4 \succ A_1$
SPFWG	No	$A_2 \succ A_5 \succ A_3 \succ A_4 \succ A_1$
PFWGBM	$p = q = 1$	$A_2 \succ A_5 \succ A_3 \succ A_4 \succ A_1$
PFWMSM	$k = 2$	$A_2 \succ A_5 \succ A_3 \succ A_4 \succ A_1$
PFWMM	$Q = (1,1,1,1)$	$A_2 \succ A_5 \succ A_3 \succ A_4 \succ A_1$
PFDWMM	$Q = (1,1,1,1)$	$A_2 \succ A_5 \succ A_3 \succ A_4 \succ A_1$

As we can see from Table 13, the ranking order of the ERP systems by the PFWG operator is slightly different with the other methods, but the best and worst ERP systems are A_2 and A_1, respectively. This verifies that the PFWMM and PFDWMM operators we developed are reasonable and valid for MCGDM problems with Pythagorean fuzzy information.

In what follows, the comparisons of proposed approaches and the other methods with regard to some characteristics are shown in Table 14. In light of Table 14, some conclusions are summarized as follows:

(1) The two methods developed by Ma and Xu [6] aggregate fuzzy information easily. The drawbacks of Ma and Xu's method are they assume that the input arguments are not correlated, that is, they fail to consider the relationships between the input arguments. Nevertheless, our developed operators can capture the correlations among all the input arguments,

and fuse fuzzy information more flexibly by the parameter vector. Furthermore, the PFWA and PFWG operators are a special case of PFWMM and PFDWMM operators, respectively, when the parameter vector $Q = (1,0,0,0)$. Therefore, our developed approaches are more general and flexible comparing with that proposed by Ma and Xu.

(2) The primary advantage of our proposed operators is that they can capture the relationships between the multi-input arguments, while the method proposed by Liang et al. [14] can only deal with a correlation between any two input arguments. In reality, interdependent relationships may exist between more than two input arguments. Apparently, the PFWGBM operator is unable to handle this situation because it only captures the relationship between any two arguments. Furthermore, we also find that the PFDMM operator can transform into the PFGBM operator when the parameter vector is set to $Q = (1,1,0,\cdots,0)$. Therefore, our approach can overcome the weakness of the PFWGBM operator because our operators can deal with any number of input arguments being interdependent.

(3) Compared with our developed operators, although the PFWMSM operator can also deal with relationships between multi-input arguments, our methods can provide a more flexible information aggregation process through setting different parameter vector Q. Similarly, we can obtain the PFMSM operator when the parameter vector is set to $Q = (\underbrace{1,1,\cdots,1}_{k},\underbrace{0,0,\cdots,0}_{n-k})$. Thus, our developed operators are more general.

In short, based on the above comparisons and discussion, we can conclude that there are advantages to the PFWMM operator and PFDWMM operator compared with the existing other operators, including (a) they can capture the relationships between the multi-input arguments; and (b) they are more robust and it is more convenient to fuse the Pythagorean fuzzy information by the parameter vector Q.

Table 14. A comparison of the different approaches.

Approaches	Captures Correlation of Two Criteria	Captures Correlation of Multiple Criteria	Makes Method Flexible by the Parameter Vector
PFWA	No	No	No
PFWG	No	No	No
SPFWA	No	No	No
SPFWG	No	No	No
PFWGBM	Yes	No	No
PFWMSM	Yes	Yes	No
PFWMM	Yes	Yes	Yes
PFDWMM	Yes	Yes	Yes

7. Conclusions

In recent years, a number of researchers have developed aggregation operators under various fuzzy environments and applied these to solve different decision-making problems. However, these aggregation operators have some drawbacks in actual applications, such as being unable to reflect the correlation of all input arguments. The MM operator has an apparent advantage in that it can deal with the relationships between all the input arguments according to the parameter vector Q. Motivated by the ideal characteristic of the MM operator, in this paper we extended the MM operator into the Pythagorean fuzzy environment to deal with MCGDM problems with relationships between any number of arguments. We proposed some aggregation operators, including the PFMM operator, PFWMM operator, PFDMM operator, and PFDWMM operator. Then, some desirable properties and special cases of the proposed operators were investigated and discussed in detail. In addition, we have used the PFWMM and PFDWMM operators to present two methods to solve MCGDM problems with PFNs. Finally, we gave an example to demonstrate the effectiveness and feasibility of the presented methods through comparing with other existing approaches.

In future research, it is necessary to verify the validity of the proposed methods by solving other MCGDM problems such as supplier selection, risk assessment, and environment evaluation. Moreover, based on the prominent characteristics of the MM operator, we shall extend the MM operator into other fuzzy contexts such as the interval-valued 2-tuple linguistic environment, the interval-valued Pythagorean fuzzy environment, and the triangle intuitionistic fuzzy context.

Author Contributions: Conceptualization and Methodology, J.Z. and Y.L.; Data Curation, J.Z.; Writing—Original Draft Preparation, J.Z.; Writing—Review& Editing, J.Z. and Y.L.; Visualization, J.Z.; Supervision, Y.L.

Funding: This research was funded by the National Natural Science Foundation of China (No. 71371156) and the Doctoral Innovation Fund Program of Southwest Jiaotong University (D-CX201729).

Conflicts of Interest: The authors declare no conflict of interest.

Abbreviations

All abbreviations used in this paper:

BM	Bonferroni mean
ERP	Enterprise resource planning
FS	Fuzzy set
IFS	Intuitionistic fuzzy set
MCDM	Multi-criteria decision-making
MCGDM	Multi-criteria group decision-making
MM	Muirhead mean
MSM	Maclaurin symmetric mean
PFDMM	Pythagorean fuzzy dual Muirhead mean
PFDWMM	Pythagorean fuzzy dual weighted Muirhead mean
PFMM	Pythagorean fuzzy Muirhead mean
PFN	Pythagorean fuzzy number
PFS	Pythagorean fuzzy set
PFWA	Pythagorean fuzzy weighted averaging
PFWG	Pythagorean fuzzy weighted geometric
PFWGBM	Pythagorean fuzzy weighted geometric Bonferroni mean
PFWMM	Pythagorean fuzzy weighted Muirhead mean
PFWMSM	Pythagorean fuzzy weighted Maclaurin symmetric mean
SPFWA	Symmetric Pythagorean fuzzy weighted averaging
SPFWG	Symmetric Pythagorean fuzzy weighted geometric

References

1. Zadeh, L.A. Fuzzy sets. *Inf. Control* **1965**, *8*, 338–353. [CrossRef]
2. Atanassov, K.T.; Rangasamy, P. Intuitionistic fuzzy sets. *Fuzzy Sets Syst.* **1986**, *20*, 87–96. [CrossRef]
3. Yager, R.R. Pythagorean fuzzy subsets. In Proceedings of the IFSA World Congress and NAFIPS Annual Meeting, Edmonton, AB, Canada, 24–28 June 2013; pp. 57–61.
4. Yager, R.R. Pythagorean membership grades in multicriteria decision making. *IEEE Trans. Fuzzy Syst.* **2014**, *22*, 958–965. [CrossRef]
5. Zhang, X.; Xu, Z. Extension of topsis to multiple criteria decision making with Pythagorean fuzzy sets. *Int. J. Intell. Syst.* **2015**, *29*, 1061–1078. [CrossRef]
6. Ma, Z.; Xu, Z. Symmetric Pythagorean fuzzy weighted geometric/averaging operators and their application in multicriteria decision-making problems. *Int. J. Intell. Syst.* **2016**, *31*, 1198–1219. [CrossRef]
7. Rahman, K.; Abdullah, S.; Ahmed, R.; Ullah, M. Pythagorean fuzzy Einstein weighted geometric aggregation operator and their application to multiple attribute group decision making. *J. Intell. Fuzzy Syst.* **2017**, *33*, 1–13. [CrossRef]
8. Garg, H. A new generalized Pythagorean fuzzy information aggregation using Einstein operations and its application to decision making. *Int. J. Intell. Syst.* **2016**, *31*, 886–920. [CrossRef]

9. Garg, H. Confidence levels based Pythagorean fuzzy aggregation operators and its application to decision-making process. *Comput. Math. Organ. Theory* **2017**, *23*, 1–26. [CrossRef]
10. Zeng, S.; Chen, J.; Li, X. A hybrid method for Pythagorean fuzzy multiple-criteria decision making. *Int. J. Inf. Technol. Decis. Mak.* **2016**, *15*, 403–422. [CrossRef]
11. Peng, X.; Yang, Y. Pythagorean fuzzy choquet integral based mabac method for multiple attribute group decision making. *Int. J. Intell. Syst.* **2016**, *31*, 989–1020. [CrossRef]
12. Wei, G.; Lu, M. Pythagorean fuzzy power aggregation operators in multiple attribute decision making. *Int. J. Intell. Syst.* **2018**, *33*, 169–186. [CrossRef]
13. Liang, D.; Zhang, Y.; Xu, Z.; Darko, A.P. Pythagorean fuzzy bonferroni mean aggregation operator and its accelerative calculating algorithm with the multithreading. *Int. J. Intell. Syst.* **2018**. [CrossRef]
14. Liang, D.; Xu, Z.; Darko, A.P. Projection model for fusing the information of Pythagorean fuzzy multicriteria group decision making based on geometric bonferroni mean. *Int. J. Intell. Syst.* **2017**, *32*, 966–987. [CrossRef]
15. Muirhead, R.F. Some methods applicable to identities and inequalities of symmetric algebraic functions of n letters. *Proc. Edinb. Math. Soc.* **1902**, *21*, 144–162. [CrossRef]
16. Bonferroni, C. Sulle medie multiple di potenze. *Bollettino dell'Unione Matematica Italiana* **1950**, *5*, 267–270.
17. Detemple, D.W.; Robertson, J.M. On generalized symmetric means of two varibles. *Angew. Chem.* **1979**, *47*, 4638–4660.
18. Qin, J.; Liu, X. 2-tuple linguistic Muirhead mean operators for multiple attribute group decision making and its application to supplier selection. *Kybernetes* **2016**, *45*, 2–29. [CrossRef]
19. Liu, P.; You, X. Interval neutrosophic Muirhead mean operators and their application in multiple attribute group decision making. *Int. J. Uncertain. Quantif.* **2017**, *7*. [CrossRef]
20. Liu, P.; Li, D. Some Muirhead mean operators for intuitionistic fuzzy numbers and their applications to group decision making. *PLoS ONE* **2017**, *12*, e0168767. [CrossRef] [PubMed]
21. Liu, P.; Li, Y.; Zhang, M.; Zhang, L.; Zhao, J. Multiple-attribute decision-making method based on hesitant fuzzy linguistic Muirhead mean aggregation operators. *Soft Comput.* **2018**, 1–12. [CrossRef]
22. Wang, J.; Zhang, R.; Zhu, X.; Xing, Y.; Buchmeister, B. Some hesitant fuzzy linguistic Muirhead means with their application to multi-attribute group decision making. *Complexity* **2018**. [CrossRef]
23. Liu, P.; Teng, F. Some Muirhead mean operators for probabilistic linguistic term sets and their applications to multiple attribute decision-making. *Appl. Soft Comput.* **2018**, *68*, 396–431. [CrossRef]
24. Liu, H.C.; Li, Z.; Zhang, J.Q.; You, X.Y. A large group decision making approach for dependence assessment in human reliability analysis. *Reliab. Eng. Syst. Saf.* **2018**, *176*, 135–144. [CrossRef]
25. Wei, G.; Lu, M. Pythagorean fuzzy Maclaurin symmetric mean operators in multiple attribute decision making. *Int. J. Intell. Syst.* **2017**. [CrossRef]

information

MDPI

Article

Multiple Attributes Group Decision-Making under Interval-Valued Dual Hesitant Fuzzy Unbalanced Linguistic Environment with Prioritized Attributes and Unknown Decision-Makers' Weights

Xiao-Wen Qi [1], Jun-Ling Zhang [2,*] and Chang-Yong Liang [3]

1 School of Business Administration, Zhejiang University of Finance & Economics, Hangzhou 310018, China; qixiaowen@zufe.edu.cn

2 School of Economics and Management, Zhejiang Normal University, Jinhua 321004, China

3 School of Management, Hefei University of Technology, Hefei 230009, China; cyliang@hfut.edu.cn

* Correspondence: zhangjunling@zjnu.cn or zhangjunling_dc@126.com; Tel.: +86-579-8229-8567

Received: 11 May 2018; Accepted: 12 June 2018; Published: 14 June 2018

Abstract: Aiming at a special type of ill-defined complicate multiple attributes group decision-making (MAGDM) problem, which exhibits hybrid complexity features of decision hesitancy, prioritized evaluative attributes, and unknown decision-makers' weights, we investigate an effective approach in this paper. To accommodate decision hesitancy, we employ a compound expression tool of interval-valued dual hesitant fuzzy unbalanced linguistic set (IVDHFUBLS) to help decision-makers elicit their assessments more comprehensively and completely. To exploit prioritization relations among evaluating attributes, we develop a prioritized weighted aggregation operator for IVDHFUBLS-based decision-making scenarios and then analyze its properties and special cases. To objectively derive unknown decision-makers' weighting vector, we next develop a hybrid model that simultaneously takes into account the overall accuracy measure of the individual decision matrix and maximizing deviation among all decision matrices. Furthermore, on the strength of the above methods, we construct an MAGDM approach and demonstrate its practicality and effectiveness using applied study on a green supplier selection problem.

Keywords: multiple attributes decision-making; group decision-making; unbalanced linguistic set; prioritized average operator; maximizing deviation model

1. Introduction

After decades of extension and exploitation research [1], multiple attributes decision-making (MADM) approaches have been widely applied to many practical problems in social and technical systems, such as supply chain management [2–5], business intelligence evaluation [6,7], emergency management [8–10], teaching evaluation [11], product design evaluation [12], energy management [13], and waste management [14], among others. Due to increasing complexity in socioeconomic scenarios, and limitedness and uncertainty in human cognition, a single decision-maker is quite often incompetent when confronted with complicated decision-making scenarios. Therefore, multiple attributes group decision-making (MAGDM) methodologies have been developed and deeply studied regarding the strength of fuzzy tools for preferences expression [15–23], such as fuzzy sets [24], intuitionistic fuzzy sets [25,26], hesitant fuzzy sets (HFS) [27,28], dual-hesitant fuzzy sets (DHFS) [29–32], etc. Especially, HFS and DHFS are capable of addressing the common phenomena of decision hesitancy, that is, decision-makers are often irresolute about possible membership degrees to a fuzzy set [27,28]; comparatively, DHFS manages to reflect decision hesitancy more completely than HFS by accommodating both membership degrees and non-membership degrees when depicting decision hesitancy [29].

Regarding ill-structured decision-making problems with higher complexity that cannot be quantified by the above-mentioned fuzzy tools, effective MAGDM approaches have also been put forward by employing the linguistic term set [33] to qualitatively express decision-makers' opinions directly [34,35]. However, decision-makers quite often approximate the most-preferred linguistic label in a certain linguistic term set but still hesitate about possible membership degrees or non-membership degrees with regard to that linguistic label [36]. Consequently, by fusing the merits of both the linguistic term set and the hesitant fuzzy set, more effective and comprehensive expression tools have been further introduced and exploited to construct multiple attributes decision-making approaches, such as hesitant fuzzy linguistic set [37,38], interval-valued hesitant fuzzy linguistic set [36,39], dual-hesitant fuzzy linguistic set [40], interval-valued dual hesitant fuzzy linguistic set [41], etc. As can be seen, nearly all the above approaches drew on the presumption that linguistic labels must be distributed in a symmetrical and balanced manner [34,35]. However, practical investigations [42,43] have indicated that decision-makers preferred non-uniform or asymmetric linguistic term sets, i.e., the unbalanced linguistic term set (ULTS) [44], to express their complicate assessments more precisely and objectively. Most recent studies [45,46] also verified that ULTS attains better adaptability and flexibility. So, coherently, to tackle complex decision-making more effectively under ill-structured scenarios with decision hesitancy, there are actual needs to develop hybrid hesitant fuzzy linguistic expression tools that are capable of inheriting advantages of both ULTS and hesitant fuzzy sets. However, thus far, to the best of our knowledge, only Qi et al. [47] developed the interval-valued dual hesitant fuzzy unbalanced linguistic set (IVDHFUBLS) and its power aggregation operators. Although IVDHFUBLS manages to be more effective and flexible in depicting complicated assessments with interval values for both membership degrees and non-membership degrees to a designated unbalanced linguistic label, their MAGDM approaches were only developed to cope with the special type of decision-making problems with mutually supportive assessments in decision matrices. Obviously, it is still substantively necessary to investigate various hesitant fuzzy unbalanced linguistic expression tools and exploit their derivative multiple attributes decision-making approaches to resolve practical complex problems.

In fact, in determination of appropriate weights for attributes in MAGDM, decision-makers are generally required to reciprocally compare evaluating attributes so that AHP-like method can be used to derive attributes weights [48]. However, quite often, due to limited expertise on ill-structured problems, decision-makers need many iterations to achieve acceptable consistency or are even unwilling to fulfill the reciprocal comparisons, while on the contrary, for difficulties of high uncertainty, Delphi-like analytical processes provide decision-makers with ways of utilizing collective knowledge to approximate fairly accurate prioritization relations among evaluating attributes [49,50]. For instance, considering four indicators to select emergency response plans for chemical spills events: response efficiency (A_1), environmental impact (A_2), social impact (A_3) and cost (A_4). If the event location L_1 was in districts with scarce any residence but freeways, decision-makers would naturally deduce the prioritization relation among the indicators as $A_1 \succ A_4 \succ A_3 \succ A_2$, while if L_1 was nearby a residence district, decision-makers would derive different prioritization as $A_1 \succ A_3 \succ A_2 \succ A_4$. In viewing of the common existence of prioritization relations among assessing attributes in multiple attributes decision-making, ref. [49,51] introduced the prioritized average (PA) operator and the prioritized ordered weighted average (POWA) operator, which provide effective ways with which to consider decision information from both assessments under attributes and prioritization relation among the attributes. Since then, prioritized operators have been extended to complicated decision environments of high uncertainty, such as prioritized operators for decision-making under intuitionistic fuzzy environments [52–55], multi-granular uncertain linguistic environments [56], hesitant fuzzy environments [50], dual hesitant fuzzy environments [57], and hesitant fuzzy linguistic environments [58]. Nevertheless, there is still a lack of investigation on prioritized operators in hesitant fuzzy unbalanced linguistic environments. Therefore, aiming at resolving these types of practical multiple attributes decision-making problems with prioritization relation among evaluating attributes, on the basis of IVDHFUBLS [47], we focus on studying prioritized average operators for IVDHFUBLS and corresponding effective MAGDM approaches.

To do so, in this paper, we first propose a fundamental prioritized aggregation operator for fusing preferences in the form of IVDHFUBLS and simultaneously considering prioritization relation among evaluative attributes, i.e., the interval-valued dual hesitant fuzzy unbalanced linguistic prioritized weighted aggregation (IVDHFUBLPWA) operator. We then investigate its desirable properties and discuss its special cases. Further, to objectively determine decision-makers' weights, which cannot be obtained up-front in complex problem scenarios, we develop a hybrid model that takes into account overall accuracy measure of the individual decision matrix and the maximizing deviation among all decision matrices. Subsequently, on the strength of the above-developed aggregation operator and decision-makers' weighting model, an effective approach is constructed to tackle practical MAGDM problems that take features of decision-makers' decision hesitancy, prioritization relationships among evaluative attributes, and unknown decision-makers' weights.

The remainder of this paper unfolds as follows. Section 2 presents a literature review to discuss the limitations of existing approaches, thereby showing the motivation of this paper. In Section 3, necessary preliminaries for the interval-valued dual hesitant fuzzy unbalanced linguistic set (IVDHFUBLS) are detailed. In Section 4, we firstly define the interval-valued dual hesitant fuzzy unbalanced linguistic prioritized weighted aggregation (IVDHFUBLPWA) operator and discuss its properties, as well as special cases; next, the hybrid model is developed for determining unknown weights for decision-makers; then, an effective MAGDM approach based on the above methods is constructed in detail. In Section 5, an illustrative example of the green supplier selection problem is given to demonstrate the effectiveness and practicality of our proposed approach. Finally, conclusions and future research directions are given in Section 6.

2. Literature Review on Hesitant Fuzzy Linguistic MADM Approaches

With support of fuzzy set and its extensions, classic MADM methodologies have been successfully extended and enhanced to accommodate complicated decision-making environments in which decision-makers have imprecise, uncertain, or vague assessments [15], while for those decision scenarios of ill-structured definition, fuzzy expression tools cannot directly apply. Zadeh [59] thus suggested employing linguistic variables to facilitate expression of judgments. However, no matter whether assigning membership degrees to given fuzzy set or utilizing linguistic labels to depict decision-makers' complicate judgments, there is a common phenomenon that decision-makers quite often hesitate among possible values [27,28]. Torra and Narukawa [27] and Torra [28] thus use hesitant fuzzy set (HFS) to describe the decision hesitancy. In viewing of same importance of membership degrees and non-membership degrees in depicturing decision hesitancy, Zhu et al. [29] further extended HFS to dual hesitant fuzzy set (DHFS).

Regarding linguistic decision-making scenarios, Rodríguez et al. [60] introduced the hesitant fuzzy linguistic term sets (HFLTSs) to allow decision-makers to directly express their uncertain opinions with possible linguistic labels, based on which the authors then developed a group decision-making model through comparative linguistic expressions [61]. Using HFLTSs, Beg and Rashid [62] endowed conventional TOPSIS with the ability to deal with decision hesitancy. From another perspective, when using linguistic variables to denote their judgments, decision-makers commonly are capable of efficiently determining the most approximate linguistic term while having decision hesitancy with regard to the one selected. Therefore, Lin et al. [37] proposed the effective compound expression tool of hesitant fuzzy linguistic set (HFLS) that employs hesitant fuzzy set to describe decision hesitancy with regard to the selected linguistic label. Wang et al. [36] then introduced the interval-valued hesitant fuzzy linguistic set (IVHFLS) to help decision-makers express their decision hesitancy with possible interval values, on the basis of which they developed a single-person MADM approach. Be aware that above hybrid linguistic expression tools only took into account possible membership degrees but neglected same importance of non-membership degrees; Yang and Ju [40] introduced the dual hesitant fuzzy linguistic set (DHFLS) by incorporating both possible membership degrees and possible non-membership degrees. However, in Yang and Ju [40], they also yet investigated a single-person MADM approach. Qi et al. [41] took a step further to study the interval-valued dual hesitant fuzzy linguistic set (IVDHFLS) and constructed a multiple attributes group decision-making approach based on a family of generalized power aggregation operators. Recently, researchers have started to extend classic MADM methodologies by utilizing the above compound hesitant fuzzy linguistic tools; for example, Wang et al. [63] developed a MADM approach based on TOPSIS and TODIM methods in which attribute values take the form of hesitant fuzzy linguistic numbers. As can be seen, recent studies have verified effectiveness of the compound hesitant fuzzy linguistic expression tools at eliciting complicate uncertain assessments under ill-structured decision situations.

Unfortunately, all the formerly discussed hesitant fuzzy linguistic decision-making models presumed that linguistic labels must be distributed in a symmetrical and balanced manner [34,35]. However, Herrera-Viedma and López-Herrera [42] revealed from their studies on information retrieval system that users (decision-makers) preferred more labels on the right side of a non-uniform or asymmetric linguistic scale, which further was verified by the experiments of olive sensory evaluation in Martínez et al. [43]. Herrera et al. [44] thus defined this special type of linguistic variable as the unbalanced linguistic term set (ULTS) and studied its operations. To a deeper extent, Meng and Pei [45] developed some weighted, unbalanced linguistic aggregation operators and applied them to a multiple attributes group decision-making problem; Dong, et al. [46] investigated group decision-making based on unbalanced linguistic preference relations and proposed a consistency reaching method. Generally speaking, ULTS attains better adaptability and flexibility than strictly symmetrical or balanced linguistic term set. However, till now, regarding MADM based on compound hesitant fuzzy linguistic expressions, only Qi et al. [47] introduced the expression tool of interval-valued dual hesitant fuzzy unbalanced linguistic set (IVDHFUBLS) and investigated its power aggregation operators. Despite merits of ULTS into IVDHFUBLS, the group decision-making approach in Qi et al. [47] only applies to decision scenarios with mutually supportive assessments in decision matrices; nevertheless, the weighting methods for both attributes and decision-makers were derived from and specific to the mutually supportive relations. For more clarity, representative hesitant fuzzy linguistic MADM methods from above discussion and their properties have been compared in Table 1.

Table 1. Representative hesitant fuzzy linguistic MADM methods and their properties.

Authors	Methodology Properties					
	Linguistic Variable		Description of Hesitancy		Prioritized Attributes	Unknown Decision-Makers' Weights
	Balanced	Unbalanced	Hesitant Fuzzy Set	Dual Hesitant Fuzzy Set		
Lin, et al. [37]	√	×	√	×	×	Single-person MADM
Wang, et al. [36]	√	×	√	×	√	Single-person MADM
Yang and Ju [40]	√	×	×	√	√	Single-person MADM
Qi, et al. [41]	√	×	×	√	×	√
Wang, et al. [63]	√	×	√	×	×	×
Qi, et al. [47]	√	√	√	√	×	Power aggregation-based method
This paper	√	√	√	√	√	Deviation-maximizing method

Furthermore, to determine unknown attributes' weights in complex decision problems, analytical hierarchy process (AHP) generally exhibits an effective way with which to obtain relative importance among attributes [64]. However, AHP method requires precisely consistent judgments for reciprocal comparisons to proceed, which quite often cannot be guaranteed for complex problems and thus result in multiple rounds of adjustments or even failure in decision-making, that is, lack of efficiency to some extent, while, in fact, decision-makers are generally capable of obtaining rather accurate prioritization relations among evaluative attributes thanks to their group intelligence and expertise [49].

For example, suppose we choose the following eight attributes to evaluate alterative response solutions to specific emergency event: response time to start emergency response solution (C_1) [9,48], reasonable organizational structure and clear awareness of responsibilities (C_2) [9,48], economic cost (C_3) [65], operability of the response solution (C_4) [9,66], monitoring and forecasting potential hazards (C_5) [9,48], reconstruction ability (C_6) [9,48], social impact (C_7) [67], and environmental impact (C_8) [68,69]. Additionally, features of the target emergency event have already been identified as follows: (i) Located on an intersection of two highways in a sandstorm desert area where no residences are nearby; (ii) Truck drivers were injured; (iii) A large amount of highly corrosive fluid materials in both trucks are leaking; (iv) Accident trucks destroyed a critical sand control dam. Then, decision-makers will efficiently arrive at the prioritization relations among the eight evaluative attributes $(C_1) \succ (C_4) \succ (C_2) \succ (C_5) \succ (C_6) \succ (C_3) \succ (C_7) \succ (C_8)$. Apparently, there is a practical need to further investigate IVDHFUBLS-based decision-making approaches that exploit the prioritization relations. Therefore, based on the prioritized average (PA) operator [49], in the following, we firstly develop PA operator for IVDHFUBLS to address prioritization relations among attributes. To gain more generality of decision-makers' weighting method rather than the problem-specific limitedness of the one devised in Qi et al. [47], we then develop a deviation-maximizing method to objectively derive decision-makers' weights. Finally, on the strength of these methods, we manage to construct a practical and effective IVDHFUBLS-based multiple attributes group decision-making approach.

3. Preliminaries for IVDHFUBLS

By fusing the merits of both unbalanced linguistic term set [44] and interval-valued dual hesitant fuzzy set (IVDHFS) [32], most recently, Qi et al. [47] introduced the effective hybrid expression tool called interval-valued dual hesitant fuzzy unbalanced linguistic set (IVDHFUBLS), as shown in following definition.

Definition 1 [47]. *Let X be a fixed set and S be a finite and continuous linguistic label set; then, an interval-valued dual hesitant fuzzy unbalanced linguistic set (IVDHFUBLS) SD on X is defined as*

$$SD = \left\{ \left\langle x, s_i, \tilde{h}(x), \tilde{g}(x) \right\rangle | x \in X \right\}, \tag{1}$$

in which s_i is an unbalanced linguistic variable from predefined unbalanced linguistic label set S, which represents decision-makers' judgments of an evaluated object x; $\tilde{h}(x) = \cup_{[\mu^L,\mu^U] \in \tilde{h}(x)} \{\tilde{\mu}\} = \cup_{[\mu^L,\mu^U] \in \tilde{h}(x)} \{[\mu^L, \mu^U]\}$ is a set of closed interval values in [0, 1], denoting possible membership degrees to which x belongs to s_i; $\tilde{g}(x) = \cup_{[\nu^L,\nu^U] \in \tilde{g}(x)} \{\tilde{\nu}\} = \cup_{[\nu^L,\nu^U] \in \tilde{g}(x)} \{[\nu^L, \nu^U]\}$ is a set of closed interval values in [0, 1], denoting possible non-membership degrees to which x belongs to s_i. In $\tilde{h}(x)$ and $\tilde{g}(x)$, $\tilde{\mu}, \tilde{\nu} \in [0,1]$ and $0 \le (\mu^U)^+ + (\nu^U)^+ \le 1$, in which $(\mu^U)^+ \in \tilde{h}^+(x) = \cup_{[\mu^L,\mu^U] \in \tilde{h}(x)} \max\{\mu^U\}$ and $(\nu^U)^+ \in \tilde{g}^+(x) = \cup_{[\nu^L,\nu^U] \in \tilde{g}(x)} \max\{\nu^U\}$ for all $x \in X$.

Generally, $sd = \left(s_i, \tilde{h}, \tilde{g}\right)$ is called an interval-valued dual hesitant fuzzy unbalanced linguistic number (IVDHFUBLN) and IVDHFUBLNs are all elements of IVDHFUBLS.

Definition 2 [47]. *Let $sd = \left(s_k, \tilde{h}, \tilde{g}\right)$, $sd_1 = \left(s_i, \tilde{h}_1, \tilde{g}_1\right)$, and $sd_2 = \left(s_j, \tilde{h}_2, \tilde{g}_2\right)$ be any three IVDHFUBLNs, $\lambda \in [0,1]$; some operations on these IVDHFUBLNs are defined by*

$$(1) \quad \lambda sd = \cup_{(s_k,\tilde{h},\tilde{g}) \in sd} \left(s_{\lambda \Delta_{t_0}^{-1}(TF_{t_0}^{t_k}(\psi(s_k)))}, \cup_{[\mu^L,\mu^U] \in \tilde{h}, [\nu^L,\nu^U] \in \tilde{g}} \left\{ \left\{ [1-(1-\mu^L)^\lambda, 1-(1-\mu^U)^\lambda] \right\}, \left\{ [(\nu^L)^\lambda, (\nu^U)^\lambda] \right\} \right\} \right);$$

$$(2) \quad sd^\lambda = \cup_{(s_k,\tilde{h},\tilde{g}) \in sd} \left(s_{(\Delta_{t_0}^{-1}(TF_{t_0}^{t_k}(\psi(s_k))))^\lambda}, \cup_{[\mu^L,\mu^U] \in \tilde{h}, [\nu^L,\nu^U] \in \tilde{g}} \left\{ \left\{ [(\mu^L)^\lambda, (\mu^U)^\lambda] \right\}, \left\{ [1-(1-\nu^L)^\lambda, 1-(1-\nu^U)^\lambda] \right\} \right\} \right);$$

(3) $sd_1 \oplus sd_2 = \cup_{(s_i,\widetilde{h}_1,\widetilde{g}_1)\in sd_1,(s_j,\widetilde{h}_2,\widetilde{g}_2)\in sd_2}\left(s_{\Delta_{t_0}^{-1}(TF_{t_0}^{t_i}(\psi(s_i)))+\Delta_{t_0}^{-1}(TF_{t_0}^{t_j}(\psi(s_j)))},\right.$
$\left.\cup_{[\mu_1^L,\mu_1^U]\in\widetilde{h}_1,[\mu_2^L,\mu_2^U]\in\widetilde{h}_2,[v_1^L,v_1^U]\in\widetilde{g}_1,[v_2^L,v_2^U]\in\widetilde{g}_2}\{\{[\mu_1^L+\mu_2^L-\mu_1^L\mu_2^L,\mu_1^U+\mu_2^U-\mu_1^U\mu_2^U]\},\{[v_1^Lv_2^L,v_1^Uv_2^U]\}\}\right);$

(4) $sd_1 \otimes sd_2 = \cup_{(s_i,\widetilde{h}_1,\widetilde{g}_1)\in sd_1,(s_j,\widetilde{h}_2,\widetilde{g}_2)\in sd_2}\left(s_{\Delta_{t_0}^{-1}(TF_{t_0}^{t_i}(\psi(s_i)))\times\Delta_{t_0}^{-1}(TF_{t_0}^{t_j}(\psi(s_j)))},\right.$
$\left.\cup_{[\mu_1^L,\mu_1^U]\in\widetilde{h}_1,[\mu_2^L,\mu_2^U]\in\widetilde{h}_2,[v_1^L,v_1^U]\in\widetilde{g}_1,[v_2^L,v_2^U]\in\widetilde{g}_2}\{\{[\mu_1^L\mu_2^L,\mu_1^U\mu_2^U]\},\{[v_1^L+v_2^L-v_1^Lv_2^L,v_1^U+v_2^U-v_1^Uv_2^U]\}\}\right).$

In Definition 2, t_k, t_i, t_j are the corresponding levels of unbalanced linguistic terms s_k, s_i, s_j in the linguistic hierarchy (LH) [44], respectively; t_0 is the maximum level of s_k, s_i, s_j in LH. Using the transformation function defined in following Definition 3, any 2-tuple linguistic representation format can be transformed into a term in LH.

Definition 3 [44]. *In linguistic hierarchies $LH = \cup_t l(t,n(t))$, whose linguistic term sets are represented by $S^{n(t)} = \left\{s_0^{n(t)},\ldots,s_{n(t)-1}^{n(t)}\right\}$, the transformation function from a linguistic label in level t to a label in consecutive level t′ is defined as $TF_{t'}^t$: $l(t,n(t)) \to l(t',n(t'))$, such that*

$$TF_{t'}^t\left(s_i^{n(t)},\alpha^{n(t)}\right) = \Delta_{t'}\left(\frac{\Delta_t^{-1}\left(s_i^{n(t)},\alpha^{n(t)}\right)(n(t')-1)}{n(t)-1}\right). \quad (2)$$

In order to compare any two IVDHFUBLNs, following Definition 4 introduces comparison rules based on a score function and accuracy function. Taking a step further, Definition 5 defines a fundamental distance measure to calculate separation degree between any two IVDHFUBLNs.

Definition 4 [47]. *Let $sd = \left(s_i,\widetilde{h},\widetilde{g}\right)$ be an IVDHFUBLN, and then a score function $S(sd)$ can be denoted as*

$$S(sd) = \Delta_{t_0}^{-1}\left(TF_{t_0}^{t_i}(\psi(s_i))\right)\times\frac{1}{2}\left(\frac{1}{l(\widetilde{h})}\sum_{[\mu^L,\mu^U]\in\widetilde{h}}\mu^L-\frac{1}{l(\widetilde{g})}\sum_{[v^L,v^U]\in\widetilde{g}}v^L+\frac{1}{l(\widetilde{h})}\sum_{[\mu^L,\mu^U]\in\widetilde{h}}\mu^U-\frac{1}{l(\widetilde{g})}\sum_{[v^L,v^U]\in\widetilde{g}}v^U\right), \quad (3)$$

and an accuracy function $P(sd)$ can be denoted as

$$P(sd) = \Delta_{t_0}^{-1}\left(TF_{t_0}^{t_i}(\psi(s_i))\right)\times\frac{1}{2}\left(\frac{1}{l(\widetilde{h})}\sum_{[\mu^L,\mu^U]\in\widetilde{h}}\mu^L+\frac{1}{l(\widetilde{g})}\sum_{[v^L,v^U]\in\widetilde{g}}v^L+\frac{1}{l(\widetilde{h})}\sum_{[\mu^L,\mu^U]\in\widetilde{h}}\mu^U+\frac{1}{l(\widetilde{g})}\sum_{[v^L,v^U]\in\widetilde{g}}v^U\right). \quad (4)$$

Here, $l(\widetilde{h})$ and $l(\widetilde{g})$ are numbers of interval values in $\widetilde{h}$ and $\widetilde{g}$, respectively, and t_i is the corresponding level of unbalanced linguistic term s_i in the LH; t_0 is the maximum level of t_i in LH. Subsequently, given any two $sd_1 = \left(s_i,\widetilde{h}_1,\widetilde{g}_1\right)$ and $sd_2 = \left(s_j,\widetilde{h}_2,\widetilde{g}_2\right)$, based on $S(sd)$ and $P(sd)$, we have following comparison rules:

(1) If $S(sd_1) < S(sd_2)$, then $sd_1 < sd_2$.
(2) If $S(sd_1) = S(sd_2)$, then

  (a) If $P(sd_1) = P(sd_2)$, then $sd_1 = sd_2$;
  (b) If $P(sd_1) < P(sd_2)$, then $sd_1 < sd_2$.

Definition 5 [47]. *Let two IVDHFUBLNs $sd_1 = \left(s_i,\widetilde{h}_1,\widetilde{g}_1\right)$ and $sd_2 = \left(s_j,\widetilde{h}_2,\widetilde{g}_2\right)$, $l_{\widetilde{h}_1}$, $l_{\widetilde{h}_2}$, $l_{\widetilde{g}_1}$, and $l_{\widetilde{g}_2}$ are the lengths of $\widetilde{h}_1$, $\widetilde{h}_2$, $\widetilde{g}_1$, and $\widetilde{g}_2$, respectively, which represent number of elements in the sets of $\widetilde{h}_1$, $\widetilde{h}_2$, $\widetilde{g}_1$, and $\widetilde{g}_2$. Suppose $I_1 = \frac{1}{n(t_i)-1}\Delta_{t_0}^{-1}\left(TF_{t_0}^{t_i}(\psi(s_i))\right)$, $I_2 = \frac{1}{n(t_j)-1}\Delta_{t_0}^{-1}\left(TF_{t_0}^{t_j}(\psi(s_j))\right)$, in whiche t_i and t_j are* *the corresponding levels of unbalanced linguistic terms s_i and s_j in the linguistic hierarchy LH, and t_0 is the maximum level of s_i and s_j in LH. Then, a distance measure d based on the normalized Euclidean distance can be defined as follows:*

Situation 1. When $l_{\tilde{h}_1} = l_{\tilde{h}_2} = l_1$ and $l_{\tilde{g}_1} = l_{\tilde{g}_2} = l_2$, then $d(sd_1, sd_2) =$

$$\left(\frac{1}{2}\left(\frac{1}{l_1}\sum_{k=1}^{l_1}\left(\left|I_1\mu_{\tilde{h}_1}^{L_j} - I_2\mu_{\tilde{h}_2}^{L_k}\right|^2 + \left|I_1\mu_{\tilde{h}_1}^{U_j} - I_2\mu_{\tilde{h}_2}^{U_k}\right|^2\right) + \frac{1}{l_2}\sum_{k=1}^{l_2}\left(\left|I_1\nu_{\tilde{g}_1}^{L_j} - I_2\nu_{\tilde{g}_2}^{L_k}\right|^2 + \left|I_1\nu_{\tilde{g}_1}^{U_j} - I_2\nu_{\tilde{g}_2}^{U_k}\right|^2\right)\right)\right)^{\frac{1}{2}}. \quad (5)$$

Situation 2. When $l_{\tilde{h}_1} \neq l_{\tilde{h}_2}$ or $l_{\tilde{g}_1} \neq l_{\tilde{g}_2}$, then $d(sd_1, sd_2) =$

$$\left(\frac{1}{2}\left(\frac{1}{l_{\tilde{h}_1} l_{\tilde{h}_2}}\sum_{j=1}^{l_{\tilde{h}_1}}\sum_{k=1}^{l_{\tilde{h}_2}}\left(\left|I_1\mu_{\tilde{h}_1}^{L_j} - I_2\mu_{\tilde{h}_2}^{L_k}\right|^2 + \left|I_1\mu_{\tilde{h}_1}^{U_j} - I_2\mu_{\tilde{h}_2}^{U_k}\right|^2\right) + \frac{1}{l_{\tilde{g}_1} l_{\tilde{g}_2}}\sum_{j=1}^{l_{\tilde{g}_1}}\sum_{k=1}^{l_{\tilde{g}_2}}\left(\left|I_1\nu_{\tilde{g}_1}^{L_j} - I_2\nu_{\tilde{g}_2}^{L_k}\right|^2 + \left|I_1\nu_{\tilde{g}_1}^{U_j} - I_2\nu_{\tilde{g}_2}^{U_k}\right|^2\right)\right)\right)^{\frac{1}{2}}. \quad (6)$$

Example 1. *Suppose we utilize an unbalanced linguistic term set S_0 for evaluation, in which $S_0 = \{N, L, M, AH, H, QH, VH, AT, T\}$. Figure 1 demonstrates S_0 and its mapping in a linguistic hierarchy. Then, we got two IVDHFUBLNs sd_1 and sd_2. Let sd_1 = (L,{[0.2,0.3],[0.4,0.5],[0.5,0.6]},{[0.1,0.2],[0.3,0.4]}) and sd_2 = (VH,{[0.1,0.4],[0.5,0.6]},{[0.2,0.3]}).*

Then, by use of Definition 5, we can directly calculate the distance between sd_1 and sd_2 without adding any elements into sd_2, and we get $d(sd_1, sd_2) = 0.4771$.

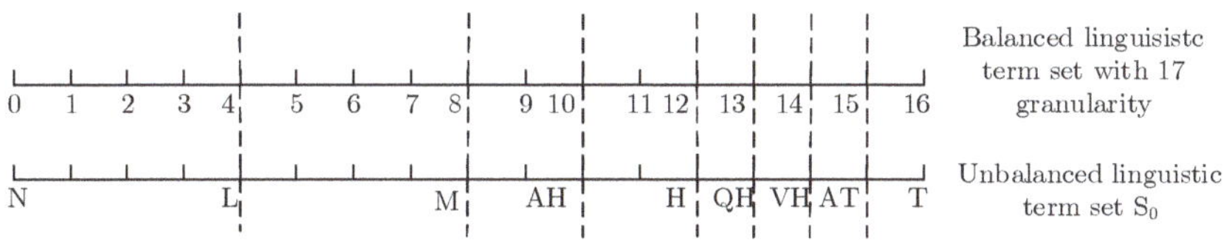

Figure 1. Unbalanced linguistic term set S_0 and its mapping in linguistic hierarchy.

4. Proposed Approach for MAGDM Based on IVDHFUBLS

When confronted with complicate practical MAGDM scenarios, decision-makers usually are inadequate in determining exact weighting information for evaluative attributes due to time limit or lack of domain knowledge, while they are capable of deriving relatively exact prioritization relation among evaluative attributes, such as the talent introduction decision-making problem [70], teaching quality evaluation problem [54], software selection problem [71], etc. In actuality, Yager [49] acutely noticed the real-world prioritization phenomena among assessing criteria and thus developed the prioritized average (PA) operator. PA operator has been verified as a fundamentally effective aggregation operator that enables classic multiple criteria decision-making methodologies to include prioritization relations among indicators as decision information in their mechanisms [54,70–72].

Therefore, firstly in this section, we develop the fundamental prioritized average aggregation operator for IVDHFUBLS and study its desirable properties. Next, considering that decision-makers' weighting information also quite often cannot be subjectively obtained in advance under complex decision-making environments, we develop a programming model based on deviation maximizing method to objectively derive weighting vector for decision-makers. Furthermore, based on the developed prioritized average aggregation operator and the programming model, we propose an algorithm for MAGDM under IVDHFUBLS environment in which prioritization relation among evaluating attributes exists and decision-makers' weighting vector is unknown.

4.1. Prioritized Average Aggregation Operator for IVDHFUBLS

Definition 6. *For a collection of IVDHFUBLNs* $sd_j (j = 1, 2, \ldots, n)$*, which are prioritized such that* $sd_j \prec sd_{j-1}$*, the IVDHFUBLPWA operator is defined as follows:*

$$IVDHFUBLPWA(sd_1, sd_2, \ldots, sd_n) = \frac{T_1}{\sum_{j=1}^n T_j} sd_1 \oplus \frac{T_2}{\sum_{j=1}^n T_j} sd_2 \oplus \ldots \oplus \frac{T_n}{\sum_{j=1}^n T_j} sd_n \tag{7}$$

$$= \bigoplus_{j=1}^{n} \left(\frac{T_j sd_j}{\sum_{j=1}^n T_j} \right) \tag{8}$$

in which $T_1 = 1$, $T_j = \prod_{k=1}^{j-1} P(sd_k) = P(sd_{j-1}) T_{j-1}$ *and* $P(sd_k)$ *is the accuracy value of* sd_k *calculated by Definition 4.*

IVDHFUBLPWA operator also can be rewritten as following Theorem 1.

Theorem 1. *Let* $sd_j = \left(s_j, \widetilde{h}_j, \widetilde{g}_j\right)$ *be a collection of IVDHFUBLNs. By noticing that aggregation results obtained from Definition 6 have been transformed to the form of interval-valued dual hesitant fuzzy balanced linguistic numbers (IVDHFBLNs), we have*

$$\begin{aligned} IVDHFUBLPWA(sd_1, sd_2, \ldots, sd_n) = \cup_{(s_j, \widetilde{h}_j, \widetilde{g}_j) \in sd_j} \Bigg(s_{\sum_{j=1}^{n} \frac{T(sd_j)}{\sum_{i=1}^n T(sd_i)} \Delta_{t_0}^{-1}(TF_{t_0}^{t_j}(\psi(s_j)))}, \\ \cup_{[\mu_j^L, \mu_j^U] \in \widetilde{h}_j, [\nu_j^L, \nu_j^U] \in \widetilde{g}_j} \left\{ \left\{ \left[1 - \prod_{j=1}^{n} (1 - \mu_j^L)^{\frac{T(sd_j)}{\sum_{i=1}^n T(sd_i)}}, 1 - \prod_{j=1}^{n} (1 - \mu_j^U)^{\frac{T(sd_j)}{\sum_{i=1}^n T(sd_i)}} \right] \right\}, \right. \\ \left. \left\{ \left[\prod_{j=1}^{n} (\nu_j^L)^{\frac{T(sd_j)}{\sum_{i=1}^n T(sd_i)}}, \prod_{j=1}^{n} (\nu_j^U)^{\frac{T(sd_j)}{\sum_{i=1}^n T(sd_i)}} \right] \right\} \right\} \Bigg). \end{aligned} \tag{9}$$

Proof.

(1) When $n = 1$, obviously, it is right.

$$IVDHFUBLPWA(sd) = \cup_{(s_0, \widetilde{h}, \widetilde{g}) \in sd} \left(s_{\Delta_{t_0}^{-1}(TF_{t_0}^{t_0}(\psi(s_0)))}, \cup_{[\mu^L, \mu^U] \in \widetilde{h}, [\nu^L, \nu^U] \in \widetilde{g}} \left\{ \left\{ [\mu^L, \mu^U] \right\}, \left\{ [\nu^L, \nu^U] \right\} \right\} \right);$$

(2) When $n = 2$, $\frac{T(sd_1)}{\sum_{i=1}^2 T(sd_i)} sd_1 = \cup_{(s_1, \widetilde{h}, \widetilde{g}) \in sd_1} \Bigg(s_{\frac{T(sd_1)}{\sum_{i=1}^2 T(sd_i)} (\Delta_{t_0}^{-1}(TF_{t_0}^{t_1}(\psi(s_1))))},$

$$\cup_{[\mu_1^L, \mu_1^U] \in \widetilde{h}_1, [\nu_1^L, \nu_1^U] \in \widetilde{g}_1} \left\{ \left\{ \left[1 - (1 - \mu_1^L)^{\frac{T(sd_1)}{\sum_{i=1}^2 T(sd_i)}}, 1 - (1 - \mu_1^U)^{\frac{T(sd_1)}{\sum_{i=1}^2 T(sd_i)}} \right] \right\}, \left\{ \left[(\nu_1^L)^{\frac{T(sd_1)}{\sum_{i=1}^2 T(sd_i)}}, (\nu_1^U)^{\frac{T(sd_1)}{\sum_{i=1}^2 T(sd_i)}} \right] \right\} \right\} \Bigg),$$

$$\frac{T(sd_2)}{\sum_{i=1}^2 T(sd_i)} sd_2 = \cup_{(s_2, \widetilde{h}, \widetilde{g}) \in sd_2} \Bigg(s_{\frac{T(sd_2)}{\sum_{i=1}^2 T(sd_i)} (\Delta_{t_0}^{-1}(TF_{t_0}^{t_2}(\psi(s_2))))},$$

$$\cup_{[\mu_2^L, \mu_2^U] \in \widetilde{h}_2, [\nu_2^L, \nu_2^U] \in \widetilde{g}_2} \left\{ \left\{ \left[1 - (1 - \mu_2^L)^{\frac{T(sd_2)}{\sum_{i=1}^2 T(sd_i)}}, 1 - (1 - \mu_2^U)^{\frac{T(sd_2)}{\sum_{i=1}^2 T(sd_i)}} \right] \right\}, \left\{ \left[(\nu_2^L)^{\frac{T(sd_2)}{\sum_{i=1}^2 T(sd_i)}}, (\nu_2^U)^{\frac{T(sd_2)}{\sum_{i=1}^2 T(sd_i)}} \right] \right\} \right\} \Bigg),$$

$$\frac{T(sd_1)}{\sum_{i=1}^2 T(sd_i)} sd_1 + \frac{T(sd_2)}{\sum_{i=1}^2 T(sd_i)} sd_2 =$$

$$\begin{aligned} \cup_{(s_1, \widetilde{h}_1, \widetilde{g}_1) \in sd_1, (s_2, \widetilde{h}_2, \widetilde{g}_2) \in sd_2} \Bigg(s_{\sum_{j=1}^{2} \frac{T(sd_j)}{\sum_{i=1}^2 T(sd_i)} (\Delta_{t_0}^{-1}(TF_{t_0}^{t_j}(\psi(s_j))))}, \cup_{[\mu_1^L, \mu_1^U] \in \widetilde{h}_1, [\mu_2^L, \mu_2^U] \in \widetilde{h}_2, [\nu_1^L, \nu_1^U] \in \widetilde{g}_1, [\nu_2^L, \nu_2^U] \in \widetilde{g}_2} \\ \left\{ \left\{ \left[1 - (1 - \mu_1^L)^{\frac{T(sd_1)}{\sum_{i=1}^2 T(sd_i)}} (1 - \mu_2^L)^{\frac{T(sd_2)}{\sum_{i=1}^2 T(sd_i)}}, 1 - (1 - \mu_1^U)^{\frac{T(sd_1)}{\sum_{i=1}^2 T(sd_i)}} (1 - \mu_2^U)^{\frac{T(sd_2)}{\sum_{i=1}^2 T(sd_i)}} \right] \right\}, \right. \\ \left. \left\{ \left[(\nu_1^L)^{\frac{T(sd_1)}{\sum_{i=1}^2 T(sd_i)}} (\nu_2^L)^{\frac{T(sd_2)}{\sum_{i=1}^2 T(sd_i)}}, (\nu_1^U)^{\frac{T(sd_1)}{\sum_{i=1}^2 T(sd_i)}} (\nu_2^U)^{\frac{T(sd_2)}{\sum_{i=1}^2 T(sd_i)}} \right] \right\} \right\} \Bigg). \end{aligned}$$

So, when $n = 2$, Theorem 1 also is right.

(3) Suppose that when $n = k$, Theorem 1 is right; then, we have

$$IVDHFULPWA(sd_1, sd_2, \ldots, sd_k) = \cup_{(s_j,\tilde{h}_j,\tilde{g}_j)\in sd_j}\left(s_{\sum\limits_{j=1}^{k} \frac{T(sd_j)}{\sum_{i=1}^{n} T(sd_i)} \Delta_{t_0}^{-1}(TF_{t_0}^{t_j}(\psi(s_j)))}, \right.$$
$$\cup_{[\mu_j^L,\mu_j^U]\in\tilde{h}_j,[v_j^L,v_j^U]\in\tilde{g}_j}\left\{\left\{\left[1-\prod_{j=1}^{k}(1-\mu_j^L)^{\frac{T(sd_j)}{\sum_{i=1}^{n}T(sd_i)}}, 1-\prod_{j=1}^{k}(1-\mu_j^U)^{\frac{T(sd_j)}{\sum_{i=1}^{n}T(sd_i)}}\right]\right\},\right.$$
$$\left.\left.\left\{\left[\prod_{j=1}^{k}(v_j^L)^{\frac{T(sd_j)}{\sum_{i=1}^{n}T(sd_i)}}, \prod_{j=1}^{k}(v_j^U)^{\frac{T(sd_j)}{\sum_{i=1}^{n}T(sd_i)}}\right]\right\}\right\}\right).$$

Then, when $n = k + 1$,

$$IVDHFUBLPWA(sd_1, sd_2, \ldots, sd_k, sd_{k+1}) =$$
$$\left(\mathop{\oplus}_{j=1}^{k} \frac{T(sd_j)}{\sum_{i=1}^{n} T(sd_i)} sd_j\right) \oplus \frac{T(sd_{k+1})}{\sum_{i=1}^{n} T(sd_i)} sd_{k+1} = \cup_{(s_j,\tilde{h}_j,\tilde{g}_j)\in sd_j}\left(s_{\sum\limits_{j=1}^{k+1} \frac{T(sd_j)}{\sum_{i=1}^{n} T(sd_i)} \Delta_{t_0}^{-1}(TF_{t_0}^{t_j}(\psi(s_j)))}, \right.$$
$$\cup_{[\mu_j^L,\mu_j^U]\in\tilde{h}_j,[v_j^L,v_j^U]\in\tilde{g}_j}\left\{\left\{\left[1-\prod_{j=1}^{k+1}(1-\mu_j^L)^{\frac{T(sd_j)}{\sum_{i=1}^{n}T(sd_i)}}, 1-\prod_{j=1}^{k+1}(1-\mu_j^U)^{\frac{T(sd_j)}{\sum_{i=1}^{n}T(sd_i)}}\right]\right\},\right.$$
$$\left.\left.\left\{\left[\prod_{j=1}^{k+1}(v_j^L)^{\frac{T(sd_j)}{\sum_{i=1}^{n}T(sd_i)}}, \prod_{j=1}^{k+1}(v_j^U)^{\frac{T(sd_j)}{\sum_{i=1}^{n}T(sd_i)}}\right]\right\}\right\}\right).$$

So, when $n = k + 1$, Theorem 1 is right too.

According to steps (1), (2), and (3), we get that Theorem 1 is right for all n. □

Theorem 2. *IVDHFUBLPWA operator holds following properties:*

(1) Commutativity: Let $(sd_1{}^*, sd_2{}^*, \ldots, sd_n{}^*)$ *be any permutation of* $(sd_1, sd_2, \ldots, sd_n)$*, then*

$$IVDHFUBLPWA(sd_1{}^*, sd_2{}^*, \ldots, sd_n{}^*) = IVDHFUBLPWA(sd_1, sd_2, \ldots, sd_n).$$

(2) Idempotency: Let $sd_j = sd$*, for all* $j = 1, 2, \ldots, n$*, then*

$$IVDHFUBLPWA(sd_1, sd_2, \ldots, sd_n) = sd.$$

(3) Boundedness: the IVDHFUBLPWA operator lies between the max and min operators,

$$sd^- \leq IVDHFUBLPWA(sd_1, sd_2, \ldots, sd_n) \leq sd^+.$$

Proof.

(1) Assume that $(sd_1{}^*, sd_2{}^*, \ldots, sd_n{}^*)$ is any permutation of $(sd_1, sd_2, \ldots, sd_n)$; then, for each sd_j, there exists one and only one $sd_k{}^*$, such that $sd_k{}^* = sd_j$ and vice versa. Additionally, also we have $T(sd_j) = T(sd_k{}^*)$. Thus, based on Theorem 1, we have

$$IVDHFUBLPWA(sd_1, sd_2, \ldots, sd_n) = \frac{\mathop{\oplus}_{j=1}^{n} T(sd_j) sd_j}{\sum_{i=1}^{n} T(sd_i)} = \frac{\mathop{\oplus}_{j=1}^{n} T(sd_k{}^*) sd_k{}^*)}{\sum_{i=1}^{n} T(sd_i)} = IVDHFULPWA(sd_1{}^*, sd_2{}^*, \ldots, sd_n{}^*).$$

(2) Since $sd_j = sd$ for all $j = 1, 2, \ldots, n$, then $IVDHFUBLPWA(sd_1, sd_2, \ldots, sd_n)$

$$= \cup_{(s_0, \widetilde{h}_j, \widetilde{g}_j) \in sd} \left(s_0, \cup_{[\mu^L, \mu^U] \in \widetilde{h}, [\nu^L, \nu^U] \in \widetilde{g}} \left\{ \left\{ \left[\mu^L, \mu^U \right] \right\}, \left\{ \left[\nu^L, \nu^U \right] \right\} \right\} \right) = sd.$$

(3) Suppose $sd^- = \left(s_0^-, \widetilde{h}^-, \widetilde{g}^- \right)$, $sd^+ = \left(s_0^+, \widetilde{h}^+, \widetilde{g}^+ \right)$, in which

$$s_0^- = \min_j \left(s_{\Delta_{t_0}^{-1}(TF_{t_0}^{t_j}(\psi(s_j)))} \right), s_0^+ = \max_j \left(s_{\Delta_{t_0}^{-1}(TF_{t_0}^{t_j}(\psi(s_j)))} \right),$$
$$\widetilde{h}^- = \cup_{[\mu_j^L, \mu_j^U] \in \widetilde{h}_j} \left\{ \left[\mu^{L-}, \mu^{U-} \right] \right\} = \cup_{[\mu_j^L, \mu_j^U] \in \widetilde{h}_j} \left\{ \left[\min_{1 \le j \le n} \mu_j^L, \min_{1 \le j \le n} \mu_j^U \right] \right\},$$
$$\widetilde{h}^+ = \cup_{[\mu_j^L, \mu_j^U] \in \widetilde{h}_j} \left\{ \left[\mu^{L+}, \mu^{U+} \right] \right\} = \cup_{[\mu_j^L, \mu_j^U] \in \widetilde{h}_j} \left\{ \left[\max_{1 \le j \le n} \mu_j^L, \max_{1 \le j \le n} \mu_j^U \right] \right\},$$
$$\widetilde{g}^- = \cup_{[\nu_j^L, \nu_j^U] \in \widetilde{g}_j} \left\{ \left[\nu^{L-}, \nu^{U-} \right] \right\} = \cup_{[\nu_j^L, \nu_j^U] \in \widetilde{g}_j} \left\{ \left[\max_{1 \le j \le n} \nu_j^L, \max_{1 \le j \le n} \nu_j^U \right] \right\},$$
$$\widetilde{g}^+ = \cup_{[\nu_j^L, \nu_j^U] \in \widetilde{g}_j} \left\{ \left[\nu^{L+}, \nu^{U+} \right] \right\} = \cup_{[\nu_j^L, \nu_j^U] \in \widetilde{g}_j} \left\{ \left[\min_{1 \le j \le n} \nu_j^L, \min_{1 \le j \le n} \nu_j^U \right] \right\}.$$

Obviously,

$$s_0^- = \min_j \left(s_{\Delta_{t_0}^{-1}(TF_{t_0}^{t_j}(\psi(s_j)))} \right) \le s_{\sum_{j=1}^{n} \frac{T(sd_j)}{\sum_{i=1}^{n} T(sd_i)} \Delta_{t_0}^{-1}(TF_{t_0}^{t_j}(\psi(s_j)))} \le \max_j \left(s_{\Delta_{t_0}^{-1}(TF_{t_0}^{t_j}(\psi(s_j)))} \right) = s_0^+.$$

Additionally, for all $j = 1, 2, \ldots, n$, we have

$$\left(1 - \prod_{j=1}^{n} (1 - \mu^{L+})^{\frac{T(sd_j)}{\sum_{i=1}^{n} T(sd_i)}} \right) + \left(1 - \prod_{j=1}^{n} (1 - \mu^{U+})^{\frac{T(sd_j)}{\sum_{i=1}^{n} T(sd_i)}} \right) \ge$$
$$\left(1 - \prod_{j=1}^{n} (1 - \mu_j^L)^{\frac{T(sd_j)}{\sum_{i=1}^{n} T(sd_i)}} \right) + \left(1 - \prod_{j=1}^{n} (1 - \mu_j^U)^{\frac{T(sd_j)}{\sum_{i=1}^{n} T(sd_i)}} \right) \ge$$
$$\left(1 - \prod_{j=1}^{n} (1 - \mu^{L-})^{\frac{T(sd_j)}{\sum_{i=1}^{n} T(sd_i)}} \right) + \left(1 - \prod_{j=1}^{n} (1 - \mu^{U-})^{\frac{T(sd_j)}{\sum_{i=1}^{n} T(sd_i)}} \right);$$

Meanwhile, we have

$$\left(\prod_{j=1}^{n} (\nu^{L-})^{\frac{T(sd_j)}{\sum_{i=1}^{n} T(sd_i)}} \right) + \left(\prod_{j=1}^{n} (\nu^{U-})^{\frac{T(sd_j)}{\sum_{i=1}^{n} T(sd_i)}} \right) \ge \left(\prod_{j=1}^{n} (\nu^{L})^{\frac{T(sd_j)}{\sum_{i=1}^{n} T(sd_i)}} \right) + \left(\prod_{j=1}^{n} (\nu^{U})^{\frac{T(sd_j)}{\sum_{i=1}^{n} T(sd_i)}} \right)$$
$$\ge \left(\prod_{j=1}^{n} (\nu^{L+})^{\frac{T(sd_j)}{\sum_{i=1}^{n} T(sd_i)}} \right) + \left(\prod_{j=1}^{n} (\nu^{U+})^{\frac{T(sd_j)}{\sum_{i=1}^{n} T(sd_i)}} \right).$$

Then

$$\left(1-\prod_{j=1}^{n}(1-\mu^{L+})^{\frac{T(sd_j)}{\sum_{i=1}^{n}T(sd_i)}}\right)+\left(1-\prod_{j=1}^{n}(1-\mu^{U+})^{\frac{T(sd_j)}{\sum_{i=1}^{n}T(sd_i)}}\right)-\left(\prod_{j=1}^{n}(\nu^{L+})^{\frac{T(sd_j)}{\sum_{i=1}^{n}T(sd_i)}}\right)$$
$$-\left(\prod_{j=1}^{n}(\nu^{U+})^{\frac{T(sd_j)}{\sum_{i=1}^{n}T(sd_i)}}\right)\geq\left(1-\prod_{j=1}^{n}(1-\mu_j^{L})^{\frac{T(sd_j)}{\sum_{i=1}^{n}T(sd_i)}}\right)+\left(1-\prod_{j=1}^{n}(1-\mu_j^{U})^{\frac{T(sd_j)}{\sum_{i=1}^{n}T(sd_i)}}\right)$$
$$-\left(\prod_{j=1}^{n}(\nu^{L})^{\frac{T(sd_j)}{\sum_{i=1}^{n}T(sd_i)}}\right)-\left(\prod_{j=1}^{n}(\nu^{U})^{\frac{T(sd_j)}{\sum_{i=1}^{n}T(sd_i)}}\right)\geq\left(1-\prod_{j=1}^{n}(1-\mu^{L-})^{\frac{T(sd_j)}{\sum_{i=1}^{n}T(sd_i)}}\right)+$$
$$\left(1-\prod_{j=1}^{n}(1-\mu^{U-})^{\frac{T(sd_j)}{\sum_{i=1}^{n}T(sd_i)}}\right)-\left(\prod_{j=1}^{n}(\nu^{L-})^{\frac{T(sd_j)}{\sum_{i=1}^{n}T(sd_i)}}\right)-\left(\prod_{j=1}^{n}(\nu^{U-})^{\frac{T(sd_j)}{\sum_{i=1}^{n}T(sd_i)}}\right).$$

According to Definition 4 and Theorem 1, we have

$$sd^{-}\leq IVDHFUBLPWA(sd_1,sd_2,\ldots,sd_n)\leq sd^{+},$$

which completes the proof. □

Theorem 3. *For a collection of IVDHFUBLNs $sd_j(j=1,2,\ldots,n)$, if there is no prioritized relationship between theme, then IVDHFUBLPWA operator reduces to the interval-valued dual hesitant fuzzy unbalanced linguistic weighted average (IVDHFUBLWA) operator, in which*

$$IVDHFUBLWA(sd_1,sd_2,\ldots,sd_n)=\frac{\overset{n}{\underset{j=1}{\oplus}}(\omega_j sd_j)}{\sum_{i=1}^{n}\omega_i} \tag{10}$$

in which $\omega=(\omega_1,\omega_2,\ldots,\omega_n)^T$ is the weighting vector for $sd_j(j=1,2,\ldots,n)$ with $\omega_j\in[0,1]$ and $\sum_{j=1}^{n}\omega_j=1$.

4.2. A Hybrid Model for Determining the Unknown Experts' Weights

Generally, when the weighting information for decision-makers cannot be subjectively acquired in advance, decision matrices given by decision-makers should be taken into account to derive the unknown weighting vector objectively.

Basically, there are two indispensible aspects with which to exploit assessments in decision matrices objectively. On one side, the accuracy function for hesitant fuzzy elements [32] can be utilized to measure the overall fuzziness of individual decision matrix given by each decision-maker; hence, the less fuzziness there is in a decision matrix, the bigger the weight that should be configured to the corresponding decision-maker. On the other side, according to deviation maximizing methodology [73], the smaller the difference between the assessments offered by one specific decision-maker with those offered by the other decision-makers, the more precise the evaluation information given that specific decision-maker; a larger weight thus should be correspondingly assigned to the decision-maker.

Therefore, firstly, we apply the accuracy function $P(sd)$ in Definition 4 to indicate information fuzziness in IVDHFUBL individual decision matrix R^k. There is less fuzzy information contained in individual decision matrix R^k than other IVDHFUBL decision matrix, so the kth decision-maker plays an important role in prioritization process and should be assigned a bigger weight. Then, specifically from this aspect, we naturally can obtain type of experts' weights $\widetilde{\lambda}^k(k=1,2,\ldots,t)$ by

$$\widetilde{\lambda}^k=\frac{\frac{1}{mn}\sum_{i=1}^{n}\sum_{j=1}^{m}P(r_{ij}^k)}{\sum_{k=1}^{t}\left(\frac{1}{mn}\sum_{i=1}^{n}\sum_{j=1}^{m}P(r_{ij}^k)\right)}. \tag{11}$$

Secondly, we here take the divergence degree measure $DD(R^k, R^l) = 1 - \frac{1}{mn}\sum\limits_{i=1}^{n}\sum\limits_{j=1}^{m} d(r_{ij}^k, r_{ij}^l)$ to calculate the deviation between IVDHFUBL decision matrix R^k given by the kth decision-maker and IVDHFUBL decision matrix R^l given by the lth decision-maker. If the overall divergence of R^k appears to be larger than other decision matrices, then the kth decision-maker should be assigned a smaller weight. On the contrary, overall divergence of evaluations in IVDHFUBL decision matrix R^k comes to be smaller than other decision matrices; then, it can be seen that the kth decision-maker should be assigned a larger weight. As a result, we establish the following programming model (M-1) for calculating the divergence-based weighting vector $\overline{\lambda}^k (k = 1, 2, \ldots, t)$ for decision-makers.

$$(\mathrm{M}-1)\begin{cases} \max F(\overline{\lambda}^k) = \sum\limits_{k=1}^{t} \frac{1}{t}\left(\sum\limits_{l=1, l\neq k}^{t} DD(R^k, R^l)\overline{\lambda}^k \right) \\ s.t \sum_{k=1}^{t} (\overline{\lambda}^k)^2 = 1, \overline{\lambda}^k \geq 0, k = 1, 2, \ldots, t \end{cases}.$$

Because of $DD(R^k, R^l) = 1 - \frac{1}{mn}\sum\limits_{i=1}^{n}\sum\limits_{j=1}^{m} d(r_{ij}^k, r_{ij}^l)$, we rewrite the above model (M-1) to following model (M-2).

$$(\mathrm{M}-2)\begin{cases} \max F(\overline{\lambda}^k) = \sum\limits_{k=1}^{t} \frac{1}{t}\left(\sum\limits_{l=1, l\neq k}^{t} \left(1 - \frac{1}{mn}\sum\limits_{i=1}^{n}\sum\limits_{j=1}^{m} d(r_{ij}^k, r_{ij}^l)\right)\overline{\lambda}^k \right) \\ s.t \sum_{k=1}^{t} (\overline{\lambda}^k)^2 = 1, \overline{\lambda}^k \geq 0, k = 1, 2, \ldots, t \end{cases},$$

in which $d(r_{ij}^k, r_{ij}^l)$ is applied according to Definition 5.

Regarding the model (M-2), we have following Theorems 4 and 5.

Theorem 4. *The optimal solution to (M-2) is*

$$\overline{\lambda}^k = \frac{\sum_{l=1, l\neq k}^{t}\left(1 - \frac{1}{mn}\sum_{i=1}^{n}\sum_{j=1}^{m} d(r_{ij}^k, r_{ij}^l)\right)}{\sum_{k=1}^{t}\left(\sum_{l=1, l\neq k}^{t}\left(1 - \frac{1}{mn}\sum_{i=1}^{n}\sum_{j=1}^{m} d(r_{ij}^k, r_{ij}^l)\right)\right)}. \tag{12}$$

Proof. To solve this model, we construct the Lagrange function as follows:

$$L(\overline{\lambda}^k, \zeta) = \sum_{k=1}^{t} \frac{1}{t}\left(\sum_{l=1, l\neq k}^{t} \left(1 - \frac{1}{mn}\sum_{i=1}^{n}\sum_{j=1}^{m} d(r_{ij}^k, r_{ij}^l)\right)\right)\overline{\lambda}^k + \frac{1}{2}\zeta\sum_{k=1}^{t}\left((\overline{\lambda}^k)^2 - 1 \right), \tag{13}$$

By differentiation on Equation (13) with respect to $\overline{\lambda}^k (k = 1, 2, \ldots, t)$ and ζ, and setting these partial derivatives equal to zero, the following set of equations is obtained:

$$\begin{cases} \frac{\partial L}{\partial \overline{\lambda}^k} = \frac{1}{t}\left(\sum\limits_{l=1, l\neq k}^{t} \left(1 - \frac{1}{mn}\sum\limits_{i=1}^{n}\sum\limits_{j=1}^{m} d(r_{ij}^k, r_{ij}^l)\right)\right) + \zeta\overline{\lambda}^k = 0 \\ \frac{\partial L}{\partial \zeta} = \sum\limits_{k=1}^{t}\left((\overline{\lambda}^k)^2 - 1 \right) = 0 \end{cases}. \tag{14}$$

By solving Equation (14), we get a simple and exact formula for determining the weights of decision-makers, as follows:

$$\overline{\lambda}^k = \frac{\frac{1}{t}\sum_{l=1, l\neq k}^{t}\left(1 - \frac{1}{mn}\sum_{i=1}^{n}\sum_{j=1}^{m} d(r_{ij}^k, r_{ij}^l)\right)}{\sqrt{\sum_{k=1}^{t}\left(\frac{1}{t}\sum_{l=1, l\neq k}^{t}\left(1 - \frac{1}{mn}\sum_{i=1}^{n}\sum_{j=1}^{m} d(r_{ij}^k, r_{ij}^l)\right)\right)^2}}. \tag{15}$$

Then, by normalizing $\overline{\lambda}^k (k = 1, 2, \ldots, t)$ be a unit, we have the optimal solution:

$$\overline{\lambda}^k = \frac{\sum_{l=1, l \neq k}^{t} \left(1 - \frac{1}{mn} \sum_{i=1}^{n} \sum_{j=1}^{m} d(r_{ij}^k, r_{ij}^l)\right)}{\sum_{k=1}^{t} \left(\sum_{l=1, l \neq k}^{t} \left(1 - \frac{1}{mn} \sum_{i=1}^{n} \sum_{j=1}^{m} d(r_{ij}^k, r_{ij}^l)\right)\right)}. \tag{16}$$

As can be seen, $\overline{\lambda}^k (k = 1, 2, \ldots, t)$ is the unique solution to (M-2) and applies to determine DMs' weights for MAGDM under IVDHFUBLS environment, which completes the proof. □

Theorem 5. *If $DD(R^k, R^l) = 1$, then it is reasonable to assign the experts $\overline{\lambda}^k (k = 1, 2, \ldots, t)$ the same weight.*

Proof. If $DD(R^k, R^l) = 1$, then we have $d(r_{ij}^k, r_{ij}^l) = 0$. By solving the programming model (M-2), we will obtain the experts weights $\overline{\lambda}^k = \frac{1}{t} (k = 1, 2, \ldots, t)$, which completes the proof. □

Now, to simultaneously consider the fuzziness of individual decision matrix and deviation measures between decision matrices, based on Equations (11) and (16), we can get the overall experts' weights $\lambda^k (k = 1, 2, \ldots, t)$ according to a hybrid model, as follows:

$$\lambda^k = \alpha \overline{\lambda}^k + \beta \widetilde{\lambda}^k (k = 1, 2, \ldots, t), \tag{17}$$

in which $\alpha + \beta = 1$, generally $\alpha = \beta = 0.5$, or α and β depend on real decision situations.

4.3. Algorithm for MAGDM Based on IVDHFUBLS with Prioritization Relation among Evaluative Attributes and Unknown Decision-Makers' Weights

Let $X = \{x_1, \ldots, x_i, \ldots, x_n\}$ be the set of response solutions, $A = \{A_1, \ldots, A_j, \ldots, A_m\}$ be the set of attributes, $E = \{E_1, \ldots, E_k, \ldots, E_t\}$ be the set of decision-makers. Suppose that, according to knowledge from decision contexts and Delphi method, decision-makers are capable of determining a prioritization relation, $A_{\sigma(1)} \succ \ldots \succ A_{\sigma(j)} \succ \ldots \succ A_{\sigma(m)}$, among evaluative attributes, which means the attribute $A_{\sigma(j-1)}$ has a higher priority level than the attribute $A_{\sigma(j)}$. Suppose $R^k = (r_{ij}^k)_{n \times m} (k = 1, 2, \ldots, t)$ constitutes the IVDHFUBL decision matrices given by all t decision-makers, among which r_{ij}^k denotes assessments presented by the kth decision-maker based on an unbalanced linguistic term set S^k with respect to alternative x_i under attribute A_j, and $r_{ij}^k = \left(s_{\alpha_{ij}}^k, \widetilde{h}_{ij}^k, \widetilde{g}_{ij}^k\right)$ and $s_{\alpha_{ij}}^k \in S^k$ take the form of IVDHFUBLNs. Then, on the strength of above-developed methods, we here construct the following Procedure I for MAGDM based on IVDHFUBLS with prioritization relation among evaluative attributes and unknown decision-makers' weights.

Procedure I. MAGDM based on IVDHFUBLS with prioritization relation among evaluative attributes and unknown decision-makers' weights.

Step I-1. Compute the weight vector $\lambda = (\lambda^1, \ldots, \lambda^k, \ldots, \lambda^t)$ for decision-makers by applying Equation (17).

Step I-2. According to the prioritization relation, $A_{\sigma(1)} \succ \ldots \succ A_{\sigma(j)} \succ \ldots \succ A_{\sigma(m)}$, among attributes, transform each individual decision matrix $R^k = \left(r_{ij}^k\right)_{n \times m} = \left(s_{\alpha_{ij}}^k, \widetilde{h}_{ij}^k, \widetilde{g}_{ij}^k\right)_{n \times m}$ to the prioritized individual decision matrix $\overline{R}^k = (\overline{r}_{ij}^k)_{n \times m} = \left(\overline{s}_{\alpha_{ij}}^k, \overline{\overline{h}}_{ij}^k, \overline{\overline{g}}_{ij}^k\right)_{n \times m}, k = 1, 2, \ldots, t,$ in which $\overline{\overline{h}}_{ij}^k = \cup_{\overline{\overline{\mu}}_{ij}^k \in \overline{\overline{h}}_{ij}^{(k)}} \left\{\overline{\overline{\mu}}_{ij}^k\right\}, \overline{\overline{g}}_{ij}^k = \cup_{\overline{\overline{v}}_{ij}^k \in \overline{\overline{g}}_{ij}^{(k)}} \left\{\overline{\overline{v}}_{ij}^k\right\}$.

Step I-3. Calculate prioritized levels in prioritized individual IVDHFUBL decision matrices: $\overline{R}^k = (\overline{r}_{ij}^k)_{n \times m} = \left(\overline{s}_{\alpha_{ij}}^k, \overline{\overline{h}}_{ij}^k, \overline{\overline{g}}_{ij}^k\right)_{n \times m}, k = 1, 2, \ldots, t.$

Calculate the score values of $\overline{r}_{ij}^k$ according to Equation (3) in Definition 4, then compute the numerical prioritized levels $T_{ij}^k(i = 1,2,\ldots,n; j = 1,2,\ldots,m; k = 1,2,\ldots,t)$ in each prioritized individual IVDHFUBL decision matrix, in which

$$T_{ij}^k = \prod_{l=1}^{j-1} S(\overline{r}_{il}^k) = (S(\overline{r}_{i(j-1)}^k))T_{i(j-1)}^k \tag{18}$$

$$T_{i1}^k = 1 \tag{19}$$

Step I-4. Obtain aggregated results in prioritized individual decision matrices, $\overline{R}^k = (\overline{r}_{ij}^k)_{n\times m} = \left(\overline{s}_{\alpha_{ij}}^k, \overline{\overline{h}}_{ij}^k, \overline{\overline{g}}_{ij}^k\right)_{n\times m}$, $k = 1,2,\ldots,t$, by applying operator IVDHFUBLPWA.

Utilize the IVDHFUBLPWA operator described in Definition 6 to aggregate $\overline{r}_{ij}^k$ so that we get the k decision-maker's decision result r_i^k on the alternative x_i, in which

$$\begin{aligned} r_i^k = IVDHFUBLPWA(\overline{r}_{i1}^k,\ldots,\overline{r}_{ij}^k,\ldots,\overline{r}_{im}^k) = \cup_{(s_{\alpha_{ij}}^k, h_{ij}^k, g_{ij}^k)\in r_{ij}^k} \Bigg(& s_{\sum_{k=1}^t \frac{T_{ij}^k}{\sum_{j=1}^m T_{ij}^k} \Delta_{t_{ij}}^{-1}(TF_{t_{ij}}^{t_{ij}^k}(\psi(s_{\alpha_{ij}}^k)))}, \\ \cup_{[\overline{\mu}_{ij}^{Lk}, \overline{\mu}_{ij}^{Uk}]\in \overline{\overline{h}}_{ij}^k, [\overline{\nu}_{ij}^{Lk}, \overline{\nu}_{ij}^{Uk}]\in \overline{\overline{g}}_{ij}^k} & \left(\left\{ \left[1 - \prod_{j=1}^m (1-\overline{\mu}_{ij}^{Lk})^{\frac{T_{ij}^k}{\sum_{j=1}^m T_{ij}^k}}, 1 - \prod_{j=1}^m (1-\overline{\mu}_{ij}^{Uk})^{\frac{T_{ij}^k}{\sum_{j=1}^m T_{ij}^k}} \right] \right\}, \right. \\ & \left. \left\{ \left[\prod_{j=1}^m (\nu_{ij}^{Lk})^{\frac{T_{ij}^k}{\sum_{j=1}^m T_{ij}^k}}, \prod_{j=1}^m (\nu_{ij}^{Uk})^{\frac{T_{ij}^k}{\sum_{j=1}^m T_{ij}^k}} \right] \right\} \right) \Bigg). \end{aligned} \tag{20}$$

Step I-5. Obtain collective results of all alternatives by applying decision-makers' weighting vector.

Given the weighting vector $\lambda = \{\lambda_1,\ldots,\lambda_k,\ldots,\lambda_t\}$ for decision-makers, which has been determined in Step 1, we now aggregate all the individual overall decision values $r_i^k(k = 1,2,\ldots,t)$ into the overall group decision values $r_i(i = 1,2,\ldots,n)$ by use of the IVDHFUBLWA operator described in Equation (10), in which

$$\begin{aligned} r_i = IVDHFUBLWA(r_i^1,\ldots,r_i^k,\ldots,r_i^t) = \cup_{(s_{\alpha_i}^k, h_i^k, g_i^k)\in r_i^k} \Bigg(& s_{\sum_{k=1}^t \lambda^k \Delta_{t_i}^{-1}(TF_{t_i}^{t_i^k}(\psi(s_{\alpha_i}^k)))}, \\ \cup_{[\overline{\mu}_i^{Lk}, \overline{\mu}_i^{Uk}]\in \overline{\overline{h}}_i^k, [\overline{\nu}_i^{Lk}, \overline{\nu}_i^{Uk}]\in \overline{\overline{g}}_i^k} & \left(\left\{ \left[1 - \prod_{k=1}^t (1-\overline{\mu}_{ij}^{Lk})^{\lambda^k}, 1 - \prod_{k=1}^t (1-\overline{\mu}_{ij}^{Uk})^{\lambda^k} \right] \right\}, \left\{ \left[\prod_{k=1}^t (\nu_i^{Lk})^{\lambda^k}, \prod_{k=1}^t (\nu_i^{Uk})^{\lambda^k} \right] \right\} \right) \Bigg). \end{aligned} \tag{21}$$

Step I-6. According to Definition 4, calculate the score value $S(r_i)$ of the group overall assessments $r_i(i = 1,2,\ldots,n)$ to alternatives $x_i(i = 1,2,\ldots,n)$, then rank all the alternatives $x_i(i = 1,2,\ldots,n)$ and select the most desirable one(s).

5. Illustrative Examples

5.1. Applied Case Study on Green Supplier Selection Problem

Due to increasing environmental concerns in socioeconomic activities, more and more companies have been urged to enhance their green images so as to maintain and improve competitiveness. As a result, leading enterprises like Dell, HP, and IBM have already turned to include green supply chains in their business processes. Obviously, in order to construct effective green supply chains, core companies generally are only willing to select suppliers who exhibit better practices regarding green supply chain management [2]. Therefore, to demonstrate the practicality and effectiveness of our proposed approach, we apply the Algorithm I to resolve the following example of green supplier selection problem.

Suppose we are evaluating three alternative suppliers, i.e., x_i, $i = 1,2,3$, according to eight attributes A_j: (1) A_1—Economic performance, (2) A_2—Regulation, (3) A_3—Perceived stakeholders' pressure, (4) A_4—Green design, (5) A_5—Environmental performance, (6) A_6—Recovery and reuse of used products, (7) A_7—Supplier/customer collaboration, and (8) A_8—Green purchasing. A panel of decision-makers, i.e., E_k, $k = 1,2,3$, have been already organized, and decision-makers also reached a consensus opinion on the prioritization relation, i.e., $(A_2) \succ (A_5) \succ (A_1) \succ (A_8) \succ (A_4) \succ (A_6) \succ (A_7) \succ (A_3)$ among the attributes.

Next, all three decision-makers $E_k(k = 1,2,3)$ were invited to provide their preferences in the form of interval-valued dual hesitant fuzzy unbalanced linguistic numbers. The corresponding linguistic variables are chosen from two unbalanced linguistic term sets S_1 and S_2, in which $S_1 = \{N, L, M, AH, H, QH, VH, AT, T\}$ and $S_2 = \{N, M, H, VH, T\}$. The relationship between unbalanced linguistic term sets S_1, S_2, and linguistic hierarchies is shown in Figure 2. Decisionmakers E_1 and E_2 evaluate the three suppliers by the unbalanced linguistic term set S_1, while E_3 utilizes the unbalanced linguistic term set S_2. Then, three interval-valued dual hesitant fuzzy unbalanced linguistic (IVDHFUBL) decision matrices, i.e., $R^k = (r_{ij}^k)_{3\times 8}(k = 1,2,3)$, have been collected, as shown in Tables 2–4.

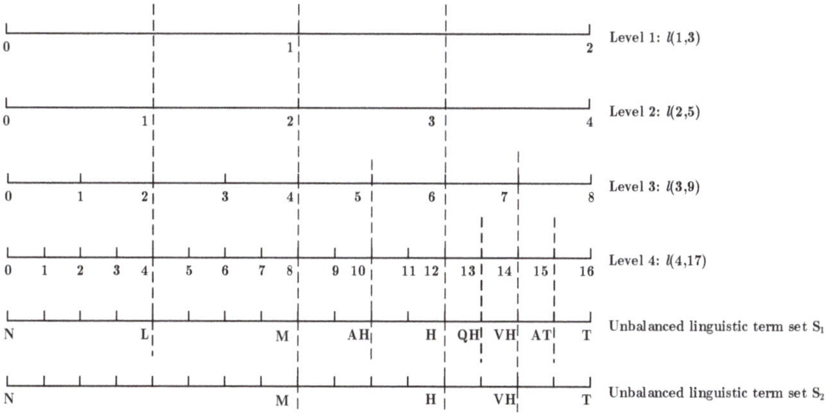

Figure 2. Unbalanced linguistic term sets (S_1 and S_2) and their mapping in linguistic hierarchies.

Table 2. The IVDHFUBL decision matrix R^1 provided by decision-maker E_1.

	A_1	A_2	A_3	A_4
x_1	(VH,{[0.2,0.3]}, {[0.2,0.4],[0.3,0.4]})	(H,{[0.5,0.6]}, {[0.2,0.3]})	(M,{[0.3,0.4]}, {[0.4,0.5],[0.5,0.6]})	(T,{[0.2,0.4]}, {[0.5,0.6]})
x_2	(AT,{[0.4,0.5],[0.5,0.6]}, {[0.2,0.3],[0.2,0.4]})	(T,{[0.6,0.7]}, {[0.1,0.2]})	(L,{[0.6,0.7], [0.7,0.8]},{[0.1,0.2]})	(QH,{[0.3,0.5]}, {[0.2,0.3]})
x_3	(L,{[0.1,0.2],[0.1,0.3]}, {[0.6,0.7]})	(H,{[0.2,0.3]}, {[0.5,0.6],[0.6,0.7]})	(AH,{[0.4,0.5]}, {[0.2,0.3]})	(VH,{[0.6,0.7]}, {[0.1,0.2],[0.2,0.3]})

	A_5	A_6	A_7	A_8
x_1	(L,{[0.4,0.5]}, {[0.1,0.2],[0.4,0.5]})	(M,{[0.1,0.2],[0.3,0.5]}, {[0.3,0.5]})	(AH,{[0.4,0.5],[0.5,0.6]}, {[0.2,0.3],[0.2,0.4]})	(L,{[0.1,0.3]}, {[0.4,0.6]})
x_2	(M,{[0.6,0.7]}, {[0.1,0.2]})	(VH,{[0.2,0.4],[0.5,0.6]}, {[0.2,0.3]})	(L,{[0.5,0.6],[0.7,0.8]}, {[0.1,0.2]})	(H,{[0.5,0.7]}, {[0.1,0.2],[0.2,0.3]})
x_3	(M,{[0.4,0.5],[0.6,0.7]}, {[0.1,0.3]})	(H,{[0.3,0.4]}, {[0.4,0.5]})	(QH,{[0.4,0.5],[0.5,0.6]}, {[0.3,0.4]})	(VH,{[0.4,0.6]}, {[0.3,0.4]})

Table 3. The IVDHFUBL decision matrix R^2 provided by decision-maker E_2.

	A_1	A_2	A_3	A_4
x_1	(M,{[0.3,0.5]}, {[0.1,0.2]})	(AH,{[0.1,0.4]}, {[0.2,0.3],[0.3,0.4]})	(H,{[0.2,0.4]}, {[0.4,0.5]})	(QH,{[0.2,0.4]}, {[0.5,0.6]})
x_2	(AT,{[0.4,0.7]}, {[0.2,0.3] })	(L,{[0.5,0.6]}, {[0.1,0.2] })	(AH,{[0.6,0.7],[0.7,0.8]}, {[0.1,0.2]})	(M,{[0.2,0.3]}, {[0.5,0.6],[0.6,0.7]})
x_3	(AT,{[0.6,0.8]}, {[0.1,0.2]})	(H,{[0.4,0.5]}, {[0.3,0.4],[0.4,0.5]})	(H,{[0.4,0.5]}, {[0.2,0.3]})	(AH,{[0.4,0.5]}, {[0.2,0.3]})
	A_5	A_6	A_7	A_8
x_1	(M,{[0.1,0.2],[0.2,0.3]}, {[0.1,0.2]})	(L,{[0.6,0.7]}, {[0.1,0.2]})	(H,{[0.3,0.4]}, {[0.2,0.3],[0.4,0.5]})	(M,{[0.5,0.7]}, {[0.2,0.3]})
x_2	(VH,{[0.6,0.7]}, {[0.1,0.2]})	(AT,{[0.2,0.3]}, {[0.5,0.7]})	(H,{[0.5,0.8]}, {[0.1,0.2]})	(AT,{[0.3,0.5]}, {[0.3,0.4]})
x_3	(QH,{[0.4,0.5]}, {[0.3,0.4]})	(L,{[0.7,0.8]}, {[0.1,0.2]})	(VH,{[0.2,0.5]}, {[0.3,0.4]})	(H,{[0.3,0.5]}, {[0.3,0.4]})

Table 4. The IVDHFUBL decision matrix R^3 provided by decision-maker E_3.

	A_1	A_2	A_3	A_4
x_1	(M,{[0.6,0.8]}, {[0.1,0.2]})	(H,{[0.4,0.5]}, {[0.4,0.5]})	(H,{[0.2,0.4],[0.3,0.4]}, {[0.2,0.3]})	(M,{[0.7,0.8]}, {[0.1,0.2]})
x_2	(T,{[0.3,0.4]}, {[0.4,0.6]})	(M,{[0.4,0.5],[0.5,0.6]}, {[0.2,0.3] })	(VH,{[0.6,0.7]}, {[0.1,0.3]})	(H,{[0.1,0.3],[0.2,0.4]}, {[0.3,0.5]})
x_3	(VH,{[0.4,0.5]}, {[0.1,0.2],[0.3,0.4]})	(VH,{[0.7,0.8]}, {[0.1,0.2]})	(H,{[0.6,0.8]}, {[0.1,0.2]})	(M,{[0.6,0.7]}, {[0.1,0.3]})
	A_5	A_6	A_7	A_8
x_1	(T,{[0.2,0.5]}, {[0.3,0.5]})	(VH,{[0.6,0.7]}, {[0.2,0.3]})	(M,{[0.1,0.2]}, {[0.5,0.8]})	(VH,{[0.3,0.4]}, {[0.1,0.3],[0.2,0.5]})
x_2	(M,{[0.4,0.6], [0.5,0.7]},{[0.1,0.2]})	(VH,{[0.3,0.6]}, {[0.1,0.3],[0.2,0.4]})	(H,{[0.4,0.6]}, {[0.3,0.4]})	(VH,{[0.7,0.8]}, {[0.1,0.2]})
x_3	(VH,{[0.6,0.7]}, {[0.1,0.2]})	(M,{[0.5,0.6]}, {[0.3,0.4]})	(H,{[0.3,0.5]}, {[0.4,0.5]})	(H,{[0.6,0.7]}, {[0.1,0.3]})

Now, we apply the proposed Procedure I to solve the above green supplier selection problem. The following are details steps in Procedure I.

Step I-1. Compute the weight vector $\overline{\lambda} = (\overline{\lambda}^1, \overline{\lambda}^2, \overline{\lambda}^3)$ for decision-makers. Firstly, by solving the programming model (M-2), we obtain deviation-based weighting vector as

$$\overline{\lambda} = (0.3259, 0.3419, 0.3286).$$

Then, by applying Equation (11), we get the accuracy-measure based experts' weighting vector $\widetilde{\lambda} = (\widetilde{\lambda}^1, \widetilde{\lambda}^2, \widetilde{\lambda}^3)$ as

$$\widetilde{\lambda} = (0.3125, 0.3195, 0.368).$$

Finally, according to Equation (17), we here suppose $\alpha = \beta = 0.5$; then, the hybrid experts' weighting vector $\lambda = (\lambda^1, \lambda^2, \lambda^3)$ is obtained as

$$\lambda = (0.3192, 0.3307, 0.3483).$$

Step I-2. Transform each individual IVDHFUBL decision matrix $R^k = (r_{ij}^k)_{3\times 8}$ into the prioritized individual IVDHFUBL decision matrix $\overline{R}^k = (\overline{r}_{ij}^k)_{3\times 8} (k = 1, 2, 3)$ according to the different priority levels of attributes, as listed in Tables 5–7.

Step I-3. Calculate prioritized levels $T_{ij}^{k}(i = 1,2,3; j = 1,2,\ldots,8; k = 1,2,3)$ in each prioritized individual IVDHFUBL decision matrix by Equations (18) and (19), and then we have

$T_{11}^{1} = 1$, $T_{12}^{1} = 0.6$, $T_{13}^{1} = 0.1125$, $T_{14}^{1} = 0.0566$, $T_{15}^{1} = 0.0099$, $T_{16}^{1} = 0.0084$, $T_{17}^{1} = 0.0028$, $T_{18}^{1} = 0.0014$; $T_{21}^{1} = 1$, $T_{22}^{1} = 0.8$, $T_{23}^{1} = 0.32$, $T_{24}^{1} = 0.2325$, $T_{25}^{1} = 0.1395$, $T_{26}^{1} = 0.0737$, $T_{27}^{1} = 0.0435$, $T_{28}^{1} = 0.0087$; $T_{31}^{1} = 1$, $T_{32}^{1} = 0.6375$, $T_{33}^{1} = 0.2391$, $T_{34}^{1} = 0.0493$, $T_{35}^{1} = 0.0367$, $T_{36}^{1} = 0.0273$, $T_{37}^{1} = 0.0164$, $T_{38}^{1} = 0.0113$.
$T_{11}^{2} = 1$, $T_{12}^{2} = 0.3438$, $T_{13}^{2} = 0.0602$, $T_{14}^{2} = 0.0165$, $T_{15}^{2} = 0.007$, $T_{16}^{2} = 0.0049$, $T_{17}^{2} = 0.001$, $T_{18}^{2} = 0.0005$; $T_{21}^{2} = 1$, $T_{22}^{2} = 0.175$, $T_{23}^{2} = 0.1225$, $T_{24}^{2} = 0.0919$, $T_{25}^{2} = 0.0646$, $T_{26}^{2} = 0.0275$, $T_{27}^{2} = 0.0219$, $T_{28}^{2} = 0.0131$; $T_{31}^{2} = 1$, $T_{32}^{2} = 0.6375$, $T_{33}^{2} = 0.4144$, $T_{34}^{2} = 0.3302$, $T_{35}^{2} = 0.1857$, $T_{36}^{2} = 0.0813$, $T_{37}^{2} = 0.0183$, $T_{38}^{2} = 0.0112$.
$T_{11}^{3} = 1$, $T_{12}^{3} = 0.675$, $T_{13}^{3} = 0.5063$, $T_{14}^{3} = 0.2152$, $T_{15}^{3} = 0.1177$, $T_{16}^{3} = 0.0529$, $T_{17}^{3} = 0.0417$, $T_{18}^{3} = 0.0167$; $T_{21}^{3} = 1$, $T_{22}^{3} = 0.375$, $T_{23}^{3} = 0.1312$, $T_{24}^{3} = 0.1116$, $T_{25}^{3} = 0.0879$, $T_{26}^{3} = 0.0428$, $T_{27}^{3} = 0.0262$, $T_{28}^{3} = 0.0167$; $T_{31}^{3} = 1$, $T_{32}^{3} = 0.7875$, $T_{33}^{3} = 0.5513$, $T_{34}^{3} = 0.3376$, $T_{35}^{3} = 0.2152$, $T_{36}^{3} = 0.0915$, $T_{37}^{3} = 0.0412$, $T_{38}^{3} = 0.0262$.

Step I-4. Utilize the IVDHFUBLPWA operator described in Definition 6 to aggregate $\bar{r}_{ij}^{k}$, so that we get the k th expert's decision result $r_i^k(k = 1,2,3)$ on alternatives $x_i(i = 1,2,3)$, in which

$r_1^1 = (s_{10.181}^{17}$,{[0.439,0.5428],[0.4392,0.5429],[0.4397,0.5438],[0.4399,0.5439]},{[0.1633,0.2744],[0.1633,0.2744],[0.1633,0.2745],[0.1633,0.2746],[0.1675,0.2744],[0.1675,0.2744],[0.1675,0.2745],[0.1675,0.2746],[0.2598,0.373],[0.2598,0.373],[0.2598,0.3731],[0.2598,0.3732],[0.2665,0.373],[0.2665,0.373],[0.2665,0.3731],[0.2665,0.3732]}});
$r_2^1 = (s_{12.6218}^{17}$,{[0.5479,0.6638],[0.5483,0.6642],[0.5517,0.6676],[0.5521,0.6681],[0.5538,0.6676],[0.5542,0.668],[0.5576,0.6714],[0.558,0.6718],[0.5579,0.6728],[0.5583,0.6732],[0.5616,0.6766],[0.562,0.677],[0.5637,0.6765],[0.5641,0.677],[0.5673,0.6802],[0.5677,0.6806]},{[0.1152,0.2172],[0.1225,0.2252],[0.1152,0.225],[0.1225,0.2333]});
$r_3^1 = (s_{9.8701}^{17}$,{[0.2778,0.383],[0.2789,0.3841],[0.2778,0.3927],[0.2789,0.3938],[0.3647,0.475],[0.3656,0.476],[0.3647,0.4832],[0.3656,0.4842]},{[0.2911,0.4718],[0.2948,0.4753],[0.3186,0.5093],[0.3227,0.5131]});
$r_1^2 = (s_{9.4101}^{17}$,{[0.1186,0.3686],[0.1431,0.3885]},{[0.1649,0.2682],[0.165,0.2683],[0.2188,0.3278],[0.2189,0.3279]});
$r_2^2 = (s_{7.2463}^{17}$,{[0.4817,0.6085],[0.483,0.6099]},{[0.1246,0.231],[0.1256,0.2326]});
$r_3^2 = (s_{12.3335}^{17}$,{[0.4365,0.5779]},{[0.2377,0.3445],[0.2646,0.3745]});
$r_1^3 = (s_{12.22}^{17}$,{[0.4134,0.5925],[0.4139,0.5925]},{[0.235,0.3833],[0.2487,0.3997]});
$r_2^3 = (s_{9.5402}^{17}$,{[0.4072,0.5418],[0.4106,0.5452],[0.4294,0.5685],[0.4327,0.5718],[0.4646,0.5954],[0.4676,0.5985],[0.4846,0.6191],[0.4876,0.6219]},{[0.1748,0.2911],[0.1777,0.2932]});
$r_3^3 = (s_{13.3978}^{17}$,{[0.6027,0.7085]},{[0.1053,0.2225],[0.1284,0.2522]}).

Step I-5. Aggregate all the individual overall decision values $r_i^k(k = 1,2,3)$ into the overall group decision values $r_i(i = 1,2,\ldots,n)$ by use of the IVDHFUBLWA operator described in Equation (10) and experts' weighting vector $\lambda = (0.3192, 0.3307, 0.3483)$ determined in Step 1. Taking r_3 as an example, we have

$r_3 = (s_{11.8957}^{17}$,{[0.4597,0.5809],[0.46,0.5811],[0.4597,0.583],[0.46,0.5833], [0.4815,0.6021],[0.4817,0.6023],[0.4815,0.6041],[0.4817,0.6043]},{[0.191,0.3273],[0.2047,0.3419],[0.1979,0.3364],[0.2121,0.3514],[0.1918,0.328],[0.2055,0.3427],[0.1987,0.3372],[0.2129,0.3522],[0.1966,0.3354],[0.2107,0.3504],[0.2037,0.3448],[0.2183,0.3601],[0.1974,0.3362],[0.2116,0.3512],[0.2046,0.3456],[0.2192,0.361]}}).

Step I-6. Calculating scores $S(r_i)$ of the alternatives $r_i(i = 1,2,3)$, we have

$$S(r_1) = 0.1019, S(r_2) = 0.2234, S(r_3) = 0.1916.$$

Accordingly, then the ranking order of all the alternatives is determined as

$$x_2 \succ x_3 \succ x_1.$$

Therefore, solution x_2 is the most desirable green supplier.

Table 5. The prioritized individual IVDHFUBL decision matrix $\overline{R}^1$.

	$A_{\sigma(1)}$	$A_{\sigma(2)}$	$A_{\sigma(3)}$	$A_{\sigma(4)}$
x_1	(H,{[0.5,0.6]}, {[0.2,0.3]})	(L,{[0.4,0.5]}, {[0.1,0.2],[0.4,0.5]})	(VH,{[0.2,0.3 0.2,0.4],[0.3,0.4]})	(L,{[0.1,0.3]}, {[0.4,0.6]})
x_2	(T,{[0.6,0.7]}, {[0.1,0.2]})	(M,{[0.6,0.7]}, {[0.1,0.2]})	(AT,{[0.4,0.5],[0.5,0.6]}, {[0.2,0.3],[0.2,0.4]})	(H,{[0.5,0.7]}, {[0.1,0.2],[0.2,0.3]})
x_3	(H,{[0.2,0.3]}, {[0.5,0.6],[0.6,0.7]})	(M,{[0.4,0.5],[0.6,0.7]}, {[0.1,0.3]})	(L,{[0.1,0.2],[0.1,0.3]}, {[0.6,0.7]})	(VH,{[0.4,0.6]}, {[0.3,0.4]})
	$A_{\sigma(5)}$	$A_{\sigma(6)}$	$A_{\sigma(7)}$	$A_{\sigma(8)}$
x_1	(T,{[0.2,0.4]}, {[0.5,0.6]})	(M,{[0.1,0.2],[0.3,0.5]}, {[0.3,0.5]})	(AH,{[0.4,0.5],[0.5,0.6]}, {[0.2,0.3],[0.2,0.4]})	(M,{[0.3,0.4]}, {[0.4,0.5],[0.5,0.6]})
x_2	(QH,{[0.3,0.5]}, {[0.2,0.3]})	(VH,{[0.2,0.4],[0.5,0.6]}, {[0.2,0.3]})	(L,{[0.5,0.6],[0.7,0.8]}, {[0.1,0.2]})	(L,{[0.6,0.7],[0.7,0.8]}, {[0.1,0.2]})
x_3	(VH,{[0.6,0.7]}, {[0.1,0.2],[0.2,0.3]})	(H,{[0.3,0.4]}, {[0.4,0.5]})	(QH,{[0.4,0.5],[0.5,0.6]}, {[0.3,0.4]})	(AH,{[0.4,0.5]}, {[0.2,0.3]})

Table 6. The prioritized individual IVDHFUBL decision matrix $\overline{R}^2$.

	$A_{\sigma(1)}$	$A_{\sigma(2)}$	$A_{\sigma(3)}$	$A_{\sigma(4)}$
x_1	(AH,{[0.1,0.4]}, {[0.2,0.3],[0.3,0.4]})	(M,{[0.1,0.2], [0.2,0.3]},{[0.1,0.2]})	(M,{[0.3,0.5]}, {[0.1,0.2]})	(M,{[0.5,0.7]}, {[0.2,0.3]})
x_2	(L,{[0.5,0.6]}, {[0.1,0.2] })	(VH,{[0.6,0.7]}, {[0.1,0.2]})	(AT,{[0.4,0.7]}, {[0.2,0.3] })	(AT,{[0.3,0.5]}, {[0.3,0.4]})
x_3	(H,{[0.4,0.5]}, {[0.3,0.4],[0.4,0.5]})	(QH,{[0.4,0.5]}, {[0.3,0.4]})	(AT,{[0.6,0.8]}, {[0.1,0.2]})	(H,{[0.3,0.5]}, {[0.3,0.4]})
	$A_{\sigma(5)}$	$A_{\sigma(6)}$	$A_{\sigma(7)}$	$A_{\sigma(8)}$
x_1	(QH,{[0.2,0.4]}, {[0.5,0.6]})	(L,{[0.6,0.7]}, {[0.1,0.2]})	(H,{[0.3,0.4]}, {[0.2,0.3],[0.4,0.5]})	(H,{[0.2,0.4]}, {[0.4,0.5]})
x_2	(M,{[0.2,0.3]}, {[0.5,0.6],[0.6,0.7]})	(AT,{[0.2,0.3]}, {[0.5,0.7]})	(H,{[0.5,0.8]}, {[0.1,0.2]})	(AH,{[0.6,0.7],[0.7,0.8]}, {[0.1,0.2]})
x_3	(AH,{[0.4,0.5]}, {[0.2,0.3]})	(L,{[0.7,0.8]}, {[0.1,0.2]})	(VH,{[0.2,0.5]}, {[0.3,0.4]})	(H,{[0.4,0.5]}, {[0.2,0.3]})

Table 7. The prioritized individual IVDHFUBL decision matrix $\overline{R}^3$.

	$A_{\sigma(1)}$	$A_{\sigma(2)}$	$A_{\sigma(3)}$	$A_{\sigma(4)}$
x_1	(H,{[0.4,0.5]}, {[0.4,0.5]})	(T,{[0.2,0.5]}, {[0.3,0.5]})	(M,{[0.6,0.8]}, {[0.1,0.2]})	(VH,{[0.3,0.4]}, {[0.1,0.3],[0.2,0.5]})
x_2	(M,{[0.4,0.5],[0.5,0.6]}, {[0.2,0.3] })	(M,{[0.4,0.6], [0.5,0.7]},{[0.1,0.2]})	(T,{[0.3,0.4]}, {[0.4,0.6]})	(VH,{[0.7,0.8]}, {[0.1,0.2]})
x_3	(VH,{[0.7,0.8]}, {[0.1,0.2]})	(VH,{[0.6,0.7]}, {[0.1,0.2]})	(VH,{[0.4,0.5]}, {[0.1,0.2],[0.3,0.4]})	(H,{[0.6,0.7]}, {[0.1,0.3]})
	$A_{\sigma(5)}$	$A_{\sigma(6)}$	$A_{\sigma(7)}$	$A_{\sigma(8)}$
x_1	(M,{[0.7,0.8]}, {[0.1,0.2]})	(VH,{[0.6,0.7]}, {[0.2,0.3]})	(M,{[0.1,0.2]}, {[0.5,0.8]})	(H,{[0.2,0.4],[0.3,0.4]}, {[0.2,0.3]})
x_2	(H,{[0.1,0.3],[0.2,0.4]}, {[0.3,0.5]})	(VH,{[0.3,0.6]}, {[0.1,0.3],[0.2,0.4]})	(H,{[0.4,0.6]}, {[0.3,0.4]})	(VH,{[0.6,0.7]}, {[0.1,0.3]})
x_3	(M,{[0.6,0.7]}, {[0.1,0.3]})	(M,{[0.5,0.6]}, {[0.3,0.4]})	(H,{[0.3,0.5]}, {[0.4,0.5]})	(H,{[0.6,0.8]}, {[0.1,0.2]})

5.2. Comparison with IVDHFUBLS-Based TOPSIS Method

Due to the fact that there are no directly related decision-making approaches based on IVDHFUBLS for comparison with our proposed Procedure I, in this section, we firstly develop a IVDHFUBLS-based TOPSIS method as shown in following Procedure II, in which conventional TOPSIS method is endowed the ability to address linguistic decision hesitancy and to accommodate group decision-making scenarios by use of the interval-valued dual hesitant fuzzy unbalanced linguistic weighted aggregation (IVDHFULWA) operator, which was defined in Section 3. We then apply the following Procedure II to solve the same problem adopted in Section 5.1 and discuss their ranking results.

Procedure II. IVDHFUBLS-based TOPSIS method for group decision-making.

Step II-1. Obtaining individual decision matrices from decision-makers, we get $R^k = \left(r_{ij}^k\right)_{n\times m} = \left(s_{\alpha_{ij}}^k, \widetilde{h}_{ij}^k, \widetilde{g}_{ij}^k\right)_{n\times m}$.

Step II-2. Aggregate individual decision matrices $R^k = (r_{ij}^k)_{n\times m}, k = 1,2,\ldots,t$, into individual overall evaluation values $r_i^k, i = 1,2,\ldots,n, k = 1,2,\ldots,t$, corresponding to each alternative x_i according to IVDHFULWA operator. Here, assume $\omega = (\frac{1}{m})_{1\times m}$, and

$$\begin{aligned} r_i^k = IVDHFULWA(r_{ij}^1,\ldots,r_{ij}^k,\ldots,r_{ij}^t) = \cup_{(s_{\alpha_{ij}}^k, h_{ij}^k, g_{ij}^k)\in r_{ij}^k} \Bigg(& s_{\sum_{j=1}^m \frac{1}{m}\Delta_{t_{ij}}^{-1}(TF_{t_{ij}}^{t_{ij}^k}(\psi(s_{\alpha_{ij}}^k)))}, \\ & \cup_{[\overline{\mu}_{ij}^{Lk},\overline{\mu}_{ij}^{Uk}]\in \overline{\widetilde{h}}_{ij}^k, [\overline{v}_{ij}^{Lk},\overline{v}_{ij}^{Uk}]\in \overline{\widetilde{g}}_{ij}^k} \left(\left\{\left[1-\prod_{j=1}^m (1-\overline{\mu}_{ij}^{Lk})^{\frac{1}{m}}, 1-\prod_{j=1}^m (1-\overline{\mu}_{ij}^{Uk})^{\frac{1}{m}}\right]\right\}, \left\{\left[\prod_{j=1}^m (v_i^{Lk})^{\frac{1}{m}}, \prod_{j=1}^m (v_i^{Uk})^{\frac{1}{m}}\right]\right\}\right)\Bigg). \end{aligned} \tag{22}$$

Step II-3. Calculate separating measure from positive and negative ideal solutions.

Determine positive ideal solution (PIS) $r^+ = (r_1^+, r_2^+,\ldots,r_i^+,\ldots,r_n^+)$ and negative ideal solution (NIS) $r^- = (r_1^-, r_2^-,\ldots,r_i^-,\ldots,r_n^-)$, in which $r_i^+ = (\{[1,1]\},\{[0,0]\})$, $r_i^- = (\{[0,0]\},\{[1,1]\})$.

Then, we calculate the separating measure from the PIS and NIS for each alternative according to the distance measure introduced in Equation (5), in which

$$d(\widetilde{r}_{ij},\widetilde{r}_i^+) = \left(\frac{1}{2}\left(\frac{1}{l_1}\sum_{k=1}^{l_1}\left(\left|I_1\mu_{\widetilde{h}_1}^{L_j}-1\right|^2+\left|I_1\mu_{\widetilde{h}_1}^{U_j}-1\right|^2\right)+\frac{1}{l_2}\sum_{k=1}^{l_2}\left(\left|I_1 v_{\widetilde{g}_1}^{L_j}-0\right|^2+\left|I_1 v_{\widetilde{g}_1}^{U_j}-0\right|^2\right)\right)\right)^{\frac{1}{2}}, \tag{23}$$

$$d(\widetilde{r}_{ij},\widetilde{r}_i^-)=\left(\frac{1}{2}\left(\frac{1}{l_1}\sum_{k=1}^{l_1}\left(\left|I_1\mu_{\widetilde{h}_1}^{L_j}-0\right|^2+\left|I_1\mu_{\widetilde{h}_1}^{U_j}-0\right|^2\right)+\frac{1}{l_2}\sum_{k=1}^{l_2}\left(\left|I_1\nu_{\widetilde{g}_1}^{L_j}-1\right|^2+\left|I_1\nu_{\widetilde{g}_1}^{U_j}-1\right|^2\right)\right)\right)^{\frac{1}{2}}. \quad (24)$$

Next, we can obtain

$$d_i^+=\sum_{j=1}^{m}d(\widetilde{r}_{ij},\widetilde{r}_i^+),\, d_i^-=\sum_{j=1}^{m}d(\widetilde{r}_{ij},\widetilde{r}_i^-).$$

Step II-4. Calculate the relative closeness to the ideal solution by

$$c_i=\frac{d_i^-}{d_i^-+d_i^+}. \quad (25)$$

Step II-5. Rank the green suppliers according to the descending order of c_i; then, we get the most desirable supplier.

Now we can apply Procedure II to the same problem adopted in Section 5.1 and compare their ranking results.

In Step II-1, we directly accept the decision matrices in Section 5.1. In Step II-2, we adopt $\omega=(\frac{1}{8},\frac{1}{8},\frac{1}{8},\frac{1}{8},\frac{1}{8},\frac{1}{8},\frac{1}{8},\frac{1}{8})^T$. Then, according to Equations (23) and (24), in Step II-3, we calculate the separating measure from the PIS and NIS for each alternative, and we get $d_1^+=2.1661$, $d_1^-=2.6042$, $d_2^+=2.0799$, $d_2^-=2.6879$, $d_3^+=1.8427$, and $d_3^-=2.7484$. Subsequently, in Step II-4, according to Equation (24), we obtain the relative closeness to the ideal solution: c_1 = 0.5459, c_2 = 0.5638, and c_3 = 0.5986. Therefore, Step II-5 generates the ranking result of $x_3 \succ x_2 \succ x_1$, which means the most desirable alternative is x_3.

By comparing the ranking results obtained by Procedure I and Procedure II, we find out that the two algorithms unanimously identify that the supplier x_1 is the worst alternative. However, the permutation of suppliers x_2 and x_3 changes in the ranking results. The reasons are that Algorithm II takes equal weights for both attributes and decision-makers and obviously is incapable of more completely including decision information in complicated decision-making scenarios. Contrariwise, Procedure I manages to exploit prioritization relations among attributes and objectively deduce relative importance among decision-makers, thus producing a different result.

In sum, when tackling ill-structured MAGDM problems, our proposed Procedure I provides decision-makers with an effective expression tool with which to depict their complicated assessments more comprehensively. Using the developed prioritized aggregation operator, Procedure I manages to exploit more efficiently group opinions on prioritization relations among evaluative attributes, rather than multiple rounds adjustments in conventional AHP-based methodologies under decision-making environments of high complexity. Additionally, the maximizing deviation model help Procedure I achieves more generality and objectivity in deriving unknown weights for decision-makers. Therefore, the proposed Procedure I performs an effective and efficient approach to complicate decision-making problems.

6. Conclusions

Focusing on the special type of ill-structured complex multiple attributes group decision-making problems, which characterize facets of decision-makers' decision hesitancy and prioritization relationships among evaluative attributes and unknown weighting information for decision-makers, we have developed an effective approach by employing IVDHFUBLS to elicit hesitant assessments more precisely and completely. To accommodate prioritization relationships among evaluative attributes, the proposed interval-valued dual hesitant fuzzy unbalanced linguistic prioritized weighted aggregation (IVDHFUBLPWA) operator is capable of simultaneously considering both assessments given by decision-makers and prioritization relationships. As for deducing unknown weights for decision-makers, the devised hybrid model succeeds in objectively determining rational decision-makers' weights by exploiting the overall accuracy measure of the individual decision matrix and maximizing the deviation among all decision matrices. Applied study on a green supplier selection problem has demonstrated the effectiveness and practicality of our approach.

Although we have constructed an effective approach for MAGDM under IVDHFUBLS environments, the approach applies to only the homogeneous format of assessments in group decision-making scenarios. However, sophisticated MAGDM approaches for tackling more complex practical problems should allow decision-makers to denote their specific preferences with various expression tools so as to attain better flexibility and adaptability. Therefore, future research should be firstly directed to investigate heterogeneous MAGDM approaches under IVDHFUBLS environments to deep depth, and more application studies on real problems as well, such as sustainable supplier selection, risk evaluation, etc.

Author Contributions: X.-W.Q. contributed to theoretical analysis and writing the whole paper. J.-L.Z. carried out the case studies. C.-Y.L. provided overall instructions of this work.

Funding: This research received no external funding.

Acknowledgments: The authors would like to thank precious suggestions by all anonymous reviewers. The authors would also like to thank great joint-support by the National Natural Science Foundation of China (Nos. 71701181, 71771075, 71331002), the Social Science Foundation of Ministry of Education of China (No. 16YJC630094), and the Natural Science Foundation of Zhejiang Province of China (LQ17G010002, LY18G010010, and LQ18G030012).

Conflicts of Interest: The authors declare no conflict of interest.

References

1. Zimmermann, H.J. *Fuzzy Set Theory and Its Applications*; Springer: Dordrecht, The Netherlands, 2001.
2. Lin, R.-J. Using fuzzy dematel to evaluate the green supply chain management practices. *J. Clean. Prod.* **2013**, *40*, 32–39. [CrossRef]
3. Qin, J.; Liu, X. 2-tuple linguistic muirhead mean operators for multiple attribute group decision making and its application to supplier selection. *Kybernetes* **2016**, *45*, 2–29. [CrossRef]
4. Junior, F.R.L.; Osiro, L.; Carpinetti, L.C.R. A comparison between fuzzy AHP and fuzzy TOPSIS methods to supplier selection. *Appl. Soft Comput.* **2014**, *21*, 194–209. [CrossRef]
5. Qi, X.; Liang, C.; Zhang, J. Generalized cross-entropy based group decision making with unknown expert and attribute weights under interval-valued intuitionistic fuzzy environment. *Comput. Ind. Eng.* **2015**, *79*, 52–64. [CrossRef]
6. Ghazanfari, M.; Rouhani, S.; Jafari, M. A fuzzy topsis model to evaluate the business intelligence competencies of port community systems. *Pol. Marit. Res.* **2014**, *21*, 86–96. [CrossRef]
7. Rouhani, S.; Ghazanfari, M.; Jafari, M. Evaluation model of business intelligence for enterprise systems using fuzzy topsis. *Expert Syst. Appl.* **2012**, *39*, 3764–3771. [CrossRef]
8. Zhang, J.; Hegde, G.; Shang, J.; Qi, X. Evaluating emergency response solutions for sustainable community development by using fuzzy multi-criteria group decision making approaches: Ivdhf-topsis and ivdhf-vikor. *Sustainability* **2016**, *8*, 291. [CrossRef]
9. Ju, Y.B.; Wang, A.H.; You, T.H. Emergency alternative evaluation and selection based on ANP, DEMATEL, and TL-TOPSIS. *Nat. Hazards* **2015**, *75*, 347–379. [CrossRef]
10. Ju, Y.; Wang, A.; Liu, X. Evaluating emergency response capacity by fuzzy AHP and 2-tuple fuzzy linguistic approach. *Expert Syst. Appl.* **2012**, *39*, 6972–6981. [CrossRef]
11. Zhang, X.; Wang, J.; Zhang, H.; Hu, J. A heterogeneous linguistic magdm framework to classroom teaching quality evaluation. *Eurasia J. Math. Sci. Technol. Educ.* **2017**, *13*, 4929–4956. [CrossRef]
12. Xu, Y.P. Model for evaluating the mechanical product design quality with dual hesitant fuzzy information. *J. Intell. Fuzzy Syst.* **2016**, *30*, 1–6. [CrossRef]
13. Başar, Ö.; Cengiz, K. Evaluation of renewable energy alternatives using hesitant fuzzy topsis and interval type-2 fuzzy AHP. In *Soft Computing Applications for Renewable Energy and Energy Efficiency*; Maria del Socorro García, C., Juan Miguel Sánchez, L., Antonio David Masegosa, A., Carlos Cruz, C., Eds.; IGI Global: Hershey, PA, USA, 2015; pp. 191–224.

14. Liu, H.-C.; You, J.-X.; Lu, C.; Chen, Y.-Z. Evaluating health-care waste treatment technologies using a hybrid multi-criteria decision making model. *Renew. Sustain. Energy Rev.* **2015**, *41*, 932–942. [CrossRef]
15. Mardani, A.; Jusoh, A.; Zavadskas, E.K. Fuzzy multiple criteria decision-making techniques and applications—Two decades review from 1994 to 2014. *Expert Syst. Appl.* **2015**, *42*, 4126–4148. [CrossRef]
16. Kahraman, C.; Onar, S.C.; Oztaysi, B. Fuzzy multicriteria decision-making: A literature review. *Int. J. Comput. Intell. Syst.* **2015**, *8*, 637–666. [CrossRef]
17. Montazer, G.A.; Saremi, H.Q.; Ramezani, M. Design a new mixed expert decision aiding system using fuzzy electre III method for vendor selection. *Expert Syst. Appl.* **2009**, *36*, 10837–10847. [CrossRef]
18. Chen, T.-Y. An interval type-2 fuzzy promethee method using a likelihood-based outranking comparison approach. *Inf. Fusion* **2015**, *25*, 105–120. [CrossRef]
19. Park, J.H.; Park, I.Y.; Kwun, Y.C.; Tan, X.G. Extension of the topsis method for decision making problems under interval-valued intuitionistic fuzzy environment. *Appl. Math. Model.* **2011**, *35*, 2544–2556. [CrossRef]
20. Zhang, X.L.; Xu, Z.S. The todim analysis approach based on novel measured functions under hesitant fuzzy environment. *Knowl.-Based Syst.* **2014**, *61*, 48–58. [CrossRef]
21. Zhang, X.L.; Xu, Z.S. Hesitant fuzzy qualiflex approach with a signed distance-based comparison method for multiple criteria decision analysis. *Expert Syst. Appl.* **2015**, *42*, 873–884. [CrossRef]
22. Chen, N.; Xu, Z.S. Hesitant fuzzy electre II approach: A new way to handle multi-criteria decision making problems. *Inf. Sci.* **2015**, *292*, 175–197. [CrossRef]
23. Peng, J.-J.; Wang, J.-Q.; Wang, J.; Yang, L.-J.; Chen, X.-H. An extension of electre to multi-criteria decision-making problems with multi-hesitant fuzzy sets. *Inf. Sci.* **2015**, *307*, 113–126. [CrossRef]
24. Turksen, I.B. Interval valued fuzzy sets based on normal forms. *Fuzzy Sets Syst.* **1986**, *20*, 191–210. [CrossRef]
25. Atanassov, K.T. Intuitionistic fuzzy sets. *Fuzzy Sets Syst.* **1986**, *20*, 87–96. [CrossRef]
26. Atanassov, K.T.; Gargov, G. Interval valued intuitionistic fuzzy sets. *Fuzzy Sets Syst.* **1989**, *31*, 343–349. [CrossRef]
27. Torra, V.; Narukawa, Y. On hesitant fuzzy sets and decision. In Proceedings of the 18th IEEE International Conference on Fuzzy Systems, Jeju Island, Korea, 20–24 August 2009; pp. 1378–1382.
28. Torra, V. Hesitant fuzzy sets. *Int. J. Intell. Syst.* **2010**, *25*, 529–539. [CrossRef]
29. Zhu, B.; Xu, Z.S.; Xia, M.M. Dual hesitant fuzzy sets. *J. Appl. Math.* **2012**, *2012*. [CrossRef]
30. Farhadinia, B. Correlation for dual hesitant fuzzy sets and dual interval-valued hesitant fuzzy sets. *Int. J. Intell. Syst.* **2014**, *29*, 184–205. [CrossRef]
31. Zhu, B.; Xu, Z.S. Some results for dual hesitant fuzzy sets. *J. Intell. Fuzzy Syst.* **2014**, *26*, 1657–1668.
32. Ju, Y.B.; Liu, X.Y.; Yang, S.H. Interval-valued dual hesitant fuzzy aggregation operators and their applications to multiple attribute decision making. *J. Intell. Fuzzy Syst.* **2014**, *27*, 1203–1218.
33. Herrera, F.; Herrera-Viedma, E.; Martínez, L. A fusion approach for managing multi-granularity linguistic term sets in decision making. *Fuzzy Sets Syst.* **2000**, *114*, 43–58. [CrossRef]
34. Tao, Z.; Liu, X.; Chen, H.; Zhou, L. Using new version of extended t-Norms and s-Norms for aggregating interval linguistic labels. *IEEE Trans. Syst. Man Cybern. Syst.* **2016**, *47*, 3284–3298. [CrossRef]
35. Martínez, L.; Herrera, F. An overview on the 2-tuple linguistic model for computing with words in decision making: Extensions, applications and challenges. *Inf. Sci.* **2012**, *207*, 1–18. [CrossRef]
36. Wang, J.Q.; Wu, J.T.; Wang, J.; Zhang, H.Y.; Chen, X.H. Interval-valued hesitant fuzzy linguistic sets and their applications in multi-criteria decision-making problems. *Inf. Sci.* **2014**, *288*, 55–72. [CrossRef]
37. Lin, R.; Zhao, X.F.; Wang, H.J.; Wei, G.W. Hesitant fuzzy linguistic aggregation operators and their application to multiple attribute decision making. *J. Intell. Fuzzy Syst.* **2014**, *27*, 49–63.
38. Zhang, N. Hesitant fuzzy linguistic information aggregation in decision making. *Int. J. Oper. Res.* **2014**, *21*, 489–507. [CrossRef]
39. Zhang, W.; Ju, Y.; Liu, X. Multiple criteria decision analysis based on shapley fuzzy measures and interval-valued hesitant fuzzy linguistic numbers. *Comput. Ind. Eng.* **2017**, *105*, 28–38. [CrossRef]
40. Yang, S.H.; Ju, Y.B. Dual hesitant fuzzy linguistic aggregation operators and their applications to multi-attribute decision making. *J. Intell. Fuzzy Syst.* **2014**, *27*, 1935–1947.
41. Qi, X.; Liang, C.; Zhang, J. Multiple attribute group decision making based on generalized power aggregation operators under interval-valued dual hesitant fuzzy linguistic environment. *Int. J. Mach. Learn. Cybern.* **2016**, *7*, 1147–1193. [CrossRef]

42. Herrera-Viedma, E.; López-Herrera, A.G. A model of an information retrieval system with unbalanced fuzzy linguistic information. *Int. J. Intell. Syst.* **2007**, *22*, 1197–1214. [CrossRef]
43. Martínez, L.; Espinilla, M.; Liu, J.; Pérez, L.G.; Sánchez, P.J. An evaluation model with unbalanced linguistic information applied to olive oil sensory evaluation. *J. Mult.-Valued Log. Soft Comput.* **2009**, *15*, 229–251.
44. Herrera, F.; Herrera-Viedma, E.; Martinez, L. A fuzzy linguistic methodology to deal with unbalanced linguistic term sets. *IEEE Trans. Fuzzy Syst.* **2008**, *16*, 354–370. [CrossRef]
45. Meng, D.; Pei, Z. On weighted unbalanced linguistic aggregation operators in group decision making. *Inf. Sci.* **2013**, *223*, 31–41. [CrossRef]
46. Dong, Y.; Li, C.C.; Herrera, F. An optimization-based approach to adjusting unbalanced linguistic preference relations to obtain a required consistency level. *Inf. Sci.* **2015**, *292*, 27–38. [CrossRef]
47. Qi, X.; Zhang, J.; Liang, C. Multiple attributes group decision-making approaches based on interval-valued dual hesitant fuzzy unbalanced linguistic set and their applications. *Complexity* **2018**, *2018*. [CrossRef]
48. Ju, Y.B.; Wang, A.H. Emergency alternative evaluation under group decision makers: A method of incorporating ds/ahp with extended topsis. *Expert Syst. Appl.* **2012**, *39*, 1315–1323. [CrossRef]
49. Yager, R.R. Prioritized aggregation operators. *Int. J. Approx. Reason.* **2008**, *48*, 263–274. [CrossRef]
50. Wei, G. Hesitant fuzzy prioritized operators and their application to multiple attribute decision making. *Knowl.-Based Syst.* **2012**, *31*, 176–182. [CrossRef]
51. Yager, R.R. Prioritized owa aggregation. *Fuzzy Optim. Decis Mak.* **2009**, *8*, 245–262. [CrossRef]
52. Yu, D.; Wu, Y.; Lu, T. Interval-valued intuitionistic fuzzy prioritized operators and their application in group decision making. *Knowl.-Based Syst.* **2012**, *30*, 57–66. [CrossRef]
53. Yu, X.H.; Xu, Z.S. Prioritized intuitionistic fuzzy aggregation operators. *Inf. Fusion* **2013**, *14*, 108–116. [CrossRef]
54. Yu, D.J. Prioritized information fusion method for triangular intuitionistic fuzzy set and its application to teaching quality evaluation. *Int. J. Intell. Syst.* **2013**, *28*, 411–435. [CrossRef]
55. Zhao, Q.Y.; Chen, H.Y.; Zhou, L.G.; Tao, Z.F.; Liu, X. The properties of fuzzy number intuitionistic fuzzy prioritized operators and their applications to multi-criteria group decision making. *J. Intell. Fuzzy Syst.* **2015**, *28*, 1835–1848.
56. Peng, D.H.; Wang, T.D.; Gao, C.Y.; Wang, H. Multigranular uncertain linguistic prioritized aggregation operators and their application to multiple criteria group decision making. *J. Appl. Math.* **2013**, *2013*. [CrossRef]
57. Ren, Z.; Wei, C. A multi-attribute decision-making method with prioritization relationship and dual hesitant fuzzy decision information. *Int. J. Mach. Learn. Cybern.* **2017**, *8*, 755–763. [CrossRef]
58. Wu, J.T.; Wang, J.Q.; Wang, J.; Zhang, H.Y.; Chen, X.H. Hesitant fuzzy linguistic multicriteria decision-making method based on generalized prioritized aggregation operator. *Sci. World J.* **2014**. [CrossRef] [PubMed]
59. Zadeh, L.A. The concept of a linguistic variable and its application to approximate reasoning-I. *Inf. Sci.* **1975**, *8*, 199–249. [CrossRef]
60. Rodríguez, R.M.; Martínez, L.; Herrera, F. Hesitant fuzzy linguistic term sets for decision making. *IEEE Trans. Fuzzy Syst.* **2012**, *20*, 109–119. [CrossRef]
61. Rodríguez, R.M.; Martínez, L.; Herrera, F. A group decision making model dealing with comparative linguistic expressions based on hesitant fuzzy linguistic term sets. *Inf. Sci.* **2013**, *241*, 28–42. [CrossRef]
62. Beg, I.; Rashid, T. Topsis for hesitant fuzzy linguistic term sets. *Int. J. Intell. Syst.* **2013**, *28*, 1162–1171. [CrossRef]
63. Wang, J.-Q.; Wu, J.-T.; Wang, J.; Zhang, H.-Y.; Chen, X.-H. Multi-criteria decision-making methods based on the hausdorff distance of hesitant fuzzy linguistic numbers. *Soft Comput.* **2016**, *20*, 1621–1633. [CrossRef]
64. Saaty, T.L. *The Analytic Hierarchy Process*; McGraw-Hill: New York, NY, USA, 1980.
65. Chinese Government. *Regulations on Natural Disaster Rscue and Assistance*; The-Ministry-of-Civil-Affairs, Chinese Government: Beijing, China, 2010. Available online: http://www.mca.gov.cn/article/zwgk/fvfg/jzjj/201008/20100800095101.shtml (accessed on 13 June 2018).
66. Chinese Government. *Emergency Plan for Natural Disaster Rescue*, Modified ed.; The-Ministry-of-Civil-Affairs, Chinese Government: Beijing, China, 2011. Available online: http://www.mca.gov.cn/article/zwgk/fvfg/jzjj/201111/20111100191129.shtml (accessed on 13 June 2018).
67. Momoh, J.A.; Zhang, Y.; Fanara, P.; Kurban, H.; Iwarere, L.J. Social impact based contingency screening and ranking. *Int. J. Crit. Infrastruct.* **2007**, *3*, 124–141. [CrossRef]

68. Kelly, C. *Quick Guide: Rapid Environmental Impact Assessment in Disaster*; Benfield Hazard Research Centre, University College London and CARE International: London, UK, 2003; pp. 1–43.
69. Kelly, C. *Guidelines for Rapid Environmental Impact Assessment in Disasters*; Benfield Greig Hazard Research Centre, University College London and CARE International: London, UK, 2005; pp. 1–86.
70. Zhou, X.Q.; Li, Q.G. Generalized hesitant fuzzy prioritized einstein aggregation operators and their application in group decision making. *Int. J. Fuzzy Syst.* **2014**, *16*, 303–316.
71. Ye, J. Prioritized aggregation operators of trapezoidal intuitionistic fuzzy sets and their application to multicriteria decision-making. *Neural Comput. Appl.* **2014**, *25*, 1447–1454. [CrossRef]
72. Chen, L.; Xu, Z. A new prioritized multi-criteria outranking method: The prioritized promethee. *J. Intell. Fuzzy Syst.* **2015**, *29*, 2099–2110. [CrossRef]
73. Wang, Y.M. Using the method of maximizing deviations to make decision for multi-indices. *Syst. Eng. Electron.* **1997**, *8*, 21–26.

information

Article

Pythagorean Fuzzy Interaction Muirhead Means with Their Application to Multi-Attribute Group Decision-Making

Yuan Xu, Xiaopu Shang * and Jun Wang

School of Economics and Management, Beijing Jiaotong University, Beijing 100044, China; 17120627@bjtu.edu.cn (Y.X.); 14113149@bjtu.edu.cn (J.W.)

* Correspondence: sxp@bjtu.edu.cn; Tel.: +86-10-5168-3854

Received: 4 June 2018; Accepted: 23 June 2018; Published: 27 June 2018

Abstract: Due to the increased complexity of real decision-making problems, representing attribute values correctly and appropriately is always a challenge. The recently proposed Pythagorean fuzzy set (PFS) is a powerful and useful tool for handling fuzziness and vagueness. The feature of PFS that the square sum of membership and non-membership degrees should be less than or equal to one provides more freedom for decision makers to express their assessments and further results in less information loss. The aim of this paper is to develop some Pythagorean fuzzy aggregation operators to aggregate Pythagorean fuzzy numbers (PFNs). Additionally, we propose a novel approach to multi-attribute group decision-making (MAGDM) based on the proposed operators. Considering the Muirhead mean (MM) can capture the interrelationship among all arguments, and the interaction operational rules for PFNs can make calculation results more reasonable, to take full advantage of both, we extend MM to PFSs and propose a family of Pythagorean fuzzy interaction Muirhead mean operators. Some desirable properties and special cases of the proposed operators are also investigated. Further, we present a novel approach to MAGDM with Pythagorean fuzzy information. Finally, we provide a numerical instance to illustrate the validity of the proposed model. In addition, we perform a comparative analysis to show the superiorities of the proposed method.

Keywords: Pythagorean fuzzy set; Muirhead mean; interaction operational laws; multi-attribute group decision-making

1. Introduction

As one of the most important branches of modern decision-making theory, multi-attribute group decision-making (MAGDM) has been widely investigated and successfully applied to many fields, owing to its high capacity of modelling the process of real decision-making problems [1–6]. With the development of management and economics, actual decision-making problems are becoming more and more diversified and complicated. Thus, one of the most significant issues is representing and denoting attribute values appropriately. Zadeh [7] originally introduced the fuzzy set (FS) theory, which makes it possible to describe vagueness and uncertainty. However, the shortcoming of the FS is that it only has a membership degree, making it insufficient to express fuzziness comprehensively. Recently, Atanassov [3] put forward the concept of an intuitionistic fuzzy set (IFS), which can express the complex fuzzy information effectively as it simultaneously has a membership degree and a non-membership degree. Considering its effective vagueness information processing capabilities, IFS has been widely investigated and applied to so many fields since its appearance. For instance, Liu and Ren [8] proposed a novel intuitionistic fuzzy entropy and based on which a novel approach to MAGDM was proposed. Ren and Wang [9] proposed a new similarity measure for interval-valued IFSs, which considers not only the impacts of membership and membership degrees but also the median point of interval-valued IFSs. Kaur and Garg [10] extended IFSs and proposed cubic intuitionistic fuzzy sets as well as their aggregation operators. P. Liu and X. Liu [11] proposed the concept of linguistic intuitionistic fuzzy sets based on the combination of IFSs and linguistic terms sets and applied them to MAGDM. Liu and Wang [12] extend partitioned Heronian mean operator to linguistic intuitionistic fuzzy sets and applied it to MAGDM. Lakshmana et al. [13] proposed a total order on the entire class of intuitionistic fuzzy numbers using an upper lower dense sequence in the interval [1]. Liu and Teng [14] proposed the concept of normal interval-valued intuitionistic fuzzy numbers and applied it to decision-making. Liu and Chen [15] introduced some intuitionistic fuzzy Heronian mean operators based on the Archimedean t-conorm and t-norm and applied them to dealing with MAGDM problems.

Recently, as an extension of the IFS, the Pythagorean fuzzy set (PFS) [16], which is also characterized by a membership degree and a non-membership degree, has been proposed. The prominent feature of the PFS that the sum of membership and non-membership degrees may be greater than one and their square sum should be less than or equal to one, makes the PFS more powerful and useful than the IFS. Since its appearance, it has drawn much attention. For example, Zhang [17] proposed a novel similarity measure for PFSs and based on which a new method to Pythagorean fuzzy MAGDM problems was developed. Zhang and Xu [18] and Ren et al. [19] respectively extended the traditional TOPSIS (technique for order preference by similarity to ideal solution) method and the TODIM (an acronym in Portuguese for interactive multi-criteria decision-making) approach to solve MAGDM in a Pythagorean fuzzy context. Aggregation operators are a central topic in MAGDM, as they can ingrate individual input data into collective ones, and rank the alternatives based on the collective value. In the past years, quite a few Pythagorean fuzzy operators have been proposed and been applied to MAGDM successfully [20–26]. However, the main shortcomings of these operators are:

(1) They cannot consider the interrelationship between Pythagorean fuzzy numbers (PFNs). In other words, these aggregation operators assume that the attributes are independent, signifying that the correlations among attribute values are not taken into consideration when aggregating them. Generally, the Bonferroni mean (BM) [27], Heronian mean (HM) [28], and Maclaurin symmetric mean (MSM) [29] are aggregation technologies that consider the interrelationships among arguments. Thus, in order to overcome the shortcoming of the aforementioned aggregation operators, some other Pythagorean fuzzy aggregation operators have been proposed. Liang et al. [30,31] proposed some Pythagorean fuzzy Bonferroni mean and geometric Bonferroni mean operators, respectively. Zhang et al. [32] investigated the generalized Bonferroni mean to aggregate Pythagorean fuzzy information and proposed a family of Pythagorean fuzzy generalized Bonferroni means. Wei and Lu [33], and Qin [34] proposed some Pythagorean fuzzy Maclaurin symmetric mean operators, respectively. These operators consider the interrelationships between any two or among multiple arguments, however, they fail to capture the interrelationships among all arguments. The Muirhead mean (MM) [35] is a useful and powerful aggregation technology that captures the interrelationships among all arguments. Moreover, it has a parameter vector that leads to flexible aggregation processes. Quite a few existing aggregation operators are some special cases of MM. The MM was introduced for crisp numbers and, up to now, MM has been investigated in intuitionistic fuzzy [36] and 2-tuple linguistic environments [37]. However, to the best of our knowledge, nothing has been done about MM in a Pythagorean fuzzy environment. Thus, in order to aggregate Pythagorean fuzzy information, it is necessary to extend the MM to a Pythagorean fuzzy environment

(2) The aforementioned aggregation operators are based on the traditional Pythagorean fuzzy operational rules introduced in [18]. However, these operations cannot be used to deal with some situations. For instance, let $p_1 = (\mu_1, v_1)$ and $p_2 = (\mu_2, v_2)$ be two PFNs, if $\mu_1 = 0$ and $\mu_2 \neq 0$, then according to the operational laws proposed by Zhang and Xu [18], we can obtain $\mu_{p_1 \oplus p_2} = 0$. It is noted that μ_2 is not accounted for at all. Similarly, if $v_1 = 0$ and $v_2 \neq 0$, then and v_2 is not accounted for at all. It is not consistent with our intuition and the reality. To overcome the drawback of the proposed operations, Wei [38] proposed the interaction operations for PFNs.

Therefore, to take full advantages of MM and Wei' [38] Pythagorean fuzzy interaction operations, we propose a family of Pythagorean fuzzy interaction Muirhead mean operators. Thus, the proposed operators not only capture the interrelationships among all input arguments, but also effectively handle situations in which a membership or non-membership degree of an attribute value is equal to one. It is worth pointing out that in [39], Zhu and Li also proposed some Pythagorean fuzzy Muirhead mean operators. However, the proposed operators in this paper are different from those proposed by Zhu and Li. The main difference is that Zhu and Li's [39] operators are based on the basic operational laws proposed in [18]. Therefore, Zhu and Li's [39] operators do not work for situations in which one membership degree or one non-membership degree is equal to one. Our operators are based on the interaction operational rules of PFNs, so that the proposed operators in this study are more powerful and flexible than Zhu and Li's operators. Further, based on the proposed aggregation operators, we propose a novel approach to MAGDM in which attribute values take the form of PFNs. The main aims and motivations of this paper are: (1) to develop a family of Pythagorean fuzzy Muirhead mean operations based on interaction operational laws; and (2) to propose a novel approach to MAGDM with Pythagorean fuzzy information. The rest of the paper is organized as follows. Section 2 recalls some basic concepts, such as PFS, MM, and the interaction operations of PFNs. Section 3 extends the MM to Pythagorean fuzzy environment and proposes the Pythagorean fuzzy interaction Muirhead mean (PFIMM) operator and the Pythagorean fuzzy interaction weighted Muirhead mean (PFIWMM) operator. Section 4 extends the DMM to aggregating Pythagorean fuzzy information and develops the Pythagorean fuzzy interaction dual Muirhead mean (PFIDMM) operator and the Pythagorean fuzzy interaction weighted dual Muirhead mean (PFIDWMM) operator. Section 5 develops a novel approach to MAGDM with Pythagorean fuzzy information based on the proposed operators. Section 6 provides a numerical example to illustrate the performance of the proposed method and the final section summarizes the whole paper.

2. Basic Concepts

In this section, we briefly review the concepts of IFS, PFS, and MM.

2.1. IFS and PFS

Definition 1 [3]. *An intuitionistic fuzzy set A with an object X is defined as follows:*

$$A = \{\langle x, \mu A(x), vA(x)\rangle \,|x \in X\} \tag{1}$$

where $\mu_A(x)$ and $v_A(x)$ represent the membership and non-membership degrees respectively, satisfying $\mu A(x) \in [0,1]$, $vA(x) \in [0,1]$ and $\mu A(x) + vA(x) \in [0,1]$, $\forall x \in X$. For convenience, $(\mu A(x), vA(x))$ is called an intuitionistic fuzzy number (IFN), which can be denoted by $\alpha = (\mu, v)$.

Yager [16] extended Atanassov's IFS and proposed the PFS.

Definition 2 [16]. *A Pythagorean fuzzy set P with an object X is defined as follows:*

$$P = \{\langle x, \mu_p(x), v_p(x)\rangle \,|x \in X\}, \tag{2}$$

where $\mu_p(x)$ and $v_p(x)$ are the membership degree the non-membership degree respectively, satisfying $\mu_p(x) \in [0,1]$, $v_p(x) \in [0,1]$ and $(\mu_P(x))^2 + (v_P(x))^2 \leq 1$, $\forall x \in X$. Then the hesitancy degree of P is defined as $\pi_P(x) = \sqrt{1 - (\mu_P(x))^2 - (v_P(x))^2}$, $\forall x \in X$. For convenience, $(\mu_p(x), v_p(x))$ is called a PFN, which can be denoted by $p = (\mu_P, v_P)$.

To compare two PFNs, Zhang and Xu [18] proposed a comparison law.

Definition 3 [18]. *Let $p = (\mu, v)$ be a PFN, then the score function of p is defined as $S(p) = \mu^2 - v^2$. For any two PFNs, $p_1 = (\mu_1, v_1)$ and $p_2 = (\mu_2, v_2)$, if $S(p_1) > S(p_2)$, then $p_1 > p_2$; if $S(p_1) = S(p_2)$, then $p_1 = p_2$.*

Moreover, Zhang and Xu [18] proposed some operations for PFNs.

Definition 4 [18]. *Let $p = (\mu, v)$, $p_1 = (\mu_1, v_1)$ and $p_2 = (\mu_2, v_2)$ be any three PFNs, and λ be a positive real number, then*

(1) $p_1 \oplus p_2 = \left(\sqrt{\mu_1^2 + \mu_2^2 - \mu_1^2\mu_2^2}, v_1 v_2\right)$,

(2) $p_1 \otimes p_2 = \left(\mu_1\mu_2, \sqrt{v_1^2 + v_2^2 - v_1^2 v_2^2}\right)$,

(3) $\lambda p = \left(\sqrt{1 - (1 - \mu^2)^\lambda}, v^\lambda\right)$,

(4) $p^\lambda = \left(\mu^\lambda, \sqrt{1 - (1 - v^2)^\lambda}\right)$.

However, the operational laws shown above cannot reflect the correlations between membership degrees and non-membership degrees. Thus, Wei [38] proposed some interaction operations for PFNs that are shown as the following.

Definition 5 [38]. *Let $p = (\mu, v)$,$p_1 = (\mu_1, v_1)$ and $p_2 = (\mu_2, v_2)$ be any of the three PFNs, and λ be* any positive real number, then

(1) $p_1 \oplus p_2 = \left(\sqrt{1 - (1 - \mu_1^2)(1 - \mu_2^2)}, \sqrt{(1 - \mu_1^2)(1 - \mu_2^2) - (1 - \mu_1^2 - v_1^2)(1 - \mu_2^2 - v_2^2)}\right)$,

(2) $p_1 \otimes p_2 = \left(\sqrt{(1 - v_1^2)(1 - v_2^2) - (1 - \mu_1^2 - v_1^2)(1 - \mu_2^2 - v_2^2)}, \sqrt{1 - (1 - v_1^2)(1 - v_2^2)}\right)$,

(3) $\lambda p = \left(\sqrt{1 - (1 - \mu^2)^\lambda}, \sqrt{(1 - \mu^2)^\lambda - (1 - \mu^2 - v^2)^\lambda}\right)$,

(4) $p^\lambda = \left(\sqrt{(1 - v^2)^\lambda - (1 - \mu^2 - v^2)^\lambda}, \sqrt{1 - (1 - v^2)^\lambda}\right)$.

2.2. The Muirhead Mean

The MM was introduced by Muirhead [35] for crisp numbers. The prominent advantage of the MM is that it can capture interrelationships among all of the aggregated arguments.

Definition 6 [35]. *Let $a_i (i = 1, 2, \cdots, n)$ be a collection of crisp numbers and $R = (r_1, r_2, \ldots, r_n) \in R^n$ be a vector of parameters, then the MM can be defined as*

$$MM^R(a_1, a_2, \ldots, a_n) = \left(\frac{1}{n!}\sum_{\vartheta \in S_n}\prod_{j=1}^{n} a_{\vartheta(j)}^{r_j}\right)^{\frac{1}{\sum_{j=1}^{n} r_j}} \tag{3}$$

where $\vartheta(j)(j = 1, 2, \cdots, n)$ is any permutation of $(1, 2, \ldots, n)$, S_n is the collection of $\vartheta(j)(j = 1, 2, \cdots, n)$.

Liu and Li [36] proposed the dual operator of MM, which is called the DMM operator.

Definition 7 [36]. *Let $a_i (i = 1, 2, \cdots, n)$ be a collection of crisp numbers and $P = (p_1, p_2, \cdots, p_n) \in R^n$ be a vector of parameters. If*

$$DMM^P(a_1, a_2, \ldots, a_n) = \frac{1}{\sum_{j=1}^{n} p_j}\left(\prod_{\vartheta \in S_n}\sum_{j=1}^{n}\left(p_j a_{\vartheta(j)}\right)\right)^{\frac{1}{n!}} \tag{4}$$

Then DMM^P is called the DMM, where $\vartheta(j)(j = 1, 2, \cdots, n)$ is any permutation of $(1, 2, \ldots, n)$ and S_n is the collection of $\vartheta(j)(j = 1, 2, \cdots, n)$.

3. The Pythagorean Fuzzy Interaction Muirhead Mean and the Pythagorean Fuzzy Interaction Weighted Muirhead Mean

In this section, we extend the MM to Pythagorean fuzzy environment and propose some new Pythagorean fuzzy aggregation operators.

3.1. The Pythagorean Fuzzy Interaction Muirhead Mean

Definition 8 . *Let $p_i (i = 1, 2, \ldots, n)$ be a collection of PFNs and $R = (r_1, r_2, \ldots, r_n) \in R^n$ be a vector of parameters. If*

$$PFIMM^R(p_1, p_2, \ldots, p_n) = \left(\frac{1}{n!} \sum_{\vartheta \in S_n} \prod_{j=1}^{n} p_{\vartheta(j)}^{r_j} \right)^{\frac{1}{\sum_{j=1}^{n} r_j}} \tag{5}$$

then $PFIMM^R$ is called the PFIMM, where $\vartheta(j)(j = 1, 2, \cdots, n)$ is any a permutation of $(1, 2, \cdots, n)$, and S_n is the collection of $\vartheta(j)(j = 1, 2, \cdots, n)$.

According to the interaction operations for PFNs presented in Definition 5, the following theorem can be obtained.

Theorem 1. *Let $p_i = (\mu_i, v_i)(i = 1, 2, \cdots, n)$ be a collection of PFNs, the aggregated value by using the PFIMM is still a PFN and*

$$PFIMM^R(p_1, p_2, \ldots, p_n) = \\ \left(\left(\left(1 - \prod_{\vartheta \in S_n} \left(1 - \prod_{j=1}^{n} \left(1 - v_{\vartheta(j)}^2\right)^{r_j} + \prod_{j=1}^{n} \left(1 - \mu_{\vartheta(j)}^2 - v_{\vartheta(j)}^2\right)^{r_j} \right)^{\frac{1}{n!}} + \prod_{\vartheta \in S_n} \prod_{j=1}^{n} \left(1 - \mu_{\vartheta(j)}^2 - v_{\vartheta(j)}^2\right)^{\frac{r_j}{n!}} \right)^{\frac{1}{\sum_{j=1}^{n} r_j}} - \prod_{\vartheta \in S_n} \prod_{j=1}^{n} \left(1 - \mu_{\vartheta(j)}^2 - v_{\vartheta(j)}^2\right)^{\frac{r_j}{n! \sum_{j=1}^{n} r_j}} \right)^{\frac{1}{2}}, \right. \\ \left. \left(1 - \left(1 - \prod_{\vartheta \in S_n} \left(1 - \prod_{j=1}^{n} \left(1 - v_{\vartheta(j)}^2\right)^{r_j} + \prod_{j=1}^{n} \left(1 - \mu_{\vartheta(j)}^2 - v_{\vartheta(j)}^2\right)^{r_j} \right)^{\frac{1}{n!}} + \prod_{\vartheta \in S_n} \left(\prod_{j=1}^{n} \left(1 - \mu_{\vartheta(j)}^2 - v_{\vartheta(j)}^2\right)^{r_j} \right)^{\frac{1}{n!}} \right)^{\frac{1}{\sum_{j=1}^{n} r_j}} \right)^{\frac{1}{2}} \right) \tag{6}$$

Proof. According to the Definition 5, we have

$$p_{\vartheta(j)}^{r_j} = \left(\sqrt{\left(1 - v_{\vartheta(j)}^2\right)^{r_j} - \left(1 - \mu_{\vartheta(j)}^2 - v_{\vartheta(j)}^2\right)^{r_j}}, \sqrt{1 - \left(1 - v_{\vartheta(j)}^2\right)^{r_j}} \right) \tag{7}$$

and,

$$\prod_{j=1}^{n} p_{\vartheta(j)}^{r_j} = \left(\sqrt{\prod_{j=1}^{n} \left(1 - v_{\vartheta(j)}^2\right)^{r_j} - \prod_{j=1}^{n} \left(1 - \mu_{\vartheta(j)}^2 - v_{\vartheta(j)}^2\right)^{r_j}}, \sqrt{1 - \prod_{j=1}^{n} \left(1 - v_{\vartheta(j)}^2\right)^{r_j}} \right) \tag{8}$$

Then,

$$\sum_{\vartheta\in S_n}\prod_{j=1}^{n}p_{\vartheta(j)}^{r_j}=\left(\sqrt{1-\prod_{\vartheta\in S_n}\left(1-\prod_{j=1}^{n}\left(1-v_{\vartheta(j)}^2\right)^{r_j}+\prod_{j=1}^{n}\left(1-\mu_{\vartheta(j)}^2-v_{\vartheta(j)}^2\right)^{r_j}\right)},\right.$$
$$\left.\sqrt{\prod_{\vartheta\in S_n}\left(1-\prod_{j=1}^{n}\left(1-v_{\vartheta(j)}^2\right)^{r_j}+\prod_{j=1}^{n}\left(1-\mu_{\vartheta(j)}^2-v_{\vartheta(j)}^2\right)^{r_j}\right)-\prod_{\vartheta\in S_n}\left(\prod_{j=1}^{n}\left(1-\mu_{\vartheta(j)}^2-v_{\vartheta(j)}^2\right)^{r_j}\right)}\right) \tag{9}$$

Further,

$$\frac{1}{n!}\sum_{\vartheta\in S_n}\prod_{j=1}^{n}p_{\vartheta(j)}^{r_j}=\left(\sqrt{1-\prod_{\vartheta\in S_n}\left(1-\prod_{j=1}^{n}\left(1-v_{\vartheta(j)}^2\right)^{r_j}+\prod_{j=1}^{n}\left(1-\mu_{\vartheta(j)}^2-v_{\vartheta(j)}^2\right)^{r_j}\right)^{\frac{1}{n!}}},\right.$$
$$\left.\sqrt{\prod_{\vartheta\in S_n}\left(1-\prod_{j=1}^{n}\left(1-v_{\vartheta(j)}^2\right)^{r_j}+\prod_{j=1}^{n}\left(1-\mu_{\vartheta(j)}^2-v_{\vartheta(j)}^2\right)^{r_j}\right)^{\frac{1}{n!}}-\prod_{\vartheta\in S_n}\prod_{j=1}^{n}\left(1-\mu_{\vartheta(j)}^2-v_{\vartheta(j)}^2\right)^{\frac{r_j}{n!}}}\right) \tag{10}$$

Moreover,

$$\left(\frac{1}{n!}\sum_{\vartheta\in S_n}\prod_{j=1}^{n}p_{\vartheta(j)}^{r_j}\right)^{\frac{1}{\sum_{j=1}^{n}r_j}}=$$
$$\left(\left(\begin{array}{l}\left(1-\prod_{\vartheta\in S_n}\left(1-\prod_{j=1}^{n}\left(1-v_{\vartheta(j)}^2\right)^{r_j}+\prod_{j=1}^{n}\left(1-\mu_{\vartheta(j)}^2-v_{\vartheta(j)}^2\right)^{r_j}\right)^{\frac{1}{n!}}+\prod_{\vartheta\in S_n}\prod_{j=1}^{n}\left(1-\mu_{\vartheta(j)}^2-v_{\vartheta(j)}^2\right)^{\frac{r_j}{n!}}\right)^{\frac{1}{\sum_{j=1}^{n}r_j}}\\-\prod_{\vartheta\in S_n}\prod_{j=1}^{n}\left(1-\mu_{\vartheta(j)}^2-v_{\vartheta(j)}^2\right)^{\frac{r_j}{n!\sum_{j=1}^{n}r_j}}\end{array}\right)^{\frac{1}{2}},\right.$$
$$\left.\left(1-\left(1-\prod_{\vartheta\in S_n}\left(1-\prod_{j=1}^{n}\left(1-v_{\vartheta(j)}^2\right)^{r_j}+\prod_{j=1}^{n}\left(1-\mu_{\vartheta(j)}^2-v_{\vartheta(j)}^2\right)^{r_j}\right)^{\frac{1}{n!}}+\prod_{\vartheta\in S_n}\prod_{j=1}^{n}\left(1-\mu_{\vartheta(j)}^2-v_{\vartheta(j)}^2\right)^{\frac{r_j}{n!}}\right)^{\frac{1}{\sum_{j=1}^{n}r_j}}\right)^{\frac{1}{2}}\right) \tag{11}$$

Hence, Equation (6) is maintained.

For convenience, let

$$\mu=\left(\begin{array}{l}\left(1-\prod_{\vartheta\in S_n}\left(1-\prod_{j=1}^{n}\left(1-v_{\vartheta(j)}^2\right)^{r_j}+\prod_{j=1}^{n}\left(1-\mu_{\vartheta(j)}^2-v_{\vartheta(j)}^2\right)^{r_j}\right)^{\frac{1}{n!}}+\prod_{\vartheta\in S_n}\prod_{j=1}^{n}\left(1-\mu_{\vartheta(j)}^2-v_{\vartheta(j)}^2\right)^{\frac{r_j}{n!}}\right)^{\frac{1}{\sum_{j=1}^{n}r_j}}\\-\prod_{\vartheta\in S_n}\prod_{j=1}^{n}\left(1-\mu_{\vartheta(j)}^2-v_{\vartheta(j)}^2\right)^{\frac{r_j}{n!\sum_{j=1}^{n}r_j}}\end{array}\right)^{\frac{1}{2}}$$

and

$$v=\left(1-\left(1-\prod_{\vartheta\in S_n}\left(1-\prod_{j=1}^{n}\left(1-v_{\vartheta(j)}^2\right)^{r_j}+\prod_{j=1}^{n}\left(1-\mu_{\vartheta(j)}^2-v_{\vartheta(j)}^2\right)^{r_j}\right)^{\frac{1}{n!}}+\prod_{\vartheta\in S_n}\prod_{j=1}^{n}\left(1-\mu_{\vartheta(j)}^2-v_{\vartheta(j)}^2\right)^{\frac{r_j}{n!}}\right)^{\frac{1}{\sum_{j=1}^{n}r_j}}\right)^{\frac{1}{2}}$$

Evidently,

$$0\le\mu_{\vartheta(j)}\le1,\ 0\le v_{\vartheta(j)}\le1,\ 0\le\mu_{\vartheta(j)}^2+v_{\vartheta(j)}^2\le1, \tag{12}$$

and,

$$0\le\prod_{j=1}^{n}\left(1-v_{\vartheta(j)}^2\right)^{r_j}\le1,\text{ and }0\le\prod_{j=1}^{n}\left(1-\mu_{\vartheta(j)}^2-v_{\vartheta(j)}^2\right)^{r_j}\le1. \tag{13}$$

Then,

$$0 \leq 1 - \left(\prod_{j=1}^{n} \left(1 - v_{\vartheta(j)}^2\right)^{r_j} - \prod_{j=1}^{n} \left(1 - \mu_{\vartheta(j)}^2 - v_{\vartheta(j)}^2\right)^{r_j} \right) \leq 1 \tag{14}$$

Further,

$$0 \leq 1 - \prod_{j=1}^{n} \left(1 - v_{\vartheta(j)}^2\right)^{r_j} + \prod_{j=1}^{n} \left(1 - \mu_{\vartheta(j)}^2 - v_{\vartheta(j)}^2\right)^{r_j} \leq 1 \tag{15}$$

and,

$$0 \leq \prod_{\vartheta \in S_n} \left(1 - \prod_{j=1}^{n} \left(1 - v_{\vartheta(j)}^2\right)^{r_j} + \prod_{j=1}^{n} \left(1 - \mu_{\vartheta(j)}^2 - v_{\vartheta(j)}^2\right)^{r_j} \right)^{\frac{1}{n!}} \leq 1,\ 0 \leq \prod_{\vartheta \in S_n} \prod_{j=1}^{n} \left(1 - \mu_{\vartheta(j)}^2 - v_{\vartheta(j)}^2\right)^{\frac{r_j}{n!}} \leq 1. \tag{16}$$

Moreover,

$$0 \leq \left(1 - \prod_{\vartheta \in S_n} \left(1 - \prod_{j=1}^{n} \left(1 - v_{\vartheta(j)}^2\right)^{r_j} + \prod_{j=1}^{n} \left(1 - \mu_{\vartheta(j)}^2 - v_{\vartheta(j)}^2\right)^{r_j} \right)^{\frac{1}{n!}} + \prod_{\vartheta \in S_n} \prod_{j=1}^{n} \left(1 - \mu_{\vartheta(j)}^2 - v_{\vartheta(j)}^2\right)^{\frac{r_j}{n!}} \right)^{\frac{1}{\sum_{j=1}^{n} r_j}} \leq 1 \tag{17}$$

and,

$$0 \leq \prod_{\vartheta \in S_n} \prod_{j=1}^{n} \left(1 - \mu_{\vartheta(j)}^2 - v_{\vartheta(j)}^2\right)^{\frac{r_j}{n! \sum_{j=1}^{n} r_j}} \leq 1 \tag{18}$$

Therefore,

$$0 \leq \left(\begin{array}{l} \left(1 - \prod_{\vartheta \in S_n} \left(1 - \prod_{j=1}^{n} \left(1 - v_{\vartheta(j)}^2\right)^{r_j} + \prod_{j=1}^{n} \left(1 - \mu_{\vartheta(j)}^2 - v_{\vartheta(j)}^2\right)^{r_j} \right)^{\frac{1}{n!}} + \prod_{\vartheta \in S_n} \prod_{j=1}^{n} \left(1 - \mu_{\vartheta(j)}^2 - v_{\vartheta(j)}^2\right)^{\frac{r_j}{n!}} \right)^{\frac{1}{\sum_{j=1}^{n} r_j}} \\ - \prod_{\vartheta \in S_n} \prod_{j=1}^{n} \left(1 - \mu_{\vartheta(j)}^2 - v_{\vartheta(j)}^2\right)^{\frac{r_j}{n! \sum_{j=1}^{n} r_j}} \end{array} \right)^{\frac{1}{2}} \leq 1$$

Therefore, $0 \leq \mu \leq 1$. Similarly, we can get $0 \leq v \leq 1$.
Then,

$$\mu^2 + v^2 = 1 - \prod_{\vartheta \in S_n} \prod_{j=1}^{n} \left(1 - \mu_{\vartheta(j)}^2 - v_{\vartheta(j)}^2\right)^{\frac{r_j}{n! \sum_{j=1}^{n} r_j}}$$

We have proved that

$$0 \leq \prod_{\vartheta \in S_n} \prod_{j=1}^{n} \left(1 - \mu_{\vartheta(j)}^2 - v_{\vartheta(j)}^2\right)^{\frac{r_j}{n!}} \leq 1$$

Thus,

$$0 \leq \prod_{\vartheta \in S_n} \prod_{j=1}^{n} \left(1 - \mu_{\vartheta(j)}^2 - v_{\vartheta(j)}^2\right)^{\frac{r_j}{n! \sum_{j=1}^{n} r_j}} \leq 1, \text{ and } 0 \leq 1 - \prod_{\vartheta \in S_n} \prod_{j=1}^{n} \left(1 - \mu_{\vartheta(j)}^2 - v_{\vartheta(j)}^2\right)^{\frac{r_j}{n! \sum_{j=1}^{n} r_j}} \leq 1$$

Therefore, $0 \leq \mu^2 + v^2 \leq 1$, which completes the proof. □

Moreover, the PFIMM has the following properties.

Theorem 2. *(Idempotency) If all of the $p_i (i = 1, 2, \cdots, n)$ are equal, i.e., $p_i = p = (\mu, v)$, then*

$$PFIMM^R(p_1, p_2, \cdots, p_n) = p \tag{19}$$

Proof. According to Theorem 1, we can get

$$\begin{aligned}
&PFIMM^R(p, p, \cdots, p) \\
&= \left(\left(\begin{array}{l} \left(1 - \prod_{\vartheta \in S_n} \left(1 - \prod_{j=1}^{n} (1 - v^2)^{r_j} + \prod_{j=1}^{n} (1 - \mu^2 - v^2)^{r_j}\right)^{\frac{1}{n!}} + \prod_{\vartheta \in S_n} \prod_{j=1}^{n} (1 - \mu^2 - v^2)^{\frac{r_j}{n!}}\right)^{\frac{1}{\sum_{j=1}^{n} r_j}} \\ - \prod_{\vartheta \in S_n} \prod_{j=1}^{n} (1 - \mu^2 - v^2)^{\frac{r_j}{n! \sum_{j=1}^{n} r_j}} \end{array} \right)^{\frac{1}{2}}, \right. \\
&\left. \left(1 - \left(1 - \prod_{\vartheta \in S_n} \left(1 - \prod_{j=1}^{n} (1 - v^2)^{r_j} + \prod_{j=1}^{n} (1 - \mu^2 - v^2)^{r_j}\right)^{\frac{1}{n!}} + \prod_{\vartheta \in S_n} \prod_{j=1}^{n} (1 - \mu^2 - v^2)^{\frac{r_j}{n!}}\right)^{\frac{1}{\sum_{j=1}^{n} r_j}}\right)^{\frac{1}{2}} \right) \\
&= \left(\sqrt{\left(1 - \prod_{\vartheta \in S_n} \left(1 - (1 - v^2)^{\sum_{j=1}^{n} r_j} + (1 - \mu^2 - v^2)^{\sum_{j=1}^{n} r_j}\right)^{\frac{1}{n!}} + \prod_{\vartheta \in S_n} (1 - \mu^2 - v^2)^{\frac{\sum_{j=1}^{n} r_j}{n!}}\right)^{\frac{1}{\sum_{j=1}^{n} r_j}} - \prod_{\vartheta \in S_n} (1 - \mu^2 - v^2)^{\frac{1}{n!}}}, \right. \\
&\left. \sqrt{1 - \left(1 - \prod_{\vartheta \in S_n} \left(1 - (1 - v^2)^{\sum_{j=1}^{n} r_j} + (1 - \mu^2 - v^2)^{\sum_{j=1}^{n} r_j}\right)^{\frac{1}{n!}} + \prod_{\vartheta \in S_n} (1 - \mu^2 - v^2)^{\frac{\sum_{j=1}^{n} r_j}{n!}}\right)^{\frac{1}{\sum_{j=1}^{n} r_j}}} \right) \\
&= \left(\sqrt{\left(1 - \left(1 - (1 - v^2)^{\sum_{j=1}^{n} r_j} + (1 - \mu^2 - v^2)^{\sum_{j=1}^{n} r_j}\right) + (1 - \mu^2 - v^2)^{\sum_{j=1}^{n} r_j}\right)^{\frac{1}{\sum_{j=1}^{n} r_j}} - \left((1 - \mu^2 - v^2)^{\sum_{j=1}^{n} r_j}\right)^{\frac{1}{\sum_{j=1}^{n} r_j}}}, \right. \\
&\left. \sqrt{1 - \left(1 - \left(1 - (1 - v^2)^{\sum_{j=1}^{n} r_j} + (1 - \mu^2 - v^2)^{\sum_{j=1}^{n} r_j}\right) + (1 - \mu^2 - v^2)^{\sum_{j=1}^{n} r_j}\right)^{\frac{1}{\sum_{j=1}^{n} r_j}}} \right) \\
&= \left(\sqrt{\left((1 - v^2)^{\sum_{j=1}^{n} r_j}\right)^{\frac{1}{\sum_{j=1}^{n} r_j}} - \left((1 - \mu^2 - v^2)^{\sum_{j=1}^{n} r_j}\right)^{\frac{1}{\sum_{j=1}^{n} r_j}}}, \sqrt{1 - \left((1 - v^2)^{\sum_{j=1}^{n} r_j}\right)^{\frac{1}{\sum_{j=1}^{n} r_j}}} \right) \\
&= \left(\sqrt{(1 - v^2) - (1 - \mu^2 - v^2)}, \sqrt{1 - (1 - v^2)} \right) = \left(\sqrt{\mu^2}, \sqrt{v^2} \right) = (\mu, v).
\end{aligned}$$

The parameter vector R of PFIMM plays an important role in the final result. In the following, we explore some special cases of PFIMM. □

Case 1: If $R = (1, 0, \ldots, 0)$, then the PFIMM is reduced to the following

$$PFIMM^{(1,0,0,\ldots,0)}(p_1, p_2, \cdots, p_n) = \left(\sqrt{1 - \prod_{j=1}^{n} (1 - \mu_i^2)^{\frac{1}{n}}}, \sqrt{\prod_{j=1}^{n} (1 - \mu_i^2)^{\frac{1}{n}} - \prod_{j=1}^{n} (1 - \mu_i^2 - v_i^2)^{\frac{1}{n}}} \right) = \frac{1}{n} \sum_{i=1}^{n} p_i \tag{20}$$

which is the Pythagorean fuzzy interaction averaging (PFIA) operator.
Case 2: If $R = (\lambda, 0, \ldots, 0)$, then the PFIMM is reduced to the following

$$\begin{aligned}
PFIMM^{(\lambda,0,0,\ldots,0)}(p_1, p_2, \ldots, p_n) = &\left(\sqrt{\left(1 - \left(1 - \prod_{j=1}^{n} (1 - v_i^2)^{\lambda} + \prod_{j=1}^{n} (1 - \mu_i^2 - v_i^2)^{\lambda}\right)^{\frac{1}{n}}\right)^{\frac{1}{\lambda}} - \prod_{j=1}^{n} (1 - \mu_i^2 - v_i^2)^{\frac{1}{n}}}, \right. \\
&\left. \sqrt{1 - \left(1 - \left(1 - \prod_{j=1}^{n} (1 - v_i^2)^{\lambda} + \prod_{j=1}^{n} (1 - \mu_i^2 - v_i^2)^{\lambda}\right)^{\frac{1}{n}} + \prod_{j=1}^{n} (1 - \mu_i^2 - v_i^2)^{\frac{\lambda}{n}}\right)^{\frac{1}{\lambda}}} \right) = \left(\frac{1}{n} \sum_{i=1}^{n} p_i^{\lambda} \right)^{\frac{1}{\lambda}},
\end{aligned} \tag{21}$$

which is the generalized Pythagorean fuzzy interaction averaging (GPFIA) operator.
Case 3: If $R = (1,1,0,0,\cdots,0)$, then the PFIMM is reduced to the following

$$
\begin{aligned}
&PFIMM^{(1,1,0,0,\ldots,0)}(p_1,p_2,\ldots,p_n) = \\
&\left(\left(\begin{array}{l} 1-\prod\limits_{\substack{i,j=1\\i\neq j}}^{n}\left(1-(1-v_i^2)\left(1-v_j^2\right)+\left(1-\mu_i^2-v_j^2\right)\left(1-\mu_j^2-v_j^2\right)\right)^{\frac{1}{n(n-1)}}+\prod\limits_{\substack{i,j=1\\i\neq j}}^{n}\left(\left(1-\mu_i^2-v_j^2\right)\left(1-\mu_j^2-v_j^2\right)\right)^{\frac{1}{n(n-1)}} \\ -\prod\limits_{\substack{i,j=1\\i\neq j}}^{n}\left(\left(1-\mu_i^2-v_j^2\right)\left(1-\mu_j^2-v_j^2\right)\right)^{\frac{1}{2n(n-1)}} \end{array}\right)^{\frac{1}{2}}\right)^{\frac{1}{2}}, \\
&\left(1-\left(1-\prod\limits_{\substack{i,j=1\\i\neq j}}^{n}\left(1-(1-v_i^2)\left(1-v_j^2\right)+(1-\mu_i^2-v_i^2)\left(1-\mu_j^2-v_j^2\right)\right)^{\frac{1}{n(n-1)}}+\prod\limits_{\substack{i,j=1\\i\neq j}}^{n}\left((1-\mu_i^2-v_i^2)\left(1-\mu_j^2-v_j^2\right)\right)^{\frac{1}{n(n-1)}}\right)^{\frac{1}{2}}\right)^{\frac{1}{2}}\Bigg) \\
&= \left(\frac{1}{n(n-1)}\sum_{\substack{i,j=1\\i\neq j}}^{n} p_ip_j\right)^{\frac{1}{2}},
\end{aligned}
\tag{22}
$$

which is the Pythagorean fuzzy interaction BM (PFIBM) operator.

Case 4: If $R = \left(\overbrace{1,1,\cdots,1}^{k},\overbrace{0,0,\cdots,0}^{n-k}\right)$, then the PFIMM is reduced to the following

$$
\begin{aligned}
&PFIMM^{\left(\overbrace{1,1,\cdots,1}^{k},\overbrace{0,0,\cdots,0}^{n-k}\right)}(p_1,p_2,\ldots,p_n) = \\
&\left(\left(\left(\begin{array}{l} 1-\prod\limits_{1\leq i_1\prec\cdots\prec i_k\leq n}\left(1-\prod\limits_{j=1}^{k}\left(1-v_{i_j}\right)^2+\prod\limits_{j=1}^{n}\left(1-\mu_{i_j}-v_{i_j}\right)^2\right)^{\frac{1}{C_n^k}}+\prod\limits_{1\leq i_1\prec\cdots\prec i_k\leq n}\prod\limits_{j=1}^{n}\left(1-\mu_{i_j}-v_{i_j}\right)^{\frac{2}{C_n^k}} \\ -\prod\limits_{1\leq i_1\prec\cdots\prec i_k\leq n}\prod\limits_{j=1}^{n}\left(1-\mu_{i_j}-v_{i_j}\right)^{\frac{2}{kC_n^k}} \end{array}\right)^{\frac{1}{k}}\right)^{\frac{1}{2}}, \\
&\left(1-\left(1-\prod\limits_{1\leq i_1\prec\cdots\prec i_k\leq n}\left(1-\prod\limits_{j=1}^{k}\left(1-v_{i_j}\right)^2+\prod\limits_{j=1}^{n}\left(1-\mu_{i_j}-v_{i_j}\right)^2\right)^{\frac{1}{C_n^k}}+\prod\limits_{1\leq i_1\prec\cdots\prec i_k\leq n}\prod\limits_{j=1}^{n}\left(1-\mu_{i_j}-v_{i_j}\right)^{\frac{2}{C_n^k}}\right)^{\frac{1}{k}}\right)^{\frac{1}{2}}\Bigg) \\
&= \left(\frac{\bigoplus\limits_{1\leq i_1\prec\ldots\prec i_k\leq n}\bigotimes\limits_{j=1}^{k} p_{i_j}}{C_n^k}\right)^{\frac{1}{k}},
\end{aligned}
\tag{23}
$$

which is the Pythagorean fuzzy interaction Maclaurin symmetric mean (PFIMSM) operator.
Case 5: If $R = (1,1,\cdots,1)$, then the PFIMM is reduced to the following

$$
PFIMM^{(1,1,\cdots,1)}(p_1,p_2,\cdots,p_n) = \left(\sqrt{\prod_{i=1}^{n}(1-v_i^2)^{\frac{1}{n}}-\prod_{i=1}^{n}(1-\mu_i^2-v_i^2)^{\frac{1}{n}}},\ \sqrt{1-\prod_{i=1}^{n}(1-v_i^2)^{\frac{1}{n}}}\right) = \left(\prod_{i=1}^{n} p_i\right)^{\frac{1}{n}}
\tag{24}
$$

which is the Pythagorean fuzzy interaction geometric averaging (PFIGA) operator.
Case 6: If $R = (1/n, 1/n, \ldots, 1/n)$, then the PFIMM is reduced to the PFIGA operator, which is shown as Equation (24).

3.2. The Pythagorean Fuzzy Interaction Weighted Muirhead Mean

Evidently, the main drawback of the PFIMM is that it cannot take the weights of arguments into consideration. Therefore, we propose the PFIWMM.

Definition 9 . *Let $p_i = (\mu_i, v_i)(i = 1, 2, \cdots, n)$ be a collection of PFNs, $w = (w_1, w_2, \cdots, w_n)^T$ be the weight vector of $p_i(i = 1, 2, \cdots, n)$, satisfying $w_i \in [0, 1]$ and $\sum_{i=1}^{n} w_i = 1$. Let $R = (r_1, r_2, \ldots, r_n) \in R^n$ be a vector of parameter. If*

$$PFIWMM^R(r_1, r_2, \ldots, r_n) = \left(\frac{1}{n!}\sum_{\vartheta \in S_n}\prod_{j=1}^{n}\left(nw_{\vartheta(j)}p_{\vartheta(j)}\right)^{r_j}\right)^{\frac{1}{\sum_{j=1}^{n} r_j}} \tag{25}$$

then we call $PFIWMM^R$ the PFIWMM operator, where $\vartheta(j) = (j = 1, 2, \ldots, n)$ is any a permutation of $(1, 2, \ldots, n)$, and S_n is the collection of all permutations of $(1, 2, \ldots, n)$.

According to Definition 5, we can get the following theorem.

Theorem 3. *Let $p_i = (\mu_i, v_i)(i = 1, 2, \cdots, n)$ be a collection of PFNs, then the aggregated value by the PFIWMM is still a PFN and*

$$\begin{aligned} &PFIWMM^R(p_1, p_2, \cdots, p_n) = \\ &\left(\left(\left(\begin{array}{c} 1 - \prod_{\vartheta \in S_n}\left(1 - \prod_{j=1}^{n}\left(1 - \left(1 - \mu_{\vartheta(j)}^2\right)^{nw_{\vartheta(j)}} + \left(1 - \mu_{\vartheta(j)}^2 - v_{\vartheta(j)}^2\right)^{nw_{\vartheta(j)}}\right)^{r_j} + \prod_{j=1}^{n}\left(1 - \mu_{\vartheta(j)}^2 - v_{\vartheta(j)}^2\right)^{nw_{\vartheta(j)}r_j}\right)^{\frac{1}{n!}} + \\ \prod_{\vartheta \in S_n}\prod_{j=1}^{n}\left(1 - \mu_{\vartheta(j)}^2 - v_{\vartheta(j)}^2\right)^{\frac{nw_{\vartheta(j)}r_j}{n!}} \end{array} \right)^{\frac{1}{\sum_{j=1}^{n} r_j}} \right. \right. \\ &\left. \left. - \prod_{\vartheta \in S_n}\prod_{j=1}^{n}\left(1 - \mu_{\vartheta(j)}^2 - v_{\vartheta(j)}^2\right)^{\frac{nw_{\vartheta(j)}r_j}{n!\sum_{j=1}^{n} r_j}} \right)^{\frac{1}{2}}, \right. \\ &\left. \left(1 - \left(\begin{array}{c} 1 - \left(\prod_{\vartheta \in S_n}\left(1 - \prod_{j=1}^{n}\left(1 - \left(1 - \mu_{\vartheta(j)}^2\right)^{nw_{\vartheta(j)}} + \left(1 - \mu_{\vartheta(j)}^2 - v_{\vartheta(j)}^2\right)^{nw_{\vartheta(j)}}\right)^{r_j} + \prod_{j=1}^{n}\left(1 - \mu_{\vartheta(j)}^2 - v_{\vartheta(j)}^2\right)^{nw_{\vartheta(j)}r_j}\right)\right)^{\frac{1}{n!}} \\ + \prod_{\vartheta \in S_n}\prod_{j=1}^{n}\left(1 - \mu_{\vartheta(j)}^2 - v_{\vartheta(j)}^2\right)^{\frac{nw_{\vartheta(j)}r_j}{n!}} \end{array} \right)^{\frac{1}{\sum_{j=1}^{n} r_j}} \right)^{\frac{1}{2}} \right) \end{aligned} \tag{26}$$

The proof of Theorem 3 is similar to that of Theorem 1, which is omitted here in order to save space.

4. The Pythagorean Fuzzy Interaction Dual Muirhead Mean and the Pythagorean Fuzzy Interaction Weighted Dual Muirhead Mean

4.1. The Pythagorean Fuzzy Interaction Dual Muirhead Mean Operator

Definition 10 . *Let $p_i = (\mu_i, v_i)(i = 1, 2, \cdots, n)$ be a collection of PFNs, and $R = (r_1, r_2, \ldots, r_n) \in R^n$ be a vector of parameters. If*

$$PFIDMM^R(p_1, p_2, \cdots, p_n) = \frac{1}{\sum_{j=1}^{n} r_j}\left(\prod_{\vartheta \in S_n}\sum_{j=1}^{n}\left(r_j p_{\vartheta(j)}\right)\right)^{\frac{1}{n!}} \tag{27}$$

then we call $PFIDMM^R$ the PFIDMM operator, where $\vartheta(j) = (j = 1, 2, \ldots, n)$ is any a permutation of $(1, 2, \ldots, n)$ and S_n is the collection of all permutations of $(1, 2, \ldots, n)$.

Theorem 4. *Let $p_i = (\mu_i, v_i)(i = 1, 2, \ldots, n)$ be a collection of all permutations of PFNs, the aggregated value by the PFIDMM is also a PFN and*

$$PFIDMM^{R}(p_1, p_2, \cdots, p_n) =$$

$$\left(\begin{array}{l} \left(1 - \left(1 - \prod\limits_{\vartheta \in S_n} \left(1 - \prod\limits_{j=1}^{n}\left(1-\mu_{\vartheta(j)}^2\right)^{r_j} + \prod\limits_{j=1}^{n}\left(1-\mu_{\vartheta(j)}^2 - v_{\vartheta(j)}^2\right)^{r_j}\right)^{\frac{1}{n!}} + \prod\limits_{\vartheta \in S_n}\prod\limits_{j=1}^{n}\left(1-\mu_{\vartheta(j)}^2 - v_{\vartheta(j)}^2\right)^{\frac{r_j}{n!}}\right)^{\frac{1}{\sum_{j=1}^{n} r_j}}\right)^{\frac{1}{2}}, \\ \left(\begin{array}{l} \left(1 - \prod\limits_{\vartheta \in S_n} \left(1 - \prod\limits_{j=1}^{n}\left(1-\mu_{\vartheta(j)}^2\right)^{r_j} + \prod\limits_{j=1}^{n}\left(1-\mu_{\vartheta(j)}^2 - v_{\vartheta(j)}^2\right)^{r_j}\right)^{\frac{1}{n!}} + \prod\limits_{\vartheta \in S_n}\prod\limits_{j=1}^{n}\left(1-\mu_{\vartheta(j)}^2 - v_{\vartheta(j)}^2\right)^{\frac{r_j}{n!}}\right)^{\frac{1}{\sum_{j=1}^{n} r_j}} \\ - \prod\limits_{\vartheta \in S_n}\prod\limits_{j=1}^{n}\left(1-\mu_{\vartheta(j)}^2 - v_{\vartheta(j)}^2\right)^{\frac{r_j}{n! \sum_{j=1}^{n} r_j}} \end{array} \right)^{\frac{1}{2}} \end{array} \right) \tag{28}$$

Proof. According to the operational laws of PFNs in Definition 5, we can get

$$r_j p_{\vartheta(j)} = \left(\sqrt{1-\left(1-\mu_{\vartheta(j)}^2\right)^{r_j}}, \sqrt{\left(1-\mu_{\vartheta(j)}^2\right)^{r_j} - \left(1-\mu_{\vartheta(j)}^2 - v_{\vartheta(j)}^2\right)^{r_j}}\right) \tag{29}$$

and,

$$\sum_{j=1}^{n}\left(r_j p_{\vartheta(j)}\right) = \left(\sqrt{1-\prod_{j=1}^{n}\left(1-\mu_{\vartheta(j)}^2\right)^{r_j}}, \sqrt{\prod_{j=1}^{n}\left(1-\mu_{\vartheta(j)}^2\right)^{r_j} - \prod_{j=1}^{n}\left(1-\mu_{\vartheta(j)}^2 - v_{\vartheta(j)}^2\right)^{r_j}}\right) \tag{30}$$

Therefore,

$$\prod_{\vartheta \in S_n}\sum_{j=1}^{n}\left(r_j p_{\vartheta(j)}\right) = \left(\begin{array}{l} \sqrt{\prod\limits_{\vartheta \in S_n}\left(1-\prod\limits_{j=1}^{n}\left(1-\mu_{\vartheta(j)}^2\right)^{r_j} + \prod\limits_{j=1}^{n}\left(1-\mu_{\vartheta(j)}^2 - v_{\vartheta(j)}^2\right)^{r_j}\right) - \prod\limits_{\vartheta \in S_n}\left(\prod\limits_{j=1}^{n}\left(1-\mu_{\vartheta(j)}^2 - v_{\vartheta(j)}^2\right)^{r_j}\right)}, \\ \sqrt{1-\prod\limits_{\vartheta \in S_n}\left(1-\prod\limits_{j=1}^{n}\left(1-\mu_{\vartheta(j)}^2\right)^{r_j} + \prod\limits_{j=1}^{n}\left(1-\mu_{\vartheta(j)}^2 - v_{\vartheta(j)}^2\right)^{r_j}\right)} \end{array} \right). \tag{31}$$

Further,

$$\left(\prod_{\vartheta \in S_n}\sum_{j=1}^{n}\left(r_j p_{\vartheta(j)}\right)\right)^{\frac{1}{n!}} = \left(\begin{array}{l} \sqrt{\prod\limits_{\vartheta \in S_n}\left(1-\prod\limits_{j=1}^{n}\left(1-\mu_{\vartheta(j)}^2\right)^{r_j} + \prod\limits_{j=1}^{n}\left(1-\mu_{\vartheta(j)}^2 - v_{\vartheta(j)}^2\right)^{r_j}\right)^{\frac{1}{n!}} - \prod\limits_{\vartheta \in S_n}\prod\limits_{j=1}^{n}\left(1-\mu_{\vartheta(j)}^2 - v_{\vartheta(j)}^2\right)^{\frac{r_j}{n!}}}, \\ \sqrt{1-\prod\limits_{\vartheta \in S_n}\left(1-\prod\limits_{j=1}^{n}\left(1-\mu_{\vartheta(j)}^2\right)^{r_j} + \prod\limits_{j=1}^{n}\left(1-\mu_{\vartheta(j)}^2 - v_{\vartheta(j)}^2\right)^{r_j}\right)^{\frac{1}{n!}}} \end{array} \right). \tag{32}$$

Therefore,

$$\frac{1}{\sum_{j=1}^{n} r_j}\left(\prod_{\vartheta \in S_n}\sum_{j=1}^{n}\left(r_j p_{\vartheta(j)}\right)\right)^{\frac{1}{n!}} =$$

$$\left(\begin{array}{l} \left(1 - \left(1 - \prod\limits_{\vartheta \in S_n} \left(1 - \prod\limits_{j=1}^{n}\left(1-\mu_{\vartheta(j)}^2\right)^{r_j} + \prod\limits_{j=1}^{n}\left(1-\mu_{\vartheta(j)}^2 - v_{\vartheta(j)}^2\right)^{r_j}\right)^{\frac{1}{n!}} + \prod\limits_{\vartheta \in S_n}\prod\limits_{j=1}^{n}\left(1-\mu_{\vartheta(j)}^2 - v_{\vartheta(j)}^2\right)^{\frac{r_j}{n!}}\right)^{\frac{1}{\sum_{j=1}^{n} r_j}}\right)^{\frac{1}{2}}, \\ \left(\begin{array}{l} \left(1 - \prod\limits_{\vartheta \in S_n} \left(1 - \prod\limits_{j=1}^{n}\left(1-\mu_{\vartheta(j)}^2\right)^{r_j} + \prod\limits_{j=1}^{n}\left(1-\mu_{\vartheta(j)}^2 - v_{\vartheta(j)}^2\right)^{r_j}\right)^{\frac{1}{n!}} + \prod\limits_{\vartheta \in S_n}\prod\limits_{j=1}^{n}\left(1-\mu_{\vartheta(j)}^2 - v_{\vartheta(j)}^2\right)^{\frac{r_j}{n!}}\right)^{\frac{1}{\sum_{j=1}^{n} r_j}} \\ - \prod\limits_{\vartheta \in S_n}\prod\limits_{j=1}^{n}\left(1-\mu_{\vartheta(j)}^2 - v_{\vartheta(j)}^2\right)^{\frac{r_j}{n! \sum_{j=1}^{n} r_j}} \end{array} \right)^{\frac{1}{2}} \end{array} \right) \tag{33}$$

Therefore, Equation (28) is kept.

In the following, we prove the aggregated value is a PFN. For convenience, let

$$\mu = \left(1 - \left(1 - \prod_{\vartheta \in S_n}\left(1 - \prod_{j=1}^{n}\left(1 - \mu_{\vartheta(j)}^2\right)^{r_j} + \prod_{j=1}^{n}\left(1 - \mu_{\vartheta(j)}^2 - v_{\vartheta(j)}^2\right)^{r_j}\right)^{\frac{1}{n!}} + \prod_{\vartheta \in S_n}\prod_{j=1}^{n}\left(1 - \mu_{\vartheta(j)}^2 - v_{\vartheta(j)}^2\right)^{\frac{r_j}{n!}}\right)^{\frac{1}{\sum_{j=1}^{n} r_j}}\right)^{\frac{1}{2}}$$

$$v = \left(\begin{array}{l} \left(1 - \prod_{\vartheta \in S_n}\left(1 - \prod_{j=1}^{n}\left(1 - \mu_{\vartheta(j)}^2\right)^{r_j} + \prod_{j=1}^{n}\left(1 - \mu_{\vartheta(j)}^2 - v_{\vartheta(j)}^2\right)^{r_j}\right)^{\frac{1}{n!}} + \prod_{\vartheta \in S_n}\prod_{j=1}^{n}\left(1 - \mu_{\vartheta(j)}^2 - v_{\vartheta(j)}^2\right)^{\frac{r_j}{n!}}\right)^{\frac{1}{\sum_{j=1}^{n} r_j}} \\ - \prod_{\vartheta \in S_n}\prod_{j=1}^{n}\left(1 - \mu_{\vartheta(j)}^2 - v_{\vartheta(j)}^2\right)^{\frac{r_j}{n! \sum_{j=1}^{n} r_j}} \end{array}\right)^{\frac{1}{2}}.$$

Evidently,

$$\mu_{\vartheta(j)} \in [0,1], v_{\vartheta(j)} \in [0,1],\ 0 \leq \mu_{\vartheta(j)}^2 + v_{\vartheta(j)}^2 \leq 1. \tag{34}$$

Therefore,

$$0 \leq \left(1 - \mu_{\vartheta(j)}^2\right)^{r_j} \leq 1, 0 \leq \left(1 - v_{\vartheta(j)}^2\right)^{r_j} \leq 1, 0 \leq \left(1 - \mu_{\vartheta(j)}^2 - v_{\vartheta(j)}^2\right)^{r_j} \leq 1 \tag{35}$$

Further,

$$0 \leq \prod_{j=1}^{n}\left(1 - \mu_{\vartheta(j)}^2 - v_{\vartheta(j)}^2\right)^{r_j} \leq 1,\ 0 \leq \prod_{j=1}^{n}\left(1 - \mu_{\vartheta(j)}^2\right)^{r_j} - \prod_{j=1}^{n}\left(1 - \mu_{\vartheta(j)}^2 - v_{\vartheta(j)}^2\right)^{r_j} \leq 1. \tag{36}$$

Thus,

$$0 \leq 1 - \prod_{j=1}^{n}\left(1 - \mu_{\vartheta(j)}^2\right)^{r_j} + \prod_{j=1}^{n}\left(1 - \mu_{\vartheta(j)}^2 - v_{\vartheta(j)}^2\right)^{r_j} \leq 1. \tag{37}$$

Further,

$$0 \leq \prod_{\vartheta \in S_n}\prod_{j=1}^{n}\left(1 - \mu_{\vartheta(j)}^2 - v_{\vartheta(j)}^2\right)^{\frac{r_j}{n!}} \leq 1, 0 \leq \prod_{\vartheta \in S_n}\left(1 - \prod_{j=1}^{n}\left(1 - \mu_{\vartheta(j)}^2\right)^{r_j} + \prod_{j=1}^{n}\left(1 - \mu_{\vartheta(j)}^2 - v_{\vartheta(j)}^2\right)^{r_j}\right)^{\frac{1}{n!}} \leq 1. \tag{38}$$

Moreover,

$$0 \leq \left(1 - \prod_{\vartheta \in S_n}\left(1 - \prod_{j=1}^{n}\left(1 - \mu_{\vartheta(j)}^2\right)^{r_j} + \prod_{j=1}^{n}\left(1 - \mu_{\vartheta(j)}^2 - v_{\vartheta(j)}^2\right)^{r_j}\right)^{\frac{1}{n!}} + \prod_{\vartheta \in S_n}\prod_{j=1}^{n}\left(1 - \mu_{\vartheta(j)}^2 - v_{\vartheta(j)}^2\right)^{\frac{r_j}{n!}}\right) \leq 1. \tag{39}$$

Therefore,

$$0 \leq 1 - \left(1 - \prod_{\vartheta \in S_n}\left(1 - \prod_{j=1}^{n}\left(1 - \mu_{\vartheta(j)}^2\right)^{r_j} + \prod_{j=1}^{n}\left(1 - \mu_{\vartheta(j)}^2 - v_{\vartheta(j)}^2\right)^{r_j}\right)^{\frac{1}{n!}} + \prod_{\vartheta \in S_n}\prod_{j=1}^{n}\left(1 - \mu_{\vartheta(j)}^2 - v_{\vartheta(j)}^2\right)^{\frac{r_j}{n!}}\right)^{\frac{1}{\sum_{j=1}^{n} r_j}} \leq 1 \tag{40}$$

i.e., $0 \leq \mu \leq 1$. Similarly, we can get $0 \leq v \leq 1$.

Moreover,

$$\mu^2 + v^2 = 1 - \prod_{\vartheta \in S_n}\prod_{j=1}^{n}\left(1 - \mu_{\vartheta(j)}^2 - v_{\vartheta(j)}^2\right)^{\frac{r_j}{n! \sum_{j=1}^{n} r_j}} \tag{41}$$

As we have proved that

$$0 \leq \prod_{\vartheta \in S_n}\prod_{j=1}^{n}\left(1 - \mu_{\vartheta(j)}^2 - v_{\vartheta(j)}^2\right)^{\frac{r_j}{n!}} \leq 1 \tag{42}$$

Therefore,

$$\prod_{\vartheta \in S_n}\prod_{j=1}^{n}\left(1-\mu_{\vartheta(j)}^2-v_{\vartheta(j)}^2\right)^{\frac{r_j}{n!\sum_{j=1}^{n} r_j}},\ 0 \le 1-\prod_{\vartheta \in S_n}\prod_{j=1}^{n}\left(1-\mu_{\vartheta(j)}^2-v_{\vartheta(j)}^2\right)^{\frac{r_j}{n!\sum_{j=1}^{n} r_j}} \le 1. \tag{43}$$

Therefore, $0 \le \mu^2 + v^2 \le 1$, which completes the proof. □

Moreover, the PFIDMM has the following properties.

Theorem 5. *(Idempotency) If all* $p_i = (i = 1, 2, \ldots, n)$ *are equal, i.e.,* $p_i = p = (\mu, v)$*, then*

$$PFIDMM^R = (p_1, p_2, \ldots, p_n) = p \tag{44}$$

In the following, we investigate some special cases of PFIDMM with respect to R.
Case 1: If $R = (1, 0, \cdots, 0)$, the PFIDMM is reduced to the following

$$PFIDMM^{(1,0,\cdots,0)}(p_1, p_2, \ldots, p_n) = \left(\sqrt{\prod_{j=1}^{n}(1-v_i^2)^{\frac{1}{n}} - \prod_{j=1}^{n}(1-\mu_i^2-v_i^2)^{\frac{1}{n}}},\ \sqrt{1-\prod_{j=1}^{n}(1-v_i^2)^{\frac{1}{n}}}\right) \tag{45}$$

which is the Pythagorean fuzzy interaction arithmetic averaging operator.
Case 2: If $R = (\lambda, 0, \cdots, 0)$, the PFIDMM is reduced to the following

$$\begin{aligned} PFIDMM^{(\lambda,0,\cdots,0)}(p_1, p_2, \ldots, p_n) = & \left(\sqrt{1-\left(1-\left(1-\prod_{j=1}^{n}(1-\mu_i^2)^{\lambda}+\prod_{j=1}^{n}(1-\mu_i^2-v_i^2)^{\lambda}\right)^{\frac{1}{n}}+\prod_{j=1}^{n}(1-\mu_i^2-v_i^2)^{\frac{\lambda}{n}}\right)^{\frac{1}{\lambda}}},\right. \\ & \left.\sqrt{\left(1-\left(1-\prod_{j=1}^{n}(1-\mu_i^2)^{\lambda}+\prod_{j=1}^{n}(1-\mu_i^2-v_i^2)^{\lambda}\right)^{\frac{1}{n}}\right)^{\frac{1}{\lambda}}-\prod_{j=1}^{n}(1-\mu_i^2-v_i^2)^{\frac{1}{n}}}\right), \end{aligned} \tag{46}$$

which is the Pythagorean fuzzy interaction generalized arithmetic averaging operator.
Case 3: If $R = (1, 1, 0, 0, \cdots, 0)$, the PFIDMM is reduced to the following

$$\begin{aligned} & PFIDMM^{(1,1,0,0,\cdots,0)}(p_1, p_2, \ldots, p_n) = \\ & \left(\left(\left(1-\left(1-\prod_{\substack{i,j=1\\ i\neq j}}^{n}\left(1-(1-\mu_i^2)\left(1-\mu_j^2\right)+(1-\mu_i^2-v_i^2)\left(1-\mu_j^2-v_j^2\right)\right)^{\frac{1}{n(n-1)}}+\prod_{\substack{i,j=1\\ i\neq j}}^{n}\left((1-\mu_i^2-v_i^2)\left(1-\mu_j^2-v_j^2\right)\right)^{\frac{1}{n(n-1)}}\right)^{\frac{1}{2}}\right)^{\frac{1}{2}}\right., \\ & \left.\left(\begin{array}{l}\left(1-\prod_{\substack{i,j=1\\ i\neq j}}^{n}\left(1-(1-\mu_i^2)\left(1-\mu_j^2\right)+\left(1-\mu_i^2-v_j^2\right)\left(1-\mu_j^2-v_j^2\right)\right)^{\frac{1}{n(n-1)}}+\prod_{\substack{i,j=1\\ i\neq j}}^{n}\left(\left(1-\mu_i^2-v_j^2\right)\left(1-\mu_j^2-v_j^2\right)\right)^{\frac{1}{n(n-1)}}\right)^{\frac{1}{2}} \\ -\prod_{\substack{i,j=1\\ i\neq j}}^{n}\left(\left(1-\mu_i^2-v_j^2\right)\left(1-\mu_j^2-v_j^2\right)\right)^{\frac{1}{2n(n-1)}}\end{array}\right)^{\frac{1}{2}}\right), \end{aligned} \tag{47}$$

which is the Pythagorean fuzzy interaction arithmetic BM operator.

Case 4: If $R = \left(\overbrace{1, 1, \cdots, 1}^{k}, \overbrace{0, 0, \cdots, 0}^{n-k}\right)$, the PFIDMM is reduced to the following

$$PFIDMM^{\left(\overbrace{1,1,\cdots,1}^{k}\overbrace{0,0,\cdots,0}^{n-k}\right)}(p_1,p_2,\ldots,p_n)=$$
$$\left(\left(1-\left(1-\prod_{1\le i_1\prec\cdots\prec i_k\le n}\left(1-\prod_{j=1}^{k}\left(1-\mu_{i_j}\right)^2+\prod_{j=1}^{n}\left(1-\mu_{i_j}-v_{i_j}\right)^2\right)^{\frac{1}{C_n^k}}+\prod_{1\le i_1\prec\cdots\prec i_k\le n}\prod_{j=1}^{n}\left(1-\mu_{i_j}-v_{i_j}\right)^{\frac{2}{C_n^k}}\right)^{\frac{1}{k}}\right)^{\frac{1}{2}},\right.$$
$$\left.\left(\begin{array}{l}\left(1-\prod_{1\le i_1\prec\cdots\prec i_k\le n}\left(1-\prod_{j=1}^{k}\left(1-\mu_{i_j}\right)^2+\prod_{j=1}^{n}\left(1-\mu_{i_j}-v_{i_j}\right)^2\right)^{\frac{1}{C_n^k}}+\prod_{1\le i_1\prec\cdots\prec i_k\le n}\prod_{j=1}^{n}\left(1-\mu_{i_j}-v_{i_j}\right)^{\frac{2}{C_n^k}}\right)^{\frac{1}{k}}\\-\prod_{1\le i_1\prec\cdots\prec i_k\le n}\prod_{j=1}^{n}\left(1-\mu_{i_j}-v_{i_j}\right)^{\frac{2}{kC_n^k}}\end{array}\right)^{\frac{1}{2}}\right),\tag{48}$$

which is the Pythagorean fuzzy interaction Maclaurin symmetric mean operator.
Case 5: If $R=(1,1,\cdots,1)$, the PFIDMM is reduced to the following

$$PFIDMM^{(1,1,\cdots,1)}(p_1,p_2,\ldots,p_n)=\left(\sqrt{1-\prod_{i=1}^{n}(1-\mu_i^2)^{\frac{1}{n}}},\sqrt{\prod_{i=1}^{n}(1-\mu_i^2)^{\frac{1}{n}}-\prod_{i=1}^{n}(1-\mu_i^2-v_i^2)^{\frac{1}{n}}}\right)\tag{49}$$

which is the Pythagorean fuzzy interaction arithmetic averaging operator.
Case 6: If $R=\left(\frac{1}{n},\frac{1}{n},\cdots,\frac{1}{n}\right)$, the PFIDMM is reduced to the Pythagorean fuzzy interaction arithmetic averaging operator, which is shown as Equation (49).

4.2. The Pythagorean Fuzzy Interaction Dual Weighted Muirhead Mean Operator

In the following, we introduce the PFIDWMM operator so as to consider the weights vector of the attribute values.

Definition 11 . *Let $p_i=(\mu_i,v_i)(i=1,2,\cdots,n)$ be a collection of PFNs, $w=(w_1,w_2,\cdots,w_n)^T$ be the weight vector of $p_i(i=1,2,\cdots,n)$, which satisfies $w_i\in[0,1]$ and $\sum_{i=1}^{n}w_i=1$, and let $R=(r_1,r_2,\ldots,r_n)\in R^n$ be a vector of parameters. If*

$$PFIDWMM^R(p_1,p_2,\ldots,p_n)=\frac{1}{\sum_{j=1}^{n}r_j}\left(\prod_{\vartheta\in S_n}\sum_{j=1}^{n}\left(r_j p_{\vartheta(j)}{}^{nw_{\vartheta(j)}}\right)\right)^{\frac{1}{n!}}\tag{50}$$

then we call $PFIDWMM^R$ the PFIDWMM operator, where $\vartheta(j)=(j=1,2,\ldots,n)$ is any a permutation of $(1,2,\ldots,n)$, and S_n is the collection of all permutations of $(1,2,\ldots,n)$.

Theorem 6. *Let $p_i=(\mu_i,v_i)(i=1,2,\cdots,n)$ be a collection of PFNs, we can see that the aggregation result from by the PFIDWMM is still a PFN, it can be obtained as follows:*

$$PFIDWMM^R(p_1,p_2,\ldots,p_n)=$$
$$\left(\left(1-\left(\begin{array}{l}1-\prod_{\vartheta\in S_n}\left(1-\prod_{j=1}^{n}\left(1-\left(1-v_{\vartheta(j)}^2\right)^{nw_{\vartheta(j)}}+\left(1-\mu_{\vartheta(j)}^2-v_{\vartheta(j)}^2\right)^{nw_{\vartheta(j)}}\right)^{r_j}+\prod_{j=1}^{n}\left(1-\mu_{\vartheta(j)}^2-v_{\vartheta(j)}^2\right)^{nw_{\vartheta(j)}r_j}\right)^{\frac{1}{n!}}\\+\prod_{\vartheta\in S_n}\prod_{j=1}^{n}\left(1-\mu_{\vartheta(j)}^2-v_{\vartheta(j)}^2\right)^{\frac{nw_{\vartheta(j)}r_j}{n!}}\end{array}\right)^{\frac{1}{\sum_{j=1}^{n}r_j}}\right)^{\frac{1}{2}},\right.$$
$$\left.\left(\begin{array}{l}\left(\begin{array}{l}1-\prod_{\vartheta\in S_n}\left(1-\prod_{j=1}^{n}\left(1-\left(1-v_{\vartheta(j)}^2\right)^{nw_{\vartheta(j)}}+\left(1-\mu_{\vartheta(j)}^2-v_{\vartheta(j)}^2\right)^{nw_{\vartheta(j)}}\right)^{r_j}+\prod_{j=1}^{n}\left(1-\mu_{\vartheta(j)}^2-v_{\vartheta(j)}^2\right)^{nw_{\vartheta(j)}r_j}\right)^{\frac{1}{n!}}\\+\prod_{\vartheta\in S_n}\prod_{j=1}^{n}\left(1-\mu_{\vartheta(j)}^2-v_{\vartheta(j)}^2\right)^{\frac{nw_{\vartheta(j)}r_j}{n!}}\end{array}\right)^{\frac{1}{\sum_{j=1}^{n}r_j}}\\-\prod_{\vartheta\in S_n}\prod_{j=1}^{n}\left(1-\mu_{\vartheta(j)}^2-v_{\vartheta(j)}^2\right)^{\frac{nw_{\vartheta(j)}r_j}{n!\sum_{j=1}^{n}r_j}}\end{array}\right)^{\frac{1}{2}}\right).\tag{51}$$

Proof. Because $p_{\vartheta(j)}^{nw_{\vartheta(j)}} = \left(\sqrt{\left(1-v_{\vartheta(j)}^{2}\right)^{nw_{\vartheta(j)}} - \left(1-\mu_{\vartheta(j)}^{2}-v_{\vartheta(j)}^{2}\right)^{nw_{\vartheta(j)}}}, \sqrt{1-\left(1-v_{\vartheta(j)}^{2}\right)^{nw_{\vartheta(j)}}}\right)$, we can replace $\mu_{\vartheta(j)}$ in Equation (28) with $\sqrt{(1-v^2)^{nw_{\vartheta(j)}} - (1-\mu^2-v^2)^{nw_{\vartheta(j)}}}$, and $v_{\vartheta(j)}$ in Equation (28) with $\sqrt{1-(1-v^2)^{nw_{\vartheta(j)}}}$, then we can get Equation (51).

Because $p_{\vartheta(j)}$ is a PFN, $p_{\vartheta(j)}^{nw_{\vartheta(j)}}$ is also a PFN. By Equation (28), we have $PFIDWMM^{R}(p_1, p_2, \ldots, p_n)$ is a PFN.

Just the same as the PFIDMM operator, the PFIDWMM operator still does not have the monotonicity and the boundedness. □

Theorem 7. *The PFIDMM operator is a special case of the PFIDWMM operator.*

Proof. When $w = \left(\frac{1}{n}, \frac{1}{n}, \cdots, \frac{1}{n}\right)$

$$
\begin{aligned}
&PFIDWMM^{R}(p_1, p_2, \ldots, p_n) = \\
&\left(\left(1-\left(\begin{array}{l}1-\prod\limits_{\vartheta\in S_n}\left(1-\prod\limits_{j=1}^{n}\left(1-\left(1-v_{\vartheta(j)}^{2}\right)^{nw_{\vartheta(j)}}+\left(1-\mu_{\vartheta(j)}^{2}-v_{\vartheta(j)}^{2}\right)^{nw_{\vartheta(j)}}\right)^{r_j}+\prod\limits_{j=1}^{n}\left(1-\mu_{\vartheta(j)}^{2}-v_{\vartheta(j)}^{2}\right)^{nw_{\vartheta(j)}r_j}\right)^{\frac{1}{n!}}\\ +\prod\limits_{\vartheta\in S_n}\prod\limits_{j=1}^{n}\left(1-\mu_{\vartheta(j)}^{2}-v_{\vartheta(j)}^{2}\right)^{\frac{nw_{\vartheta(j)}r_j}{n!}}\end{array}\right)^{\frac{1}{\sum_{j=1}^{n} r_j}}\right)^{\frac{1}{2}},\right.\\
&\left.\left(\begin{array}{l}\left(\begin{array}{l}1-\prod\limits_{\vartheta\in S_n}\left(1-\prod\limits_{j=1}^{n}\left(1-\left(1-v_{\vartheta(j)}^{2}\right)^{nw_{\vartheta(j)}}+\left(1-\mu_{\vartheta(j)}^{2}-v_{\vartheta(j)}^{2}\right)^{nw_{\vartheta(j)}}\right)^{r_j}+\prod\limits_{j=1}^{n}\left(1-\mu_{\vartheta(j)}^{2}-v_{\vartheta(j)}^{2}\right)^{nw_{\vartheta(j)}r_j}\right)^{\frac{1}{n!}}\\ \qquad+\prod\limits_{\vartheta\in S_n}\prod\limits_{j=1}^{n}\left(1-\mu_{\vartheta(j)}^{2}-v_{\vartheta(j)}^{2}\right)^{\frac{nw_{\vartheta(j)}r_j}{n!}}\end{array}\right)^{\frac{1}{\sum_{j=1}^{n} r_j}}\\ -\prod\limits_{\vartheta\in S_n}\prod\limits_{j=1}^{n}\left(1-\mu_{\vartheta(j)}^{2}-v_{\vartheta(j)}^{2}\right)^{\frac{nw_{\vartheta(j)}r_j}{n!\sum_{j=1}^{n} r_j}}\end{array}\right)^{\frac{1}{2}}\right)\\
&=\left(\left(1-\left(1-\prod\limits_{\vartheta\in S_n}\left(1-\prod\limits_{j=1}^{n}\left(1-\left(1-v_{\vartheta(j)}^{2}\right)+\left(1-\mu_{\vartheta(j)}^{2}-v_{\vartheta(j)}^{2}\right)\right)^{r_j}+\prod\limits_{j=1}^{n}\left(1-\mu_{\vartheta(j)}^{2}-v_{\vartheta(j)}^{2}\right)^{r_j}\right)^{\frac{1}{n!}}+\prod\limits_{\vartheta\in S_n}\prod\limits_{j=1}^{n}\left(1-\mu_{\vartheta(j)}^{2}-v_{\vartheta(j)}^{2}\right)^{\frac{r_j}{n!}}\right)^{\frac{1}{\sum_{j=1}^{n} r_j}}\right)^{\frac{1}{2}},\right.\\
&\left.\left(\begin{array}{l}\left(1-\prod\limits_{\vartheta\in S_n}\left(1-\prod\limits_{j=1}^{n}\left(1-\left(1-v_{\vartheta(j)}^{2}\right)+\left(1-\mu_{\vartheta(j)}^{2}-v_{\vartheta(j)}^{2}\right)\right)^{r_j}+\prod\limits_{j=1}^{n}\left(1-\mu_{\vartheta(j)}^{2}-v_{\vartheta(j)}^{2}\right)^{r_j}\right)^{\frac{1}{n!}}+\prod\limits_{\vartheta\in S_n}\prod\limits_{j=1}^{n}\left(1-\mu_{\vartheta(j)}^{2}-v_{\vartheta(j)}^{2}\right)^{\frac{r_j}{n!}}\right)^{\frac{1}{\sum_{j=1}^{n} r_j}}\\ -\prod\limits_{\vartheta\in S_n}\prod\limits_{j=1}^{n}\left(1-\mu_{\vartheta(j)}^{2}-v_{\vartheta(j)}^{2}\right)^{\frac{r_j}{n!\sum_{j=1}^{n} r_j}}\end{array}\right)^{\frac{1}{2}}\right)\\
&=\left(\left(1-\left(1-\prod\limits_{\vartheta\in S_n}\left(1-\prod\limits_{j=1}^{n}\left(1-\mu_{\vartheta(j)}^{2}\right)^{r_j}+\prod\limits_{j=1}^{n}\left(1-\mu_{\vartheta(j)}^{2}-v_{\vartheta(j)}^{2}\right)^{r_j}\right)^{\frac{1}{n!}}+\prod\limits_{\vartheta\in S_n}\prod\limits_{j=1}^{n}\left(1-\mu_{\vartheta(j)}^{2}-v_{\vartheta(j)}^{2}\right)^{\frac{r_j}{n!}}\right)^{\frac{1}{\sum_{j=1}^{n} r_j}}\right)^{\frac{1}{2}},\right.\\
&\left.\left(\begin{array}{l}\left(1-\prod\limits_{\vartheta\in S_n}\left(1-\prod\limits_{j=1}^{n}\left(1-\mu_{\vartheta(j)}^{2}-v_{\vartheta(j)}^{2}\right)^{r_j}+\prod\limits_{j=1}^{n}\left(1-\mu_{\vartheta(j)}^{2}-v_{\vartheta(j)}^{2}\right)^{r_j}\right)^{\frac{1}{n!}}+\prod\limits_{\vartheta\in S_n}\prod\limits_{j=1}^{n}\left(1-\mu_{\vartheta(j)}^{2}-v_{\vartheta(j)}^{2}\right)^{\frac{r_j}{n!}}\right)^{\frac{1}{\sum_{j=1}^{n} r_j}}\\ -\prod\limits_{\vartheta\in S_n}\prod\limits_{j=1}^{n}\left(1-\mu_{\vartheta(j)}^{2}-v_{\vartheta(j)}^{2}\right)^{\frac{r_j}{n!\sum_{j=1}^{n} r_j}}\end{array}\right)^{\frac{1}{2}}\right)\\
&= PFIDMM^{R}(p_1, p_2, \ldots, p_n).
\end{aligned}
$$

5. A Novel Approach to MAGDM with Pythagorean Fuzzy Information

Based on the proposed operators, this section provides a novel approach to MAGDM problems in which attribute values take the form of PFNs and the weights of attributes take the form of crisp numbers. The description of a typical MAGDM problem with Pythagorean fuzzy information is shown as follows. Let $X = \{x_1, x_2, \cdots, x_m\}$ be a set of alternatives and $G = \{G_1, G_2, \cdots, G_n\}$ be a set of attributes with the weights vector being $w = (w_1, w_2, \cdots, w_n)^T$, satisfying $w_i \in [0,1]$ and

$\sum_{i=1}^{n} w_i = 1$. For attribute $G_j (j = 1, 2, \cdots, n)$ of alternative $x_i (i = 1, 2, \ldots, m)$, a PFN $p_{ij} = (\mu_{ij}, v_{ij})$ $(i = 1, 2, \cdots, m; j = 1, 2, \cdots, n)$ is utilized to represent decision makers' preference information, in which μ_{ij} denotes that degree that alternative x_i satisfies the criteria G_j and v_{ij} represents the degree that alternative x_i dissatisfies the criteria G_j. Therefore, we can get a Pythagorean fuzzy decision matrix finally, which and be denoted by $P = (p_{ij})_{m \times n}$. In the followings, we introduce an algorithm to solve this problem based on the proposed operators.

Step 1. Standardized the original decision matrix. In real decision-making problems, there exists two kinds of attributes: benefit attributes and cost attributes. Therefore, the original decision matrix should be normalized by

$$p_{ij} = \begin{cases} (\mu_{ij}, v_{ij}) & G_j \in I_1 \\ (v_{ij}, \mu_{ij}) & G_j \in I_2 \end{cases} \tag{52}$$

where I_1 represents benefit attributes and I_2 represents cost attributes.

Step 2. For alternative $x_i (i = 1, 2, \ldots, m)$, utilize the PFIWMM operator

$$p_i = PFIWMM^R(p_{i1}, p_{i2}, \cdots, p_{in}) \tag{53}$$

or the PFIDWMM operator

$$p_i = PFIDWMM^R(p_{i1}, p_{i2}, \cdots, p_{in}) \tag{54}$$

to aggregate all the attributes values, so that a series of comprehensive preference value can be obtained.

Step 3. Rank the overall values $p_i (i = 1, 2, \ldots, m)$ based on their scores according to Definition 3.

Step 4. Rank the corresponding alternatives according to the rank of overall values and select the best alternative.

6. Numerical Example

In the following, we provide a numerical example that is adopted from [21] to illustrate the application of the proposed method. In order to know the best airline in Taiwan, the civil aviation administration of Taiwan (CAAT) organizes several experts to form a committee to assess the four major domestic airlines. The four airlines are the UNI Air (x_1), Transasia (x_2), Mandarin (x_3), and Daily Air (x_4).The alternatives are assessed from four attributes: (1) the booking and ticketing service (G_1); (2) the check-in and boarding process (G_2); (3) the cabin service (G_3); (4) the responsiveness (G_4). Weight vector of the attributes is $w = (0.15, 0.25, 0.35, 0.25)^T$. Experts are required to utilize a PFN $p_{ij} = (\mu_{ij}, v_{ij})$ to express their assessments for attributes $G_j (j = 1, 2, 3, 4)$ of airline $x_i (i = 1, 2, 3, 4)$, and a Pythagorean fuzzy decision matrix $P = (p_{ij})_{4 \times 4} (i, j = 1, 2, 3, 4)$ is shown in Table 1. In the following, we will solve this problem based on the proposed method.

Table 1. The Pythagorean fuzzy decision matrix.

	G_1	G_2	G_3	G_4
x_1	(0.9, 0.3)	(0.7, 0.6)	(0.5, 0.8)	(0.6, 0.3)
x_2	(0.4, 0.7)	(0.9, 0.2)	(0.8, 0.1)	(0.5, 0.3)
x_3	(0.8, 0.4)	(0.7, 0.5)	(0.6, 0.2)	(0.7, 0.4)
x_4	(0.7, 0.2)	(0.8, 0.2)	(0.8, 0.4)	(0.6, 0.6)

6.1. The Decision-Making Process

Step 1. As all of the attribute values are the same type, the original decision matrix does not need to be standardized.

Step 2. For each alternative, utilize Equation (53) to aggregate the assessments. Here, we assume $R = (1, 1, 1, 1)$. Therefore, we can obtain

$$p_1 = (0.3895, 0.2816)\ p_2 = (0.3526, 0.1679)\ p_3 = (0.3415, 0.1979)\ p_4 = (0.3588, 0.1878).$$

Step 3. Based on Definition 3, we can calculate the score function $S(p_i)(i = 1, 2, 3, 4)$ as follows

$$s(p_1) = 0.0724\ s(p_2) = 0.0961\ s(p_3) = 0.0775\ s(p_4) = 0.0935.$$

Therefore, the ranking order of the overall values is $p_2 > p_4 > p_3 > p_1$.

Step 4. According to the ranking order of the overall values, we can get the ranking order of the corresponding alternatives. That is $x_2 \succ x_4 \succ x_3 \succ x_1$. Therefore, x_2 is the best alternative, which means Transasia is the best airline of Taiwan.

In [14], the ranking results by using the Pythagorean fuzzy weighted averaging (PFWA) operator, the symmetric Pythagorean fuzzy weighted averaging (SPFWA) operator and the symmetric Pythagorean fuzzy weighted geometric (SPFWG) operator are also $x_2 \succ x_4 \succ x_3 \succ x_1$, which proves the validity of the proposed method.

In step 2, if we utilize the PFIWDMM operator to aggregate the decision makers' preference information, we can obtain

$$p_1 = (0.3622, 0.3159)\ p_2 = (0.3295, 0.2096)\ p_3 = (0.3439, 0.1938)\ p_4 = (0.3524, 0.1996).$$

Therefore, the scores of the overall values are

$$s(p_1) = 0.0314\ s(p_2) = 0.0647\ s(p_3) = 0.0807\ s(p_4) = 0.0843$$

Thus, the ranking order of the alternatives is $x_4 \succ x_3 \succ x_2 \succ x_1$. In Ref [11], the ranking result by utilizing the Pythagorean fuzzy weighted geometric is also $x_4 \succ x_3 \succ x_2 \succ x_1$, which also illustrate the validity of the proposed approach.

6.2. Further Discussion

The prominent advantage of the proposed aggregation operators is that the interrelationship among all PFNs can be taken into consideration. Moreover, it has a parameter vector that leads to flexible aggregation operators. To show the validity and superiorities of the proposed operators, we conduct a comparative analysis. We solve the same problem by some existing MAGDM approaches including the SPFWA and the SPFWG operators in [22], the Pythagorean fuzzy ordered weighted averaging weighted averaging distance (PFOWAWAD) operator in [22], the Pythagorean fuzzy point (PFP) operator and generalized Pythagorean fuzzy point ordered weighted averaging (GPFPOWA) in [23], the Pythagorean fuzzy Einstein ordered weighted averaging (PFEOWA) operator in [24], the Pythagorean fuzzy Einstein ordered weighted geometric (PFEOWG) operator in [25,26], the Pythagorean fuzzy weighted Bonferroni mean (PFWBM) operator in [30], the Pythagorean fuzzy weighted geometric Bonferroni mean (PFWGBM) operator in [31], the generalized Pythagorean fuzzy weighted Bonferroni mean (GPFWBM) operator and generalized Pythagorean fuzzy Bonferroni geometric mean (GPFBGM) operator in [32], the dual generalized Pythagorean fuzzy weighted Bonferroni mean (DGPFWBM) operator and dual generalized Pythagorean fuzzy weighted Bonferroni geometric mean (DGPFWBGM) operator in [32], the Pythagorean fuzzy weighted Maclaurin symmetric mean (PFWMSM) operator in [33], the generalized Pythagorean fuzzy weighted Maclaurin symmetric mean (GPFWMSM) operator in [34], the Pythagorean fuzzy interaction ordered weighted averaging (PFIOWA) operator and the Pythagorean fuzzy interaction ordered weighted geometric (PFIOWG) operator in [38], the Pythagorean fuzzy weighted Muirhead mean (PFWMM) operator, and Pythagorean fuzzy weighted dual Muirhead mean (PFWDMM) operator [39]. Details can be found in Table 2.

The approaches in [11–26] are based on a simple weighted averaging operator. The weaknesses of these approaches are (1) they assume that all the input arguments are independent, which is somewhat inconsistent with reality; (2) they cannot consider the interrelationship among input arguments; (3) they cannot capture the interrelationship between membership degree and non-membership degrees. However, on the contrary, the method in the present paper can capture the interrelationship among input arguments. In addition, it provides a feasible aggregation process as it has a parameter vector R. Quite a few existing aggregation operators are special cases of the proposed operators. Moreover, the method is based on the interaction operations for the PFNs. Thus, the proposed method can consider the relationship among membership and non-membership degrees. In other words, the proposed method can effectively handle situations in which a membership degree or a non-membership degree is zero. Thus, the proposed method is more powerful and flexible than the methods in [21–26].

Table 2. Comparison of different aggregation operators.

Approaches	Whether Captures Interrelationship of Two Attributes	Whether Captures Interrelationship of Multiple Attributes	Whether Captures Interrelationship of All Attributes	Whether Captures Relationship of Membership and Non-Membership Degrees	Whether Makes the Method Flexible by the Parameter Vector
SPFWA [21]	No	No	No	No	No
SPFWG [21]	No	No	No	No	No
PFOWAWAD [22]	No	No	No	No	No
PFP [23]	No	No	No	No	No
GPFPOWA [23]	No	No	No	No	No
PFEOWA [24]	No	No	No	No	No
PFEOWG [25,26]	No	No	No	No	No
PFWBM [30]	Yes	No	No	No	No
PFWGBM [31]	Yes	No	No	No	No
GPFWBM [32]	Yes	No	No	No	No
GPFWBGM [32]	Yes	No	No	No	No
DGPFWBM [32]	Yes	Yes	Yes	No	Yes
DGPFWBGM [32]	Yes	Yes	Yes	No	Yes
PFWMSM [33]	Yes	Yes	No	No	No
GPFWMSM [34]	Yes	Yes	No	No	No
PFIOWA [38]	No	No	No	Yes	No
PFIOWG [38]	No	No	No	Yes	No
PFWMM [39]	Yes	Yes	Yes	No	Yes
PFWDMM [39]	Yes	Yes	Yes	No	Yes
PFIWMM	Yes	Yes	Yes	Yes	Yes
PFIWDMM	Yes	Yes	Yes	Yes	Yes

Approaches in [30,31] are based on BM, so that they consider the interrelationships between arguments. However, the main flaw is that they can only capture the interrelationship between any two arguments. Approaches based on GPFWBM and GPFWBGM operators are better than approaches in [32], as the former approaches can capture the interrelationship between any three approaches. Approaches in [33,34] can consider the interrelationship among multiple arguments; however, all the methods [30–34] fail to reflect the interrelationship among all input arguments. Additionally, these methods do not consider the interrelationship among membership degree and non-membership degree. The proposed method in this paper not only captures the interrelationship between all input arguments but also takes the relationship between membership and non-membership degrees.

The approaches in [32] based on the DGPFWBM and GPFWBGM operators are much better than the methods in [30–34], as they can consider the interrelationship among all arguments. Additionally, they have vectors of the parameters, leading to a flexible and feasible aggregation process. However, the main drawback of these operators is that they do not consider the relationship between membership degree and non-membership degree. The proposed method in this paper takes the interrelationships of all arguments into consideration and simultaneously considers the relationship between membership and non-membership degrees. Thus, our method in this paper is more powerful than the method based on the DGPFWBM or GPFWBGM operators.

Compared with the approach based on the PFIOWA and PFIOWG operators, the merit of the proposed approach is that it can reflect the membership and non-membership degrees, as it is based on the interaction operations for PFNs. However, it cannot reflect of the interrelationship among PFNs. Moreover, it is not as flexible as the proposed method. In addition, the Pythagorean fuzzy Muirhead mean operators in [39] are based on basic operational laws, so that the relationship among membership and non-membership degrees is overlooked. In other words, the operators in [39] do not work for the situations in which one membership or non-membership degree is equal to one.

All in all, the proposed method in this paper can reflect the interrelationships among all input arguments. In addition, it works for situations in which a membership degree or a non-membership degree is zero, leading to less information loss and consequently making decision-making results more reasonable. Therefore, the proposed method is more powerful and flexible than others.

It is noted that there exists a vector of parameter R in the proposed method. The parameter vector R plays a significant role in the final ranking results. Some existing Pythagorean fuzzy aggregation operators are special cases of the proposed operators. By assigning different parameter vectors in the proposed operators, different overall values as well as the final ranking results can be obtained. Thus, in the following, we investigate the influence of the vector of parameters R on the score functions and the ranking results. We assign different values to R in the PFIWMM and PFIWDMM operators, and the score function and ranking orders are presented in Tables 3 and 4.

Table 3. Ranking results by utilizing the different parameter vector R in the Pythagorean fuzzy interaction weighted Muirhead mean (PFIWMM) operator.

Parameter Vector R	The Scores of $s(p_i)(i = 1,2,3,4)$	Ranking Results
$R = (1,0,0,0)$	$s(p_1) = 0.0874$ $s(p_2) = 0.2680$ $s(p_3) = 0.1966$ $s(p_4) = 0.2440$	$x_2 \succ x_4 \succ x_3 \succ x_1$
$R = (1,1,0,0)$	$s(p_1) = 0.1040$ $s(p_2) = 0.1706$ $s(p_3) = 0.1337$ $s(p_4) = 0.1625$	$x_2 \succ x_4 \succ x_3 \succ x_1$
$R = (1,1,1,0)$	$s(p_1) = 0.0868$ $s(p_2) = 0.1232$ $s(p_3) = 0.0984$ $s(p_4) = 0.1190$	$x_2 \succ x_4 \succ x_3 \succ x_1$
$R = (1,1,1,1)$	$s(p_1) = 0.0724$ $s(p_2) = 0.0961$ $s(p_3) = 0.0775$ $s(p_4) = 0.0935$	$x_2 \succ x_4 \succ x_3 \succ x_1$
$R = (2,0,0,0)$	$s(p_1) = 0.1423$ $s(p_2) = 0.2989$ $s(p_3) = 0.1888$ $s(p_4) = 0.2646$	$x_2 \succ x_4 \succ x_3 \succ x_1$

As we can see in Table 3, by assigning different vector R to the PFIWMM operator, different scores of the overall assessments can be obtained. However, the ranking results are always the same. In addition, the more interrelationships between PFNs are taken into consideration, the smaller the value of score functions will become. Similarly, as we can see in Table 4, different scores of the overall assessments are obtained with different parameter vector R in the PFIWDMM operator. Similar to the PFIWMM operator, the more interrelationships among attributes are taken into account, the smaller the scores of the overall assessments. However, no matter what the parameter vector is, the ranking result is always the same. Therefore, the parameter vector can be viewed as the decision makers' risk preference.

Table 4. Ranking results by utilizing the different parameter vector R in the PFIWDMM operator.

Parameter Vector R	The Scores of $s(p_i)(i = 1,2,3,4)$	Ranking Results
$R = (1,0,0,0)$	$s(p_1) = 0.2103$ $s(p_2) = 0.2530$ $s(p_3) = 0.2990$ $s(p_4) = 0.3116$	$x_4 \succ x_3 \succ x_2 \succ x_1$
$R = (1,1,0,0)$	$s(p_1) = 0.0818$ $s(p_2) = 0.1290$ $s(p_3) = 0.1577$ $s(p_4) = 0.1647$	$x_4 \succ x_3 \succ x_2 \succ x_1$
$R = (1,1,1,0)$	$s(p_1) = 0.0465$ $s(p_2) = 0.0862$ $s(p_3) = 0.1068$ $s(p_4) = 0.1116$	$x_4 \succ x_3 \succ x_2 \succ x_1$
$R = (1,1,1,1)$	$s(p_1) = 0.0314$ $s(p_2) = 0.0647$ $s(p_3) = 0.0807$ $s(p_4) = 0.0843$	$x_4 \succ x_3 \succ x_2 \succ x_1$
$R = (2,0,0,0)$	$s(p_1) = 0.0807$ $s(p_2) = 0.1966$ $s(p_3) = 0.3076$ $s(p_4) = 0.3078$	$x_4 \succ x_3 \succ x_2 \succ x_1$

7. Conclusions

In the field of aggregation operators, more and more operators have been proposed. However, some operators do not take the correlations among attributes into consideration, which cannot satisfy the needs of real decision-making problems. The MM operator can consider the interaction relationships among any number of attributes with a parameter R. In this paper, we extend the MM operator to PFNs and propose some new Pythagorean fuzzy operators, including the PFIMM, PFIWMM, and PFIDWMM operators. These operators can reflect the correlations among all Pythagorean fuzzy elements. Further, we propose a novel approach to MAGDM by using these operators. Moreover, in order to show the application of the proposed method in this paper, we provide a numerical example and the advantages of the new operator are more obvious by comparing the new operator with the existing ones. Finally, we give the parameter vector R some different values to discuss the advantages of the new approach on the ranking results of the numerical example. In further works, we will apply the proposed method in more practical decision-making problems, such as low carbon supplier selection, hospital-based post-acute care, risk management, medical diagnosis, and resource evaluation, etc. In addition, we will investigate more aggregation operators for fusing Pythagorean fuzzy information.

Author Contributions: Conceptualization, Y.X.; Formal Analysis, J.W.; All the authors have participated in writing the manuscript and have revised the final version. All authors read and approved the final manuscript.

Funding: This work was partially supported by the National Science Foundation of China (Grant number 61702023, 71532002), Humanities and Social Science Foundation of Ministry of Education of China (Grant number 17YJC870015), the Fundamental Research Funds for the Central Universities of China (Grant number 2018JBM304).

Conflicts of Interest: The authors declare that there is no conflict of interest regarding the publication of this paper.

References

1. Liu, P.D.; Liu, J.L.; Chen, S.M. Some intuitionistic fuzzy Dombi Bonferroni mean operators and their application to multi-attribute group decision making. *J. Oper. Res. Soc.* **2018**, *69*, 1–24. [CrossRef]
2. Li, L.; Zhang, R.T.; Wang, J.; Shang, X.P.; Bai, K.Y. A novel approach to multi-attribute group decision-making with q-rung picture linguistic information. *Symmetry* **2018**, *10*, 172. [CrossRef]
3. Liu, P.D.; Liu, J.L.; Merigó, J.M. Partitioned Heronian means based on linguistic intuitionistic fuzzy numbers for dealing with multi-attribute group decision making. *Appl. Soft Comput.* **2018**, *62*, 395–422. [CrossRef]
4. Liu, P.D.; Chen, S.M. Multiattribute group decision making based on intuitionistic 2-tuple linguistic information. *Inf. Sci.* **2018**, *430*, 599–619. [CrossRef]
5. Xing, Y.P.; Zhang, R.T.; Xia, M.M.; Wang, J. Generalized point aggregation operators for dual hesitant fuzzy information. *J. Intell. Fuzzy Syst.* **2017**, *33*, 515–527. [CrossRef]
6. Liu, P.D.; Wang, P. Some q-rung orthopair fuzzy aggregation operators and their applications to multiple-attribute decision making. *Int. J. Intell. Syst.* **2018**, *33*, 259–280. [CrossRef]

7. Zadeh, L.A. Fuzzy sets. *Inf. Control* **1965**, *8*, 338–353. [CrossRef]
8. Atanassov, K.T. Intuitionistic fuzzy sets. *Fuzzy Sets Syst.* **1986**, *20*, 87–96. [CrossRef]
9. Liu, M.; Ren, H. A new intuitionistic fuzzy entropy and application in multi-attribute decision making. *Information* **2014**, *5*, 587–601. [CrossRef]
10. Ren, H.; Wang, G. An interval-valued intuitionistic fuzzy MADM method based on a new similarity measure. *Information* **2015**, *6*, 880–894. [CrossRef]
11. Kaur, G.; Garg, H. Multi-Attribute Decision-Making Based on Bonferroni Mean Operators under Cubic Intuitionistic Fuzzy Set Environment. *Entropy* **2018**, *20*, 65. [CrossRef]
12. Liu, P.; Liu, X. Multi-attribute group decision making methods based on linguistic intuitionistic fuzzy power Bonferroni mean operators. *Complexity* **2017**, *2017*, 1–15.
13. Liu, P.; Wang, P. Some improved linguistic intuitionistic fuzzy aggregation operators and their applications to multiple-attribute decision making. *Int. J. Inf. Technol. Decis. Mak.* **2017**, *16*, 817–850. [CrossRef]
14. Lakshmana, V.; Jeevaraj, S.; Sivaraman, G. Total ordering for intuitionistic fuzzy numbers. *Complexity* **2016**, *21*, 54–66. [CrossRef]
15. Liu, P.; Teng, F. Multiple criteria decision making method based on normal interval-valued intuitionistic fuzzy generalized aggregation operator. *Complexity* **2016**, *21*, 277–290. [CrossRef]
16. Liu, P.; Chen, S.M. Group decision making based on Heronian aggregation operators of intuitionistic fuzzy numbers. *IEEE Trans. Cybern.* **2017**, *47*, 2514–2530. [CrossRef] [PubMed]
17. Yager, R.R. Pythagorean membership grades in multi-criteria decision making. *IEEE Trans. Fuzzy Syst.* **2014**, *22*, 958–965. [CrossRef]
18. Zhang, X.L. A novel approach based on similarity measure for Pythagorean fuzzy multiple criteria group decision making. *Int. J. Intell. Syst.* **2016**, *31*, 593–611. [CrossRef]
19. Zhang, X.L.; Xu, Z.S. Extension of TOPSIS to multiple criteria decision making with Pythagorean fuzzy sets. *Int. J. Intell. Syst.* **2015**, *29*, 1061–1078. [CrossRef]
20. Ren, P.J.; Xu, Z.S.; Gou, X.J. Pythagorean fuzzy TODIM approach to multi-criteria decision making. *Appl. Soft Comput.* **2016**, *42*, 245–259. [CrossRef]
21. Ma, Z.M.; Xu, Z.S. Symmetric Pythagorean fuzzy weighted geometric/averaging operators and their application in multi-criteria decision-making problems. *Int. J. Intell. Syst.* **2016**, *31*, 1198–1219. [CrossRef]
22. Zeng, S.Z.; Chen, J.P.; Li, X.S. A hybrid method for Pythagorean fuzzy multiple-criteria decision making. *Int. J. Inf. Technol. Decis. Mak.* **2016**, *15*, 403–422. [CrossRef]
23. Peng, X.D.; Yuan, H.Y. Fundamental properties of Pythagorean fuzzy aggregation operators. *Fund. Inform.* **2016**, *147*, 415–446. [CrossRef]
24. Garg, H. Generalized Pythagorean fuzzy geometric aggregation operators using Einstein t-norm and t-conorm for multi-criteria decision making process. *Int. J. Intell. Syst.* **2017**, *32*, 597–630. [CrossRef]
25. Garg, H. A new generalized Pythagorean fuzzy information aggregation using Einstein operations and its application to decision making. *Int. J. Intell. Syst.* **2016**, *31*, 886–920. [CrossRef]
26. Rahman, K.; Abdullah, S.; Ahmed, R.; Ullah, M. Pythagorean fuzzy Einstein weighted geometric aggregation operator and their application to multiple attribute group decision making. *J. Intell. Fuzzy Syst.* **2017**, *33*, 635–647. [CrossRef]
27. Bonferroni, C. Sulle medie multiple di potenze. *Boll. Unione Mat. Ital.* **1950**, *5*, 267–270.
28. Sykora, S. Mathematical means and averages: Generalized Heronian means. *Stan's Libr.* **2009**, *3*. [CrossRef]
29. Maclaurin, C. A second letter to Martin Folkes, Esq.; concern-ing the roots of equations, with the demonstration of other rules in algebra, Phil. *Philos. Trans. R. Soc. Lond. Ser. A* **1729**, *36*, 59–96.
30. Liang, D.; Xu, Z.; Darko, A.P. Projection model for fusing the information of Pythagorean fuzzy multicriteria group decision making based on geometric Bonferroni mean. *Int. J. Intell. Syst.* **2017**, *32*, 966–987. [CrossRef]
31. Liang, D.; Zhang, Y.; Xu, Z.; Darko, A.P. Pythagorean fuzzy Bonferroni mean aggregation operator and its accelerative calculating algorithm with the multithreading. *Int. J. Intell. Syst.* **2018**, *33*, 615–633. [CrossRef]
32. Zhang, R.; Wang, J.; Zhu, X.; Yu, M. Some generalized Pythagorean fuzzy Bonferroni mean aggregation operators with their application to multiattribute group decision-making. *Complexity* **2017**, *6*, 1–16. [CrossRef]
33. Wei, G.; Lu, M. Pythagorean fuzzy Maclaurin symmetric mean operators in multiple attribute decision making. *Int. J. Intell. Syst.* **2018**, *33*, 1043–1070. [CrossRef]
34. Qin, J. Generalized Pythagorean Fuzzy Maclaurin Symmetric Means and Its Application to Multiple Attribute SIR Group Decision Model. *Int. J. Intell. Syst.* **2017**, *20*, 1–15. [CrossRef]

35. Muirhead, R.F. Some methods applicable to identities and inequalities of symmetric algebraic functions of n letters. *Proc. Edinb. Math. Soc.* **1902**, *21*, 144–162. [CrossRef]
36. Liu, P.D.; Li, D.F. Some Muirhead mean operators for intuitionistic fuzzy numbers and their applications to group decision making. *PLoS ONE* **2017**, *12*, e0168767. [CrossRef] [PubMed]
37. Qin, J.D.; Liu, X.W. 2-tuple linguistic Muirhead mean operators for multiple attribute group decision making and its application to supplier selection. *Kybernetes* **2016**, *45*, 2–29. [CrossRef]
38. Wei, G.W. Pythagorean fuzzy interaction aggregation operators and their application to multiple attribute decision making. *J. Intell. Fuzzy Syst.* **2017**, *33*, 2119–2132. [CrossRef]
39. Zhu, J.H.; Li, Y.L. Pythagorean fuzzy Muirhead mean operators and their application in multiple-criteria group decision-Making. *Information* **2018**, *9*, 142. [CrossRef]

information

MDPI

Article

A Green Supplier Assessment Method for Manufacturing Enterprises Based on Rough ANP and Evidence Theory

Lianhui Li [1,*] and Hongguang Wang [2]

1 College of Mechatronic Engineering, North Minzu University, Yinchuan 750021, China
2 The 713th Research Institute of China Shipbuilding Industry Corporation, Zhengzhou 450052, China; silverystream@foxmail.com
* Correspondence: lilianhui@nmu.edu.cn; Tel.: +86-951-2066393

Received: 10 May 2018; Accepted: 28 June 2018; Published: 2 July 2018

Abstract: Within the context of increasingly serious global environmental problems, green supplier assessment has become one of the key links in modern green supply chain management. In the actual work of green supplier assessment, the information of potential suppliers is often ambiguous or even absent, and there are interrelationships and feedback-like effects among assessment indexes. Additionally, the thinking of experts in index importance judgment is always ambiguous and subjective. To handle the uncertainty and incompleteness in green supplier assessment, we propose a green supplier assessment method based on rough ANP and evidence theory. The uncertain index value is processed by membership degree. Trapezoidal fuzzy number is adopted to express experts' judgment on the relative importance of the indexes, and rough boundary interval is used to integrate the judgment opinions of multiple experts. The ANP structure is built to deal with the interrelationship and feedback-like effects among indexes. Then, the index weight is calculated by ANP method. Finally, the green suppliers are assessed by a trust interval, based on evidence theory. The feasibility and effectiveness of the proposed method is verified by an application of a bearing cage supplier assessment.

Keywords: green supplier; rough ANP; trapezoidal fuzzy number; rough boundary interval; evidence theory; trust interval

1. Introduction

With the increasing awareness of global environmental protection and the increasing number of related environmental regulations, manufacturing enterprises are facing more stringent environmental requirements. Nowadays, green supply chains have become inevitable choices for manufacturing enterprises who wish to deal with environmental problems. Green supply chain management includes many links, such as green supplier assessment, green product design, green production, green marketing and waste recovery [1–3]. Green supplier assessment is in the upstream of the whole supply chain, and its effect on environmental protection and cost saving can be transmitted to every part of the downstream through the supply chain.

In the process of green supply chain management, various factors make the relationship between suppliers and manufacturing enterprises complicated and vague. However, competitors constantly adjust their strategies, and the supply chain must constantly improve to adapt to the complex environment which changes rapidly. In this context, green supplier assessment plays a very important role in reducing costs, and improving product quality and market competitiveness. Through effective assessment and supervisions of suppliers, problems can be found and solved in time, and the green and healthy development of the entire supply chain can be promoted [4–9]. It can be seen that green supplier assessment plays a decisive role in green supply chain management, which directly determines the competitiveness of the entire supply chain. The rise of the internet has provided convenience for manufacturing enterprises to assess green suppliers, but enterprises cannot quickly choose suppliers which meet their needs in the face of so many uneven suppliers. Considering finances and the effective utilization of resources, how to assess green supplier quickly and effectively becomes the key problem in modern green supply chain.

As seen in the present literature, there are many significant works on green supplier assessment. On the whole, the existing research mainly includes the following aspects.

(1) Supplier assessment models based on Analytic Hierarchy Process (AHP) [10] or Analytic Network Process (ANP) [11]. Noci [12] used AHP to evaluate supplier's environmental efficiency. Lee et al. [13] used the Delphi technique to distinguish the evaluation criterion difference between the traditional supplier and the green supplier, and then used Fuzzy Analytic Hierarchy Process (FAHP) to solve the green supplier selection process. Hsu and Hu [14] contained the interdependence between components of decision structure and used ANP for green supplier selection which reflected a more realistic result.
(2) Supplier assessment model based on mathematical Programming. Yeh and Chuang [15] put forward a mathematical programming model of green partner selection, which includes four goals: cost, time, product quality, and green score. They adopted two multi-objective genetic algorithms (MOGA) to find a set of Pareto optimal solutions, and used weighted summation to generate more solutions. Yousefi et al. [16] used Dynamic Data Envelopment Analysis (D-DEA) and scenario-based robust model for supplier selection. In this supplier selection model, the shortcomings of the DEA model (the benchmarks were determined based on previous performance) were overcome, and the disadvantages of the D-DEA model (the decision unit couldn't get a unified efficiency score) were avoided.
(3) Supplier assessment model based on Technique for Order Preference by Similarity to an Ideal Solution (TOPSIS). Awasthi et al. [17] proposed a three-step method for green supplier evaluation: identification standard, expert score, evaluate expert score by fuzzy TOPSIS. The fuzzy TOPSIS method integrated profit and cost standard, and this method is suitable for the situation of lack of partial quantitative information. Kannan and Jabbour [18] used fuzzy TOPSIS to solve the problem of green supplier selection, and applies three types of fuzzy TOPSIS method to sort green suppliers.
(4) Other hybrid models for supplier assessment. Gandhi et al. [5] proposed a combined approach using AHP and Decision-Making and Trial Evaluation Laboratory (DEMATEL) for evaluating success factors in implementation of green supply chain management and gave a case study in Indian manufacturing industries. Chatterjee et al. [19] combined DEMATEL and ANP in a rough context, and then proposed a rough DEMATEL-ANP (R'AMATEL) method to evaluate the performance of suppliers for green supply chain implementation in electronics industry. Wu et al. [20] used the Continuous Ordered Weighted Averaging (COWA) operator to transform the trapezoid fuzzy number into the exact real number to select the green supplier, and make a sensitivity analysis according to the degree of risk of decision-maker to rank the suppliers. Luo and Peng [21] proposed a multi-level supplier evaluation and selection model. In this model, AHP is used to determine the weights, and then TOPSIS is used for supplier evaluation. Kuo and Lin [22] integrated ANP and DEA and proposed a green supplier evaluation method. The interdependence between standards were considered by ANP, which allowed users to choose their own weight preferences to limit weights, expanded DEA method, and allowed more flexible number of decision units. Shi et al. [23] used the improved attribute reduction algorithm based on rough set to reduce the index of the green supplier evaluation index system, and then evaluated the data by RBF neural network training. Akman [24] identified the suppliers that should be included in the green supplier development plan through the C mean clustering algorithm and

the VIKOR method. This method can be used to solve the problem of supplier classification and evaluation.

The study of green supplier assessment, which is of great theoretical and practical significance, has been a hot topic all along. However, the existing research has obvious shortcomings and the research gaps are mainly as follows:

(1) The information of the suppliers to be assessed is not clear enough in the actual work of green supplier assessment. There is often no information sharing between manufacturing enterprise and suppliers, and the information between them is often ambiguous or even absent. The deterministic assessing model can no longer meet the needs of the increasingly complex decision-making environment.
(2) Green supplier assessment is a complex decision problem and its indexes are interrelated. When calculating the index weight, the core idea of traditional AHP is to divide the index system into isolated and hierarchical levels. Only the upper level elements' dominating effect on the lower level elements is considered and the elements in the same level are deemed to be independent of each other. However, the relationships among the indexes are often interdependent and sometimes provide feedback-like effects in green supplier assessment. Therefore, traditional AHP cannot solve the complex relationships among indexes to obtain the weight in green supplier assessment.
(3) The accurate number is used to describe the relative importance of the indexes in the expression of experts' judgment in most of the existing research, which cannot reflect the ambiguities and subjectivity of the actual thinking. It is more reasonable to use fuzzy numbers [25] to express the experts' judgment. After introducing fuzzy numbers to express experts' judgment, analyzing and processing the imprecise and inconsistent information becomes a difficult problem.

To fill the research gaps in green supplier assessment, a green supplier assessment method for manufacturing enterprises based on rough ANP and evidence theory is proposed. We process the uncertain index value by membership degree, adopt trapezoidal fuzzy number to express experts' judgment on the relative importance of the indexes, use rough boundary interval to integrate the judgment opinions of multiple experts, set up the ANP structure to deal with the interrelationship and the feedback-like effects among indexes and then calculate the index weight by ANP method, and finally solve the incomplete information problem by evidence theory and assess the green suppliers by trust interval.

The rest of this paper is organized as follows. Section 2 establishes the index system of green supplier assessment; Section 3 uses membership degree method to process the uncertain index value of suppliers to be assessed; Section 4 adopts rough ANP to calculate the index weight; Section 5 gives the green supplier assessment procedure based on evidence theory; Section 6 provides an application case of bearing cage supplier assessment and discusses the feasibility and effectiveness of the proposed method for green supplier assessment. We conclude this paper in Section 7.

2. Index System

The first and very important segment of green supplier assessment is the establishment of a complete and overall index system. The attribute of the supplier's product is the main representative of its ability, and the comprehensive ability of supplier can provide a strong support to its product. The comprehensive ability of supplier mainly includes internal competitiveness, external competitiveness and cooperation ability. Internal competitiveness of a supplier can be subdivided into its innovation capacity, manufacturing capacity and agility capacity. Furthermore, a supplier is not isolated and is inevitably restricted by its external competitiveness. External competitiveness of a supplier mainly includes its economic environment, geographical environment, social environment and legal environment. Additionally, cooperation ability of a supplier is affected by its technical compatibility degree, cultural compatibility degree, information platform compatibility degree, and reputation.

As shown in Figure 1, we establish the index system of green supplier assessment.

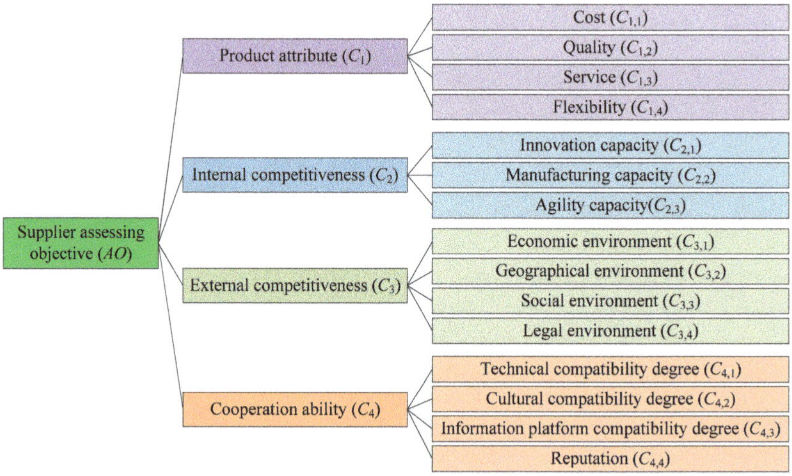

Figure 1. The index system.

The green supplier assessment objective (AO) includes four first-level indexes: product attribute (C_1), internal competitiveness (C_2), external competitiveness (C_3) and cooperation ability (C_4).

- C_1 is decomposed into four second-level indexes: cost ($C_{1,1}$), quality ($C_{1,2}$), service ($C_{1,3}$) and flexibility ($C_{1,4}$). $C_{1,1}$ and $C_{1,2}$ belong to quantitative type, and $C_{1,3}$ and $C_{1,4}$ belong to qualitative type.
- C_2 is decomposed into three second-level indexes: innovation capacity ($C_{2,1}$), manufacturing capacity ($C_{2,2}$) and agility capacity ($C_{2,3}$). $C_{2,1}$, $C_{2,2}$ and $C_{2,3}$ all belong to qualitative type.
- C_3 is decomposed into four second-level indexes: economic environment ($C_{3,1}$), geographical environment ($C_{3,2}$), social environment ($C_{3,3}$) and legal environment ($C_{3,4}$). $C_{3,1}$, $C_{3,2}$, $C_{3,3}$ and $C_{3,4}$ all belong to qualitative type.
- C_4 is decomposed into four second-level indexes: technical compatibility degree ($C_{4,1}$), cultural compatibility degree ($C_{4,2}$), information platform compatibility degree ($C_{4,3}$) and reputation ($C_{4,4}$). $C_{4,1}$, $C_{4,2}$, $C_{4,3}$ and $C_{4,4}$ all belong to qualitative type.

3. Index Value Processing

For different types of indexes, different methods are used to get their values. To quantitative type index (i.e., $C_{1,1}$ and $C_{1,2}$), its value is obtained directly. To qualitative type index (i.e., $C_{1,3}$, $C_{1,4}$, $C_{2,1}$, $C_{2,2}$, $C_{2,3}$ $C_{3,1}$, $C_{3,2}$, $C_{3,3}$, $C_{3,4}$, $C_{4,1}$, $C_{4,2}$, $C_{4,3}$ and $C_{4,4}$), its value, which is a score, is given by manager. If an index value can be accurately determined, it is a point value. If an index value is relatively fuzzy, it is an interval value. If an index value is completely unknown, it is a null value.

The suppliers to be assessed are $x_1, x_2, \ldots, x_M$. For the supplier x_r ($r = 1, 2, \ldots, M$), the value on the index $C_{j,l}$($j = 1, 2, \ldots, N$ and $l = 1, 2, \ldots, n_j$) is represented as $v_{r,(j,l)}$. Then, the normalized index value $v'_{r,(j,l)}$ is calculated as follows. If the index belongs to benefit-type, $v'_{r,(j,l)} = v_{r,(j,l)}/\max\left\{v_{1,(j,l)}, v_{2,(j,l)}, \ldots, v_{M,(j,l)}\right\}$. If the index belongs to cost-type, $v'_{r,(j,l)} = \min\left\{v_{1,(j,l)}, v_{2,(j,l)}, \ldots, v_{M,(j,l)}\right\}/v_{r,(j,l)}$. Here, the interval index value is replaced with its left and right ends.

We set five comment levels which are very bad (G_1), bad (G_2), middle (G_3), good (G_4), very good (G_5). Furtherly, G_1 and G_5 are the comment level corresponding to the lowest normalized index value $v'_{(j,l)}(G_1) = \min\left\{v'_{1,(j,l)}, v'_{2,(j,l)}, \ldots, v'_{M,(j,l)}\right\}$ and the highest normalized index value $v'_{(j,l)}(G_5) = \max\left\{v'_{1,(j,l)}, v'_{2,(j,l)}, \ldots, v'_{M,(j,l)}\right\}$, respectively. Similarly, the interval index value is replaced with its left and right ends. Then, a number sequence $v'_{(j,l)}(G_1), v'_{(j,l)}(G_2), v'_{(j,l)}(G_3), v'_{(j,l)}(G_4), v'_{(j,l)}(G_5)$ is obtained.

It is assumed that the corresponding numbers of the five comment levels are $\pi_1 = 0.1$, $\pi_2 = 0.3$, $\pi_3 = 0.5$, $\pi_4 = 0.7$ and $\pi_5 = 0.9$. β_u represents the membership degree of the index value to the comment level G_u. On the index $C_{j,l}$, the utility value of the supplier x_r is represented as $\tau_{r,(j,l)}$. The normalized index value of the supplier x_r is a point-value χ_1, an interval value $[\chi_1, \chi_2]$ or a null value. When $v'_p \le \chi_1 \le v'_{p+1}$ or $v'_p \le \chi_1 \le \chi_2 \le v'_{p+1} (p = 1, 2, 3, 4)$, $\tau_{r,(j,l)} = \beta_p \pi_p + \beta_{p+1}\pi_{p+1}$. When $v'_p \le \chi_1 \le v'_{p+1}$ and $v'_{p+1} \le \chi_2 \le v'_{p+2} (p = 1, 2, 3)$, $\tau_{r,(j,l)} = \beta_p \pi_p + \beta_{p+1}\pi_{p+1} + \beta_{p+2}\pi_{p+2}$. When $v'_p \le \chi_1 \le v'_{p+1}$ and $v'_o \le \chi_2 \le v'_{o+1}$ ($p = 1, 2, 3, 4, o = 1, 2, 3, 4$ and $o > p + 1$), $\tau_{r,(j,l)} = \beta_p \pi_p + \beta_{p+1}\pi_{p+1} + \ldots + \beta_o \pi_o + \beta_{o+1}\pi_{o+1}$.

4. Index Weight Calculating

In ANP [11], the system elements are divided into two parts: (1) The first is called the control layer, including the problem objective and decision criteria. All decision criteria are considered independent of each other and are governed only by the problem objective. There can be no decision criteria in the control layer, but at least one objective; (2) The second part is the network layer, which is composed of all the elements that are controlled by the control layer, and its internal network structure is interacted.

Therefore, we set up the ANP structure of green supplier assessment as shown in Figure 2. The control layer only has one element: the green supplier assessment objective (*AO*), and the network layer has four element groups: product attribute (C_1), internal competitiveness (C_2), external competitiveness (C_3) and cooperation ability (C_4). Each element group affects each other and contains different elements. The elements in the same element group also affect each other. For example, internal competitiveness (C_2) is affected by product attribute (C_1), external competitiveness (C_3) and cooperation ability (C_4), and innovation capacity ($C_{2,1}$), manufacturing capacity ($C_{2,2}$) and agility capacity ($C_{2,3}$) also affect each other.

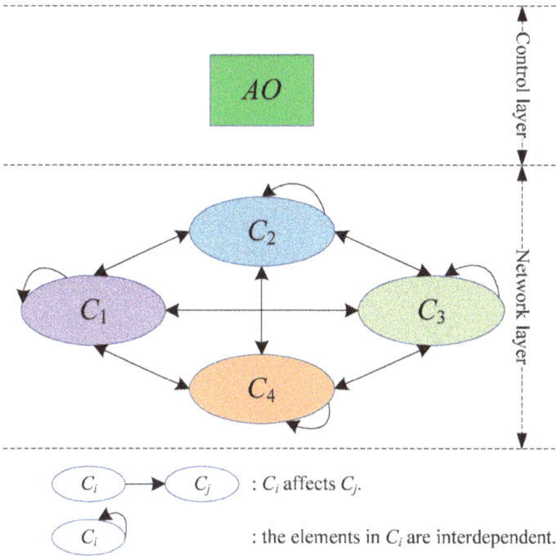

Figure 2. The ANP structure of green supplier assessment.

According to the ANP structure of green supplier assessment shown in Figure 2, the control layer has the element *AO* and the network layer has the element groups $C_1, C_2, \ldots, C_N$ (here, $N = 4$). The element group C_i ($i = 1, 2, \ldots, N$) contains the elements $C_{i,1}, C_{i,2}, \ldots, C_{i,n_i}$. The control layer element *AO* is taken as the criterion and the element $C_{j,l}$ ($l = 1, 2, \ldots, n_j$) in C_j ($j = 1, 2, \ldots, N$) is taken as the sub-criterion. Based on the influence of the elements in C_i on $C_{j,l}$, the indirect dominance comparison of the elements in C_i are conducted.

Here, the influence of the elements in C_i on $C_{j,l}$ are assessed according to the personal experience and subjective judgment of experts, so using exact numbers to describe the influence of the elements in C_i on $C_{j,l}$ is unreasonable. In contrast, the fuzzy number can reflect the inherent uncertainty of the expert's preference. At the same time, when integrating the opinions of multiple experts, the assessment of the influence of the elements in C_i on $C_{j,l}$ by experts is obviously with indiscernibility. The rough boundary interval in rough sets theory [26–29] can describe the indiscernibility as a set boundary area instead of a membership function, which can better integrate the assessment of multiple experts.

Based the index weight obtaining method in Reference [30], we design a rough ANP method to determine the index weight in green supplier assessment. The specific process of the designed rough ANP is as follows.

Step 1: Under the control layer element AO, we conduct the indirect dominance comparison of the elements in C_i according to their influence on $C_{j,l}$.

There are q experts participating in the indirect dominance comparison of the elements in C_i. The fuzzy reciprocal judgement matrix $E^{k,i,(j,l)} = (e_{g,h}^{k,i,(j,l)})_{n_i \times n_i}$ given by the expert k ($k = 1, 2, \ldots, q$) is as follows:

$$E^{k,i,(j,l)} = \begin{array}{c|cccc} C_{j,l} & C_{i,1} & C_{i,2} & \cdots & C_{i,n_i} \\ \hline C_{i,1} & 1 & e_{1,2}^{k,i,(j,l)} & \cdots & e_{1,n_i}^{k,i,(j,l)} \\ C_{i,2} & e_{2,1}^{k,i,(j,l)} & 1 & \cdots & e_{2,n_i}^{k,i,(j,l)} \\ \vdots & \vdots & \vdots & & \vdots \\ C_{i,n_i} & e_{n_i,1}^{k,i,(j,l)} & e_{n_i,2}^{k,i,(j,l)} & \cdots & 1 \end{array} \tag{1}$$

where $e_{g,h}^{k,i,(j,l)}$ represents the indirect dominance score of the element $C_{i,h}$ compared to the element $C_{i,g}$ giver by the expert k, here $g,h = 1, 2, \ldots, n_i$ and $g \neq h$. $e_{g,h}^{k,i,(j,l)}$ is a trapezoidal fuzzy number and $e_{g,h}^{k,i,(j,l)} = (a_{g,h}^{k,i,(j,l)}, b_{g,h}^{k,i,(j,l)}, c_{g,h}^{k,i,(j,l)}, d_{g,h}^{k,i,(j,l)})$. $a_{g,h}^{k,i,(j,l)}$, $b_{g,h}^{k,i,(j,l)}$, $c_{g,h}^{k,i,(j,l)}$ and $d_{g,h}^{k,i,(j,l)}$ are all positive real numbers and $a_{g,h}^{k,i,(j,l)} \leq b_{g,h}^{k,i,(j,l)} \leq c_{g,h}^{k,i,(j,l)} \leq d_{g,h}^{k,i,(j,l)}$. Then the consistency of the matrix given by each expert is verified. If it is qualified, the next step will be carried out; otherwise, this step will be returned.

Then, $E^{k,i,(j,l)}$ is split into $A^{k,i,(j,l)} = (a_{g,h}^{k,i,(j,l)})_{n_i \times n_i}$, $B^{k,i,(j,l)} = (b_{g,h}^{k,i,(j,l)})_{n_i \times n_i}$, $C^{k,i,(j,l)} = (c_{g,h}^{k,i,(j,l)})_{n_i \times n_i}$ and $D^{k,i,(j,l)} = (d_{g,h}^{k,i,(j,l)})_{n_i \times n_i}$. Based on $A^{1,i,(j,l)}, A^{2,i,(j,l)}, \ldots, A^{q,i,(j,l)}$, the rough group decision matrix $A^{i,(j,l)} = (a_{g,h}^{i,(j,l)})_{n_i \times n_i}$ is constructed where $a_{g,h}^{i,(j,l)} = \{a_{g,h}^{1,i,(j,l)}, a_{g,h}^{2,i,(j,l)}, \ldots, a_{g,h}^{q,i,(j,l)}\}$, $g,h = 1, 2, \ldots, n_i$ and $g \neq h$.

The rough boundary interval of $a_{g,h}^{k,i,(j,l)} \in a_{g,h}^{i,(j,l)}$ ($k = 1, 2, \ldots, q$) is $RN(a_{g,h}^{k,i,(j,l)}) = \left[a_{g,h}^{k,i,(j,l),-}, a_{g,h}^{k,i,(j,l),+}\right]$ where $a_{g,h}^{k,i,(j,l),-}, a_{g,h}^{k,i,(j,l),+}$ are the rough lower limit and rough upper limit of $a_{g,h}^{k,i,(j,l)}$ in the set $a_{g,h}^{i,(j,l)}$.

Thus, the rough boundary interval of $a_{g,h}^{i,(j,l)}$ can be expressed as $RN(a_{g,h}^{i,(j,l)}) = \left\{\left[a_{g,h}^{1,i,(j,l),-}, a_{g,h}^{1,i,(j,l),+}\right], \left[a_{g,h}^{2,i,(j,l),-}, a_{g,h}^{2,i,(j,l),+}\right], \cdots, \left[a_{g,h}^{q,i,(j,l),-}, a_{g,h}^{q,i,(j,l),+}\right]\right\}$. According to the calculation rule of rough boundary interval, the mean form of $RN(a_{g,h}^{i,(j,l)})$ can be obtained as $Avg\_RN(a_{g,h}^{i,(j,l)}) = \left[a_{g,h}^{i,(j,l),-}, a_{g,h}^{i,(j,l),+}\right] = \left[\sum_{k=1}^{q} a_{g,h}^{k,i,(j,l),-}/q, \sum_{k=1}^{q} a_{g,h}^{k,i,(j,l),+}/q\right]$ where $a_{g,h}^{i,(j,l),-}, a_{g,h}^{i,(j,l),+}$ are the rough lower limit and rough upper limit of the set $a_{g,h}^{i,(j,l)}$.

The rough judgement matrix is constructed as $EA^{i,(j,l)} = (Avg\_RN(a_{g,h}^{i,(j,l)}))_{n_i \times n_i}$. Then, $EA^{i,(j,l)}$ is split into the rough lower limit matrix $EA^{i,(j,l),-} = (a_{g,h}^{i,(j,l),-})_{n_i \times n_i}$ and the rough upper limit matrix $EA^{i,(j,l),+} = (a_{g,h}^{i,(j,l),+})_{n_i \times n_i}$.

The eigenvectors corresponding to the maximum eigenvalues of $EA^{i,(j,l),-}$ and $EA^{i,(j,l),+}$ are $VA^{i,(j,l),-} = \left[va_1^{i,(j,l),-}, va_2^{i,(j,l),-}, \cdots, va_{n_i}^{i,(j,l),-}\right]^T$ and $VA^{i,(j,l),+} = \left[va_1^{i,(j,l),+}, va_2^{i,(j,l),+}, \cdots, va_{n_i}^{i,(j,l),+}\right]^T$ respectively, where $va_h^{i,(j,l),-}, va_h^{i,(j,l),+}$ are the value of $VA^{i,(j,l),-}$ and $VA^{i,(j,l),+}$ on the h ($h = 1, 2, \ldots, n_i$) dimension. Then, a set $GA^{i,(j,l)} = \left\{ga_1^{i,(j,l)}, ga_2^{i,(j,l)}, \ldots, ga_{n_i}^{i,(j,l)}\right\}$ can be obtained where $ga_h^{i,(j,l)} = \left(\left|va_h^{i,(j,l),-}\right| + \left|va_h^{i,(j,l),+}\right|\right)/2$. Similarly, we also get the other sets: $GB^{i,(j,l)} = \left\{gb_1^{i,(j,l)}, gb_2^{i,(j,l)}, \ldots, gb_{n_i}^{i,(j,l)}\right\}$, $GC^{i,(j,l)} = \left\{gc_1^{i,(j,l)}, gc_2^{i,(j,l)}, \ldots, gc_{n_i}^{i,(j,l)}\right\}$ and $GD^{i,(j,l)} = \left\{gd_1^{i,(j,l)}, gd_2^{i,(j,l)}, \ldots, gd_{n_i}^{i,(j,l)}\right\}$.

Thereupon, the eigenvector $w^{i,(j,l)} = [w_1^{i,(j,l)}, w_2^{i,(j,l)}, \ldots, w_{n_i}^{i,(j,l)}]^T$ is obtained, where $w_h^{i,(j,l)} = \frac{[(gd_h^{i,(j,l)})^2 + gd_h^{i,(j,l)} \cdot gc_h^{i,(j,l)} + (gc_h^{i,(j,l)})^2] - [(ga_h^{i,(j,l)})^2 + ga_h^{i,(j,l)} \cdot gb_h^{i,(j,l)} + (gb_h^{i,(j,l)})^2]}{3(gd_h^{i,(j,l)} + gc_h^{i,(j,l)} - ga_h^{i,(j,l)} - gb_h^{i,(j,l)})}$, $h = 1, 2, \ldots, n_i$. Then, the eigenvector $w^{i,(j,l)} = [w_1^{i,(j,l)}, w_2^{i,(j,l)}, \ldots, w_{n_i}^{i,(j,l)}]^T$ is normalized as $\omega^{i,(j,l)} = [\omega_1^{i,(j,l)}, \omega_2^{i,(j,l)}, \ldots, \omega_{n_i}^{i,(j,l)}]^T$ where $\omega_h^{i,(j,l)} = w_h^{i,(j,l)} / \sum\limits_{h=1}^{n_i} w_h^{i,(j,l)}$.

Step 2: We represent $\Omega^{i,j} = (\omega_h^{i,(j,l)})_{n_i \times n_j}$ as follows:

$$\Omega^{i,j} = \begin{bmatrix} \omega_1^{i,(j,1)} & \omega_1^{i,(j,2)} & \cdots & \omega_1^{i,(j,n_j)} \\ \omega_2^{i,(j,1)} & \omega_2^{i,(j,2)} & \cdots & \omega_2^{i,(j,n_j)} \\ \vdots & \vdots & & \vdots \\ \omega_{n_i}^{i,(j,1)} & \omega_{n_i}^{i,(j,2)} & \cdots & \omega_{n_i}^{i,(j,n_j)} \end{bmatrix} \tag{2}$$

where the column vector $\omega^{i,(j,l)}$ is the normalized influence degree sorting vector of the elements $C_{i,1}, C_{i,2}, \ldots, C_{i,n_i}$ in C_i on the element $C_{j,l}$ in C_j. If the elements in C_j is not affected by the elements in C_i, $\Omega^{i,j} = 0$.

So we get the hyper-matrix Ω under the control layer element AO as follows:

$$\Omega = \begin{array}{cc} & \begin{array}{ccccccccc} & C_1 & & & C_2 & & & & C_N \\ C_{1,1} & \cdots & C_{1,n_1} & C_{2,1} & \cdots & C_{2,n_2} & \cdots & C_{N,1} & \cdots & C_{N,n_N} \end{array} \\ \begin{array}{cc} C_1 & \begin{array}{c} C_{1,1} \\ \vdots \\ C_{1,n_1} \end{array} \\ C_2 & \begin{array}{c} C_{2,1} \\ \vdots \\ C_{2,n_2} \end{array} \\ \vdots & \vdots \\ C_N & \begin{array}{c} C_{N,1} \\ \vdots \\ C_{N,n_N} \end{array} \end{array} & \begin{bmatrix} \Omega^{1,1} & \Omega^{1,2} & \cdots & \Omega^{1,N} \\ \Omega^{2,1} & \Omega^{2,2} & \cdots & \Omega^{2,N} \\ \vdots & \vdots & & \vdots \\ \Omega^{N,1} & \Omega^{N,2} & \cdots & \Omega^{N,N} \end{bmatrix} \end{array} \tag{3}$$

Step 3: The sub-block $\Omega^{i,j}$ of Ω is column normalized, but Ω isn't column normalized. To solve this problem, we conduct the indirect dominance comparison of the element groups $C_1, C_2, \ldots, C_N$ according to their influence on C_j ($j = 1, 2, \ldots, N$) under the control layer element AO. Here, we adopt a similar approach to Step 1 and get the relative importance matrix $\Psi = (\psi^{i,j})_{N\times N}$ of element groups as follows:

$$\Psi = \begin{bmatrix} \psi^{1,1} & \psi^{1,2} & \cdots & \psi^{1,N} \\ \psi^{2,1} & \psi^{2,2} & \cdots & \psi^{2,N} \\ \vdots & \vdots & & \vdots \\ \psi^{N,1} & \psi^{N,2} & \cdots & \psi^{N,N} \end{bmatrix} \quad (4)$$

where the column vector $\psi^{\cdot j} = [\psi^{1,j}, \psi^{2,j}, \ldots, \psi^{N,j}]^{\mathrm{T}}$ is the normalized influence degree sorting vector of the element groups $C_1, C_2, \ldots, C_N$ on C_j.

The weighted form of the hyper-matrix Ω is $\overline{\Omega}$, where $\overline{\Omega}^{i,j} = \psi^{i,j}\Omega^{i,j}$ $(i = 1,2,\ldots,N,\ j = 1,2,\ldots,N)$.

Step 4: We do the square operation of the weighted hyper-matrix $\overline{\Omega}$ until the result converges to a stable limit hyper-matrix $\overline{\Omega}^{\infty}$ as follows:

$$\overline{\Omega}^{\infty} = \lim_{t\to\infty} \overline{\Omega}^{t} \quad (5)$$

Any column of $\overline{\Omega}^{\infty}$ is the limit relative ranking vector of all elements in the network layer. So we can get the weight vector of the index system shown in Figure 1 as follows:

$$\theta = [\theta_{1,1}, \theta_{1,2}, \ldots, \theta_{1,n_1}, \theta_{2,1}, \theta_{2,2}, \ldots, \theta_{2,n_2}, \theta_{3,1}, \theta_{3,2}, \ldots, \theta_{3,n_3}, \theta_{4,1}, \theta_{4,2}, \ldots, \theta_{4,n_4}]^{\mathrm{T}} \quad (6)$$

where $\theta_{j,1}, \theta_{j,2}, \ldots, \theta_{j,n_j}$ are the weights of $C_{j,1}, C_{j,2}, \ldots, C_{j,n_j}$, $j = 1, 2, \ldots, N$.

Furtherly, the weight of C_j ($j = 1, 2, \ldots, N$) is obtained as follows:

$$\theta_j = \sum_{l=1}^{n_j} \theta_{j,l} \quad (7)$$

5. Green Supplier Assessment

5.1. Related Concepts of Evidence Theory

We assume that there are M suppliers to be assessed. Based on evidence theory [31,32], the set of the suppliers to be assessed is defined as the identification framework $\Theta = \{x_1, x_2, \ldots, x_M\}$. All possible sets in Θ are represented by the power set 2^{Θ}. If each element in Θ is incompatible with each other, the number of elements in 2^{Θ} is 2^M. Then, a set function $mass : 2^{\Theta} \to [0,1]$, which satisfies $mass(\phi) = 0$ and $\sum_{\varphi\subset\Theta} mass(\varphi) = 1$, is defined. The set function $mass$ is known as the basic probability distribution function on Θ. Here, φ represents a supplier to be assessed. $mass(\varphi)$ is the basic probability distribution value of φ and represents the trust degree for Φ, Any supplier to be assessed satisfying the condition "$mass(\varphi) > 0$" is called a focal element.

For $\varphi, \varphi_1, \varphi_2, \ldots, \varphi_n \subseteq \Theta$, the fusion rule of the basic probability distribution functions $mass_1, mass_2, \ldots, mass_n$ on Θ is as follows:

$$mass(\varphi) = \frac{1}{K} \sum_{\varphi_1\cap\varphi_2\cap\ldots\cap\varphi_n=\varnothing} mass_1(\varphi_1) \cdot mass_2(\varphi_2) \cdot \ldots \cdot mass_n(\varphi_n) \quad (8)$$

where $mass = mass_1 \oplus mass_2 \oplus \ldots \oplus mass_n$.

The normalization constant K is defined as follows:

$$K = \sum_{\varphi_1\cap\varphi_2\cap\ldots\cap\varphi_n\neq\varnothing} mass_1(\varphi_1) \cdot mass_2(\varphi_2) \cdot \ldots \cdot mass_n(\varphi_n) \quad (9)$$

The total trust degree of φ on Θ can be expressed as the belief function $Bel(\varphi) = \sum_{\phi\subset\Theta,\phi\subseteq\varphi} mass(\phi)$, and the uncertainty degree of φ on Θ can be expressed as the plausible function $Pl(\varphi) = \sum_{\phi\subset\Theta,\phi\cap\varphi\neq\varnothing} mass(\phi)$. For a supplier φ on Θ, $Bel(\varphi)$ represents the sum of the possibility measurement for all subsets of φ, and $Pl(\varphi)$ represents the sum of the uncertainty measurement for all subsets of φ. The confirmation degree of φ is represented by the trust interval $[Bel(\varphi), Pl(\varphi)]$.

Therefore, $Bel(\varphi)$ reflects the sum of exact reliability which the evidences support φ, and $Pl(\varphi)$ reflects the sum of reliability which the evidences do not negate φ. As a result, $Bel(\varphi)$ and $Pl(\varphi)$ can be considered as the minimum and maximum probability bounds respectively, so $[Bel(\varphi), Pl(\varphi)]$ can form the trust interval.

Based on the above analysis, assessing the suppliers by trust interval is more reliable than by maximum belief function or maximum plausible function [30,31]. We assume that the supplier φ_s is better than supplier φ_y with a degree of $\varepsilon_{s,y}$ $(0 \leq \varepsilon_{s,y} \leq 1)$. If the trust intervals of φ_s and φ_y are $[Bel(\varphi_s), Pl(\varphi_s)]$ and $[Bel(\varphi_y), Pl(\varphi_y)]$ respectively, $\varepsilon_{s,y}$ is defined as follows:

$$\varepsilon_{s,y} = \frac{\max\{0, Pl(\varphi_s) - Bel(\varphi_y)\} - \max\{0, Bel(\varphi_s) - Pl(\varphi_y)\}}{(Pl(\varphi_s) - Bel(\varphi_s)) + (Pl(\varphi_y) - Bel(\varphi_y))} \tag{10}$$

The decision rules based on trust interval are as follows:

- If $\varepsilon_{s,y} > 0.5$, φ_s is better than φ_y (recorded as $]\varphi_s \rhd \varphi_y$);
- If $\varepsilon_{s,y} < 0.5$, φ_s is worse than φ_y (recorded as $\varphi_s \lhd \varphi_y$);
- If $\varepsilon_{s,y} = 0.5$, φ_s and φ_y are equal (recorded as $\varphi_s = \varphi_y$);
- For three suppliers φ_s, φ_y and φ_z, if $\varepsilon_{s,y} > 0.5$ and $\varepsilon_{y,z} > 0.5$, φ_s is better than φ_y and φ_y is better than φ_z, so $\varphi_s \rhd \varphi_y \rhd \varphi_z$.

5.2. Green Supplier Assessing Procedure

On the index $C_{j,l}$, the weighted basic probability distribution value of the focal element φ_s $(s < 2^N)$ is represented as $mass'_{C_{j,l}}(\varphi_s)$. In this paper, we use $mass'_{C_{j,l}}(\varphi_s)$ as the evidence input of green supplier assessment.

The utility value of each focal element except Θ can be calculated through index value processing. The special focal element Θ can indicate the uncertainty of the expert on an index. If we don't consider the influence of Θ, the green supplier assessment problem will become a simple probability distribution problem, and the advantages of evidence theory will not be applied. Meanwhile, the trust degree of expert in each index is different and the uncertainty of index is reflected by the probability distribution of Θ. Thus the probability distribution value of Θ on different indexes should also be treated differently.

In the green supplier assessment problem, the index weight is obviously not fixed in the case of different requirements. When costs need to be reduced, C_{11} is more important than other indexes and its weight must be higher than other indexes, and the basic probability distribution value of Θ on C_{11} should be smaller than other indexes. Therefore, the index weight calculated by rough ANP is introduced to adjust the preference of experts and solve the probability distribution problem of Θ on different indexes. Then, the weighted basic probability distribution value of each focal element on each index is obtained as $mass'_{C_{j,l}}(\varphi_s)$.

We take a weighted normalization treatment for the basic probability distribution values of all focal elements and calculate $mass'_{C_{j,l}}(\varphi_s)$, as follows:

$$\begin{cases} mass'_{C_{j,l}}(\varphi_s) = \theta_{j,l}\dfrac{\tau_{s,(j,l)}}{\sum_{s=1}^{l-1}\tau_{s,(j,l)}}, & \varphi_s \neq \Theta \\ mass'_{C_{j,l}}(\varphi_s) = 1 - \theta_{j,l}, & \varphi_s = \Theta \end{cases} \tag{11}$$

Based on the two-level index system shown in Figure 1, we establish a two-order fusion evidence theory model for green supplier assessment. The procedure is shown in Figure 3.

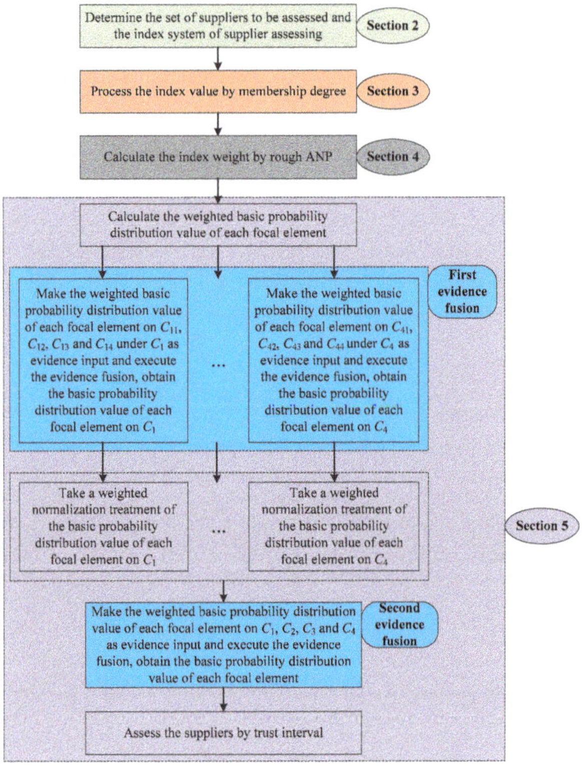

Figure 3. The green supplier assessment procedure.

6. Case Study

6.1. An Application Case of Bearing Cage Supplier Assessment

For a bearing manufacturing enterprise, there are three bearing cage suppliers to be assessed. The set of suppliers to be assessed is $\{\varphi_1, \varphi_2, \varphi_3\}$. The best bearing cage supplier need to be selected after green supplier assessment. The index value of quantitative type (C_{11} and C_{12}) is obtained directly from the enterprise resources planning system (ERP) of the bearing manufacturing enterprise, and the index value of qualitative type (other indexes) is obtained by the method of expert's scoring (i.e., 0.1, 0.3, 0.5, 0.7 and 0.9). The index values of three bearing cage suppliers are shown in Table 1. The units of the index value on $C_{1,1}$ and $C_{1,2}$ are RMB and mm (error value) respectively.

Then, the normalized index values of three bearing cage suppliers are shown in Table 2.

Five comment levels: very bad (G_1), bad (G_2), middle (G_3), good (G_4) and very good (G_5) are set. Taking the normalized index values $v'_{1,(1,1)} = 0.0023$, $v'_{2,(1,1)} = 0.0669$ and $v'_{3,(1,1)} = 1.0000$ for an example, we can get $v'_{(1,1)}(G_1) = 0.0023$ and $v'_{(1,1)}(G_5) = 1.0000$, so $v'_{(1,1)}(G_2) = 0.2517$, $v'_{(1,1)}(G_3) = 0.5011$ and $v'_{(1,1)}(G_4) = 0.7506$. The corresponding numbers of the five comment levels are $\pi_1 = 0.1$, $\pi_2 = 0.3$, $\pi_3 = 0.5$, $\pi_4 = 0.7$ and $\pi_5 = 0.9$. To the normalized index value $v'_{2,(1,1)} = 0.0669$, the membership degrees are $\beta_1 = 0.7410$ and $\beta_2 = 0.2590$, so $\tau_{2,(1,1)} = \beta_1\pi_1 + \beta_2\pi_2 = 0.0648$. The utility values of three bearing cage suppliers are shown in Table 3.

Table 1. The index values of three bearing cage suppliers.

Supplier	$C_{1,1}$	$C_{1,2}$	$C_{1,3}$	$C_{1,4}$	$C_{2,1}$	$C_{2,2}$	$C_{2,3}$	$C_{3,1}$	$C_{3,2}$	$C_{3,3}$	$C_{3,4}$	$C_{4,1}$	$C_{4,2}$	$C_{4,3}$	$C_{4,4}$
φ_1	6.4×10^1	0.01	0.7	0.1	[0.1, 0.3]	0.5	0.7	0.5	0.5	0.7	0.7	/	[0.7, 0.9]	0.5	0.7
φ_2	1.9×10^3	0.01	0.9	0.7	0.5	0.7	0.9	/	0.7	[0.5, 0.7]	0.9	0.7	0.5	0.7	0.9
φ_3	2.8×10^4	0.03	0.1	0.7	0.7	0.1	0.3	0.1	0.5	0.1	0.3	0.1	0.5	/	0.1

Table 2. The normalized index values of three bearing cage suppliers.

Supplier	$C_{1,1}$	$C_{1,2}$	$C_{1,3}$	$C_{1,4}$	$C_{2,1}$	$C_{2,2}$	$C_{2,3}$	$C_{3,1}$	$C_{3,2}$	$C_{3,3}$	$C_{3,4}$	$C_{4,1}$	$C_{4,2}$	$C_{4,3}$	$C_{4,4}$
φ_1	0.0023	1.0000	0.7778	0.1429	[0.1429, 0.4286]	0.7143	0.7778	1.0000	0.7143	1.0000	0.7778	/	[0.7778, 1.0000]	0.7143	0.7778
φ_2	0.0669	1.0000	1.0000	1.0000	0.7143	1.0000	1.0000	/	1.0000	[0.7143, 1.0000]	1.0000	1.0000	0.5556	1.0000	1.0000
φ_3	1.0000	0.3333	0.1111	1.0000	1.0000	0.1429	0.3333	0.2000	0.7143	0.1429	0.3333	0.1429	0.5556	/	0.1111

Table 3. The utility values of three bearing cage suppliers.

Supplier	$C_{1,1}$	$C_{1,2}$	$C_{1,3}$	$C_{1,4}$	$C_{2,1}$	$C_{2,2}$	$C_{2,3}$	$C_{3,1}$	$C_{3,2}$	$C_{3,3}$	$C_{3,4}$	$C_{4,1}$	$C_{4,2}$	$C_{4,3}$	$C_{4,4}$
φ_1	0.1000	0.9000	0.7778	0.1000	0.2857	0.6333	0.6334	0.9000	0.1000	0.9000	0.6334	/	0.8889	0.1000	0.7000
φ_2	0.1518	0.9000	0.9000	0.9000	0.6333	0.9000	0.9000	/	0.9000	0.8572	0.9000	0.9000	0.1000	0.9000	0.9000
φ_3	0.9000	0.1000	0.1000	0.9000	0.9000	0.1000	0.1000	0.1000	0.1000	0.1000	0.1000	0.1000	0.1000	/	0.1000

Next, the index weight is calculated by rough ANP. There are four experts: expert 1, expert 2, expert 3 and expert 4.

Taking the indirect dominance comparison of the elements in C_1 according to their influence on $C_{2,1}$ under the control layer element AO for an example, the fuzzy reciprocal judgement matrices given by the four experts are shown as follows:

$$E^{1,1,(2,1)} = \begin{array}{c|cccc} C_{2,1} & C_{1,1} & C_{1,2} & C_{1,3} & C_{1,4} \\ \hline C_{1,1} & (1,1,1,1) & (1,11/9,13/7,7/3) & (1,11/9,13/7,7/3) & (3/2,13/7,3,4) \\ C_{1,2} & (3/7,7/13,9/11,1) & (1,1,1,1) & (1,11/9,13/7,7/3) & (1,11/9,13/7,7/3) \\ C_{1,3} & (3/7,7/13,9/11,1) & (3/7,7/13,9/11,1) & (1,1,1,1) & (1,11/9,13/7,7/3) \\ C_{1,4} & (1/4,1/3,7/13,2/3) & (3/7,7/13,9/11,1) & (3/7,7/13,9/11,1) & (1,1,1,1) \end{array},$$

$$E^{2,1,(2,1)} = \begin{array}{c|cccc} C_{2,1} & C_{1,1} & C_{1,2} & C_{1,3} & C_{1,4} \\ \hline C_{1,1} & (1,1,1,1) & (1,1,1,1) & (1,11/9,13/7,7/3) & (1,11/9,13/7,7/3) \\ C_{1,2} & (1,1,1,1) & (1,1,1,1) & (1,11/9,13/7,7/3) & (1,11/9,13/7,7/3) \\ C_{1,3} & (3/7,7/13,9/11,1) & (3/7,7/13,9/11,1) & (1,1,1,1) & (1,1,1,1) \\ C_{1,4} & (3/7,7/13,9/11,1) & (3/7,7/13,9/11,1) & (1,1,1,1) & (1,1,1,1) \end{array},$$

$$E^{3,1,(2,1)} = \begin{array}{c|cccc} C_{2,1} & C_{1,1} & C_{1,2} & C_{1,3} & C_{1,4} \\ \hline C_{1,1} & (1,1,1,1) & (3/2,13/7,3,4) & (1,1,1,1) & (1,11/9,13/7,7/3) \\ C_{1,2} & (1/4,1/3,7/13,2/3) & (1,1,1,1) & (1/4,1/3,7/13,2/3) & (3/7,7/13,9/11,1) \\ C_{1,3} & (1,1,1,1) & (3/2,13/7,3,4) & (1,1,1,1) & (1,11/9,13/7,7/3) \\ C_{1,4} & (3/7,7/13,9/11,1) & (1,11/9,13/7,7/3) & (3/7,7/13,9/11,1) & (1,1,1,1) \end{array},$$

$$E^{4,1,(2,1)} = \begin{array}{c|cccc} C_{2,1} & C_{1,1} & C_{1,2} & C_{1,3} & C_{1,4} \\ \hline C_{1,1} & (1,1,1,1) & (1,11/9,13/7,7/3) & (3/2,13/7,3,4) & (7/3,3,17/3,9) \\ C_{1,2} & (3/7,7/13,9/11,1) & (1,1,1,1) & (1,11/9,13/7,7/3) & (3/2,13/7,3,4) \\ C_{1,3} & (1/4,1/3,7/13,2/3) & (3/7,7/13,9/11,1) & (1,1,1,1) & (1,11/9,13/7,7/3) \\ C_{1,4} & (1/9,3/17,1/3,3/7) & (1/4,1/3,7/13,2/3) & (3/7,7/13,9/11,1) & (1,1,1,1) \end{array}.$$

We check the consistency of $E^{1,1,(2,1)}$, $E^{2,1,(2,1)}$, $E^{3,1,(2,1)}$ and $E^{4,1,(2,1)}$ and all of them are qualified. Then, $E^{k,1,(2,1)}(k = 1, 2, 3, 4)$ is split into $A^{k,1,(2,1)} = (a_{g,h}^{k,1,(2,1)})_{4\times4}$, $B^{k,1,(2,1)} = (b_{g,h}^{k,1,(2,1)})_{4\times4}$, $C^{k,1,(2,1)} = (c_{g,h}^{k,1,(2,1)})_{4\times4}$ and $D^{k,1,(2,1)} = (d_{g,h}^{k,1,(2,1)})_{4\times4}$. For example, $A^{1,1,(2,1)} = (a_{g,h}^{1,1,(2,1)})_{4\times4}$ is as follows:

$$A^{1,1,(2,1)} = \begin{bmatrix} 1 & 1 & 1 & 3/2 \\ 3/7 & 1 & 1 & 1 \\ 3/7 & 3/7 & 1 & 1 \\ 1/4 & 3/7 & 3/7 & 1 \end{bmatrix}.$$

Based on $A^{1,1,(2,1)}$, $A^{2,1,(2,1)}$, $A^{3,1,(2,1)}$, $A^{4,1,(2,1)}$, the rough group decision matrix $A^{1,(2,1)} = (a_{g,h}^{1,(2,1)})_{4\times4}$ is constructed as follows:

$$A^{1,(2,1)} = \begin{bmatrix} \{1,1,1,1\} & \{1,1,3/2,1\} & \{1,1,1,3/2\} & \{3/2,1,1,7/3\} \\ \{3/7,1,1/4,3/7\} & \{1,1,1,1\} & \{1,1,1/4,1\} & \{1,1,3/7,3/2\} \\ \{3/7,3/7,1,1/4\} & \{3/7,3/7,3/2,3/7\} & \{1,1,1,1\} & \{1,1,1,1\} \\ \{1/4,3/7,3/7,1/9\} & \{3/7,3/7,1,1/4\} & \{3/7,1,3/7,3/7\} & \{1,1,1,1\} \end{bmatrix}.$$

For the partition $a_{1,4}^{1,1,(2,1)} = 3/2$ in the element $a_{1,4}^{1,(2,1)} = \{3/2,1,1,7/3\}$, its upper approximation set is $\{3/2, 7/3\}$ and lower approximation set is $\{3/2, 1, 1\}$, then $\underline{L}(a_{1,4}^{1,1,(2,1)}) = (3/2 + 1 + 1)/3 = 1.17$, $\overline{L}(a_{1,4}^{1,1,(2,1)}) =$

$(3/2+7/3)/2 = 1.92$ and $RN(a_{1,4}^{1,1,(2,1)}) = [1.17, 1.92]$. Similarly, $RN(a_{1,4}^{2,1,(2,1)}) = RN(a_{1,4}^{3,1,(2,1)}) = [1, 1.46]$, $RN(a_{1,4}^{4,1,(2,1)}) = [1.46, 2.33]$. So $RN(a_{1,4}^{1,(2,1)}) = \{[1.17, 1.92], [1, 1.46], [1, 1.46], [1.46, 2.33]\}$.

According to the calculation rule of rough boundary interval, the mean form of $RN(a_{1,4}^{1,(2,1)})$ is obtained as $Avg\_RN(a_{1,4}^{1,(2,1)}) = [1.16, 1.79]$. Then, the rough judgement matrix $EA^{1,(2,1)} = (Avg\_RN(a_{g,h}^{1,(2,1)}))_{4\times4}$ is constructed as follows:

$$EA^{1,(2,1)} = \begin{bmatrix} [1,1] & [1.03, 1.22] & [1.03, 1.22] & [1.16, 1.79] \\ [0.38, 0.74] & [1,1] & [0.67, 0.95] & [0.72, 1.17] \\ [0.38, 0.74] & [0.50, 0.90] & [1,1] & [1,1] \\ [0.23, 0.38] & [0.38, 0.74] & [0.46, 0.68] & [1,1] \end{bmatrix}.$$

$EA^{1,(2,1)}$ is split into the rough lower limit matrix $EA^{1,(2,1),-} = (a_{g,h}^{1,(2,1),-})_{4\times4}$ and the rough upper limit matrix $EA^{1,(2,1),+} = (a_{g,h}^{1,(2,1),+})_{4\times4}$. The eigenvectors corresponding to the maximum eigenvalues of $EA^{1,(2,1),-}$ and $EA^{1,(2,1),+}$ are $VA^{1,(2,1),-} = [0.71, 0.44, 0.45, 0.30]^T$ and $VA^{1,(2,1),+} = [0.65, 0.49, 0.47, 0.34]^T$ respectively. Then, we get a set $GA^{1,(2,1)} = \{0.68, 0.47, 0.46, 0.32\}$. Similarly, we also get the other sets: $GB^{1,(2,1)} = \{0.73, 0.51, 0.66, 0.58\}$, $GC^{1,(2,1)} = \{0.82, 0.67, 0.73, 0.69\}$ and $GD^{1,(2,1)} = \{0.95, 0.77, 0.83, 0.75\}$.

Thereupon, we obtain the normalized eigenvector $\omega^{1,(2,1)} = [0.30, 0.23, 0.25, 0.22]^T$.

Similarly, we obtain the normalized eigenvector $\omega^{1,(2,2)} = [0.28, 0.41, 0.17, 0.14]^T$ and $\omega^{1,(2,3)} = [0.33, 0.34, 0.13, 0.20]^T$. So we obtain $\Omega^{1,2} = (\omega_h^{1,(2,l)})_{4\times4}$ as follows:

$$\Omega^{1,2} = \begin{bmatrix} 0.30 & 0.28 & 0.33 \\ 0.23 & 0.41 & 0.34 \\ 0.25 & 0.17 & 0.13 \\ 0.22 & 0.14 & 0.20 \end{bmatrix}.$$

After the similar calculating, we get the hyper-matrix Ω under the control layer element G as follows:

$$\Omega = \begin{bmatrix}
0.31 & 0.19 & 0.22 & 0.33 & 0.30 & 0.28 & 0.33 & 0.18 & 0.30 & 0.41 & 0.38 & 0.22 & 0.19 & 0.33 & 0.40 \\
0.24 & 0.25 & 0.33 & 0.31 & 0.23 & 0.41 & 0.34 & 0.32 & 0.33 & 0.22 & 0.19 & 0.31 & 0.17 & 0.18 & 0.10 \\
0.29 & 0.25 & 0.24 & 0.18 & 0.25 & 0.17 & 0.13 & 0.23 & 0.10 & 0.20 & 0.22 & 0.40 & 0.33 & 0.27 & 0.29 \\
0.16 & 0.31 & 0.21 & 0.18 & 0.22 & 0.14 & 0.20 & 0.27 & 0.27 & 0.17 & 0.21 & 0.07 & 0.31 & 0.22 & 0.21 \\
\\
0.24 & 0.35 & 0.48 & 0.20 & 0.13 & 0.11 & 0.51 & 0.43 & 0.29 & 0.35 & 0.36 & 0.28 & 0.51 & 0.40 & 0.35 \\
0.25 & 0.30 & 0.31 & 0.41 & 0.29 & 0.40 & 0.24 & 0.16 & 0.13 & 0.45 & 0.40 & 0.33 & 0.30 & 0.49 & 0.35 \\
0.51 & 0.35 & 0.21 & 0.39 & 0.58 & 0.49 & 0.25 & 0.41 & 0.58 & 0.20 & 0.24 & 0.39 & 0.19 & 0.11 & 0.30 \\
\\
0.33 & 0.28 & 0.31 & 0.25 & 0.17 & 0.29 & 0.18 & 0.10 & 0.33 & 0.30 & 0.15 & 0.44 & 0.32 & 0.30 & 0.33 \\
0.32 & 0.20 & 0.25 & 0.31 & 0.44 & 0.20 & 0.25 & 0.30 & 0.18 & 0.19 & 0.25 & 0.18 & 0.28 & 0.39 & 0.33 \\
0.25 & 0.28 & 0.09 & 0.25 & 0.16 & 0.20 & 0.32 & 0.27 & 0.18 & 0.31 & 0.24 & 0.26 & 0.30 & 0.11 & 0.20 \\
0.10 & 0.24 & 0.35 & 0.19 & 0.23 & 0.31 & 0.25 & 0.33 & 0.31 & 0.20 & 0.36 & 0.12 & 0.10 & 0.20 & 0.14 \\
\\
0.22 & 0.30 & 0.18 & 0.41 & 0.15 & 0.29 & 0.19 & 0.33 & 0.23 & 0.25 & 0.28 & 0.38 & 0.41 & 0.25 & 0.33 \\
0.28 & 0.20 & 0.41 & 0.13 & 0.30 & 0.17 & 0.22 & 0.36 & 0.18 & 0.19 & 0.30 & 0.21 & 0.12 & 0.28 & 0.16 \\
0.40 & 0.19 & 0.19 & 0.17 & 0.30 & 0.21 & 0.08 & 0.17 & 0.22 & 0.25 & 0.10 & 0.19 & 0.17 & 0.19 & 0.31 \\
0.10 & 0.31 & 0.22 & 0.29 & 0.25 & 0.33 & 0.51 & 0.14 & 0.37 & 0.31 & 0.32 & 0.22 & 0.30 & 0.28 & 0.20
\end{bmatrix}.$$

Then we conduct the indirect dominance comparison of the element groups C_1, C_2, C_3, C_4 according to their influence on $C_j (j = 1, 2, 3, 4)$ under the control layer element AO. As a result, we get the relative importance matrix $\Psi = (\psi^{i,j})_{4\times4}$ of element groups as follows:

$$\Psi = \begin{bmatrix} 0.51 & 0.38 & 0.42 & 0.29 \\ 0.12 & 0.28 & 0.20 & 0.21 \\ 0.23 & 0.19 & 0.15 & 0.10 \\ 0.14 & 0.15 & 0.23 & 0.40 \end{bmatrix}.$$

The weighted form of the hyper-matrix Ω is $\overline{\Omega}$, where $\overline{\Omega}^{i,j} = \psi^{i,j}\Omega^{i,j}$ $(i = 1,2,3,4$ and $j = 1,2,3,4)$. The square operation of the weighted hyper-matrix $\overline{\Omega}$ is done continuously and $\overline{\Omega}^4$ converges to a stable limit hyper-matrix. So we get the weight vector of the index system shown in Figure 1 as follows: θ = [0.12, 0.11, 0.10, 0.09, 0.06, 0.06, 0.07, 0.05, 0.04, 0.04, 0.04, 0.06, 0.05, 0.05, 0.06]^T. In the case of ensuring the weight proportional relationship among $C_{1,1}, C_{1,2}, C_{1,3}, C_{1,4}$, we expand the weight vector of $C_{1,1}, C_{1,2}, C_{1,3}, C_{1,4}$ to [0.95, 0.87, 0.79, 0.71]^T. Similarly, the weight vector of $C_{2,1}, C_{2,2}, C_{2,3}$ is expanded to [0.81, 0.81, 0.95]^T, the weight vector of $C_{3,1}, C_{3,2}, C_{3,3}, C_{3,4}$ is expanded to [0.95, 0.76, 0.76, 0.76]^T, and the weight vector of $C_{4,1}, C_{4,2}, C_{4,3}, C_{4,4}$ is expanded to [0.95, 0.79, 0.79, 0.95]^T. Furtherly, the weight vector of C_1, C_2, C_3, C_4 is obtained as [0.42, 0.19, 0.17, 0.22]^T. The relative weight vector of the index C_1, C_2, C_3, C_4 is [0.95, 0.43, 0.38, 0.50]^T.

Based on evidence theory, the set of the three suppliers to be assessed is defined as the identification framework: $\Theta = \{\varphi_1, \varphi_2, \varphi_3\}$. According to Equation (11), the weighted basic probability distribution values of all focal elements are calculated based on the utility values of three bearing cage suppliers shown in Table 3 and the above relative weight vector as follows:

(1) $mass'_{C_{1,1}}(\varphi_1) = 0.0825$, $mass'_{C_{1,1}}(\varphi_2) = 0.1252$, $mass'_{C_{1,1}}(\varphi_3) = 0.7423$, $mass'_{C_{1,1}}(\Theta) = 0.0500$;
(2) $mass'_{C_{1,2}}(\varphi_1) = 0.4121$, $mass'_{C_{1,2}}(\varphi_2) = 0.4121$, $mass'_{C_{1,2}}(\varphi_3) = 0.0458$, $mass'_{C_{1,2}}(\Theta) = 0.1300$;
(3) $mass'_{C_{1,3}}(\varphi_1) = 0.3456$, $mass'_{C_{1,3}}(\varphi_2) = 0.3999$, $mass'_{C_{1,3}}(\varphi_3) = 0.0444$, $mass'_{C_{1,3}}(\Theta) = 0.2100$;
(4) $mass'_{C_{1,4}}(\varphi_1) = 0.0374$, $mass'_{C_{1,4}}(\varphi_2) = 0.3363$, $mass'_{C_{1,4}}(\varphi_3) = 0.3363$, $mass'_{C_{1,4}}(\Theta) = 0.2900$;
(5) $mass'_{C_{2,1}}(\varphi_1) = 0.1272$, $mass'_{C_{2,1}}(\varphi_2) = 0.2820$, $mass'_{C_{2,1}}(\varphi_3) = 0.4008$, $mass'_{C_{2,1}}(\Theta) = 0.1900$;
(6) $mass'_{C_{2,2}}(\varphi_1) = 0.3141$, $mass'_{C_{2,2}}(\varphi_2) = 0.4463$, $mass'_{C_{2,2}}(\varphi_3) = 0.0496$, $mass'_{C_{2,2}}(\Theta) = 0.1900$;
(7) $mass'_{C_{2,3}}(\varphi_1) = 0.3684$, $mass'_{C_{2,3}}(\varphi_2) = 0.5234$, $mass'_{C_{2,3}}(\varphi_3) = 0.0582$, $mass'_{C_{2,3}}(\Theta) = 0.0500$;
(8) $mass'_{C_{3,1}}(\varphi_1) = 0.8550$, $mass'_{C_{3,1}}(\varphi_3) = 0.0950$, $mass'_{C_{3,1}}(\Theta) = 0.0500$;
(9) $mass'_{C_{3,2}}(\varphi_1) = 0.0691$, $mass'_{C_{3,2}}(\varphi_2) = 0.6218$, $mass'_{C_{3,2}}(\varphi_3) = 0.0691$, $mass'_{C_{3,2}}(\Theta) = 0.2400$;
(10) $mass'_{C_{3,3}}(\varphi_1) = 0.3683$, $mass'_{C_{3,3}}(\varphi_2) = 0.3508$, $mass'_{C_{3,3}}(\varphi_3) = 0.0409$, $mass'_{C_{3,3}}(\Theta) = 0.2400$;
(11) $mass'_{C_{3,4}}(\varphi_1) = 0.2947$, $mass'_{C_{3,4}}(\varphi_2) = 0.4188$, $mass'_{C_{3,4}}(\varphi_3) = 0.0465$, $mass'_{C_{3,4}}(\Theta) = 0.2400$;
(12) $mass'_{C_{4,1}}(\varphi_2) = 0.8550$, $mass'_{C_{4,1}}(\varphi_3) = 0.0950$, $mass'_{C_{4,1}}(\Theta) = 0.0500$;
(13) $mass'_{C_{4,2}}(\varphi_1) = 0.6449$, $mass'_{C_{4,2}}(\varphi_2) = 0.0726$, $mass'_{C_{4,2}}(\varphi_3) = 0.0726$, $mass'_{C_{4,2}}(\Theta) = 0.2100$;
(14) $mass'_{C_{4,3}}(\varphi_1) = 0.0790$, $mass'_{C_{4,3}}(\varphi_2) = 0.7110$, $mass'_{C_{4,3}}(\Theta) = 0.2100$;
(15) $mass'_{C_{4,4}}(\varphi_1) = 0.3912$, $mass'_{C_{4,4}}(\varphi_2) = 0.5029$, $mass'_{C_{4,4}}(\varphi_3) = 0.0559$, $mass'_{C_{4,4}}(\Theta) = 0.0500$.

Then we make $\{mass'_{C_{1,1}}(\varphi_s), mass'_{C_{1,2}}(\varphi_s), mass'_{C_{1,3}}(\varphi_s), mass'_{C_{1,4}}(\varphi_s)\}$, $\{mass'_{C_{2,1}}(\varphi_s), mass'_{C_{2,2}}(\varphi_s), mass'_{C_{2,3}}(\varphi_s)\}$, $\{mass'_{C_{3,1}}(\varphi_s), mass'_{C_{3,2}}(\varphi_s), mass'_{C_{3,3}}(\varphi_s), mass'_{C_{3,4}}(\varphi_s)\}$ and $\{mass'_{C_{4,1}}(\varphi_s), mass'_{C_{4,2}}(\varphi_s), mass'_{C_{4,3}}(\varphi_s), mass'_{C_{4,4}}(\varphi_s)\}$ as the evidence input and execute the first evidence fusion respectively. Here, $mass_{C_1} = mass'_{C_{1,1}} \oplus mass'_{C_{1,2}} \oplus mass'_{C_{1,3}} \oplus mass'_{C_{1,4}}$, $mass_{C_2} = mass'_{C_{2,1}} \oplus mass'_{C_{2,2}} \oplus mass'_{C_{2,3}}$, $mass_{C_3} = mass'_{C_{3,1}} \oplus mass'_{C_{3,2}} \oplus mass'_{C_{3,3}} \oplus mass'_{C_{3,4}}$, $mass_{C_4} = mass'_{C_{4,1}} \oplus mass'_{C_{4,2}} \oplus mass'_{C_{4,3}} \oplus mass'_{C_{4,4}}$. The basic probability distribution values $mass_{C_1}(\varphi_s)$, $mass_{C_2}(\varphi_s)$, $mass_{C_3}(\varphi_s)$, $mass_{C_4}(\varphi_s)$ of all focal elements are calculated as follows:

(1) $mass_{C_1}(\varphi_1) = 0.0491$, $mass_{C_1}(\varphi_2) = 0.6871$, $mass_{C_1}(\varphi_3) = 0.1938$, $mass_{C_1}(\Theta) = 0.0700$;
(2) $mass_{C_2}(\varphi_1) = 0.2305$, $mass_{C_2}(\varphi_2) = 0.7000$, $mass_{C_2}(\varphi_3) = 0.0345$, $mass_{C_2}(\Theta) = 0.0350$;
(3) $mass_{C_3}(\varphi_1) = 0.8165$, $mass_{C_3}(\varphi_2) = 0.1122$, $mass_{C_3}(\varphi_3) = 0.0033$, $mass_{C_3}(\Theta) = 0.0679$;
(4) $mass_{C_4}(\varphi_1) = 0.0104$, $mass_{C_4}(\varphi_2) = 0.9841$, $mass_{C_4}(\varphi_3) = 0.0008$, $mass_{C_4}(\Theta) = 0.0046$.

With the consideration of the relative weight vector of the index C_1, C_2, C_3, C_4, we normalize the basic probability distribution values $\{mass_{C_1}(\varphi_1), mass_{C_1}(\varphi_2), mass_{C_1}(\varphi_3), mass_{C_1}(\Theta)\}$, $\{mass_{C_2}(\varphi_1), mass_{C_2}(\varphi_2), mass_{C_2}(\varphi_3), mass_{C_2}(\Theta)\}$, $\{mass_{C_3}(\varphi_1), mass_{C_3}(\varphi_2), mass_{C_3}(\varphi_3), mass_{C_3}(\Theta)\}$ and $\{mass_{C_3}(\varphi_1), mass_{C_3}(\varphi_2), mass_{C_3}(\varphi_3), mass_{C_3}(\Theta)\}$. The weighted basic probability distribution values $mass'_{C_1}(\varphi_s)$, $mass'_{C_2}(\varphi_s)$, $mass'_{C_3}(\varphi_s)$, $mass'_{C_4}(\varphi_s)$ of all focal elements are calculated as follows:

(1) $mass'_{C_1}(\varphi_1) = 0.0466$, $mass'_{C_1}(\varphi_2) = 0.6527$, $mass'_{C_1}(\varphi_3) = 0.1841$, $mass'_{C_1}(\Theta) = 0.1165$;
(2) $mass'_{C_2}(\varphi_1) = 0.0991$, $mass'_{C_2}(\varphi_2) = 0.3010$, $mass'_{C_2}(\varphi_3) = 0.0148$, $mass'_{C_2}(\Theta) = 0.5851$;
(3) $mass'_{C_3}(\varphi_1) = 0.3103$, $mass'_{C_3}(\varphi_2) = 0.0426$, $mass'_{C_3}(\varphi_3) = 0.0013$, $mass'_{C_3}(\Theta) = 0.6458$;
(4) $mass'_{C_4}(\varphi_1) = 0.0052$, $mass'_{C_4}(\varphi_2) = 0.4921$, $mass'_{C_4}(\varphi_3) = 0.0004$, $mass'_{C_4}(\Theta) = 0.5023$.

Then, we make $mass'_{C_1}(\varphi_s)$, $mass'_{C_2}(\varphi_s)$, $mass'_{C_3}(\varphi_s)$, $mass'_{C_4}(\varphi_s)$ $\widetilde{m}_4(A_i)$ as the evidence input and execute the second evidence fusion. The basic probability distribution values $mass(\varphi_s)$ are calculated as follows:

(1) $mass(\varphi_1) = 0.0042$;
(2) $mass(\varphi_2) = 0.4160$;
(3) $mass(\varphi_3) = 0.0006$;
(4) $mass(\Theta) = 0.5792$;

Therefore, the belief function $Bel(\varphi_s)$ and plausible function $Pl(\varphi_s)$ of the three suppliers are calculated, and then the trust interval $[Bel(\varphi_s), Pl(\varphi_s)]$ are obtained as follows:

(1) $[Bel(\varphi_1), Pl(\varphi_1)] = [0.0042, 0.5834]$;
(2) $[Bel(\varphi_2), Pl(\varphi_2)] = [0.4160, 0.9952]$;
(3) $[Bel(\varphi_3), Pl(\varphi_3)] = [0.0006, 0.5798]$.

According to the decision rules based on trust interval in Section 5.1, we obtain the results as follows:

(1) $P(x_1 > x_2) = 0$, so $\varphi_1 \lhd \varphi_2$;
(2) $P(x_1 > x_3) = 1$, so $\varphi_3 \lhd \varphi_1$.

Finally, we get the green supplier assessing results that are $\varphi_3 \lhd \varphi_1 \lhd \varphi_2$ and the best bearing cage supplier is φ_2.

6.2. Discussion

In this paper, the index system, which contains four first-level indexes and fifteen second-level indexes, is established. The indexes in the index system are interrelated and sometimes provide feedback-like effects. Since the suppliers to be assessed are independent, we calculate the index weight by rough ANP. Then we process the uncertain index value by membership degree and get the utility value of a supplier on each index. At last, we solve the information incomplete problem in green supplier assessing by evidence theory.

From the case study in Section 6, the green supplier assessment result is $\varphi_3 \lhd \varphi_1 \lhd \varphi_2$. Based on the Overall view of Table 1, it is also known that φ_2 is the best, φ_3 is the worst and φ_1 is middle. The details are as follows:

- To φ_2, the performance is the best on ten of the fifteen indexes (i.e., quality ($C_{1,2}$), service ($C_{1,3}$), flexibility ($C_{1,4}$), manufacturing capacity ($C_{2,2}$), agility capacity ($C_{2,3}$), geographical environment ($C_{3,2}$), legal environment ($C_{3,4}$), technical compatibility degree ($C_{4,1}$), information platform compatibility degree ($C_{4,3}$) and reputation ($C_{4,4}$)).
- To φ_3, the performance is the worst on twelve of the fifteen indexes (i.e., cost ($C_{1,1}$), quality ($C_{1,2}$), service ($C_{1,3}$), manufacturing capacity ($C_{2,2}$), agility capacity ($C_{2,3}$), economic environment ($C_{3,1}$), geographical environment ($C_{3,2}$), social environment ($C_{3,3}$), legal environment ($C_{3,4}$), technical compatibility degree ($C_{4,1}$), cultural compatibility degree ($C_{4,2}$) and reputation ($C_{4,4}$)).
- To φ_1, the performance is the worst on four indexes (i.e., flexibility ($C_{1,4}$), innovation capacity ($C_{2,1}$), geographical environment ($C_{3,2}$) and information platform compatibility degree ($C_{4,3}$)), and the performance is the best on five indexes (i.e., cost ($C_{1,1}$), quality ($C_{1,2}$), economic environment ($C_{3,1}$), social environment ($C_{3,3}$), cultural compatibility degree ($C_{4,2}$) and reputation ($C_{4,4}$)).

Thus, it can be seen that the assessing results are in accordance with the actual situation.

According to the utility value in Table 3, we compare the results of the proposed method with those based on Fuzzy Synthetic Evaluation (FSE) [33–35] and Fuzzy Analytic Hierarchy Process (FAHP) [36–38] as shown in Table 4.

Table 4. The comparison of assessing results of the proposed method, FSE and FAHP.

Supplier	Ranking of Three Suppliers		
	Proposed Method	FSE	FAHP
φ_1	2	3	3
φ_2	1	1	1
φ_3	3	2	2

As shown in Table 4, the results of the three methods, which all shows φ_2 is the best supplier, are generally consistent. However, there are differences in the ranking between φ_1 and φ_3. By FSE method and FAHP method, the complex and interrelated index system is simplified and the index interrelationship information is partially lost. By the proposed method, the weights of the indexes are processed by rough ANP and the index interrelationship information is successfully translated into the hyper-matrix through comparison among indexes.

In addition, green supplier assessment based on rough ANP and evidence theory can provide group decision-making information for enterprise managers, and the index weight can accurately reflect which index has the greatest impact on the supplier selection, thus providing the decision-making basis for the enterprise to reduce the cost and improve the competitiveness. The above analysis and the comparison in Table 4 verify the feasibility and effectiveness of the proposed method for green supplier assessing.

7. Conclusions

In the context of increasingly serious global environmental problems, an ideal manufacturer requires efficient, green suppliers. To handle the uncertainty and incompleteness in green supplier assessment, we propose a green supplier assessment method based on rough ANP and evidence theory. To the best of our knowledge, this is the first attempt to deal with green supplier assessment in a manufacturing enterprise using the hybrid method of rough ANP and evidence theory. The most prominent advantage of the proposed method is that it overcomes the shortcomings of traditional AHP by considering the dependencies and uncertainty across the indexes and processing experts' judgment on the relative importance of the indexes by fuzzy number and rough boundary interval. It can provide a simple and effective way for weight calculating. By comparing our method with FSE and FAHP approach, we have shown that the proposed method provides a systematic and optimal tool of decision-making for green supplier assessment.

The proposed method provides a way for simplified modeling of complex Multi-criteria Decision Making (MCDM) problems. The decision-making systems are becoming more and more complex nowadays, filled with imprecise and vague information. Evidence theory is adept in capturing such kind of uncertain information, and it provides us with a flexible and effective tool to deal with the green supplier selection problem under uncertain environment. Although the model has been verified on a small case including three potential suppliers and fifteen indexes, it is capable for solving more similar complex problems.

The method proposed in this paper could help us to reduce the risks of making poor investment decisions when dealing with complex networks of green suppliers. In future studies, we will demonstrate on the application of large-scale data sets and the consideration of experts' reliability.

Author Contributions: Conceptualization, L.L. and H.W.; Methodology, L.L.; Validation, L.L.; Formal Analysis, L.L.; Investigation, L.L.; Writing-Original Draft Preparation, L.L. and H.W.

Funding: This research was funded by Ningxia Natural Science Fund, Grant No. NZ17113 and Ningxia first-class discipline and scientific research projects (electronic science and technology), Grant No. NXYLXK2017A07.

Conflicts of Interest: The authors declare no conflict of interest.

References

1. Jiang, W.; Huang, C. A Multi-criteria Decision-making Model for Evaluating Suppliers in Green SCM. *Int. J. Comput. Commun. Control* **2018**, *13*, 337–352. [CrossRef]
2. Mangla, S.K.; Kumar, P.; Barua, M.K. Flexible Decision Modeling for Evaluating the Risks in Green Supply Chain Using Fuzzy AHP and IRP Methodologies. *Glob. J. Flex. Syst. Manag.* **2015**, *16*, 19–35. [CrossRef]
3. Banaeian, N.; Mobli, H.; Fahimnia, B.; Nielsen, I.E.; Omid, M. Green Supplier Selection Using Fuzzy Group Decision Making Methods: A Case Study from the Agri-Food Industry. *Comput. Oper. Res.* **2016**, *89*, 337–347. [CrossRef]
4. Yu, F.; Yang, Y.; Chang, D. Carbon footprint based green supplier selection under dynamic environment. *J. Clean. Prod.* **2018**, *170*, 880–889. [CrossRef]
5. Gandhi, S.; Mangla, S.K.; Kumar, P.; Kumar, D. A combined approach using AHP and DEMATEL for evaluating success factors in implementation of green supply chain management in Indian manufacturing industries. *Int. J. Logist. Res. Appl.* **2016**, *19*, 537–561. [CrossRef]
6. Mangla, S.K.; Kumar, P.; Barua, M.K. Risk analysis in green supply chain using fuzzy AHP approach: A case study. *Resour. Conserv. Recycl.* **2015**, *104*, 375–390. [CrossRef]
7. Mangla, S.K.; Kumar, P.; Barua, M.K. Prioritizing the responses to manage risks in green supply chain: An Indian plastic manufacturer perspective. *Sustain. Prod. Consum.* **2015**, *1*, 67–86. [CrossRef]
8. Tang, X.; Wei, G. Models for Green Supplier Selection in Green Supply Chain Management with Pythagorean 2-Tuple Linguistic Information. *IEEE Access* **2018**, *6*, 18042–18060. [CrossRef]
9. Mangla, S.K.; Govindan, K.; Luthra, S. Critical success factors for reverse logistics in Indian industries: A structural model. *J. Clean. Prod.* **2016**, *129*, 608–621. [CrossRef]
10. Saaty, T.L. *The Analytic Hierarchy Process*; McGraw-Hill: New York, NY, USA, 1980.
11. Saaty, T.L. Fundamentals of the analytic network process—Dependence and feedback in decision-making with a single network. *J. Syst. Sci. Syst. Eng.* **2004**, *13*, 129–157. [CrossRef]
12. Noci, G. Designing 'green' vendor rating systems for the assessment of a supplier's environmental performance. *Eur. J. Purch. Supply Manag.* **1997**, *3*, 103–114. [CrossRef]
13. Lee, A.H.I.; Kang, H.Y.; Hsu, C.F.; Hung, H.C. A green supplier selection model for high-tech industry. *Expert Syst. Appl.* **2009**, *36*, 7917–7927. [CrossRef]
14. Hsu, C.W.; Hu, A.H. Applying hazardous substance management to supplier selection using analytic network process. *J. Clean. Prod.* **2009**, *17*, 255–264. [CrossRef]
15. Yeh, W.C.; Chuang, M.C. Using multi-objective genetic algorithm for partner selection in green supply chain problems. *Expert Syst. Appl.* **2011**, *38*, 4244–4253. [CrossRef]
16. Yousefi, S.; Shabanpour, H.; Fisher, R.; Saen, R.F. Evaluating and ranking sustainable suppliers by robust dynamic data envelopment analysis. *Measurement* **2016**, *83*, 72–85. [CrossRef]

17. Awasthi, A.; Chauhan, S.S.; Goyal, S.K. A fuzzy multicriteria approach for evaluating environmental performance of suppliers. *Int. J. Prod. Econ.* **2010**, *126*, 370–378. [CrossRef]
18. Kannan, D.; Jabbour, C.J.C. Selecting green suppliers based on GSCM practices: Using fuzzy TOPSIS applied to a Brazilian electronics company. *Eur. J. Oper. Res.* **2014**, *233*, 432–447. [CrossRef]
19. Chatterjee, K.; Pamucar, D.; Zavadskas, E.K. Evaluating the performance of suppliers based on using the R'AMATEL-MAIRCA method for green supply chain implementation in electronics industry. *J. Clean. Prod.* **2018**, *184*, 101–129. [CrossRef]
20. Wu, J.; Cao, Q.W.; Li, H. A Method for Choosing Green Supplier Based on COWA Operator under Fuzzy Linguistic Decision-Making. *J. Ind. Eng. Eng. Manag.* **2010**, *24*, 61–65.
21. Luo, X.X.; Peng, S.H. Research on the Vendor Evaluation and Selection Based on AHP and TOPSIS in Green Supply Chain. *Soft Sci.* **2011**, *25*, 53–56.
22. Kuo, R.J.; Lin, Y.J. Supplier selection using analytic network process and data envelopment analysis. *Int. J. Prod. Res.* **2012**, *50*, 2852–2863. [CrossRef]
23. Shi, L. Green Supplier Evaluation of RS-RBF Neural Network Model. *Sci. Technol. Manag. Res.* **2012**, *32*, 198–201.
24. Akman, G. Evaluating suppliers to include green supplier development programs via fuzzy c-means and VIKOR methods. *Comput. Ind. Eng.* **2015**, *86*, 69–82. [CrossRef]
25. Pamučar, D.; Petrović, I.; Ćirović, G. Modification of the Best-Worst and MABAC methods: A novel approach based on interval-valued fuzzy-rough numbers. *Expert Syst. Appl.* **2018**, *91*, 89–106. [CrossRef]
26. Li, L.; Mo, R.; Chang, Z.; Zhang, H. Priority evaluation method for aero-engine assembly task based on balanced weight and improved TOPSIS. *Comput. Integr. Manuf. Syst.* **2015**, *21*, 1193–1201. (In Chinese)
27. Wang, X.; Xiong, W. Rough AHP approach for determining the importance ratings of customer requirements in QFD. *Comput. Integr. Manuf. Syst.* **2010**, *16*, 763–771. (In Chinese)
28. Vasiljević, M.; Fazlollahtabar, H.; Stević, Z.; Vesković, S. A rough multicriteria approach for evaluation of the supplier criteria in automotive industry. *Decis. Mak. Appl. Manag. Eng.* **2018**, *1*, 82–96. [CrossRef]
29. Pamučar, D.; Stević, Ž.; Zavadskas, E.K. Integration of interval rough AHP and interval rough MABAC methods for evaluating university web pages. *Appl. Soft Comput.* **2018**, *67*, 141–163. [CrossRef]
30. Zhu, G.N.; Hu, J.; Qi, J.; Gu, C.C.; Peng, Y.H. An integrated AHP and VIKOR for design concept evaluation based on rough number. *Adv. Eng. Inform.* **2015**, *29*, 408–418. [CrossRef]
31. Shafer, G. *A Mathematical Theory of Evidence*; Princeton University Press: Princeton, NJ, USA, 1976.
32. Sarabi-Jamab, A.; Araabi, B.N. How to decide when the sources of evidence are unreliable: A multi-criteria discounting approach in the Dempster–Shafer theory. *Inf. Sci.* **2018**, *448*, 233–248. [CrossRef]
33. Li, R.; Jin, Y. The early-warning system based on hybrid optimization algorithm and fuzzy synthetic evaluation model. *Inf. Sci.* **2017**, *435*, 296–319. [CrossRef]
34. Haider, H.; Hewage, K.; Umer, A.; Ruparathna, R.; Chhipi-Shrestha, G.; Culver, K.; Holland, M.; Kay, J.; Sadiq, R. Sustainability Assessment Framework for Small-sized Urban Neighbourhoods: An Application of Fuzzy Synthetic Evaluation. *Sustain. Cities Soc.* **2017**, *36*, 21–32. [CrossRef]
35. Zhu, J. Evaluation of supplier strength based on fuzzy synthetic assessment method. In Proceedings of the International Conference on Test and Measurement, Hong Kong, China, 5–6 December 2009; pp. 247–250.
36. Deepika, M.; Kannan, A.S.K. Global supplier selection using intuitionistic fuzzy Analytic Hierarchy Process. In Proceedings of the International Conference on Electrical, Electronics, and Optimization Techniques, Chennai, India, 3–5 March 2016; pp. 2390–2395.
37. Torng, C.; Tseng, K.W. Using Fuzzy Analytic Hierarchy Process to Construct Green Suppliers Assessment Criteria and Inspection Exemption Guidelines. In *Proceedings of the International Conference on Human-Computer Interaction*; Springer: Berlin/Heidelberg, Germany, 2013; pp. 729–732.
38. Labib, A.W. A supplier selection model: A comparison of fuzzy logic and the analytic hierarchy process. *Int. J. Prod. Res.* **2011**, *49*, 6287–6299. [CrossRef]

information

MDPI

Article

Dombi Aggregation Operators of Linguistic Cubic Variables for Multiple Attribute Decision Making

Xueping Lu * and Jun Ye

Department of Electrical and Information Engineering, Shaoxing University, 508 Huancheng West Road, Shaoxing 312000, China; yejun@usx.edu.cn

* Correspondence: luxueping@usx.edu.cn

Received: 8 July 2018; Accepted: 23 July 2018; Published: 26 July 2018

Abstract: A linguistic cubic variable (LCV) is comprised of interval linguistic variable and single-valued linguistic variable. An LCV contains decision-makers' uncertain and certain linguistic judgments simultaneously. The advantage of the Dombi operators contains flexibility due to its changeable operational parameter. Although the Dombi operations have been extended to many studies to solve decision-making problems; the Dombi operations are not used for linguistic cubic variables (LCVs) so far. Hence, the Dombi operations of LCVs are firstly presented in this paper. A linguistic cubic variable Dombi weighted arithmetic average (LCVDWAA) operator and a linguistic cubic variable Dombi weighted geometric average (LCVDWGA) operator are proposed to aggregate LCVs. Then a multiple attribute decision making (MADM) method is developed in LCV setting on the basis of LCVDWAA and LCVDWGA operators. Finally, two illustrative examples about the optimal choice problems demonstrate the validity and the application of this method.

Keywords: multiple attribute decision making; linguistic cubic variable; Dombi operations; linguistic cubic variable Dombi weighted arithmetic average (LCVDWAA) operator; linguistic cubic variable Dombi weighted geometric average (LCVDWGA) operator

1. Introduction

With the development of society and science, decision-making problems become more and more complex, and they involve more and more fields, such as manufacturing domain [1,2], hospital service quality management [3], evaluation of the supplier criterions [4], and disaster assessment [5]. Since Zadeh [6] firstly proposed that linguistic variable (LV) could evaluate the assessment for objects, many scholars put forward various linguistic aggregation operators and developed corresponding methods to handle decision-making problems with linguistic information in diversified fields [7–12]. So far language variables have come in many forms, which are classified into two types: certain linguistic evaluations and uncertain linguistic evaluations. One of the forms can be used in one decision-making problem and one linguistic variable can represent the evaluation of a decision maker. With respect to an attribute over an alternative, one of the decision makers could give an uncertain evaluation, but another could give a certain evaluation. The pre-proposed LV forms could not express uncertain evaluation and certain evaluation simultaneously. In this study we will use a new linguistic evaluation form which was defined as linguistic cubic variable (LCV) in Ye [13]. An LCV is composed of a certain linguistic variable and an uncertain linguistic variable. An LCV can represent a group of linguistic evaluations over an attribute. Thus, the multiple attribute group decision making (MAGDM) process will be much simpler. We will introduce the development of LV below.

Since single-value linguistic variable was proposed, decision makers thought only one linguistic variable could not accurately provide judgments in some uncertain environments. Xu [14] proposed that decision makers could express their opinions with a linguistic interval. The interval linguistic variable was defined as an uncertain linguistic variable (ULV). They introduced a ULV hybrid aggregation (ULHA) operator and a ULV ordered weighted averaging operator (ULOWA) for MADM under an uncertain environment. Further, some other UL operators were introduced, such as UL hybrid geometric mean operator (ULHGM) [15], UL ordered weighted averaging operator (IULOWA) [16], UL Bonferroni mean operator (ULBM) [17], UL harmonic mean operator (ULHM) [18], and UL power geometric operator (ULPG) [19]. Later, intuitionistic fuzzy sets and linguistic variables were integrated. The concept of linguistic intuitionistic fuzzy numbers was proposed. Liu [20,21] and Li et al. [22] introduced several linguistic intuitionistic MADM methods. Recently, a changeable uncertain linguistic number was defined as a neutrosophic linguistic number (NLN). Some aggregation operators of NLNs were developed to handle MADM problems with NLN information in References [23–25].

Linguistic variable forms, as introduced above, are either certain linguistic evaluations or uncertain linguistic evaluations used to describe evaluation information in the same decision-making problem simultaneously. However, in reality, uncertain linguistic evaluations and certain linguistic evaluations may exist simultaneously. Thus, Ye [13] combined a linguistic variable with a cubic set [26] and proposed a hybrid linguistic form. The hybrid linguistic form was defined as a linguistic cubic variable (LCV). An uncertain linguistic variable and a certain linguistic variable composed an LCV. Meanwhile Ye [13] developed an LCV weighted geometric averaging (LCVWGA) operator and an LCV weighted arithmetic averaging (LCVWAA) operator and further developed a MADM method on the basis of the LCVWGA operator or LCVWAA operator.

Information aggregation operators are effective and powerful tools to handle decision-making problems. Researchers have developed various operators to aggregate evaluation information. Dombi [27] firstly proposed Dombi T-conorm and T-norm operations in 1982. The operations are developed into many information aggregations to deal with various application problems; for instance, Dombi hesitant fuzzy information aggregation operators [5] for disaster assessment, or intuitionistic fuzzy set Dombi Bonferroni mean operators [28] for MADM problems. Then, the advantage of the Dombi operators contains flexibility due to its changeable operational parameter. Up to now, the Dombi operations have not been extended to LCVs. Hence, aggregation operators based the Dombi operations will be developed to handle LCV decision-making problems. So Dombi operational laws of LCVs are proposed in this study. Then an LCV Dombi weighted arithmetic average (LCVDWAA) operator and an LCV Dombi weighted geometric average (LCVDWGA) operator are presented. Further the decision-making approach on basis of the LCVDWAA or LCVDWGA operator is developed for LCV MADM problems.

The remainder of this paper is organized by following six sections. Some concepts of LCVs are introduced in Section 2, and Section 3 defines several Dombi operations of LCVs. LCVDWAA and LCVDWGA operators and some of their properties are presented in Section 4. The MADM approach based on the LCVDWAA or LCVDWGA operator is introduced in Section 5. In Section 6, two application examples are illustrated, and we discuss the validity, the influence of the operational parameter, and the sensitivity of weights. Section 7 gives the conclusions and expectations of the research.

2. Several Concepts of LCVs

Definition 1 [13]. *Set $L = \{L_0, L_1, L_2, \ldots, L_T\}$ as a linguistic term set, in which T is even. A linguistic cubic variable V is constructed by $V = (L, L_M)$, where $L = [L_G, L_H]$ is a ULV and L_M is an LV for $H \geq G$ and L_G, L_H, $L_M \in L$. If $G \leq M \leq H$, $V = ([L_G, L_H], L_M)$ is an internal LCV. If $M < G$ or $M > H$, $V = ([L_G, L_H], L_M)$ is an external LCV.*

Definition 2 [13]. *Set $V = ([L_G, L_H], L_M)$ as an LCV in $L = \{L_0, L_1, L_2, \ldots, L_T\}$ for $L_G, L_H, L_M \in L$. Then the expected value of the LCV is calculated as below:*

$$E(V) = (G + H + M)/3T \qquad for\ E(V) \in [0,1] \quad . \tag{1}$$

Definition 3 [13]. *Set* $V_1 = ([L_{G1}, L_{H1}], L_{M1})$ *and* $V_2 = ([L_{G2}, L_{H2}], L_{M2})$ *as two LCVs, their expected values are* $E(V_1)$ *and* $E(V_2)$*, then their relations are as follows:*

(a) If $E(V_1) \succ E(V_2)$, then $V_1 \succ V_2$;
(b) If $E(V_1) \prec E(V_2)$, then $V_1 \prec V_2$;
(c) If $E(V_1) = E(V_2)$, then $V_1 = V_2$.

3. Some Dombi Operations of LCVs

Dombi T-conorm operation and T-norm operation between two real numbers will be introduced in this section. Then some Dombi operations of LCVs will be proposed.

Definition 4 [27]. *Let Y and X be any two real numbers. If* $(Y, X) \in [0, 1] \times [0, 1]$*, the Dombi T-norm and Dombi T-conorm between them are defined as Equations (2) and (3):*

$$D(Y,X) = \frac{1}{1 + \left\{\left(\frac{1-Y}{Y}\right)^{\rho} + \left(\frac{1-X}{X}\right)^{\rho}\right\}^{\frac{1}{\rho}}}, \tag{2}$$

$$D^{C}(Y,X) = 1 - \frac{1}{1 + \left\{\left(\frac{Y}{1-Y}\right)^{\rho} + \left(\frac{X}{1-X}\right)^{\rho}\right\}^{\frac{1}{\rho}}}. \tag{3}$$

If $\rho > 0$, the above equations satisfy $D(Y, X) \in [0, 1]$ and $D^c(Y, X) \in [0, 1]$.

According to the above Dombi operations, the following Dombi operational laws of LCVs are defined.

Definition 5. *Let* $V_1 = ([L_{G1}, L_{H1}], L_{M1})$ *and* $V_2 = ([L_{G2}, L_{H2}], L_{M2})$ *be two LCVs, where* $(G_1, H_1, M_1, G_2, H_2, M_2) \in [0, T]$*,* $\rho > 0$*, then their Dombi operations are proposed as follows:*

$$\begin{aligned} V_1 \oplus V_2 &= ([L_{G1}, L_{H1}], L_{M1}) \oplus ([L_{G2}, L_{H2}], L_{M2}) \\ &= \left(\left[L_{T\times(1-\frac{1}{1+\{(\frac{G1/T}{1-G1/T})^{\rho}+(\frac{G2/T}{1-G2/T})^{\rho}\}^{\frac{1}{\rho}}})}, L_{T\times(1-\frac{1}{1+\{(\frac{H1/T}{1-H1/T})^{\rho}+(\frac{H2/T}{1-H2/T})^{\rho}\}^{\frac{1}{\rho}}})}\right], L_{T\times(1-\frac{1}{1+\{(\frac{M1/T}{1-M1/T})^{\rho}+(\frac{M2/T}{1-M2/T})^{\rho}\}^{\frac{1}{\rho}}})}\right) \\ &= \left(\left[L_{T-\frac{T}{1+\{(\frac{G1}{T-G1})^{\rho}+(\frac{G2}{T-G2})^{\rho}\}^{\frac{1}{\rho}}}}, L_{T-\frac{T}{1+\{(\frac{H1}{T-H1})^{\rho}+(\frac{H2}{T-H2})^{\rho}\}^{\frac{1}{\rho}}}}\right], L_{T-\frac{T}{1+\{(\frac{M1}{T-M1})^{\rho}+(\frac{M2}{1-T-M2})^{\rho}\}^{\frac{1}{\rho}}}}\right); \end{aligned} \tag{4}$$

$$\begin{aligned} V_1 \otimes V_2 &= ([L_{G1}, L_{H1}], L_{M1}) \otimes ([L_{G2}, L_{H2}], L_{M2}) \\ &= \left(\left[L_{\frac{T}{1+\{(\frac{1-G1/T}{G1/T})^{\rho}+(\frac{1-G2/T}{G2/T})^{\rho}\}^{\frac{1}{\rho}}}}, L_{\frac{T}{1+\{(\frac{1-H1/T}{H1/T})^{\rho}+(\frac{1-H2/T}{H2/T})^{\rho}\}^{\frac{1}{\rho}}}}\right], L_{\frac{T}{1+\{(\frac{1-M1/T}{M1/T})^{\rho}+(\frac{1-M2/T}{M2/T})^{\rho}\}^{\frac{1}{\rho}}}}\right) \\ &= \left(\left[L_{\frac{T}{1+\{(\frac{T-G1}{G1})^{\rho}+(\frac{T-G2}{G2})^{\rho}\}^{\frac{1}{\rho}}}}, L_{\frac{T}{1+\{(\frac{T-H1}{H1})^{\rho}+(\frac{T-H2}{H2})^{\rho}\}^{\frac{1}{\rho}}}}\right], L_{\frac{T}{1+\{(\frac{T-M1}{M1})^{\rho}+(\frac{T-M2}{M2})^{\rho}\}^{\frac{1}{\rho}}}}\right); \end{aligned} \tag{5}$$

$$KV_1 = K([L_{G1}, L_{H1}], L_{M1})$$
$$= \left(\left[L_{T-\frac{T}{1+\{K(\frac{\frac{G1}{T}}{1-\frac{G1}{T}})^{\rho}\}^{\frac{1}{\rho}}}}, L_{T-\frac{T}{1+\{K(\frac{\frac{H1}{T}}{1-\frac{H1}{T}})^{\rho}\}^{\frac{1}{\rho}}}}\right], L_{T-\frac{T}{1+\{K(\frac{\frac{M1}{T}}{1-\frac{M1}{T}})^{\rho}\}^{\frac{1}{\rho}}}}\right) \tag{6}$$
$$= \left(\left[L_{T-\frac{T}{1+\{K(\frac{G1}{T-G1})^{\rho}\}^{\frac{1}{\rho}}}}, L_{T-\frac{T}{1+\{K(\frac{H1}{T-H1})^{\rho}\}^{\frac{1}{\rho}}}}\right], L_{T-\frac{T}{1+\{K(\frac{M1}{T-M1})^{\rho}\}^{\frac{1}{\rho}}}}\right);$$

$$V_1{}^K = ([L_{G1}^K, L_{H1}^K], L_{M1}^K)$$
$$= \left(\left[L_{\frac{T}{1+\{K(\frac{1-G1/T}{G1/T})^{\rho}\}^{1/\rho}}}, L_{\frac{T}{1+\{K(\frac{1-H1/T}{H1/T})^{\rho}\}^{1/\rho}}}\right], L_{\frac{T}{1+\{K(\frac{1-M1/T}{M1/T})^{\rho}\}^{1/\rho}}}\right) \tag{7}$$
$$= \left(\left[L_{\frac{T}{1+\{K(\frac{T-G1}{G1})^{\rho}\}^{1/\rho}}}, L_{\frac{T}{1+\{K(\frac{T-H1}{H1})^{\rho}\}^{1/\rho}}}\right], L_{\frac{T}{1+\{K(\frac{T-M1}{M1})^{\rho}\}^{1/\rho}}}\right).$$

Due to functions $(G_1/T, H_1/T, M_1/T, G_2/T, H_2/T, M_2/T) \in [0, 1]$, they satisfy the parameter requirements of Dombi operations. *If* $p > 0$, $(Y, X) \in [0, 1] \times [0, 1]$, *then* $D(Y, X) \in [0, 1]$ *and* $D^c(Y, X) \in [0, 1]$. Thus we can get $T^*D^c(Y, X) \in [0, T]$ and $T^*D(Y, X) \in [0, T]$. Obviously the results of Equations (4)–(7) are also LCVs according to the Dombi operations as in Equations (2) and (3). In Equations (4)–(7), we presented Dombi operations of LCVs in the first step and simplified the equations in the second step.

Example 1. *Let* V_1 *and* V_2 *be two LCVs in the linguistic term set* $L = \{L_i \mid i \in [0, 8]\}$. *Assume that* $V_1 = ([L_4, L_6], L_5)$, $V_2 = ([L_2, L_7], L_2)$, $k = 0.5$, *and* $\rho = 1$. *According to Equations (4)–(7), the results are calculated respectively as follows:*

$$V_1 \oplus V_2 = ([L_4, L_6], L_5) \oplus ([L_2, L_7], L_2)$$
$$= \left(\left[L_{8-\frac{8}{1+\{(\frac{4}{8-4})^1+(\frac{2}{8-2})^1\}^{\frac{1}{1}}}}, L_{8-\frac{8}{1+\{(\frac{6}{8-6})^1+(\frac{7}{8-7})^1\}^{\frac{1}{1}}}}\right], L_{8-\frac{8}{1+\{(\frac{5}{8-5})^1+(\frac{2}{8-2})^1\}^{\frac{1}{1}}}}\right)$$
$$= ([L_{4.5714}, L_{7.2727}], L_{5.3333});$$

$$V_1 \otimes V_2 = ([L_4, L_6], L_5) \otimes ([L_2, L_7], L_2)$$
$$= \left(\left[L_{\frac{8}{1+\{(\frac{8-4}{4})^1+(\frac{8-2}{2})^1\}^{\frac{1}{1}}}}, L_{\frac{8}{1+\{(\frac{8-6}{6})^1+(\frac{8-7}{7})^{\rho}\}^{\frac{1}{1}}}}\right], L_{\frac{8}{1+\{(\frac{8-5}{5})^1+(\frac{8-2}{2})^1\}^{\frac{1}{1}}}}\right)$$
$$= ([L_{1.6}, L_{5.4193}], L_{1.7391});$$

$$KV_1 = K([L_4, L_6], L_5)$$
$$= \left(\left[L_{8-\frac{8}{1+\{0.5(\frac{4}{8-4})^1\}^{\frac{1}{1}}}}, L_{8-\frac{8}{1+\{0.5(\frac{6}{8-6})^1\}^{\frac{1}{1}}}}\right], L_{8-\frac{8}{1+\{0.5(\frac{5}{8-5})^1\}^{\frac{1}{1}}}}\right)$$
$$= ([L_{2.6667}, L_{4.8}], L_{3.6363});$$

$$
\begin{aligned}
V_1{}^K &= \left(\left[L_4^K, L_6^K\right], L_5^K\right) \\
&= \left(\left[L_{\frac{8}{1+\{0.5(\frac{8-4}{4})^1\}^{1/1}}}, L_{\frac{8}{1+\{0.5(\frac{8-6}{6})^1\}^{1/1}}}\right], L_{\frac{8}{1+\{0.5(\frac{8-5}{5})^1\}^{1/1}}}\right) \\
&= \left(\left[L_{5.3333}, L_{6.8571}\right], L_{6.1538}\right).
\end{aligned}
$$

4. Dombi Weighted Aggregation Operators of LCVs

4.1. Dombi Weighted Arithmetic Average Operator of LCVs

Definition 6. *Let $V = \{V_1, V_2, V_3, \ldots, V_n\}$ be an LCV set, then the Dombi weighted arithmetic average operator of LCV_S can be defined as follows:*

$$
LCVDWAA(V_1, V_2, \ldots, V_n) = \overset{n}{\underset{i=1}{\oplus}} w_i V_i \tag{8}
$$

where the weight vector w_i satisfies $\sum\limits_{i=1}^{i=n} w_i = 1$ and $w_i \in [0, 1]$.

The following Theorem 1 can be induced and proved according to Definitions 4 and 6.

Theorem 1. *Let $V_i = ([L_{Gi}, L_{Hi}], L_{Mi})$ (i = 1, 2, ..., n) be a set of LCVs and the corresponding weight vector is $w = (w_1, w_2, \ldots, w_n)$, where $\sum\limits_{i=1}^{i=n} w_i = 1$ and $w_i \in [0, 1]$, then we can calculate Equation (8) on basis of the predefined operational laws and get the following formula:*

$$
\begin{aligned}
&LCVDWAA(V_1, V_2, \ldots, V_n) \\
&= \left(\left[L_{T-\frac{T}{1+\{\sum\limits_{i=1}^{i=n} w_i(\frac{Gi}{T-Gi})^{\rho}\}^{1/\rho}}}, L_{T-\frac{T}{1+\{\sum\limits_{i=1}^{i=n} w_i(\frac{Hi}{T-Hi})^{\rho}\}^{1/\rho}}}\right], L_{T-\frac{T}{1+\{\sum\limits_{i=1}^{i=n} w_i(\frac{Mi}{T-Mi})^{\rho}\}^{1/\rho}}}\right).
\end{aligned} \tag{9}
$$

Proof:

(1) If n = 2, by the Equations (4) and (6) we can get:

$$
\begin{aligned}
&LCVDWAA(V_1, V_2) = w_1 V_1 \oplus w_2 V_2 \\
&= \left(\left[L_{T-\frac{T}{1+\{w_1(\frac{G1}{T-G1})^{\rho}\}^{\frac{1}{\rho}}}}, L_{T-\frac{T}{1+\{w_1(\frac{H1}{T-H1})^{\rho}\}^{\frac{1}{\rho}}}}\right], L_{T-\frac{T}{1+\{w_1(\frac{M1}{T-M1})^{\rho}\}^{\frac{1}{\rho}}}}\right) \oplus \\
&\quad \left(\left[L_{T-\frac{T}{1+\{w_2(\frac{G2}{T-G2})^{\rho}\}^{\frac{1}{\rho}}}}, L_{T-\frac{T}{1+\{w_2(\frac{H2}{T-H2})^{\rho}\}^{\frac{1}{\rho}}}}\right], L_{T-\frac{T}{1+\{w2(\frac{M2}{T-M2})^{\rho}\}^{\frac{1}{\rho}}}}\right)
\end{aligned}
$$

$$= \left(\left[L_{T - \frac{T}{1+\{(\frac{T-\frac{T}{1+\{w_1(\frac{G1}{T-G1})^{\rho}\}^{\frac{1}{\rho}}}}{T-(T-\frac{T}{1+\{w_1(\frac{G1}{T-G1})^{\rho}\}^{\frac{1}{\rho}}})})^{\rho}+(\frac{T-\frac{T}{1+\{w_2(\frac{G2}{T-G2})^{\rho}\}^{\frac{1}{\rho}}}}{T-(T-\frac{T}{1+\{w_2(\frac{G2}{T-G2})^{\rho}\}^{\frac{1}{\rho}}})})^{\rho}\}^{\frac{1}{\rho}}}}, L_{T - \frac{T}{1+\{(\frac{T-\frac{T}{1+\{w_1(\frac{H1}{T-H1})^{\rho}\}^{\frac{1}{\rho}}}}{T-(T-\frac{T}{1+\{w_1(\frac{H1}{T-H1})^{\rho}\}^{\frac{1}{\rho}}})})^{\rho}+(\frac{T-\frac{T}{1+\{w_2(\frac{H2}{T-H2})^{\rho}\}^{\frac{1}{\rho}}}}{T-(T-\frac{T}{1+\{w_2(\frac{H2}{T-H2})^{\rho}\}^{\frac{1}{\rho}}})})^{\rho}\}^{\frac{1}{\rho}}}} \right], L_{T - \frac{T}{1+\{(\frac{T-\frac{T}{1+\{w_1(\frac{M1}{T-M1})^{\rho}\}^{\frac{1}{\rho}}}}{T-(T-\frac{T}{1+\{w_1(\frac{M1}{T-M1})^{\rho}\}^{\frac{1}{\rho}}})})^{\rho}+(\frac{T-\frac{T}{1+\{w_2(\frac{M2}{T-M2})^{\rho}\}^{\frac{1}{\rho}}}}{T-(T-\frac{T}{1+\{w_2(\frac{M2}{T-M2})^{\rho}\}^{\frac{1}{\rho}}})})^{\rho}\}^{\frac{1}{\rho}}}} \right)$$

$$= \left(\left[L_{T-\frac{T}{1+\{w_1(\frac{G1}{T-G1})^{\rho}+w_2(\frac{G2}{T-G2})^{\rho}\}^{1/\rho}}}, L_{T-\frac{T}{1+\{w_1(\frac{H1}{T-H1})^{\rho}+w_2(\frac{H2}{T-H2})^{\rho}\}^{1/\rho}}} \right], L_{T-\frac{T}{1+\{w_1(\frac{M1}{T-M1})^{\rho}+w_2(\frac{M2}{T-M2})^{\rho}\}^{1/\rho}}} \right)$$

$$= \left(\left[L_{T-\frac{T}{1+\{\sum_{i=1}^{i=2} w_i(\frac{Gi}{T-Gi})^{\rho}\}^{1/\rho}}}, L_{T-\frac{T}{1+\{\sum_{i=1}^{i=2} w_i(\frac{Hi}{T-Hi})^{\rho}\}^{1/\rho}}} \right], L_{T-\frac{T}{1+\{\sum_{i=1}^{i=2} w_i(\frac{Mi}{T-Mi})^{\rho}\}^{1/\rho}}} \right).$$

(2) Assume $n = k$, the result is as follows:

$$LCVDWAA(V_1, V_2, \ldots, V_k) = \bigoplus_{i=1}^{k} w_i V_i$$
$$= \left(\left[L_{T-\frac{T}{1+\{\sum_{i=1}^{i=k} w_i(\frac{Gi}{T-Gi})^{\rho}\}^{1/\rho}}}, L_{T-\frac{T}{1+\{\sum_{i=1}^{i=k} w_i(\frac{Hi}{T-Hi})^{\rho}\}^{1/\rho}}} \right], L_{T-\frac{T}{1+\{\sum_{i=1}^{i=k} w_i(\frac{Mi}{T-Mi})^{\rho}\}^{1/\rho}}} \right).$$

(3) If $n = k + 1$, we have:

$$LCVDWAA(V_1, V_2, \ldots, V_k, V_{k+1}) = LCVDWAA(V_1, V_2, \ldots, V_k) \oplus w_{k+1} V_{k+1}$$
$$= \left(\left[L_{T-\frac{T}{1+\{\sum_{i=1}^{i=k} w_i(\frac{Gi}{T-Gi})^{\rho}\}^{1/\rho}}}, L_{T-\frac{T}{1+\{\sum_{i=1}^{i=k} w_i(\frac{Hi}{T-Hi})^{\rho}\}^{1/\rho}}} \right], L_{T-\frac{T}{1+\{\sum_{i=1}^{i=k} w_i(\frac{Mi}{T-Mi})^{\rho}\}^{1/\rho}}} \right)$$
$$\oplus \left(\left[L_{T-\frac{T}{1+\{w_{k+1}(\frac{G_{k+1}}{T-G_{k+1}})^{\rho}\}^{\frac{1}{\rho}}}}, L_{T-\frac{T}{1+\{w_{k+1}(\frac{H_{k+1}}{T-H_{k+1}})^{\rho}\}^{\frac{1}{\rho}}}} \right], L_{T-\frac{T}{1+\{w_{k+1}(\frac{M_{k+1}}{T-M_{k+1}})^{\rho}\}^{\frac{1}{\rho}}}} \right)$$
$$= \left(\left[L_{T-\frac{T}{1+\{\sum_{i=1}^{i=k+1} w_i(\frac{Gi}{T-Gi})^{\rho}\}^{1/\rho}}}, L_{T-\frac{T}{1+\{\sum_{i=1}^{i=k+1} w_i(\frac{Hi}{T-Hi})^{\rho}\}^{1/\rho}}} \right], L_{T-\frac{T}{1+\{\sum_{i=1}^{i=k+1} w_i(\frac{Mi}{T-Mi})^{\rho}\}^{1/\rho}}} \right).$$

Thus, we have proved that Equation (9) is correct for any n. The properties of the LCVDWAA operator are as follows:

(1) Idempotency: If there is LCVs collection $V_i = ([L_{Gi}, L_{Hi}], L_{Mi})$ for $V_i = V$ ($i = 1, 2, \ldots, n$) then LCVDWAA ($V_1, V_2, \ldots, V_n$) = V.

(2) Commutativity: Assume that the LCV set ($V'_1, V'_2, V'_3, \ldots, V'_n$) is any permutation of ($V_1, V_2, \ldots, V_n$). Then, there is LCVDWAA ($V'_1, V'_2, \ldots, V'_n$) = LCVDWAA ($V_1, V_2, \ldots, V_n$).

(3) Boundedness: If there is LCVs collection $V_i = ([L_{Gi}, L_{Hi}], L_{Mi})$ $(i = 1, 2, \ldots, n)$ $V_{\min} = ([L_{\min_i(Gi)}, L_{\min_i(Hi)}], L_{\min_i(Mi)})$, $V_{\max} = ([L_{\max_i(Gi)}, L_{\max_i(Hi)}], L_{\max_i(Mi)})$. Then, $V_{\min} \leq LCVDWAA(V_1, V_2, \ldots, V_n) \leq V_{\max}$. □

Proof:

(1) Let $V_i = ([L_{Gi}, L_{Hi}], L_{Mi}) = ([L_G, L_H], L_M)$, then we can get the result:

$$\begin{aligned}
&LCVDWAA(V_1, V_2, \ldots, V_n) \\
&= \left(\left[L_{T - \frac{T}{1+\{\sum_{i=1}^{i=n} w_i (\frac{Gi}{T-Gi})^{\rho}\}^{1/\rho}}}, L_{T - \frac{T}{1+\{\sum_{i=1}^{i=n} w_i (\frac{Hi}{T-Hi})^{\rho}\}^{1/\rho}}} \right], L_{T - \frac{T}{1+\{\sum_{i=1}^{i=n} w_i (\frac{Mi}{T-Mi})^{\rho}\}^{1/\rho}}} \right) \\
&= \left(\left[L_{T - \frac{T}{1+\{(\frac{G}{T-G})^{\rho} \sum_{i=1}^{i=n} w_i\}^{1/\rho}}}, L_{T - \frac{T}{1+\{(\frac{H}{T-H})^{\rho} \sum_{i=1}^{i=n} w_i\}^{1/\rho}}} \right], L_{T - \frac{T}{1+\{(\frac{M}{T-M})^{\rho} \sum_{i=1}^{i=n} w_i\}^{1/\rho}}} \right) \\
&= \left(\left[L_{T - \frac{T}{1+\{(\frac{G}{T-G})^{\rho}\}^{1/\rho}}}, L_{T - \frac{T}{1+\{(\frac{H}{T-H})^{\rho}\}^{1/\rho}}} \right], L_{T - \frac{T}{1+\{(\frac{M}{T-M})^{\rho}\}^{1/\rho}}} \right) \\
&= ([L_G, L_H], L_M) = V.
\end{aligned}$$

(2) The proof is obvious.

(3) Since $\min_i(Gi) \leq Gi \leq \max_i(Gi), \min_i(Hi) \leq Hi \leq \max_i(Hi), \min_i(Mi) \leq Mi \leq \max_i(Mi)$. Then the following inequalities can be induced as:

$$\begin{aligned}
T - \frac{T}{1+\left\{\sum_{i=1}^{i=n} w_i \left(\frac{\min_i(Gi)}{T-\min_i(Gi)}\right)^{\rho}\right\}^{1/\rho}} = \min_i(Gi) \leq T - \frac{T}{1+\left\{\sum_{i=1}^{i=n} w_i \left(\frac{Gi}{T-Gi}\right)^{\rho}\right\}^{1/\rho}} \leq \max_i(Gi) = T - \frac{T}{1+\left\{\sum_{i=1}^{i=n} w_i \left(\frac{\max_i(Gi)}{T-\max_i(Gi)}\right)^{\rho}\right\}^{1/\rho}} \\
T - \frac{T}{1+\left\{\sum_{i=1}^{i=n} w_i \left(\frac{\min_i(Hi)}{T-\min_i(Hi)}\right)^{\rho}\right\}^{1/\rho}} = \min_i(Hi) \leq T - \frac{T}{1+\left\{\sum_{i=1}^{i=n} w_i \left(\frac{Hi}{T-Hi}\right)^{\rho}\right\}^{1/\rho}} \leq \max_i(Hi) = T - \frac{T}{1+\left\{\sum_{i=1}^{i=n} w_i \left(\frac{\max_i(Hi)}{T-\max_i(Hi)}\right)^{\rho}\right\}^{1/\rho}} \\
T - \frac{T}{1+\left\{\sum_{i=1}^{i=n} w_i \left(\frac{\min_i(Mi)}{T-\min_i(Mi)}\right)^{\rho}\right\}^{1/\rho}} = \min_i(Mi) \leq T - \frac{T}{1+\left\{\sum_{i=1}^{i=n} w_i \left(\frac{Mi}{T-Mi}\right)^{\rho}\right\}^{1/\rho}} \leq \max_i(Mi) = T - \frac{T}{1+\left\{\sum_{i=1}^{i=n} w_i \left(\frac{\max_i(Mi)}{T-\max_i(Mi)}\right)^{\rho}\right\}^{1/\rho}}
\end{aligned}$$

Hence, $V_{\min} \leq LCVDWAA(V_1, V_2, \ldots, V_n) \leq V_{\max}$ holds. □

4.2. Dombi Weighted Geometric Average Operator of LCVs

Definition 7. *Let $V = \{V_1, V_2, \ldots, V_n\}$ be an LCV set, then the Dombi weighted geometric average operator of the LCV_S can be defined as:*

$$LCVDWGA(V_1, V_2, \ldots, V_n) = \bigotimes_{i=1}^{n} w_i V_i \tag{10}$$

where the weight vector w_i satisfies $\sum_{i=1}^{i=n} w_i = 1$ and $w_i \in [0, 1]$.

According to Definitions 5 and 7, the following theorem can be induced and proved.

Theorem 2. *Let* $V_i = ([L_{Gi}, L_{Hi}], L_{Mi})$ *(i = 1, 2, ... , n) be a set of LCVs and the corresponding weight vector is* $w = (w_1, w_2, \ldots, w_n)$*, where* $\sum_{i=1}^{i=n} w_i = 1$ *and* $w_i \in [0,1]$*, we can calculate Equation (10) on basis of the predefined operational laws and have:*

$$LCVDWGA(V_1, V_2, \ldots, V_n) = \left(\left[L_{\frac{T}{1+\{\sum_{i=1}^{i=n} w_i(\frac{T-G_i}{G_i})^{\rho}\}^{1/\rho}}}, L_{\frac{T}{1+\{\sum_{i=1}^{i=n} w_i(\frac{T-H_i}{H_i})^{\rho}\}^{1/\rho}}} \right], L_{\frac{T}{1+\{\sum_{i=1}^{i=n} w_i(\frac{T-M_i}{M_i})^{\rho}\}^{1/\rho}}} \right). \quad (11)$$

Theorem 2 is the same proof as Theorem 1. Hence, we do not prove it repeatedly.

The LCVDWGA operator also has Properties (1)–(3) as follows:

(1) Idempotency: If there is LCVs collection $V_i = ([L_{Gi}, L_{Hi}], L_{Mi})$ for $V_i = V$ $(i = 1, 2, \ldots, n)$. Then LCVDWGA $(V_1, V_2, \ldots, V_n) = V$.
(2) Commutativity: If the LCV set $(V'_1, V'_2, \ldots, V'_n)$ is any permutation of $(V_1, V_2, \ldots, V_n)$. Then, there is LCVDWGA $(V'_1, V'_2, \ldots, V'_n)$ = LCVDWGA $(V_1, V_2, \ldots, V_n)$.
(3) Boundedness: If there is LCVs collection $V_i = ([L_{Gi}, L_{Hi}], L_{Mi})$ $(i = 1,2, \ldots, n)$ $V_{\min} = ([L_{\min_i(Gi)}, L_{\min_i(Hi)}], L_{\min_i(Mi)})$, $V_{\max} = ([L_{\max_i(Gi)}, L_{\max_i(Hi)}], L_{\max_i(Mi)})$. Then, $V_{\min} \le LCVDWGA(V_1, V_2, \ldots, V_n) \le V_{\max}$.

The proofs of the above properties are omitted which are similar with the properties of the LCVDWAA operator.

5. MADM Method on Basis of the LCVDWAA or LCVDWGA Operator

If a MADM problem is described by LCV information, $V = \{V_1, V_2, \ldots, V_m\}$ and $P = \{P_1, P_2, \ldots, P_n\}$ are the sets of alternatives and attributes, respectively. $w = \{w_1, w_2, \ldots, w_n\}$ is the set of weight, where w_j is corresponding to the importance of attribute P_j with $w_j \in [0,1]$ and $\sum_{j=1}^{n} w_j = 1$. The LCV V_{ij} is the evaluation of the alternatives $V_i (i = 1, 2, \ldots, m)$ over the attributes $P_j (j = 1, 2, \ldots, n)$. Each LCV includes uncertain linguistic argument and certain linguistic argument. Thus, all the LCVs given by decision makers are constructed as an LCV decision matrix $V = (V_{ij})_{m \times n}$, where $V_{ij} = ([L_{Gij}, L_{Hij}], L_{Mij})$ is an LCV $(i = 1, 2, \ldots, m; j = 1, 2, \ldots, n)$ and $L_{Gij}, L_{Hij}, L_{Mij}$ is from the linguistic term set $L = \{L_k | k \in [0, T]\}$ with even number T.

On basis of the LCVDWAA or LCVWDGA operator, the steps of MADM method are as follows.

Step 1. According to Equation (9) or Equation (11), we can get the collective LCV of each alterative $V_i = LCVDWAA(V_{i1}, V_{i2}, \ldots, V_{in})$ or $V_i = LCVDWGA(V_{i1}, V_{i2}, \ldots, V_{in})$ $(i = 1, 2, \ldots, m)$.

Step 2. The expected values $E(V_i)(i = 1, 2, \ldots, m)$ of each collective LCV $V_i (i = 1, 2, \ldots, m)$ are calculated according to Equation (1).

Step 3. According to the expected values of $E(V_i)(i = 1, 2, \ldots, m)$, we give the rank order of all the alternatives. The best alternative $V_i (i = 1, 2, \ldots, m)$ is with the greatest value of $E(V_i)$.

6. Illustrative Examples and Discussions

Two application examples are illustrated below, then we discuss the validity of this proposed MADM approach and the influence of the operational parameter.

6.1. Illustrative Examples

Example 2 [13]. *A company needs to hire a soft engineer. There are four candidates (alternatives)* V_1, V_2, V_3, *and* V_4. *The decision makers will further evaluate them over four attributes. The four attributes are soft skills, past experience, personality, and self-confidence, in order. The corresponding weight vector of the attributes is w = (0.35, 0.25, 0.2, 0.2). The decision makers evaluate the four candidates by using the linguistic cubic values,* *which are obtained from the linguistic term set L = {L_i | i ∈ [0, 8]}, where L = {L_0 = extremely poor, L_1 = very poor, L_2 = poor, L_3 = slightly poor, L_4 = fair, L_5 = slightly good, L_6 = good, L_7 = very good, L_8 = extremely good}. The linguistic cubic decision matrix V is described as follows:*

$$V = (V_{ij})_{4\times4} = \begin{bmatrix} ([L_4, L_6], L_5) & ([L_4, L_6], L_4) & ([L_4, L_7], L_6) & ([L_5, L_6], L_6) \\ ([L_3, L_5], L_4) & ([L_5, L_7], L_6) & ([L_4, L_6], L_4) & ([L_6, L_7], L_6) \\ ([L_4, L_7], L_5) & ([L_6, L_7], L_7) & ([L_5, L_7], L_5) & ([L_5, L_7], L_7) \\ ([L_6, L_7], L_7) & ([L_5, L_7], L_6) & ([L_4, L_6], L_5) & ([L_5, L_6], L_5) \end{bmatrix}$$

Now we employ the LCVDWAA operator to solve this MADM problem.

Step 1. According to Equation (9) for ρ = 1 and T = 8, we can get the following collective LCVs for four alternatives:

$$\begin{aligned} V_1 &= LCVDWAA(V_{11}, V_{12}, \ldots, V_{14}) \\ &= \left(\left[L_{T - \frac{T}{1+\{\sum_{i=1}^{i=4} w_i (\frac{G_{1i}}{T-G_{1i}})^\rho\}^{1/\rho}}}, L_{T - \frac{T}{1+\{\sum_{i=1}^{i=4} w_i (\frac{H_{1i}}{T-H_{1i}})^\rho\}^{1/\rho}}} \right], L_{T - \frac{T}{1+\{\sum_{i=1}^{i=4} w_i (\frac{M_{1i}}{T-M_{1i}})^\rho\}^{1/\rho}}} \right) \\ &= \left(\left[L_{8 - \frac{8}{1+\sum_{i=1}^{i=4} w_i (\frac{G_{1i}}{T-G_{1i}})}}, L_{8 - \frac{8}{1+\sum_{i=1}^{i=4} w_i (\frac{H_{1i}}{T-H_{1i}})}} \right], L_{8 - \frac{8}{1+\sum_{i=1}^{i=4} w_i (\frac{M_{1i}}{T-M_{1i}})}} \right) \\ &= ([L_{4.2500}, L_{6.3333}], L_{5.3626}) \\ &= ([L_{G1}, L_{H1}], L_{M1}), \end{aligned}$$

$V_2 = LCVDWAA(V_{21}, V_{22}, \ldots, V_{24}) = ([L_{4.7033}, L_{6.5000}], L_{5.2414})$,
$V_3 = LCVDWAA(V_{31}, V_{32}, \ldots, V_{34}) = ([L_{5.1084}, L_{7.0000}], L_{6.4211})$, and
$V_4 = LCVDWAA(V_{41}, V_{42}, \ldots, V_{44}) = ([L_{5.3333}, L_{6.7500}], L_{6.3562})$.

Step 2. The expected values $E(V_i)(i = 1, 2, \ldots, m)$ of each collective LCV $V_i(i = 1, 2, 3, 4)$ are calculated according to Equation (1). The results are as follows:

$$E(V_1) = (G_1 + H_1 + M_1)/3T = 0.6644, E(V_2) = (G_2 + H_2 + M_2)/3T = 0.6852,$$
$$E(V_3) = (G_3 + H_3 + M_3)/3T = 0.7721, E(V_4) = (G_4 + H_4 + M_4)/3T = 0.7683.$$

Step 3. According to the above expected values and the rank principle, the rank order of the four candidates is $V_3 \succ V_4 \succ V_2 \succ V_1$.

Alternatively, we use LCVDWGA operator for this MADM problem with the same decision steps.

Step 1. We aggregate the LCVs for four candidates according to Equation (11) for ρ = 1 and T = 8.

$$\begin{aligned} V_1 &= LCVDWGA(V_{11}, V_{12}, \ldots, V_{14}) \\ &= \left(\left[L_{\frac{T}{1+\{\sum_{i=1}^{i=4} w_i (\frac{T-G1i}{G1i})^\rho\}^{1/\rho}}}, L_{\frac{T}{1+\{\sum_{i=1}^{i=4} w_i (\frac{T-H1i}{H1i})^\rho\}^{1/\rho}}} \right] \right), L_{\frac{T}{1+\{\sum_{i=1}^{i=4} w_i (\frac{T-M1i}{M1i})^\rho\}^{1/\rho}}} \\ &= \left(\left[L_{\frac{8}{1+\sum_{i=1}^{i=4} w_i (\frac{T-G1i}{G1i})}}, L_{\frac{8}{1+\sum_{i=1}^{i=4} w_i (\frac{T-H1i}{H1i})}} \right] \right), L_{\frac{8}{1+\sum_{i=1}^{i=4} w_i (\frac{T-M1i}{M1i})}} \\ &= ([L_{4.1677}, L_{6.1765}], L_{5.0209}) \end{aligned}$$

$V_2 = LCVDWGA(V_{21}, V_{22}, \ldots, V_{24}) = ([L_{4.0000}, L_{5.9659}], L_{4.7059})$,
$V_3 = LCVDWGA(V_{31}, V_{32}, \ldots, V_{34}) = ([L_{4.7809}, L_{7.0000}], L_{5.7377})$, and
$V_4 = LCVDWGA(V_{41}, V_{42}, \ldots, V_{44}) = ([L_{5.0420}, L_{6.5625}], L_{5.8252})$.

Step 2. The expected values $E(V_i)(i = 1, 2, \ldots, m)$ of each collective LCV $V_i(i = 1, 2, 3, 4)$ are calculated according to Equation (1). The results are as follows:

$E(V_1) = 0.6402, E(V_2) = 0.6113, E(V_3) = 0.7299, E(V_4) = 0.7262.$

Step 3. According to the above expected values and the rank principle, the rank order of the four candidates is $V_3 \succ V_4 \succ V_1 \succ V_2$.

By following the same steps above, we apply the LCVDWAA operator and LCVDWGA operator to Example 2 with parameter ρ from 1 to 100, the ranking results are shown as following Tables 1 and 2.

Table 1. Ranking orders of the LCVDWAA [1] operator, $\rho \in$ [1–5,10,15,20,30,50,100].

ρ	$E(V_1)$ [2], $E(V_2)$ [3], $E(V_3)$ [4], $E(V_4)$ [5]	Ranking Order	The Best Candidate
1	0.6644, 0.6852, 0.7721, 0.7683	$V_3 \succ V_4 \succ V_2 \succ V_1$	V_3
2	0.6766, 0.7139, 0.7875, 0.7843	$V_3 \succ V_4 \succ V_2 \succ V_1$	V_3
3	0.6876, 0.7332, 0.7978, 0.7954	$V_3 \succ V_4 \succ V_2 \succ V_1$	V_3
4	0.6969, 0.7459, 0.8049, 0.8031	$V_3 \succ V_4 \succ V_2 \succ V_1$	V_3
5	0.7045, 0.7545, 0.8099, 0.8085	$V_3 \succ V_4 \succ V_2 \succ V_1$	V_3
10	0.7253, 0.7731, 0.8214, 0.8207	$V_3 \succ V_4 \succ V_2 \succ V_1$	V_3
15	0.7336, 0.7794, 0.8254, 0.8250	$V_3 \succ V_4 \succ V_2 \succ V_1$	V_3
20	0.7377, 0.7825, 0.8274, 0.8271	$V_3 \succ V_4 \succ V_2 \succ V_1$	V_3
30	0.7419, 0.7856, 0.8294, 0.8292	$V_3 \succ V_4 \succ V_2 \succ V_1$	V_3
50	0.7451, 0.7881, 0.8310, 0.8309	$V_3 \succ V_4 \succ V_2 \succ V_1$	V_3
100	0.7476, 0.7899, 0.8322, 0.8321	$V_3 \succ V_4 \succ V_2 \succ V_1$	V_3

[1] LCVDWAA = linguistic cubic variable Dombi weighted arithmetic average; [2] $E(V_1)$ = expected value of V_1; [3] $E(V_2)$ = expected value of V_2; [4] $E(V_3)$ = expected value of V_3; [5] $E(V_4)$ = expected value of V_4.

Table 2. Ranking orders of the LCVDWGA [1] operator, $\rho \in$ [1–5,10,15,20,30,50,100].

ρ	$E(V_1)$, $E(V_2)$, $E(V_3)$, $E(V_4)$	Ranking Order	The Best Candidate
1	0.6402, 0.6113, 0.7299, 0.7262	$V_3 \succ V_4 \succ V_1 \succ V_2$	V_3
2	0.6300, 0.5827, 0.7143, 0.7076	$V_3 \succ V_4 \succ V_1 \succ V_2$	V_3
3	0.6219, 0.5633, 0.7039, 0.6929	$V_3 \succ V_4 \succ V_1 \succ V_2$	V_3
4	0.6155, 0.5503, 0.6968, 0.6816	$V_3 \succ V_4 \succ V_1 \succ V_2$	V_3
5	0.6106, 0.5414, 0.6918, 0.6730	$V_3 \succ V_4 \succ V_1 \succ V_2$	V_3
10	0.5980, 0.5213, 0.6800, 0.6509	$V_3 \succ V_4 \succ V_1 \succ V_2$	V_3
15	0.5932, 0.5143, 0.6756, 0.6424	$V_3 \succ V_4 \succ V_1 \succ V_2$	V_3
20	0.5907, 0.5107, 0.6734, 0.6381	$V_3 \succ V_4 \succ V_1 \succ V_2$	V_3
30	0.5883, 0.5071, 0.6711, 0.6337	$V_3 \succ V_4 \succ V_1 \succ V_2$	V_3
50	0.5863, 0.5043, 0.6693, 0.6303	$V_3 \succ V_4 \succ V_1 \succ V_2$	V_3
100	0.5848, 0.5021, 0.6680, 0.6276	$V_3 \succ V_4 \succ V_1 \succ V_2$	V_3

[1] LCVDWGA = linguistic cubic variable Dombi weighted geometric average.

Example 3. *Customers want to buy an air-conditioner; they choose three brands as alternatives V_1, V_2, V_3. Further, they need to evaluate the three alternatives from three attributes which are as follows: (i) P_1 is cooling effect; (ii) P_2 is heating effect; and (iii) P_3 is appearance design. Their importance lies in the weight vector w = (1/2,1/3,1/6). The customers give their evaluations over the three attributes by the linguistic cubic values V_{ij} based on the uniform linguistic term set L as Example 2. The LCVs provided by the customers constitute the decision matrix V.*

$$V = (V_{ij})_{3\times 3} = \begin{bmatrix} ([L_2, L_7], L_3) & ([L_4, L_7], L_2) & ([L_2, L_7], L_1) \\ ([L_2, L_7], L_5) & ([L_2, L_7], L_3) & ([L_2, L_7], L_3) \\ ([L_2, L_5], L_5) & ([L_1, L_6], L_4) & ([L_2, L_5], L_2) \end{bmatrix}$$

By using the same steps, we apply the LCVDWAA operator or LCVDWGA operator to this MADM problem. The ranking results based on the LCVDWAA operator with parameters ρ from 1 to 5 are shown in the Table 3. Similarly, the ranking orders on basis of the LCVDWGA operator are shown in Table 4.

Table 3. Ranking orders of the LCVDWAA operator, $\rho \in [1–5,10,15,20,30,50,100]$.

ρ	$E(V_1)$, $E(V_2)$, $E(V_3)$	Ranking Order	The Best Alterative
1	0.5117, 0.5795, 0.4804	$V_2 \succ V_1 \succ V_3$	V_2
2	0.5280, 0.5932, 0.4927	$V_2 \succ V_1 \succ V_3$	V_2
3	0.5404, 6005, 0.5016	$V_2 \succ V_1 \succ V_3$	V_2
4	0.5490, 0.6052, 0.5084	$V_2 \succ V_1 \succ V_3$	V_2
5	0.5551, 0.6085, 0.5135	$V_2 \succ V_1 \succ V_3$	V_2
10	0.5688, 0.6165, 0.5267	$V_2 \succ V_1 \succ V_3$	V_2
15	0.5736, 0.6193, 0.5317	$V_2 \succ V_1 \succ V_3$	V_2
20	0.5761, 0.6208, 0.5342	$V_2 \succ V_1 \succ V_3$	V_2
30	0.5785, 0.6222, 0.5367	$V_2 \succ V_1 \succ V_3$	V_2
50	0.5804, 0.6233, 0.5387	$V_2 \succ V_1 \succ V_3$	V_2
100	0.5819, 0.6242, 0.5402	$V_2 \succ V_1 \succ V_3$	V_2

Table 4. Ranking orders of the LCVDWGA operator, $\rho \in [1–5,10,15,20,30,50,100]$.

ρ	$E(V_1)$, $E(V_2)$, $E(V_3)$	Ranking Order	The Best Alterative
1	0.4750, 0.4826, 0.4393	$V_2 \succ V_1 \succ V_3$	V_2
2	0.4598, 0.4353, 0.4144	$V_1 \succ V_2 \succ V_3$	V_1
3	0.4488, 0.4078, 0.3946	$V_1 \succ V_2 \succ V_3$	V_1
4	0.4414, 0.3907, 0.3810	$V_1 \succ V_2 \succ V_3$	V_1
5	0.4364, 0.3796, 0.3718	$V_1 \succ V_2 \succ V_3$	V_1
10	0.4262, 0.3563, 0.3523	$V_1 \succ V_2 \succ V_3$	V_1
15	0.4229, 0.3486, 0.3459	$V_1 \succ V_2 \succ V_3$	V_1
20	0.4213, 0.3447, 0.3427	$V_1 \succ V_2 \succ V_3$	V_1
30	0.4197, 0.3409, 0.3395	$V_1 \succ V_2 \succ V_3$	V_1
50	0.4185, 0.3379, 0.3370	$V_1 \succ V_2 \succ V_3$	V_1
100	0.4176, 0.3356, 0.3352	$V_1 \succ V_2 \succ V_3$	V_1

6.2. Discussion

6.2.1. Validity of the Method

Ye [13] firstly proposed the concept of LCVs, and then used the LCVWAA operator and LCVWGA operator to handle the MADM problem of Example 2. As shown in Table 5, the ranking orders using the LCVDWAA operator and LCVDWGA operator with parameters ρ from 1 to 100 are the same as those using the LCVWAA operator [13] and LCVWGA operator [13], respectively. In Example 3, the ranking results based on the LCVDWAA operator are the same as those based on the LCVWAA operator [13] when parameter ρ ranges from 1 to 100. Then, the ranking orders based on the LCVDWGA operator are the same as those based on the LCVWGA operator [13] when parameter ρ is equal to 1.

Table 5. Ranking results of different aggregation operators with different parameters.

Example	MADM [1] Method	Ranking Order	The Best Alterative
2	LCVDWAA (ρ = 1 to 100)	$V_3 \succ V_4 \succ V_2 \succ V_1$	V_3
	LCVWAA [2] [13]	$V_3 \succ V_4 \succ V_2 \succ V_1$	V_3
	LCVDWGA (ρ = 1 to 100)	$V_3 \succ V_4 \succ V_1 \succ V_2$	V_3
	LCVWGA [3] [13]	$V_3 \succ V_4 \succ V_1 \succ V_2$	V_3
3	LCVDWAA (ρ = 1 to 100)	$V_2 \succ V_1 \succ V_3$	V_2
	LCVWAA [13]	$V_2 \succ V_1 \succ V_3$	V_2
	LCVDWGA (ρ = 1)	$V_2 \succ V_1 \succ V_3$	V_2
	LCVDWGA (ρ = 2 to 100)	$V_1 \succ V_2 \succ V_3$	V_1
	LCVWGA [13]	$V_2 \succ V_1 \succ V_3$	V_2

[1] MADM = multiple attribute decision making; [2] LCVWAA = linguistic cubic variable weighted arithmetic average;
[3] LCVWGA = linguistic cubic variable weighted geometric average.

6.2.2. The Influence of the Parameter ρ

As shown in Table 5, parameter value of ρ has no effect on the ranking results in Example 2. In Example 3, the ranking results are not sensitive to parameters ρ for the LCVDWAA operator. However, corresponding to the LCVDWGA operator, the ranking results are more sensitive to parameter ρ. When $\rho = 1$, the ranking orders of the LCVDWGA operator are the same as those of the LCVWAA [13], LCVWGA [13], and LCVDWAA operators (ρ = 1 to 100), and the best alternative is V_2. While the ranking orders of the LCVDWGA operator are obviously changed when ρ is from 2 to 100, the best alternative is V_1. From Tables 1–4, we can see that parameter value of ρ is greater and the expected values of $E(V_i)$ are greater in the LCVDWAA operator. While in the LCVDWGA operator the value of parameter ρ is greater and the expected values of $E(V_i)$ are smaller.

In any case, by using LCVDWAA or LCVDWGA operator to aggregate decision-making information, the presented approach is valid to handle MADM problems with LCV information. Especially the LCVDWGA operator is more flexible in actual applications.

6.2.3. The Sensitivity Analysis of Weights

In order to demonstrate the sensitivity of weights, we change the weights of the attributes in Examples 2 and 3. $W = (0.25, 0.25, 0.25, 0.25)$ and $w = (1/3, 1/3, 1/3)$ are used as the weight vectors in Examples 2 and 3, respectively. Then we apply LCVDWAA operator and LCVDWGA operator to the two applications again and change the parameter value of ρ from 1 to 100. The ranking results of Example 2 are shown in Figure 1 and the ranking results of Example 3 are shown in Figure 2. The curves of collective expected values $E(V_{i)}$ were shown in Figures 1 and 2. The curves clearly show that LCVDWAA and LCVDWGA have different effects on the expected value. Additionally, we find that the ranking results are identical with Table 5 when the weights are changed. Especially as Figure 2b shows, the best alternative is V_2 when ρ is equal to 1, while the best alternative is V_1 when ρ ranges from 2 to 100. It fits perfectly with Table 5. Thus, we can think that the LCVDWAA and LCVDWGA are not sensitive to the changes of weights.

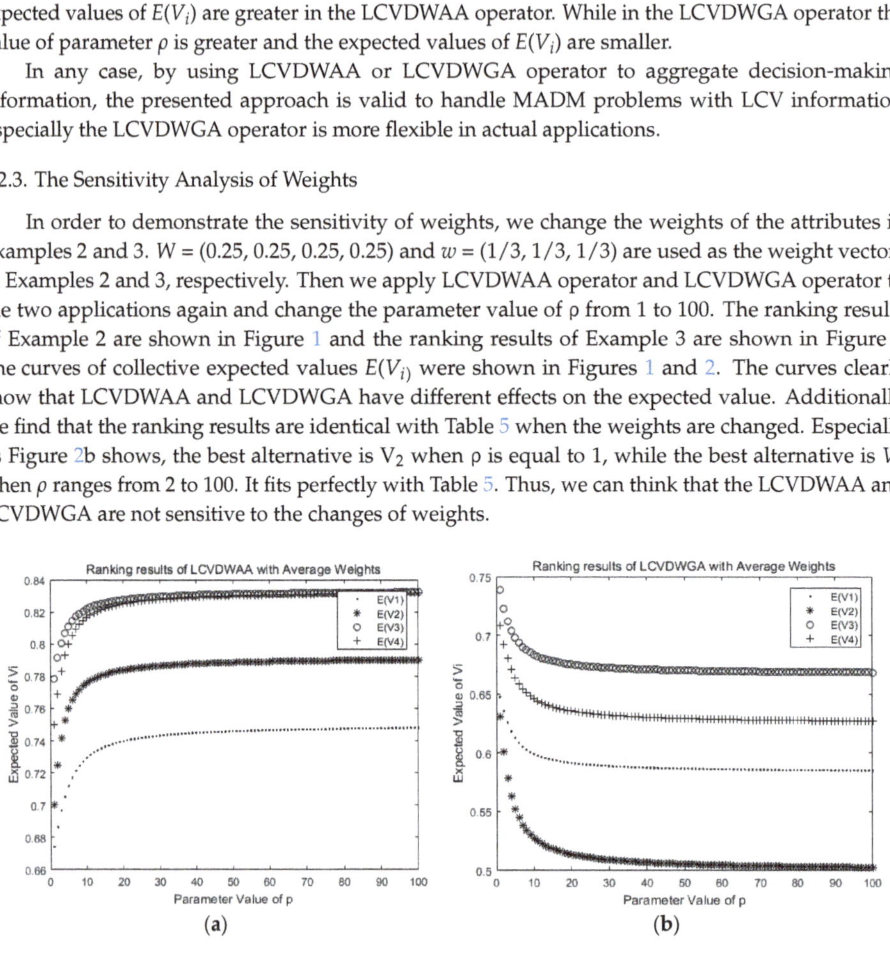

Figure 1. Ranking results with average weights in Example 2. LCVDWAA = linguistic cubic variable Dombi weighted arithmetic average; LCVDWGA = linguistic cubic variable Dombi weighted geometric average. (**a**) LCVDWAA operator; (**b**) LCVDWGA operator.

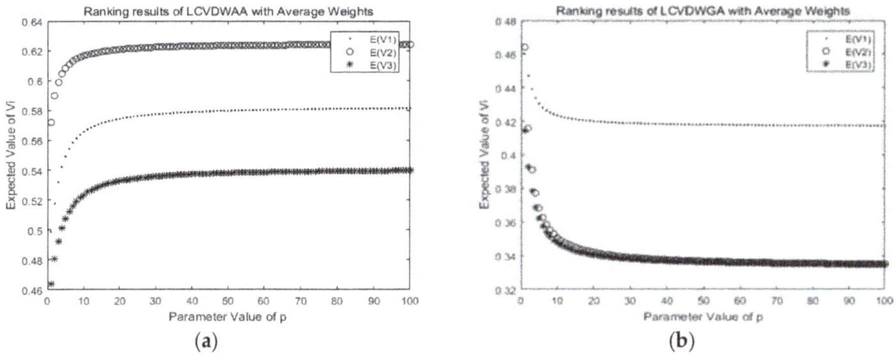

Figure 2. Ranking results with average weights in Example 3. (**a**) LCVDWAA operator; (**b**) LCVDWGA operator.

7. Conclusions

The concept of LCV was proposed by Ye [13] recently. There are few studies on LCV information aggregation operators and MADM methods about LCV information problems. In this paper, the Dombi operations were extended to an LCV environment. We proposed an LCVDWAA operator and an LCVDWGA operator, then discussed their properties. Further, based on the LCVDWAA or LCVDWGA operator, a MADM method was developed. Finally, the proposed approach was applied to two application examples. This MADM method is very simple. There is only one decision-making matrix with LCVs information in a MAGDM problem. The results demonstrated this approach is feasible and valid as the method proposed in Ye [13]. Compared with the method proposed in Ye [13], this approach not only can handle decision-making problems effectively, but also can affect the ranking order based on the LCVDWAA or LCVDWGA operator by the changeable parameter ρ. In an actual decision-making process, we can specify various parameter values based on the decision makers' preferences and requirements. However, the flexibility of the LCVDWAA or LCVDWGA operator was not fully reflected in the two examples. In order to observe the sensitivity of weights, we changed the weight vectors of the two examples and changed parameter values from 1 to 100. We found that the results were not changed when the weight was averaged. Although the operators were not sensitive to the changes of weights, there were some changes in the ranking results when we changed the weight vectors to extreme cases in the study. Thus, the ranking results are determined by weights and parameter values together for the same decision-making matrix. In future work, we can continue to develop more flexible aggregation operators of LCVs and use them to solve MADM problems in various fields.

Author Contributions: J.Y. proposed the LCVDWAA and LCVDWGA operators and the MADM method. X.L. provided the applications and comparative analysis. All authors wrote the paper together.

Funding: This research received no external funding.

Conflicts of Interest: The authors declare no conflict of interest.

References

1. Chatterjee, P.; Mondal, S.; Boral, S. A novel hybrid method for non-traditional machining process selection using factor relationship and Multi-Attributive Border Approximation Method. *Facta Univ. Ser. Mech. Eng.* **2017**, *15*, 439–456. [CrossRef]
2. Petković, D.; Madić, M.; Radovanović, M. Application of the Performance Selection Index Method for Solving Machining MCDM Problems. *Facta Univ. Ser. Mech. Eng.* **2017**, *15*, 97–106. [CrossRef]

3. Roy, J.; Adhikary, K.; Kar, S. A rough strength relational DEMATEL model for analysing the key success factors of hospital service quality. *Decis. Mak. Appl. Manag. Eng.* **2018**, *1*, 121–142. [CrossRef]
4. Vasiljević, M.; Fazlollahtabar, H.; Stević, Ž. A rough multicriteria approach for evaluation of the supplier criteria in automotive industry. *Decis. Mak. Appl. Manag. Eng.* **2018**, *1*, 82–96. [CrossRef]
5. He, X. Disaster assessment based on Dombi hesitant fuzzy information aggregation operators. *Nat. Hazards* **2017**, *90*, 1153–1175. [CrossRef]
6. Zadeh, L.A. The concept of a linguistic variable and its application to approximate reasoning-I. *Inf. Sci.* **1975**, *8*, 199–249. [CrossRef]
7. Yuen, K.K.F.; Lau, H.C.W. A linguistic possibility probability aggregation model for decision analysis with imperfect knowledge. *Appl. Soft Comput. J.* **2009**, *9*, 575–589. [CrossRef]
8. Porcel, C.; Herrera-Viedma, E. Dealing with incomplete information in a fuzzy linguistic recommender system to disseminate information in university digital libraries. *Knowl. Based Syst.* **2010**, *23*, 32–39. [CrossRef]
9. Cabrerizo, F.J.; Pérez, I.J.; Herrera-Viedma, E. Managing the consensus in group decision making in an unbalanced fuzzy linguistic context with incomplete information. *Knowl. Based Syst.* **2010**, *23*, 169–181. [CrossRef]
10. Lu, M.; Wei, G.; Alsaadi, F.E.; Hayat, T.; Alsaedi, A. Bipolar 2-tuple linguistic aggregation operators in multiple attribute decision making. *J. Intell. Fuzzy Syst.* **2017**, *33*, 1197–1207. [CrossRef]
11. Gou, X.; Xu, Z.; Lei, Q. New operational laws and aggregation method of intuitionistic fuzzy information. *J. Intell. Fuzzy Syst.* **2016**, *30*, 129–141. [CrossRef]
12. Wei, G. Pythagorean fuzzy interaction aggregation operators and their application to multiple attribute decision making. *J. Intell. Fuzzy Syst.* **2017**, *33*, 2119–2132. [CrossRef]
13. Ye, J. Multiple attribute decision making method based on linguistic cubic variables. *J. Intell. Fuzzy Syst.* **2018**, *34*, 2351–2361. [CrossRef]
14. Xu, Z.S. Uncertain linguistic aggregation operators based approach to multiple attribute group decision making under uncertain linguistic environment. *Inf. Sci.* **2004**, *168*, 171–184. [CrossRef]
15. Wei, G.W. Uncertain linguistic hybrid geometric mean operator and its application to group decision making under uncertain linguistic environment. *Int. J. Uncertain. Fuzziness Knowl. Based Syst.* **2009**, *17*, 251–267. [CrossRef]
16. Xu, Z.S. Induced uncertain linguistic OWA operators applied to group decision making. *Inf. Fusion* **2006**, *7*, 231–238. [CrossRef]
17. Wei, G.W.; Zhao, X.F.; Lin, R.; Wang, H.J. Uncertain linguistic Bonferroni mean operators and their application to multiple attribute decision making. *Appl. Math. Model.* **2013**, *37*, 5277–5285. [CrossRef]
18. Park, J.H.; Gwak, M.G.; Kwun, Y.C. Uncertain linguistic harmonic mean operators and their applications to multiple attribute group decision making. *Computing* **2011**, *93*, 47–64. [CrossRef]
19. Zhang, H. Uncertain linguistic power geometric operators and their use in multi attribute group decision making. *Math. Probl. Eng.* **2015**, *2015*, 948380. [CrossRef]
20. Liu, P.; Qin, X. Power average operators of linguistic intuitionistic fuzzy numbers and their application to multiple-attribute decision making. *J. Intell. Fuzzy Syst.* **2017**, *32*, 1029–1043. [CrossRef]
21. Liu, P. Maclaurin symmetric mean operators of linguistic intuitionistic fuzzy numbers and their application to multiple-attribute decision-making. *J. Exp. Theor. Artif. Intell.* **2017**, *29*, 1173–1202. [CrossRef]
22. Li, Z.; Liu, P. An extended VIKOR method for decision making problem with linguistic intuitionistic fuzzy numbers based on some new operational laws and entropy. *J. Intell. Fuzzy Syst.* **2017**, *33*, 1919–1931. [CrossRef]
23. Ye, J. Some aggregation operators of interval neutrosophic linguistic numbers for multiple attribute decision making. *J. Intell. Fuzzy Syst.* **2014**, *27*, 2231–2241.
24. Ye, J. Multiple Attribute Decision-Making Methods Based on the Expected Value and the Similarity Measure of Hesitant Neutrosophic Linguistic Numbers. *Cogn. Comput.* **2017**, *10*, 454–463. [CrossRef]
25. Liu, P.; Shi, L. Some Neutrosophic Uncertain Linguistic Number Heronian Mean Operators and Their Application to Multi-Attribute Group Decision Making. *Neural Comput. Appl.* **2015**, *28*, 1079–1093. [CrossRef]
26. Jun, Y.B.; Kim, C.S.; Yang, K.O. Cubic sets. *Ann. Fuzzy Math. Inform.* **2012**, *4*, 83–98.

27. Dombi, J. A general class of fuzzy operators, the demorgan class of fuzzy operators and fuzziness measures induced by fuzzy operators. *Fuzzy Sets Syst.* **1982**, *8*, 149–163. [CrossRef]
28. Liu, P.D.; Liu, J.L.; Chen, S.M. Some intuitionistic fuzzy Dombi Bonferroni mean operators and their application to multi-attribute group decision making. *J. Oper. Res. Soc.* **2017**, *69*, 1–24. [CrossRef]

information

MDPI

Article

Operations and Aggregation Methods of Single-Valued Linguistic Neutrosophic Interval Linguistic Numbers and Their Decision Making Method

Jun Ye * and Wenhua Cui

Department of Electrical Engineering and Automation, Shaoxing University, 508 Huancheng West Road, Shaoxing 312000, China; wenhuacui@usx.edu.cn

* Correspondence: yehjun@aliyun.com; Tel.: +86-575-88327323

Received: 17 July 2018; Accepted: 30 July 2018; Published: 1 August 2018

Abstract: To comprehensively describe uncertain/interval linguistic arguments and confident linguistic arguments in the decision making process by a linguistic form, this study first presents the concept of a single-valued linguistic neutrosophic interval linguistic number (SVLN-ILN), which is comprehensively composed of its uncertain/interval linguistic number (determinate linguistic argument part) and its single-valued linguistic neutrosophic number (confident linguistic argument part), and its basic operations. Then, the score function of SVLN-ILN based on the attitude index and confident degree/level is presented for ranking SVLN-ILNs. After that, SVLN-ILN weighted arithmetic averaging (SVLN-ILNWAA) and SVLN-ILN weighted geometric averaging (SVLN-ILNWGA) operators are proposed to aggregate SVLN-ILN information and their properties are investigated. Further, a multi-attribute decision-making (MADM) method based on the proposed SVLN-ILNWAA or SVLN-ILNWGA operator and the score function is established under consideration of decision makers' preference attitudes (pessimist, moderate, and optimist). Lastly, an actual example is given to show the applicability of the established MADM approach with decision makers' attitudes.

Keywords: single-valued linguistic neutrosophic interval linguistic number; score function; weighted aggregation operator; decision making

1. Introduction

Multi-attribute decision-making (MADM) explicitly evaluates multiple conflicting attributes in decision making to help people make optimal decisions [1–4]. There usually exists uncertainty and vagueness in MADM problems. In this situation, it may prove difficult for decision makers (DMs) to express their evaluation values of attributes, especially qualitative attributes, by numerical values. Then, the expression of linguistic terms (LTs) is very fit for human thinking and expressing habits. For instance, when the quality of some product is evaluated by LTs, we use LTs "good", "very good", and so on to easily express it. Hence, linguistic decision making methods have been wildly used for MADM problems with linguistic information. Firstly, Zadeh [5] presented the concept of a linguistic variable (LV) for its fuzzy reasoning application. Then, Herrera et al. [6] and Herrera and Herrera-Viedma [7] solved linguistic decision making problems using a linguistic decision analysis. After that, many scholars [8–14] introduced different linguistic aggregation operators for (group) decision making problems. Owing to the uncertainty and vagueness in the linguistic decision environment, uncertain/interval linguistic numbers (ILNs) and various uncertain linguistic aggregation operators have been also presented for uncertain linguistic (group) decision making problems [15–20]. Based on a neutrosophic number (i.e., a changeable interval number with indeterminacy), Ye [21] put forward the concept of a neutrosophic linguistic number (NLN), its basic operational laws, and two NLN weighted aggregation operators for multi-attribute group decision making (MAGDM). To represent the hybrid linguistic information of the partial uncertain and partial certain arguments, Ye [22] introduced linguistic cubic numbers (LCNs) and their operations and two weighted aggregation operators for MADM problems with LCN information. To independently depict the truth, falsity, and indeterminacy linguistic arguments in real life for an evaluated object, Fang and Ye [23] presented linguistic neutrosophic numbers (LNNs), their operations, and two weighted aggregation operators for MAGDM in an LNN setting.

In uncertain linguistic MADM problems, it may prove difficult for DMs to give accurate LT values for an attribute from a predefined LT set, but can assign a certain interval linguistic range to it. However, ILN only indicates interval/uncertain LT values of DMs for an attribute, but cannot reflect the confident degree of their judgment. Although Wang et al. [24] proposed the concept of the intuitionistic interval number (IIN) composed of its interval judgment (its uncertain argument) and its intuitionistic fuzzy judgment (its confident judgment) to express the hybrid information of both, IIN cannot express its linguistic argument information in a linguistic evaluation setting. However, how to express the hybrid information of a single-valued LNN and an ILN simultaneously is a difficult problem because there is no research in existing literature. For instance, suppose we give both the ILN $[l_4, l_6]$ (the uncertain/interval linguistic argument) and the single-valued LNN $<l_5, l_3, l_1>$ (the confident linguistic judgment) from the given LT set $L = \{l_s \mid s \in [0, 8]\}$ regarding an evaluated object. It is obvious that IIN [24] cannot express the hybrid information of both the ILN and the single-valued LNN. To comprehensively describe an uncertain linguistic argument and a confident linguistic judgment in the decision making process, we need the single-valued linguistic neutrosophic ILN (SVLN-ILN), which consists of an ILN and a single-valued LNN (SVLNN), where an SVLNN inflects the confident level/degree of decision makers indicated by the truth, indeterminacy, and falsity LT values corresponding to its ILN judgment for an evaluated object, in order to solve the gap. Therefore, the purposes of this study are as follows: (1) to propose the SVLN-ILN concept for expressing the hybrid information of both the ILN and the single-valued LNN, its operations, and score function with both attitude index and confident level/degree for ranking SVLN-ILNs; (2) to present SVLN-ILN weighted arithmetic averaging (SVLN-ILNWAA) and SVLN-ILN weighted geometric averaging (SVLN-ILNWGA) operators; and (3) to establish an MADM method using the SVLN-ILNWAA or SVLN-ILNWGA operator and the score function to handle MADM problems in SVLN-ILN setting and DMs' attitudes (pessimist, moderate, and optimist).

This study is constructed as per the following structural framework. Section 2 proposes the SVLN-ILN concept composed of ILN and SVLNN, the basic operations of SVLN-ILNs, and the score function of SVLN-ILN for ranking SVLN-ILNs. In Section 3, the SVLN-ILNWAA and SVLN-ILNWGA operators are given to aggregate SVLN-ILNs, and then their properties are discussed. In Section 4, a MADM method with DMs' attitudes is established based on the SVLN-ILNWAA or SVLN-ILNWGA operator and the score function under the SVLN-ILN setting. Section 5 presents an actual example to show the applicability of the proposed MADM method in the SVLN-ILN setting. Lastly, conclusions and future work are indicated in Section 6.

2. Single-Valued Linguistic Neutrosophic Interval Linguistic Numbers

Based on the extension of IINs [24], this section presents the concept of SVLN-ILN, which contains the hybrid information of both SVLNN and ILN, the basic operations of SVLN-ILNs, and the score function of SVLN-ILN.

Definition 1. *Let a LT set be $L = \{l_s \mid s \in [0, z]\}$, where $z + 1$ is an odd number/cardinality. A SVLN-ILN g in L is constructed as $g = \langle [l_a, l_b]; l_T, l_I, l_F \rangle$, where $[l_a, l_b]$ is the ILN part of g and l_a and l_b are linguistic lower* *and upper bounds of l_s for $l_a \le l_s \le l_b$ and $l_s \in L$, and then $<l_T, l_I, l_F>$ is the SVLNN part of g. Here, the truth linguistic function $T_g(l_s)$ of g can be defined as*

$$T_g(l_s) = \begin{cases} l_T, & l_a \le l_s \le l_b \\ l_0, & otherwise \end{cases}$$

The indeterminacy linguistic function $I_g(l_s)$ of g can be defined as

$$I_g(l_s) = \begin{cases} l_I, & l_a \le l_s \le l_b \\ l_z, & otherwise \end{cases}$$

The falsity linguistic function $F_g(l_s)$ of g can be defined as

$$F_g(l_s) = \begin{cases} l_F, & l_a \le l_s \le l_b \\ l_z, & otherwise \end{cases}$$

where $l_0 \le l_T \le l_z$, $l_0 \le l_I \le l_z$, and $l_0 \le l_F \le l_z$.

For instance, $g = <[l_4, l_6]; l_5, l_2, l_3>$ is an SVLN-ILN, where $[l_4, l_6]$ is the ILN part of g, and then $<l_5, l_2, l_3>$ is the SVLNN part depicted by the truth linguistic value l_5, the indeterminacy linguistic value l_2, and the falsity linguistic value l_3, independently. In a decision making problem, the SVLN-ILN indicates both DMs' interval linguistic judgment (uncertain linguistic judgment) and confident linguistic judgment for an evaluated object.

To express the semantics conveniently, we adopt a linguistic transformation/scale function $f(l_s) = s$ for $s \in [0, z]$, which produces the mapping f: $l_s \to s$, i.e., the mapping from a LT in $L = \{l_s \mid s \in [0, z]\}$ to a numerical value.

Definition 2. *Suppose $g_1 = \langle [l_{a_1}, l_{b_1}]; l_{T_1}, l_{I_1}, l_{F_1} \rangle$ and $g_2 = \langle [l_{a_2}, l_{b_2}]; l_{T_2}, l_{I_2}, l_{F_2} \rangle$ are two SVLN-ILNs in L. If their arguments/expected values are $m_1 = [f(l_{a_1}) + f(l_{b_1})]/2 = (a_1 + b_1)/2$ and $m_2 = [f(l_{a_2}) + f(l_{b_2})]/2 = (a_2 + b_2)/2$, and a positive scalar is $p > 0$, their basic operations can be defined below:*

$$g_1 \oplus g_2 = \left\langle \left[l_{a_1+a_2-\frac{a_1 \cdot a_2}{z}}, l_{b_1+b_2-\frac{b_1 \cdot b_2}{z}} \right]; l_{\frac{m_1 T_1 + m_2 T_2}{m_1+m_2}}, l_{\frac{m_1 I_1 + m_2 I_2}{m_1+m_2}}, l_{\frac{m_1 F_1 + m_2 F_2}{m_1+m_2}} \right\rangle \quad (1)$$

$$g_1 \otimes g_2 = \left\langle \left[l_{\frac{a_1 \cdot a_2}{z}}, l_{\frac{b_1 \cdot b_2}{z}} \right]; l_{\frac{T_1 \cdot T_2}{z}}, l_{I_1+I_2-\frac{I_1 \cdot I_2}{z}}, l_{F_1+F_2-\frac{F_1 \cdot F_2}{z}} \right\rangle \quad (2)$$

$$pg_1 = \left\langle \left[l_{z-z(1-\frac{a_1}{z})^p}, l_{z-z(1-\frac{b_1}{z})^p} \right]; l_{T_1}, l_{I_1}, l_{F_1} \right\rangle \quad (3)$$

$$g_1^p = \left\langle \left[l_{z(\frac{a_1}{z})^p}, l_{z(\frac{b_1}{z})^p} \right]; l_{z(\frac{T_1}{z})^p}, l_{z-z(1-\frac{I_1}{z})^p}, l_{z-z(1-\frac{F_1}{z})^p} \right\rangle \quad (4)$$

Clearly, the above calculated results are still SVLN-ILNs.

Example 1. *Suppose $g_1 = <[l_4, l_6]; l_5, l_2, l_3>$ and $g_2 = <[l_2, l_6]; l_6, l_1, l_2>$ are two SVLN-ILNs in the LT set $L = \{l_0, l_1, \ldots, l_8\}$ for $z = 8$ and $p = 2$. Then, their arguments are $m_1 = [f(l_4) + f(l_6)]/2 = (4 + 6)/2 = 5$ and $m_2 = [f(l_2) + f(l_6)]/2 = (2 + 6)/2 = 4$, respectively.*

Thus, using Equations (1)–(4), the operational results are yielded as follows:

$$(1)\quad g_1 \oplus g_2 = \left\langle \left[l_{a_1+a_2-\frac{a_1 \cdot a_2}{z}}, l_{b_1+b_2-\frac{b_1 \cdot b_2}{z}} \right]; l_{\frac{m_1 T_1+m_2 T_2}{m_1+m_2}}, l_{\frac{m_1 I_1+m_2 I_2}{m_1+m_2}}, l_{\frac{m_1 F_1+m_2 F_2}{m_1+m_2}} \right\rangle$$
$$= \left\langle \left[l_{4+2-\frac{4\times 2}{8}}, l_{6+6-\frac{6\times 6}{8}} \right]; l_{\frac{5\times 5+4\times 6}{5+4}}, l_{\frac{5\times 2+4\times 1}{5+4}}, l_{\frac{5\times 3+4\times 2}{5+4}} \right\rangle$$
$$= \langle [l_{5.0000}, l_{7.5000}]; l_{5.4444}, l_{1.5556}, l_{2.5556} \rangle;$$

$$(2)\quad g_1 \otimes g_2 = \left\langle \left[l_{\frac{a_1 \cdot a_2}{z}}, l_{\frac{b_1 \cdot b_2}{z}} \right]; l_{\frac{T_1 \cdot T_2}{z}}, l_{I_1+I_2-\frac{I_1 \cdot I_2}{z}}, l_{F_1+F_2-\frac{F_1 \cdot F_2}{z}} \right\rangle$$
$$= \left\langle \left[l_{\frac{4\times 2}{8}}, l_{\frac{6\times 6}{8}} \right]; l_{\frac{5\times 6}{8}}, l_{2+1-\frac{2\times 1}{8}}, l_{3+2-\frac{3\times 2}{8}} \right\rangle = \langle [l_{1.0000}, l_{4.5000}]; l_{3.7500}, l_{2.7500}, l_{4.2500} \rangle;$$

$$(3)\quad pg_1 = \left\langle \left[l_{z-z(1-\frac{a_1}{z})^p}, l_{z-z(1-\frac{b_1}{z})^p} \right]; l_{T_1}, l_{I_1}, l_{F_1} \right\rangle$$
$$= \left\langle \left[l_{8-8(1-\frac{4}{8})^2}, l_{8-8(1-\frac{6}{8})^2} \right]; l_5, l_2, l_3 \right\rangle = \langle [l_{6.0000}, l_{7.5000}]; l_5, l_2, l_3 \rangle;$$

$$(4)\quad g_1^p = \left\langle \left[l_{z(\frac{a_1}{z})^p}, l_{z(\frac{b_1}{z})^p} \right]; l_{z(\frac{T_1}{z})^p}, l_{z-z(1-\frac{I_1}{z})^p}, l_{z-z(1-\frac{F_1}{z})^p} \right\rangle$$
$$= \left\langle \left[l_{8(\frac{4}{8})^2}, l_{8(\frac{6}{8})^2} \right]; l_{8(\frac{5}{8})^2}, l_{8-8(1-\frac{2}{8})^2}, l_{8-8(1-\frac{3}{8})^2} \right\rangle = \langle [l_{2.0000}, l_{4.5000}]; l_{3.1250}, l_{3.5000}, l_{4.8750} \rangle.$$

For comparison between SVLN-ILNs, both ILN and SVLNN in an SVLN-ILN g = <$[l_a, l_b]$; l_T, l_I, l_F> should be considered as the score function containing both the attitude index of ILN and the score value of SVLNN (the confidence level/degree) regarding DMs in the decision making process.

Based on the extension of attitude index for an interval number [25], the attitude index of an ILN $[l_a, l_b]$ is defined as follows:

$$A = \frac{f(l_a) + f(l_b)}{2z} + (2\alpha - 1)\frac{f(l_b) - f(l_a)}{2z} = \frac{a+b}{2z} + (2\alpha - 1)\frac{b-a}{2z} \quad (5)$$

where $\alpha \in [0, 1]$ is the attitude coefficient.

Then, the score value of SVLNN is given as follows:

$$S = \frac{2z + f(l_T) - f(l_I) - f(l_F)}{3z} = \frac{2z + T - I - F}{3z} \quad (6)$$

Thus, the score function of a SVLN-ILN can be given by the definition below.

Definition 3. *Based on the combination of both the attitude index of ILN and the score value of SVLN (the confidence level) for a SVLN-ILN g = <$[l_a, l_b]$; l_T, l_I, l_F>, the new score function of a SVLN-ILN can be given as*

$$Y(g) = A \times S = \left(\frac{a+b}{2z} + (2\alpha - 1)\frac{b-a}{2z} \right) \times \left(\frac{2z + T - I - F}{3z} \right) \text{ for } Y(g)\ \ [0, 1] \quad (7)$$

In the score function (7), both the attitude coefficient α and the confident level/score value of S can indicate the pessimistic/moderate/optimistic degree and confident degree of DMs. On the one hand, when DM believes that the linguistic evaluation value of an attribute is in an ILN $[l_a, l_b]$, his/her linguistic evaluation value tends to the lower bound l_a for a pessimistic DM, conversely, his/her linguistic evaluation value tends to the upper bound l_b for an optimistic DM, while his/her linguistic evaluation value tends to the moderate value $[f(l_a) + f(l_b)]/2$ for a moderate DM. Obviously, the DM's attitude is increasingly more optimistic with increasing α from 0 to 1. Especially when α = 0, 0.5, and 1, the three attitude coefficients reflect the pessimistic, moderate, and optimistic attitudes of DM, respectively. On the other hand, the score value of $S \in [0, 1]$ also indicates the confident level/degree of DM. Then, the DM's confident degree is increasingly more high with increasing S from 0 to 1. Especially S = 1 for $f(l_T)$ = z, $f(l_I)$ = 0, and $f(l_F)$ = 0 in SVLNN is quite confident; while S = 0 for $f(l_T)$ = 0, $f(l_I)$ = z, and $f(l_F)$ = z in SVLNN is quite unconfident.

Example 2. *Suppose* $g = <[l_6, l_7]; l_6, l_2, l_3>$ *is the SVLN-ILN in the LT set* $L = \{l_s \mid s \in [0, 8]\}$ *for* $z = 8$. *Then, the pessimistic, moderate, and optimistic attitudes of DM are given by* $\alpha = 0, 0.5, 1$, *respectively.*

Thus, by Equation (7), we calculate the score value of the SVLN-ILN below:

$$Y(g) = \left(\frac{a+b}{2z} + (2\alpha - 1)\frac{b-a}{2z}\right)\left(\frac{2z+T-I-F}{3z}\right)$$
$$= \left(\frac{6+8}{2\times 8} + (2\alpha - 1)\frac{8-6}{2\times 8}\right)\left(\frac{2\times 8+6-2-3}{3\times 8}\right) = \begin{cases} 0.5313 \text{ for } \alpha = 0, \\ 0.6198 \text{ for } \alpha = 0.5, \\ 0.7083 \text{ for } \alpha = 1. \end{cases}$$

Clearly, the score values of SVLN-ILN can be changed with the pessimistic, moderate, and optimistic attitudes of DM (i.e., $\alpha = 0, 0.5$, and 1, respectively).

Definition 4. *Suppose* $g_1 = \langle [l_{a_1}, l_{b_1}]; l_{T_1}, l_{I_1}, l_{F_1} \rangle$ *and* $g_2 = \langle [l_{a_2}, l_{b_2}]; l_{T_2}, l_{I_2}, l_{F_2} \rangle$ *are two SVLN-ILNs in L, we give the following ranking relations:*

(i) If $Y(g_1) > Y(g_2)$, then $g_1 \succ g_2$;
(ii) If $Y(g_1) < Y(g_2)$, then $g_1 \prec g_2$;
(iii) If $Y(g_1) = Y(g_2)$, then $g_1 = g_2$.

Example 3. *If* $g_1 = <[l_5, l_7]; l_7, l_2, l_1>$ *and* $g_2 = <[l_6, l_8]; l_5, l_3, l_4>$ *are two SVLN-ILNs in the LT set* $L = \{l_s \mid s \in [0, 8]\}$ *for* $z = 8$, *they are ranked by DM with the moderate attitude* $\alpha = 0.5$.

By applying Equation (7), there exists $Y(g_1) = 0.6250 > Y(g_2) = 0.5104$, then $g_1 \succ g_2$.

3. Weighted Aggregation Operators of SVLN-ILNs

3.1. SVLN-ILNWAA Operator

Definition 5. *Suppose* $g_k = \langle [l_{a_k}, l_{b_k}]; l_{T_k}, l_{I_k}, l_{F_k} \rangle$ *(k =1, 2, ..., n) is a group of SVLN-ILNs in L. Then, the SVLN-ILNWAA operator can be given as follows:*

$$SVLN - ILNWAA(g_1, g_2, \ldots, g_n) = \sum_{k=1}^{n} \omega_k g_k \tag{8}$$

where $\omega_k \in [0, 1]$ is the weight of g_k (k =1, 2, ..., n) and $\sum_{k=1}^{n} \omega_k = 1$.

Thus, the following theorem can be given based on Equations (1), (3), and (8).

Theorem 1. *Suppose* $g_k = \langle [l_{a_k}, l_{b_k}]; l_{T_k}, l_{I_k}, l_{F_k} \rangle$ *(k =1, 2, ..., n) is a group of SVLN-ILNs in L. Thus, the aggregation result regarding Equation (8) is also a SVLN-ILN, which is yielded by the aggregation form:*

$$SVLN - ILNWAA(g_1, g_2, \ldots, g_n) = \sum_{k=1}^{n} \omega_k g_k$$
$$= \left\langle \left[l_{z - z\prod_{k=1}^{n}\left(1-\frac{a_k}{z}\right)^{\omega_k}}, l_{z - z\prod_{k=1}^{n}\left(1-\frac{b_k}{z}\right)^{\omega_k}} \right]; l_{\frac{\sum_{k=1}^{n}\omega_k T_k m_k}{\sum_{k=1}^{n}\omega_k m_k}}, l_{\frac{\sum_{k=1}^{n}\omega_k I_k m_k}{\sum_{k=1}^{n}\omega_k m_k}}, l_{\frac{\sum_{k=1}^{n}\omega_k F_k m_k}{\sum_{k=1}^{n}\omega_k m_k}} \right\rangle \tag{9}$$

where $m_k = (a_k + b_k)/2$ for $a_k, b_k \in [0, z]$ and $k = 1, 2, \ldots, n$.

Then, Theorem 1 can be proofed by the mathematical induction.

Proof:

If $k = 2$, by Equation (3), we have

$$\omega_1 g_1 = \left\langle \left[l_{z-z(1-\frac{a_1}{z})^{\omega_1}}, l_{z-z(1-\frac{b_1}{z})^{\omega_1}} \right]; l_{T_1}, l_{I_1}, l_{F_1} \right\rangle,$$

$$\omega_2 g_2 = \left\langle \left[l_{z-z(1-\frac{a_2}{z})^{\omega_2}}, l_{z-z(1-\frac{b_2}{z})^{\omega_2}} \right]; l_{T_2}, l_{I_2}, l_{F_2} \right\rangle.$$

By Equation (1), there exists the following result:

$$\begin{aligned} \sum_{k=1}^{2} \omega_k g_k &= \left\langle \begin{array}{l} \left[l_{z-z(1-\frac{a_1}{z})^{\omega_1} + z-z(1-\frac{a_2}{z})^{\omega_2} - \frac{[z-z(1-\frac{a_1}{z})^{\omega_1}] \times [z-z(1-\frac{a_2}{z})^{\omega_2}]}{z}}, l_{z-z(1-\frac{b_1}{z})^{\omega_1} + z-z(1-\frac{b_2}{z})^{\omega_2} - \frac{[z-z(1-\frac{b_1}{z})^{\omega_1}] \times [z-z(1-\frac{b_2}{z})^{\omega_2}]}{z}} \right]; \\ l_{\frac{T_1(\omega_1 a_1 + \omega_1 b_1)/2 + T_2(\omega_2 a_2 + \omega_2 b_2)/2}{(\omega_1 a_1 + \omega_1 b_1)/2 + (\omega_2 a_2 + \omega_2 b_2)/2}}, l_{\frac{I_1(\omega_1 a_1 + \omega_1 b_1)/2 + I_2(\omega_2 a_2 + \omega_2 b_2)/2}{(\omega_1 a_1 + \omega_1 b_1)/2 + (\omega_2 a_2 + \omega_2 b_2)/2}}, l_{\frac{F_1(\omega_1 a_1 + \omega_1 b_1)/2 + F_2(\omega_2 a_2 + \omega_2 b_2)/2}{(\omega_1 a_1 + \omega_1 b_1)/2 + (\omega_2 a_2 + \omega_2 b_2)/2}} \end{array} \right\rangle \\ &= \left\langle \left[l_{z-z\prod_{k=1}^{2}(1-\frac{a_k}{z})^{\omega_k}}, l_{z-z\prod_{k=1}^{2}(1-\frac{b_k}{z})^{\omega_k}} \right]; l_{\frac{\sum_{k=1}^{2}\omega_k T_k m_k}{\sum_{k=1}^{2}\omega_k m_k}}, l_{\frac{\sum_{k=1}^{2}\omega_k I_k m_k}{\sum_{k=1}^{2}\omega_k m_k}}, l_{\frac{\sum_{k=1}^{2}\omega_k F_k m_k}{\sum_{k=1}^{2}\omega_k m_k}} \right\rangle. \end{aligned} \tag{10}$$

If $k = n$, Equation (9) exists as the following form:

$$\begin{aligned} &SVLN - ILNWAA(g_1, g_2, \ldots, g_n) = \sum_{k=1}^{n} \omega_k g_k \\ &= \left\langle \left[l_{z-z\prod_{k=1}^{n}(1-\frac{a_k}{z})^{\omega_k}}, l_{z-z\prod_{k=1}^{n}(1-\frac{b_k}{z})^{\omega_k}} \right]; l_{\frac{\sum_{k=1}^{n}\omega_k T_k m_k}{\sum_{k=1}^{n}\omega_k m_k}}, l_{\frac{\sum_{k=1}^{n}\omega_k I_k m_k}{\sum_{k=1}^{n}\omega_k m_k}}, l_{\frac{\sum_{k=1}^{n}\omega_k F_k m_k}{\sum_{k=1}^{n}\omega_k m_k}} \right\rangle. \end{aligned}$$

Thus, if $k = n + 1$, by Equations (1), (3), and (10), we yield the result:

$$\begin{aligned} &SVLN - ILNWAA(g_1, g_2, \ldots, g_n, g_{k+1}) = \sum_{k=1}^{n+1} \omega_k g_k = \sum_{k=1}^{n} \omega_k g_k \oplus \omega_{n+1} g_{n+1} \\ &= \left\langle \left[l_{z-z\prod_{k=1}^{n}(1-\frac{a_k}{z})^{\omega_k}}, l_{z-z\prod_{k=1}^{n}(1-\frac{b_k}{z})^{\omega_k}} \right]; l_{\frac{\sum_{k=1}^{n}\omega_k T_k m_k}{\sum_{k=1}^{n}\omega_k m_k}}, l_{\frac{\sum_{k=1}^{n}\omega_k I_k m_k}{\sum_{k=1}^{n}\omega_k m_k}}, l_{\frac{\sum_{k=1}^{n}\omega_k F_k m_k}{\sum_{k=1}^{n}\omega_k m_k}} \right\rangle \oplus \omega_{n+1} g_{n+1} \\ &= \left\langle \begin{array}{l} \left[\begin{array}{l} l_{z-z\prod_{k=1}^{n}(1-\frac{a_k}{z})^{\omega_k} + z-z(1-\frac{a_{n+1}}{z})^{\omega_{n+1}} - \frac{[z-z\prod_{k=1}^{n}(1-\frac{a_k}{z})^{\omega_k}] \times [z-z(1-\frac{a_{n+1}}{z})^{\omega_{n+1}}]}{z}}, \\ l_{z-z\prod_{k=1}^{n}(1-\frac{b_k}{z})^{\omega_k} + z-z(1-\frac{b_{n+1}}{z})^{\omega_{n+1}} - \frac{[z-z\prod_{k=1}^{n}(1-\frac{b_k}{z})^{\omega_k}] \times [z-z(1-\frac{b_{n+1}}{z})^{\omega_{n+1}}]}{z}} \end{array} \right]; \\ l_{\frac{\sum_{k=1}^{n}\omega_k T_k m_k + \omega_{n+1} T_{n+1} m_{n+1}}{\sum_{k=1}^{n}\omega_k m_k + \omega_{n+1} m_{n+1}}}, l_{\frac{\sum_{k=1}^{n}\omega_k I_k m_k + \omega_{n+1} I_{n+1} m_{n+1}}{\sum_{k=1}^{n}\omega_k m_k + \omega_{n+1} m_{n+1}}}, l_{\frac{\sum_{k=1}^{n}\omega_k F_k m_k + \omega_{n+1} F_{n+1} m_{n+1}}{\sum_{k=1}^{n}\omega_k m_k + \omega_{n+1} m_{n+1}}} \end{array} \right\rangle \\ &= \left\langle \left[l_{z-z\prod_{k=1}^{n+1}(1-\frac{a_k}{z})^{\omega_k}}, l_{z-z\prod_{k=1}^{n+1}(1-\frac{b_k}{z})^{\omega_k}} \right]; l_{\frac{\sum_{k=1}^{n+1}\omega_k T_k m_k}{\sum_{k=1}^{n+1}\omega_k m_k}}, l_{\frac{\sum_{k=1}^{n+1}\omega_k I_k m_k}{\sum_{k=1}^{n+1}\omega_k m_k}}, l_{\frac{\sum_{k=1}^{n+1}\omega_k F_k m_k}{\sum_{k=1}^{n+1}\omega_k m_k}} \right\rangle. \end{aligned}$$

Corresponding to the above results, Equation (9) can hold for any k. This proof is completed. □

To illustrate the operational process of the SVLN-ILNWAA operator, we give the following example.

Example 4. *Suppose* $g_1 = \langle [l_5, l_6]; l_5, l_2, l_1 \rangle$, $g_2 = \langle [l_5, l_7]; l_6, l_3, l_1 \rangle$, *and* $g_3 = \langle [l_6, l_7]; l_7, l_3, l_3 \rangle$ *are three SVLN-ILNs in the LT set* $L = \{l_s \mid s \in [0, 8]\}$ *for* $z = 8$, *then their weigh vector is* $\omega = (0.32, 0.25, 0.43)$.

Thus, there are $m_1 = (5 + 6)/2 = 5.5$, $m_2 = (5 + 7)/2 = 6$, and $m_3 = (6 + 7)/2 = 6.5$.

Using Equation (9), their operational result of the SVLN-ILNWAA operator is given below:

$$SVLN-ILNWAA(g_1,g_2,g_3)=\left\langle\left[l_{z-z\prod\limits_{k=1}^{3}(1-\frac{a_k}{z})^{\omega_k}},l_{z-z\prod\limits_{k=1}^{3}(1-\frac{b_k}{z})^{\omega_k}}\right];l_{\frac{\sum_{k=1}^{3}\omega_kT_km_k}{\sum_{k=1}^{n}\omega_km_k}},l_{\frac{\sum_{k=1}^{3}\omega_kI_km_k}{\sum_{k=1}^{n}\omega_km_k}},l_{\frac{\sum_{k=1}^{3}\omega_kF_km_k}{\sum_{k=1}^{n}\omega_km_k}}\right\rangle$$
$$=\left\langle\begin{array}{l}\left[l_{8-8\times(1-5/8)^{0.32}\times(1-5/8)^{0.25}\times(1-6/8)^{0.43}},l_{8-8\times(1-6/8)^{0.32}\times(1-7/8)^{0.25}\times(1-7/8)^{0.43}}\right];\\ l_{\frac{0.32\times5\times5.5+0.25\times6\times6+0.43\times7\times6.5}{0.32\times5.5+0.25\times6+0.43\times6.5}},l_{\frac{0.32\times2\times5.5+0.25\times3\times6+0.43\times3\times6.5}{0.32\times5.5+0.25\times6+0.43\times6.5}},l_{\frac{0.32\times1\times5.5+0.25\times1\times6+0.43\times3\times6.5}{0.32\times5.5+0.25\times6+0.43\times6.5}}\end{array}\right\rangle$$
$$=\langle[l_{5.4800},l_{6.7517}];l_{6.1709},l_{2.7093},l_{1.9232}\rangle.$$

Obviously, their operational result of the SVLN-ILNWAA operator is also an SVLN-ILN and all the LT values in it still belong to *L*.

Theorem 2. *Suppose* $g_k=\langle[l_{a_k},l_{b_k}];l_{T_k},l_{I_k},l_{F_k}\rangle$ *(k =1, 2, ..., n) is a group of SVLN-ILNs in L. Thus, the SVLN-ILNWAA operator implies these properties:*

(1) Idempotency: Set g_k (k = 1, 2, ..., n) as a group of SVLN-ILNs in L. If $g_k = g$ for k = 1, 2, ..., n, then there exists $SVLN-ILNWAA(g_1,g_2,\cdots,g_n)=g$.

(2) Boundedness: Suppose g_k (k = 1, 2, ..., n) is a group of SVLN-ILNs in L. Let the minimum SVLN-ILN be $g^-=\left\langle\left[\min\limits_k f(l_{a_k}),\min\limits_k(l_{b_k})\right];\min\limits_k f(l_{T_k}),\max\limits_k f(l_{I_k}),\max\limits_k f(l_{F_k})\right\rangle$ and the maximum SVLN-ILN be $g^+=\left\langle\left[\max\limits_k f(l_{a_k}),\max\limits_k f(l_{b_k})\right],\max\limits_k(l_{T_k}),\min\limits_k(l_{I_k}),\min\limits_k(l_{F_k})\right\rangle$. Then, $g^-\le SVLN-ILNWAA(g_1,g_2,\cdots,g_n)\le g^+$ can hold.

(3) Monotonicity: Suppose g_k (k = 1, 2, ..., n) is a group of SVLN-ILNs in L. If $g_k\le g_k^*$ for k = 1, 2, ..., n, then $SVLN-ILNWAA(g_1,g_2,\cdots,g_n)\le SVLN-ILNWAA(g_1^*,g_2^*,\cdots,g_n^*)$ can hold.

Proof:

(1) Because $g_k = g$ for k = 1, 2, ..., n, there is the following result:

$$SVLN-ILNWAA(g_1,g_2,\ldots,g_n)=\sum_{k=1}^{n}\omega_kg_k$$
$$=\left\langle\left[l_{z-z\prod\limits_{k=1}^{n}(1-\frac{a_k}{z})^{\omega_k}},l_{z-z\prod\limits_{k=1}^{n}(1-\frac{b_k}{z})^{\omega_k}}\right];l_{\frac{\sum_{k=1}^{n}\omega_kT_km_k}{\sum_{k=1}^{n}\omega_km_k}},l_{\frac{\sum_{k=1}^{n}\omega_kI_km_k}{\sum_{k=1}^{n}\omega_km_k}},l_{\frac{\sum_{k=1}^{n}\omega_kF_km_k}{\sum_{k=1}^{n}\omega_km_k}}\right\rangle$$
$$=\left\langle\left[l_{z-z(1-\frac{a}{z})^{\sum_{k=1}^{n}\omega_k}},l_{z-z(1-\frac{b}{z})^{\sum_{k=1}^{n}\omega_k}}\right];l_T,l_I,l_F\right\rangle=\left\langle\left[l_{z-z(1-\frac{a}{z})},l_{z-z(1-\frac{b}{z})}\right];l_T,l_I,l_F\right\rangle$$
$$=\langle[l_a,l_b];l_T,l_I,l_F\rangle=g.$$

(2) Because g^- is the minimum SVLN-ILN and g^+ is the maximum SVLN-ILN, $g^-\le g_k\le g^+$ holds. Hence, $\sum\limits_{k=1}^{n}\omega_jg^-\le\sum\limits_{k=1}^{n}\omega_kg_k\le\sum\limits_{k=1}^{n}\omega_kg^+$ can hold. There exists $g^-\le\sum\limits_{k=1}^{n}\omega_kg_k\le g^+$ according to the property (1), that is, $g^-\le SVLN-ILNWAA(g_1,g_2,\cdots,g_n)\le g^+$.

(3) For $g_k\le g_k^*$ (k = 1, 2, ..., n), $\sum\limits_{k=1}^{n}\omega_kg_k\le\sum\limits_{k=1}^{n}\omega_kg_k^*$ can hold, that is, $SVLN-ILNWAA(g_1,g_2,\cdots,g_n)\le SVLN-ILNWAA(g_1^*,g_2^*,\cdots,g_n^*)$.

Thus, the proof of these properties is finished. □

Especially when $\omega_k = 1/n$ for k = 1, 2, ..., n, the SVLN-ILNWAA operator reduces to the SVLN-ILN arithmetic average operator.

3.2. SVLN-ILNWGA Operator

Definition 6. *Suppose* $g_k = \langle [l_{a_k}, l_{b_k}]; l_{T_k}, l_{I_k}, l_{F_k} \rangle$ *(k = 1, 2, ..., n) is a group of SVLN-ILNs in L. Then, we give the following definition of the SVLN-ILNWGA operator:*

$$SVLN - ILNWGA(g_1, g_2, \cdots, g_n) = \prod_{k=1}^{n} g_k^{\omega_k} \tag{11}$$

where $\omega_k \in [0, 1]$ is the weight of g_k ($k = 1, 2, \ldots, n$) and $\sum_{k=1}^{n} \omega_k = 1$.

Then, we can give the following theorem based on Equations (2), (4), and (11).

Theorem 3. *Suppose* $g_k = \langle [l_{a_k}, l_{b_k}]; l_{T_k}, l_{I_k}, l_{F_k} \rangle$ *(k =1, 2, ..., n) is a group of SVLN-ILNs in L. Thus, the aggregation result of Equation (11) is also an SVLN-ILN, which is obtained by the aggregation form:*

$$\begin{aligned} & SVLN - ILNWGA(g_1, g_2, \cdots, g_n) = \prod_{k=1}^{n} g_k^{\omega_k} \\ & = \left\langle \left[l_{z\prod_{k=1}^{n}(\frac{a_k}{z})^{\omega_k}}, l_{z\prod_{k=1}^{n}(\frac{b_k}{z})^{\omega_k}} \right]; l_{z\prod_{k=1}^{n}(\frac{T_k}{z})^{\omega_k}}, l_{z-z\prod_{k=1}^{n}(1-\frac{I_k}{z})^{\omega_k}}, l_{z-z\prod_{k=1}^{n}(1-\frac{F_k}{z})^{\omega_k}} \right\rangle. \end{aligned} \tag{12}$$

Similar to the proof of Theorem 1, Theorem 3 can also be proved below.

Proof:

If $k = 2$, by Equation (4) we get

$$g_1^{\omega_1} = \left\langle \left[l_{z(\frac{a_1}{z})^{\omega_1}}, l_{z(\frac{b_1}{z})^{\omega_1}} \right]; l_{z(\frac{T_1}{z})^{\omega_1}}, l_{z-z(1-\frac{I_1}{z})^{\omega_1}}, l_{z-z(1-\frac{F_1}{z})^{\omega_1}} \right\rangle,$$

$$g_2^{\omega_2} = \left\langle \left[l_{z(\frac{a_2}{z})^{\omega_2}}, l_{z(\frac{b_2}{z})^{\omega_2}} \right]; l_{z(\frac{T_2}{z})^{\omega_2}}, l_{z-z(1-\frac{I_2}{z})^{\omega_2}}, l_{z-z(1-\frac{F_2}{z})^{\omega_2}} \right\rangle.$$

Using Equation (2), there exists the following result:

$$\begin{aligned} \prod_{k=1}^{2} g_k^{\omega_k} &= \left\langle \begin{array}{l} \left[l_{\frac{z(\frac{a_1}{z})^{\omega_1} \times z(\frac{a_2}{z})^{\omega_2}}{z}}, l_{\frac{z(\frac{b_1}{z})^{\omega_1} \times z(\frac{b_2}{z})^{\omega_2}}{z}} \right]; l_{\frac{z(\frac{T_1}{z})^{\omega_1} \times z(\frac{T_2}{z})^{\omega_2}}{z}}, \\ l_{z-z(\frac{I_1}{z})^{\omega_1} + z - z(\frac{I_2}{z})^{\omega_2} - \frac{[z-z(1-\frac{I_1}{z})^{\omega_1}] \times [z-z(1-\frac{I_2}{z})^{\omega_2}]}{z}}, \\ l_{z-z(\frac{F_1}{z})^{\omega_1} + z - z(\frac{F_2}{z})^{\omega_2} - \frac{[z-z(1-\frac{F_1}{z})^{\omega_1}] \times [z-z(1-\frac{F_2}{z})^{\omega_2}]}{z}} \end{array} \right\rangle \\ &= \left\langle \left[l_{z\prod_{k=1}^{2}(\frac{a_k}{z})^{\omega_k}}, l_{z\prod_{k=1}^{2}(\frac{b_k}{z})^{\omega_k}} \right]; l_{z\prod_{k=1}^{2}(\frac{T_k}{z})^{\omega_k}}, l_{z-z(1-\frac{I_1}{z})^{\omega_1} \times (1-\frac{I_2}{z})^{\omega_2}}, l_{z-z(1-\frac{F_1}{z})^{\omega_1} \times (1-\frac{F_2}{z})^{\omega_2}} \right\rangle \\ &= \left\langle \left[l_{z\prod_{k=1}^{2}(\frac{a_k}{z})^{\omega_k}}, l_{z\prod_{k=1}^{2}(\frac{b_k}{z})^{\omega_k}} \right]; l_{z\prod_{k=1}^{2}(\frac{T_k}{z})^{\omega_k}}, l_{z-z\prod_{k=1}^{2}(1-\frac{I_k}{z})^{\omega_k}}, l_{z-z\prod_{k=1}^{2}(1-\frac{F_k}{z})^{\omega_1}} \right\rangle. \end{aligned} \tag{13}$$

If $k = n$, Equation (12) exists as the following result:

$$\begin{aligned} & SVLN - ILNWGA(g_1, g_2, \cdots, g_n) = \prod_{k=1}^{n} g_k^{\omega_k} \\ & = \left\langle \left[l_{z\prod_{k=1}^{n}(\frac{a_k}{z})^{\omega_k}}, l_{z\prod_{k=1}^{n}(\frac{b_k}{z})^{\omega_k}} \right]; l_{z\prod_{k=1}^{n}(\frac{T_k}{z})^{\omega_k}}, l_{z-z\prod_{k=1}^{n}(1-\frac{I_k}{z})^{\omega_k}}, l_{z-z\prod_{k=1}^{n}(1-\frac{F_k}{z})^{\omega_k}} \right\rangle. \end{aligned}$$

Thus if $k = n + 1$, by Equations (2), (4), and (13), we yield the following result:

$$SVLN - ILNWGA(g_1, g_2, \ldots, g_n, g_{n+1}) = \prod_{k=1}^{n+1} g_k^{\omega_k} = \prod_{k=1}^{n} g_k^{\omega_k} \otimes g_{n+1}^{\omega_{n+1}}$$

$$= \left\langle \begin{array}{l} \left[l_{\frac{z\prod_{k=1}^{n}(\frac{a_k}{z})^{\omega_k} \times z(\frac{a_{n+1}}{z})^{\omega_{n+1}}}{z}}, l_{\frac{z\prod_{k=1}^{n}(\frac{b_k}{z})^{\omega_k} \times z(\frac{b_{n+1}}{z})^{\omega_{n+1}}}{z}} \right]; l_{\frac{z\prod_{k=1}^{n}(\frac{T_k}{z})^{\omega_k} \times z(\frac{T_{n+1}}{z})^{\omega_{n+1}}}{z}}, \\ l_{(z - z\prod_{k=1}^{n}(1-\frac{I_k}{z})^{\omega_k}) + (z - z(1-\frac{I_{n+1}}{z})^{\omega_{n+1}}) - (z - z\prod_{k=1}^{n}(1-\frac{I_k}{z})^{\omega_k}) \times (z - z(1-\frac{I_{n+1}}{z})^{w_{n+1}})}, \\ l_{(z - z\prod_{k=1}^{n}(1-\frac{F_k}{z})^{\omega_k}) + (z - z(1-\frac{F_{n+1}}{z})^{\omega_{n+1}}) - (z - z\prod_{k=1}^{n}(1-\frac{F_k}{z})^{\omega_k}) \times (z - z(1-\frac{F_{n+1}}{z})^{w_{n+1}})} \end{array} \right\rangle$$

$$= \left\langle \begin{array}{l} \left[l_{z\prod_{k=1}^{n+1}(\frac{a_k}{z})^{\omega_k}}, l_{z\prod_{k=1}^{n+1}(\frac{b_k}{z})^{\omega_k}} \right]; l_{z\prod_{k=1}^{n+1}(\frac{T_k}{z})^{\omega_k}}, \\ l_{(z - z\prod_{k=1}^{n}(1-\frac{I_k}{z})^{\omega_k} \times (1-\frac{I_{n+1}}{z})^{\omega_{n+1}})}, l_{(z - z\prod_{k=1}^{n}(1-\frac{F_k}{z})^{\omega_k}(1-\frac{F_{n+1}}{z})^{\omega_{n+1}})} \end{array} \right\rangle$$

$$= \left\langle \left[l_{z\prod_{k=1}^{n+1}(\frac{a_k}{z})^{\omega_k}}, l_{z\prod_{k=1}^{n+1}(\frac{b_k}{z})^{\omega_k}} \right]; l_{z\prod_{k=1}^{n+1}(\frac{T_k}{z})^{\omega_k}}, l_{z - z\prod_{k=1}^{n+1}(1-\frac{I_k}{z})^{\omega_k}}, l_{z - z\prod_{k=1}^{n+1}(1-\frac{F_k}{z})^{\omega_k}} \right\rangle.$$

Based on the above results, Equation (12) exists for any k. This proof is finished. □

Example 5. *Consider Example 4 to calculate the aggregation result of the SVLN-ILNWGA operator.*

By Equation (12), the calculation process is shown as follows:

$$SVLN - ILNWGA(g_1, g_2, g_3) = \left\langle \left[l_{z\prod_{k=1}^{3}(\frac{a_k}{z})^{\omega_k}}, l_{z\prod_{k=1}^{3}(\frac{b_k}{z})^{\omega_k}} \right]; l_{z\prod_{k=1}^{3}(\frac{T_k}{z})^{\omega_k}}, l_{z - z\prod_{k=1}^{3}(1-\frac{I_k}{z})^{\omega_k}}, l_{z - z\prod_{k=1}^{3}(1-\frac{F_k}{z})^{\omega_k}} \right\rangle$$

$$= \left\langle \begin{array}{l} \left[l_{8 \times (5/8)^{0.32} \times (5/8)^{0.25} \times (6/8)^{0.43}}, l_{8 \times (6/8)^{0.32} \times (7/8)^{0.25} \times (7/8)^{0.43}} \right]; \\ l_{8 \times (5/8)^{0.32} \times (6/8)^{0.25} \times (7/8)^{0.43}}, l_{8 - 8 \times (1-2/8)^{0.32} \times (1-3/8)^{0.25} \times (1-3/8)^{0.43}}, l_{8 - 8 \times (1-1/8)^{0.32} \times (1-1/8)^{0.25} \times (1-3/8)^{0.43}} \end{array} \right\rangle$$

$$= \langle [l_{5.4078}, l_{6.6631}]; l_{6.0478}, l_{2.6996}, l_{1.9429} \rangle.$$

Obviously, their operational result of the SVLN-ILNWGA operator is also a SVLN-ILN and all the LT values in it still belong to L.

Theorem 4. *Suppose* $g_k = \langle [l_{a_k}, l_{b_k}]; l_{T_k}, l_{I_k}, l_{F_k} \rangle$ *(k =1, 2, ... , n) is a group of SVLN-ILNs in L. Thus, the SVLN-ILNWGA operator indicates these properties:*

(1) Idempotency: Suppose g_k (k = 1, 2, ... , n) is a group of SVLN-ILNs in L. If $g_k = g$ for k = 1, 2, ... , n, then there exists $SVLN - ILNWGA(g_1, g_2, \cdots, g_n) = g$.

(2) Boundedness: Suppose g_k (k = 1, 2, ... , n) is a group of SVLN-ILNs in L. Let the minimum SVLN-ILN be $g^- = \left\langle \left[\min_k f(l_{a_k}), \min_k(l_{b_k}) \right]; \min_k f(l_{T_k}), \max_k f(l_{I_k}), \max_k f(l_{F_k}) \right\rangle$ and the maximum SVLN-ILN be $g^+ = \left\langle \left[\max_k f(l_{a_k}), \max_k f(l_{b_k}) \right], \max_k(l_{T_k}), \min_k(l_{I_k}), \min_k(l_{F_k}) \right\rangle$. Then, $g^- \leq SVLN - ILNWGA(g_1, g_2, \cdots, g_n) \leq g^+$ can hold.

(3) Monotonicity: Suppose g_k (k = 1, 2, ... , n) is a group of SVLN-ILNs in L. If $g_k \leq g_k^*$ for k = 1, 2, ... , n, then $SVLN - ILNWGA(g_1, g_2, \cdots, g_n) \leq SVLN - ILNWGA(g_1^*, g_2^*, \cdots, g_n^*)$ can hold.

Similar to the proof of Theorem 2, these properties of the SVLN-ILNWGA operator can be also proved, and then the proof of these properties is not repeated here.

4. MADM Method Based on the SVLN-ILNWAA or SVLN-ILNWGA Operator

In the SVLN-ILN setting, we present a MADM method using the SVLN-ILNWAA or SVLN-ILNWGA operator and the score function to handle SVLN-ILN decision making problems corresponding to the pessimistic, moderate, and optimistic attitudes of DMs.

In an SVLN-ILN MADM problem, suppose $G = \{G_1, G_2, \ldots, G_m\}$ and $Q = \{Q_1, Q_2, \ldots, Q_n\}$ are represented as a set of alternatives and a set of attributes, respectively. The attribute weigh vector of Q_k ($k = 1, 2, \ldots, n$) is $\omega = (\omega_1, \omega_2, \ldots, \omega_n)$ with $\sum_{k=1}^{n} \omega_k = 1$. Then, the attributes Q_k ($k = 1, 2, \ldots, n$) over the alternatives G_j ($j = 1, 2, \ldots, m$) will be evaluated by DMs, which are expressed by SVLN-ILNs from some predefined LT set $L = \{l_s \mid s \in [0, z]\}$ regarding an even number z. In the linguistic evaluation, DM can assign an ILN as the uncertain linguistic argument and an SVLNN as the confident linguistic argument in each SVLN-ILN, so as to give the SVLN-ILN evaluation value of each attribute Q_k ($k = 1, 2, \ldots, n$) over the alternatives G_j ($j = 1, 2, \ldots, m$) regarding the LTs. Hence, all SVLN-ILNs can be established as a SVLN-ILN decision matrix $G = (g_{jk})_{m\times n}$, where $g_{jk} = \left\langle [l_{a_{jk}}, l_{b_{jk}}]; l_{T_{jk}}, l_{I_{jk}}, l_{F_{jk}} \right\rangle$ ($j = 1, 2, \ldots, m$; $k = 1, 2, \ldots, n$) is a SVLN-ILN.

Hence, the MADM method using the SVLN-ILNWAA or SVLN-ILNWGA operator and the score function is indicated as the following decision procedure:

Step 1: Compute the aggregated SVLN-ILN g_i = SVLN-ILNWAA($g_{j1}, g_{j2}, \ldots, g_{jn}$) or g_j = SVLN-ILNWGA($g_{j1}, g_{j2}, \ldots, g_{jn}$) ($j = 1, 2, \ldots, m$) based on Equation (9) or Equation (12) for G_j ($j = 1, 2, \ldots, m$).

Step 2: Calculate the score value of $Y(g_j)$ for each g_j ($j = 1, 2, \ldots, m$) by Equation (7).

Step 3: Rank the alternatives regarding the score values in a descending order and choose the best one.

Step 4: End.

5. Actual Example and Discussion

In this section, an actual example is provided to illustrate the applicability of the established MADM method in the SVLN-ILN setting, and then discuss that DMs' attitudes can affect the ranking orders of alternatives and the optimal choice.

5.1. Actual Example

In linguistic decision making environment, let us consider that some software company wants to hire a software engineer, which is adapted from the literature [22]. Then, the human resources department preliminarily chooses the four candidates (alternatives) G_1, G_2, G_3, and G_4 from all applicants, and then they require further evaluation by the four attributes: soft skill (Q_1), past experience (Q_2), personality (Q_3), and self-confidence (Q_4). A group of experts or DMs is requested to choose the best candidate by the interview. Then, the weigh vector ω = (0.35, 0.25, 0.2, 0.2) indicates the importance of the four attributes. Thus, the DMs assess the four possible candidates G_j (j = 1, 2, 3, 4) over the four attributes Q_k (k = 1, 2, 3, 4) by SVLN-ILNs from the given LT set $L = \{l_s \mid s \in [0, z]\}$, where $L = \{l_0$: extremely poor, l_1: very poor, l_2: poor, l_3: slightly poor, l_4: fair, l_5: slightly good, l_6: good, l_7: very good, l_8: extremely good$\}$. Thus, all the evaluated SVLN-ILNs can be constructed as the SVLN-ILN decision matrix:

$$G = (g_{jk})_{4\times 4} = \begin{bmatrix} \langle [l_4, l_6]; l_5, l_1, l_1 \rangle & \langle [l_4, l_6]; l_7, l_1, l_2 \rangle & \langle [l_4, l_7]; l_6, l_1, l_2 \rangle & \langle [l_5, l_6]; l_6, l_2, l_3 \rangle \\ \langle [l_3, l_5]; l_7, l_2, l_3 \rangle & \langle [l_5, l_7]; l_6, l_3, l_5 \rangle & \langle [l_4, l_6]; l_4, l_1, l_1 \rangle & \langle [l_6, l_7]; l_6, l_1, l_2 \rangle \\ \langle [l_4, l_7]; l_6, l_1, l_4 \rangle & \langle [l_6, l_7]; l_7, l_3, l_3 \rangle & \langle [l_5, l_7]; l_5, l_2, l_1 \rangle & \langle [l_5, l_7]; l_7, l_3, l_4 \rangle \\ \langle [l_6, l_7]; l_7, l_4, l_2 \rangle & \langle [l_5, l_7]; l_6, l_2, l_1 \rangle & \langle [l_4, l_6]; l_5, l_3, l_4 \rangle & \langle [l_5, l_6]; l_5, l_4, l_2 \rangle \end{bmatrix}.$$

On the one hand, the established MADM method based on the SVLN-ILNVWAA operator is used for the MADM problem with SVLN-ILN information. Thus, the decision procedure is presented below:

Step 1: Compute the aggregated value of g_1 for G_1 by Equation (9), which is shown as follows:

$$
\begin{aligned}
g_1 &= SVLN - ILNWAA(g_{11}, g_{12}, g_{13}, g_{14}) \\
&= \left\langle \left[l_{z - z\prod_{k=1}^{4}\left(1-\frac{a_{1k}}{z}\right)^{\omega_k}}, l_{z - z\prod_{k=1}^{4}\left(1-\frac{b_{1k}}{z}\right)^{\omega_k}} \right]; l_{\frac{\sum_{k=1}^{4}\omega_k T_{1k} m_{1k}}{\sum_{k=1}^{4}\omega_k m_{1k}}}, l_{\frac{\sum_{k=1}^{4}\omega_k I_{1k} m_{1k}}{\sum_{k=1}^{4}\omega_k m_{1k}}}, l_{\frac{\sum_{k=1}^{4}\omega_k F_{1k} m_{1k}}{\sum_{k=1}^{4}\omega_k m_{1k}}} \right\rangle \\
&= \left\langle \begin{array}{l} \left[l_{8-8\times(1-4/8)^{0.35}\times(1-4/8)^{0.25}\times(1-4/8)^{0.2}\times(1-5/8)^{0.2}}, l_{8-8\times(1-6/8)^{0.35}\times(1-6/8)^{0.25}\times(1-7/8)^{0.2}\times(1-6/8)^{0.2}} \right]; \\ l_{\frac{0.35\times5\times(4+6)/2+0.25\times7\times(4+6)/2+0.2\times6\times(4+7)/2+0.2\times6\times(5+6)/2}{0.35\times(4+6)/2+0.25\times(4+6)/2+0.2\times(4+7)/2+0.2\times(5+6)/2}}, \\ l_{\frac{0.35\times1\times(4+6)/2+0.25\times1\times(4+6)/2+0.2\times1\times(4+7)/2+0.2\times2\times(5+6)/2}{0.35\times(4+6)/2+0.25\times(4+6)/2+0.2\times(4+7)/2+0.2\times(5+6)/2}}, \\ l_{\frac{0.35\times1\times(4+6)/2+0.25\times2\times(4+6)/2+0.2\times2\times(4+7)/2+0.2\times3\times(5+6)/2}{0.35\times(4+6)/2+0.25\times(4+6)/2+0.2\times(4+7)/2+0.2\times(5+6)/2}} \end{array} \right\rangle \\
&= \langle [l_{4.2236}, l_{6.2589}]; l_{5.9038}, l_{1.2115}, l_{1.8750} \rangle.
\end{aligned}
$$

In the similar calculation manner, we can obtain other aggregated values of g_j for G_j (j = 2, 3, 4): g_2 = <[$l_{4.4962}$, $l_{6.3127}$], $l_{5.8846}$, $l_{1.8462}$, $l_{2.9423}$>, g_3 = <[$l_{5.002}$, l_7], $l_{6.2731}$, $l_{2.1513}$, $l_{3.1218}$>, and g_4 = <[$l_{5.2427}$, $l_{6.6805}$], $l_{6.0298}$, $l_{3.3191}$, $l_{2.0851}$>.

Step 2: Calculate the score value of $Y(g_1)$ by Equation (7) for $\alpha = 0.5$ (considering the moderate attitude of DMs):

$$
\begin{aligned}
Y(g_1) &= \left(\frac{a_1+b_1}{2z} + (2\alpha-1)\frac{b_1-a_1}{2z}\right) \times \left(\frac{2z+T_1-I_1-F_1}{3z}\right) \\
&= \left(\frac{4.2236+6.2589}{2\times8} + (2\times0.5-1)\frac{6.2589-4.2236}{2\times8}\right) \times \left(\frac{2\times8+5.9038-1.2115-1.8750}{3\times8}\right) \\
&= 0.5137.
\end{aligned}
\tag{14}
$$

In the similar calculation manner, we can obtain other score values of g_j for G_j (j = 2, 3, 4): $Y(g_2) = 0.4812$, $Y(g_3) = 0.5313$, and $Y(g_4) = 0.5162$.

Step 3: Rank the four alternatives as $G_3 \succ G_4 \succ G_1 \succ G_2$ based on the score values, and then choose G_3 as the best candidate among the four candidates.

On the other hand, the established MADM method based on the SVLN-ILNWGA operator is also used for the MADM problem with SVLN-ILN information. Thus, the decision procedure is also presented as follows:

Step 1′: Compute the aggregated value of g_1 for G_1 by Equation (12):

$$
\begin{aligned}
g_1 &= SVLN - ILNWGA(g_{11}, g_{12}, g_{13}, g_{14}) = \left\langle \left[l_{z\prod_{k=1}^{4}\left(\frac{a_{1k}}{z}\right)^{\omega_k}}, l_{z\prod_{k=1}^{4}\left(\frac{b_{1k}}{z}\right)^{\omega_k}} \right]; l_{z\prod_{k=1}^{4}\left(\frac{T_{1k}}{z}\right)^{\omega_k}}, l_{z - z\prod_{k=1}^{4}\left(1-\frac{I_{1k}}{z}\right)^{\omega_k}}, l_{z - z\prod_{k=1}^{4}\left(1-\frac{F_{1k}}{z}\right)^{\omega_k}} \right\rangle \\
&= \left\langle \begin{array}{l} \left[l_{8\times(4/8)^{0.35}\times(4/8)^{0.25}\times(4/8)^{0.2}\times(5/8)^{0.2}}, l_{8\times(6/8)^{0.35}\times(6/8)^{0.25}\times(7/8)^{0.2}\times(6/8)^{0.2}} \right]; \\ l_{8\times(5/8)^{0.35}\times(7/8)^{0.25}\times(6/8)^{0.2}\times(6/8)^{0.2}}, l_{8-8\times(1-1/8)^{0.35}\times(1-1/8)^{0.25}\times(1-1/8)^{0.2}\times(1-2/8)^{0.2}}, l_{8-8\times(1-1/8)^{0.35}\times(1-2/8)^{0.25}\times(1-2/8)^{0.2}\times(1-3/8)^{0.2}} \end{array} \right\rangle \\
&= \langle [l_{4.1826}, l_{6.1879}]; l_{5.8503}, l_{1.2125}, l_{1.8941} \rangle.
\end{aligned}
$$

In the similar calculation manner, we can obtain other aggregated values of g_j for G_j (j = 2, 3, 4): g_2 = <[$l_{4.1474}$, $l_{6.0334}$], $l_{5.8393}$, $l_{1.9027}$, $l_{3.1183}$>, g_3 = <[$l_{4.84}$, l_7], $l_{6.2007}$, $l_{2.1662}$, $l_{3.2696}$>, and g_4 = <[$l_{5.0968}$, $l_{6.5814}$], $l_{5.8872}$, $l_{3.3712}$, $l_{2.25}$>.

Step 2′: Calculate the score value of $Y(g_1)$ by Equation (7) for $\alpha = 0.5$ (considering the moderate attitude of DMs):

$$
\begin{aligned}
Y(g_1) &= \left(\frac{a_1+b_1}{2z} + (2\alpha-1)\frac{b_1-a_1}{2z}\right)\left(\frac{2z+T_1-I_1-F_1}{3z}\right) \\
&= \left(\frac{4.1826+6.1879}{2\times8} + (2\times0.5-1)\frac{6.1879-4.1826}{2\times8}\right)\left(\frac{2\times8+5.8503-1.2125-1.8941}{3\times8}\right) \\
&= 0.5062.
\end{aligned}
$$

In the similar calculation manner, we can obtain other score values of g_j for G_j (j = 2, 3, 4): $Y(g_2) = 0.4459$, $Y(g_3) = 0.5169$, and $Y(g_4) = 0.4947$.

Step 3′: Rank the four alternatives as $G_3 \succ G_1 \succ G_4 \succ G_2$ based on the score values, and then choose G_3 as the best candidate among the four candidates.

Clearly, the best candidate G_3 is identical although there exists a little difference between two kinds of ranking orders obtained by using the SVLN-ILNWAA and SVLN-ILNWGA operators under the DMs' moderate attitude.

5.2. Results and Discussion

Let us consider that the pessimistic, moderate, and optimistic attitudes of DMs may affect their ranking orders. Based on the above similar computational steps, all the decision results based on the SVLN-ILNWAA and SVLN-ILNWGA operators and the DMs' attitudes are shown in Tables 1 and 2.

Table 1. Decision results corresponding to the single-valued linguistic neutrosophic interval linguistic number weighted arithmetic averaging (SVLN-ILNWAA) operator and the decision makers' (DMs') attitudes.

DMs' Attitude	Score Value	Ranking Order
Pessimist ($\alpha = 0$)	$Y(g_1) = 0.4139$, $Y(g_2) = 0.4004$, $Y(g_3) = 0.4429$, $Y(g_4) = 0.4540$	$G_4 \succ G_3 \succ G_1 \succ G_2$
Moderate ($\alpha = 0.5$)	$Y(g_1) = 0.5137$, $Y(g_2) = 0.4812$, $Y(g_3) = 0.5313$, $Y(g_4) = 0.5162$	$G_3 \succ G_4 \succ G_1 \succ G_2$
Optimist ($\alpha = 1$)	$Y(g_1) = 0.6134$, $Y(g_2) = 0.5621$, $Y(g_3) = 0.6198$, $Y(g_4) = 0.5785$	$G_3 \succ G_1 \succ G_4 \succ G_2$

Table 2. Decision results corresponding to the SVLN-ILN weighted geometric averaging (SVLN-ILNWGA) operator and the DMs' attitudes.

DMs' Attitude	Score Value	Ranking Order
Pessimist ($\alpha = 0$)	$Y(g_1) = 0.4083$, $Y(g_2) = 0.3633$, $Y(g_3) = 0.4226$, $Y(g_4) = 0.4318$	$G_4 \succ G_3 \succ G_1 \succ G_2$
Moderate ($\alpha = 0.5$)	$Y(g_1) = 0.5062$, $Y(g_2) = 0.4459$, $Y(g_3) = 0.5169$, $Y(g_4) = 0.4947$	$G_3 \succ G_1 \succ G_4 \succ G_2$
Optimist ($\alpha = 1$)	$Y(g_1) = 0.6041$, $Y(g_2) = 0.5285$, $Y(g_3) = 0.6112$, $Y(g_4) = 0.5576$	$G_3 \succ G_1 \succ G_4 \succ G_2$

Obviously, two kinds of ranking orders based on the SVLN-ILNWGA and SVLN-ILNWGA operators in Tables 1 and 2 are identical under the DMs' pessimistic or optimistic attitudes, but the DMs' attitudes can affect the ranking orders. Then, the best candidate is G_4 for pessimist or G_3 for optimist and the worst one is G_2 in all ranking orders. Hence, the established SVLN-ILN MADM method shows its sensitivity and flexibility regarding the DMs' attitudes, which depend on their preference.

As the decision information in this study uses the SVLN-ILN that is composed of ILN (uncertain/interval linguistic argument part) and SVLNN (confident linguistic argument part) for the first time, the SVLN-ILN MADM method is established for the first time because there is no other study in existing literature. Therefore, existing various linguistic MADM methods cannot carry out such a decision making problem with SVLN-ILN information in this paper.

Generally, this study indicates a new concept of SVLN-ILN and a new SVLN-ILN MADM method, and then DMs can choose one of the SVLN-ILNWAA and the SVLN-ILNWGA operators to apply the established MADM method to MADM problems with SVLN-ILN information and their preference attitude or actual requirements.

6. Conclusions

This study proposed the SVLN-ILN concept to express the hybrid information of both a single-valued LNN and an ILN, the operational laws of SVLN-ILNs, and the score function of SVLN-ILN, along with the attitude index and confident degree for ranking SVLN-ILNs. Then, the SVLN-ILNWAA and SVLN-ILNWGA operators were presented in order to aggregate SVLN-ILN information, and then their advantage is that all the LT values in their aggregated SVLN-ILN can still belong to the predefined LT set, rather than beyond the LT set in some linguistic operations [8,9]. It is well known that the two weighted aggregation operators are not only the most basic and simplest operations, but also two main mathematical tools in MADM problems. Hence, an MADM method was established based on the SVLN-ILNWAA or SVLN-ILNWGA operator and the score function so as to handle MADM problems with SVLN-ILN information and DMs' attitudes, which existing MADM methods cannot handle. By an actual example, the decision results illustrated the applicability of the established MADM method in the SVLN-ILN setting.

This study proposed for the first time the expression and score problems of hybrid information of both the SVLN number and ILN using the SVLN-ILN and the weighted aggregation problems of SVLN-ILNs to realize MADM problems with both interval/uncertain linguistic arguments (linguistic uncertainty) and linguistic neutrosophic arguments (confident level/degree). Then, the established SVLN-ILN MADM method contains much more linguistic information (interval/uncertain linguistic arguments and confident linguistic arguments) and indicates its flexibility for DMs' preference attitudes along with pessimist, moderate, and optimist in linguistic the decision making process, which are the main advantages in this study. From the viewpoint of scientific potential impact, the proposed technologies will be extended to medical diagnosis, hospital service quality evaluation, selection of suppliers, machining process selection, and so on.

Author Contributions: J.Y. proposed the SVLN-ILN concept, the SVLN-ILNWAA and SVLN-ILNWGA operators, the score function, and the MADM method. W.C. provided the calculation of examples and comparative analysis. All authors wrote the paper together.

Funding: This research received no external funding.

Acknowledgments: This study was supported by the National Natural Science Foundation of China (Nos. 71471172, 61703280).

Conflicts of Interest: The authors declare no conflict of interest.

References

1. Chatterjee, P.; Mondal, S.; Boral, S.; Banerjee, A.; Chakraborty, S. A novel hybrid method for non-traditional machining process selection using factor relationship and multi-attributive border approximation method. *Facta Univ. Ser. Mech. Eng.* **2017**, *15*, 439–456. [CrossRef]
2. Petković, D.; Madić, M.; Radovanović, M.; Gečevska, V. Application of the performance selection index method for solving machining mcdm problems. *Facta Univ. Ser. Mech. Eng.* **2017**, *15*, 97–106. [CrossRef]
3. Roy, J.; Adhikary, K.; Kar, S.; Pamučar, D. A rough strength relational DEMATEL model for analysing the key success factors of hospital service quality. *Decis. Mak. Appl. Manag. Eng.* **2018**, *1*, 121–142. [CrossRef]
4. Vasiljevic, M.; Fazlollahtabar, H.; Stević, Ž.; Vesković, S. A rough multicriteria approach for evaluation of the supplier criteria in automotive industry. *Decis. Mak. Appl. Manag. Eng.* **2018**, *1*, 82–96. [CrossRef]
5. Zadeh, L.A. The concept of a linguistic variable and its application to approximate reasoning Part I. *Inf. Sci.* **1975**, *8*, 199–249. [CrossRef]
6. Herrera, F.; Herrera-Viedma, E.; Verdegay, L. A model of consensus in group decision making under linguistic assessments. *Fuzzy Sets Syst.* **1996**, *79*, 73–87. [CrossRef]
7. Herrera, F.; Herrera-Viedma, E. Linguistic decision analysis: Steps for solving decision problems under linguistic information. *Fuzzy Sets Syst.* **2000**, *115*, 67–82. [CrossRef]
8. Xu, Z.S. A method based on linguistic aggregation operators for group decision making with linguistic preference relations. *Inf. Sci.* **2004**, *166*, 19–30. [CrossRef]
9. Xu, Z.S. A note on linguistic hybrid arithmetic averaging operator in multiple attribute group decision making with linguistic information. *Group Decis. Negot.* **2006**, *15*, 593–604. [CrossRef]
10. Merigó, J.M.; Casanovas, M.; Martínez, L. Linguistic aggregation operators for linguistic decision making based on the Dempster-Shafer theory of evidence. *Int. J. Uncertain. Fuzziness Knowl. Based Syst.* **2010**, *18*, 287–304. [CrossRef]

11. Xu, Y.J.; Merigó, J.M.; Wang, H.M. Linguistic power aggregation operators and their application to multiple attribute group decision making. *Appl. Math. Model.* **2012**, *36*, 5427–5444. [CrossRef]
12. Merigó, J.M.; Casanovas, M.; Palacios-Marqués, D. Linguistic group decision making with induced aggregation operators and probabilistic information. *Appl. Soft Comput.* **2014**, *24*, 669–678. [CrossRef]
13. Merigó, J.M.; Palacios-Marqués, D.; Zeng, S.Z. Subjective and objective information in linguistic multi-criteria group decision making. *Eur. J. Oper. Res.* **2016**, *248*, 522–531. [CrossRef]
14. Yu, D.J.; Li, D.F.; Merigó, J.M.; Fang, L.C. Mapping development of linguistic decision making studies. *J. Intell. Fuzzy Syst.* **2016**, *30*, 2727–2736. [CrossRef]
15. Xu, Z.S. Uncertain linguistic aggregation operators based approach to multiple attribute group decision making under uncertain linguistic environment. *Inf. Sci.* **2004**, *168*, 171–184. [CrossRef]
16. Xu, Z.S. Induced uncertain linguistic OWA operators applied to group decision making. *Inf. Fusion* **2006**, *7*, 231–238. [CrossRef]
17. Wei, G.W. Uncertain linguistic hybrid geometric mean operator and its application to group decision making under uncertain linguistic environment. *Int. J. Uncertain. Fuzziness Knowl. Based Syst.* **2009**, *17*, 251–267. [CrossRef]
18. Park, J.H.; Gwak, M.G.; Kwun, Y.C. Uncertain linguistic harmonic mean operators and their applications to multiple attribute group decision making. *Computing* **2011**, *93*, 47. [CrossRef]
19. Wei, G.W.; Zhao, X.F.; Lin, R.; Wang, H.J. Uncertain linguistic Bonferroni mean operators and their application to multiple attribute decision making. *Appl. Math. Model.* **2013**, *37*, 5277–5285. [CrossRef]
20. Zhang, H. Uncertain linguistic power geometric operators and their use in multiattribute group decision making. *Math. Probl. Eng.* **2015**, *2015*, 948380. [CrossRef]
21. Ye, J. Aggregation operators of neutrosophic linguistic numbers for multiple attribute group decision making. *SpringerPlus* **2016**, *5*, 1691. [CrossRef] [PubMed]
22. Ye, J. Multiple attribute decision-making method based on linguistic cubic variables. *J. Intell. Fuzzy Syst.* **2018**, *34*, 2351–2361. [CrossRef]
23. Fang, Z.B.; Ye, J. Multiple attribute group decision-making method based on linguistic neutrosophic numbers. *Symmetry* **2017**, *9*, 111. [CrossRef]
24. Wang, J.Q.; Han, Z.Q.; Zhang, H.Y. Multi-criteria group decision-making method based on intuitionistic interval fuzzy information. *Group Decis. Negot.* **2014**, *23*, 715–733. [CrossRef]
25. Wan, S.P. Method of attitude index for interval multi-attribute decision-making. *Control Decis.* **2009**, *24*, 35–38.

information

MDPI

Article

Dual Generalized Nonnegative Normal Neutrosophic Bonferroni Mean Operators and Their Application in Multiple Attribute Decision Making

Jiongmei Mo and Han-Liang Huang *

School of Mathematics and Statistics, Minnan Normal University, Zhangzhou 363000, Fujian, China; mojiongmei123@126.com

* Correspondence: huanghl@mnnu.edu.cn

Received: 19 July 2018; Accepted: 27 July 2018; Published: 6 August 2018

Abstract: For multiple attribute decision making, ranking and information aggregation problems are increasingly receiving attention. In a normal neutrosophic number, the ranking method does not satisfy the ranking principle. Moreover, the proposed operators do not take into account the correlation between any aggregation arguments. In order to overcome the deficiencies of the existing ranking method, based on the nonnegative normal neutrosophic number, this paper redefines the score function, the accuracy function, and partial operational laws. Considering the correlation between any aggregation arguments, the dual generalized nonnegative normal neutrosophic weighted Bonferroni mean operator and dual generalized nonnegative normal neutrosophic weighted geometric Bonferroni mean operator were investigated, and their properties are presented. Here, these two operators are applied to deal with a multiple attribute decision making problem. Example results show that the proposed method is effective and superior.

Keywords: multiple attribute decision making; nonnegative normal neutrosophic number; aggregation operator

1. Introduction

During the decision making process, the evaluation information given by decision makers is often incomplete, indeterminate, and inconsistent. To deal with this uncertain information, fuzzy set (FS) was proposed by Zadeh [1] in 1965. On the basis of FS, intuitionistic fuzzy set (IFS) was introduced by Atanassov [2] in 1986. However, IFS can not deal with all types of indeterminate and inconsistent information. Hence, considering the indeterminacy-membership based on IFS, Smarandache [3] developed the neutrosophic set (NS) in 1995. In NS, the truth-membership function, indeterminacy-membership function, and false-membership function are independent of each other. In real life, normal distribution is widely applied. Nevertheless, FS, IFS, and NS do not take the normal distribution into account. Therefore, the normal fuzzy number (NFN) was firstly introduced by Yang and Ko [4] in 1996, and NFN can deal with normal fuzzy information. Based on IFS and NFN, normal intuitionistic fuzzy number (NIFN) was defined by Wang and Li [5] in 2002. Further, combining NFN with NS, Liu [6] proposed the normal neutrosophic number (NNN).

With the development of society, many achievements have been made in the research of multiple attribute decision making (MADM) [7–10]. Chatterjee et al. [7] proposed a novel hybrid method encompassing factor relationship (FARE) and multi-attributive border approximation area comparison (MABAC) methods. Petković et al. [8] introduced the performance selection index (PSI) method for solving machining MADM problems. Roy et al. [9] developed a rough strength relational-decision making and trial evaluation laboratory model. Badi et al. [10] used a new combinative distance-based assessment (CODAS) method to handle MADM problems. Lee et al. [11] developed fuzzy entropy, which determined for an FS, by using the distance measure. Based on IFS, some authors [12–14] investigated the distance-based technique for order preference by similarity to an ideal solution (TOPSIS)method, entropy, and similarity measures of IFS, and applied them to MADM. Atanassov and Gargov [15] extended IFS to interval valued intuitionistic fuzzy set (IVIFS). Huang [16] proposed a (T, S)-based IVIF composition matrix and its application. Chen et al. [17] and Biswas et al. [18] introduced a linear programming methodology and integrated the TOPSIS approach for MADM in IVIFS. Because NS—a generalization of IFS and FS—can better describe uncertain information, NS now attracts great attention. An outranking method, COPRAS method, and entropy with NS for MADM has been developed in [19–21]. Wang et al. [22,23] introduced interval neutrosophic set (INS) and single-valued neutrosophic set (SVNN). Zhang et al. [24,25] and Tian et al. [26] proposed an outranking approach and weighted correlation coefficient: cross-entropy with INS for MADM. Further, Huang [27] and Ye [28] presented a new distance measure—cross-entropy and its application to MADM in SVNN.

Presently, information aggregation operators are attracting an increasing amount of attention for dealing with MADM. Many aggregation operators have been developed in intuitionistic fuzzy MADM [29–31]. Wang and Li [32,33] and Wang et al. [34] developed some intuitionistic normal aggregation operators and proposed some MADM methods based on these operators, while Wang et al. [5] developed some aggregation operators for NIFN. For NS and INS, some aggregation operators were proposed, such as power aggregation operators [35], generalized weighted power averaging operator [36], order weighted aggregation operators [37], generalized weighted power averaging operator [38], etc. Liu [39–41] developed Frank operators, generalized weighted power averaging operators, and Heronian mean operators for application with NNN; Şahin [42] introduced generalized prioritized aggregation operators with NNN.

However, these operators do not consider the relationship between attributes. Considering the interrelation between attributes, Bonferroni mean (BM) operator was first defined by Bonferroni [43]. Liu [6] introduced normal neutrosophic weighted Bonferroni mean (NNWBM) operator and normal neutrosophic weighted geometric Bonferroni mean (NNWGBM) operator. However, these operators only take into account correlations between any two aggregation arguments, and they do not consider the connections among any three or more than three aggregation arguments. The score function and accuracy function of the NNN and their ranking method were proposed also. However, the score function and accuracy function do not satisfy the ranking principle and are counterintuitive (see Example 1 for details).

The main contribution of this paper is (1) the proposal of a score function and accuracy function that satisfy the ranking principle, and (2) the extension of the operators in Liu [6]. First, we introduce the nonnegative normal neutrosophic number (NNNN). Then, a new score function and accuracy function are defined to solve the problem of the original function. Furthermore, considering the connections between any two or more than two aggregation arguments, the operator in [6] is generalized and some new operators are defined. For MADM, it is more reasonable to consider the relationship between each attribute, and the example in this paper further illustrates the advantages of the proposed MADM method compared with Liu [6]. The example in this paper further shows that when the relationship between more aggregation arguments is considered, the aggregation result is more stable; when the parameter value is larger, the aggregation result is more sensitive.

The structure of this paper as follows. Section 2 reviews the NNN, some operational laws, the score function, accuracy function, and the ranking method. Section 3 proposes the basic concept of the NNNN, and the new score function and accuracy function are introduced. Some generalized aggregation operators are developed, which are the dual generalized nonnegative normal neutrosophic weighted Bonferroni mean (DGNNNWBM) operator and dual generalized nonnegative normal neutrosophic weighted geometric Bonferroni mean (DGNNNWGBM) operator. Their properties are discussed. In Section 4, based on the DGNNNWBM operator and DGNNNWGBM operator, a MADM method is established. Section 5 gives a numerical example to explain the application of the proposed MADM method, and compares it with the method presented in [6]. Section 6 concludes this paper.

2. Preliminaries

Yang and Ko (1996) introduced the concept of the normal fuzzy number (NFN).

Definition 1. *[4] $A = (a, \sigma)$ is an NFN if its membership function is defined by:*

$$A(x) = e^{-(\frac{x-a}{\sigma})^2} (x \in X, \sigma > 0),$$

where X is the set of real numbers, and the set of NFNs is denoted as $\tilde{N}$.

The neutrosophic number (NN) and single-valued neutrosophic number (SVNN) were proposed in 1995 and 2005.

Definition 2. *[3] Let X be a universe of discourse, with a generic element in X denoted by x. An NN A in X is*

$$A(x) = \langle x | (T_A(x), I_A(x), F_A(x)) \rangle,$$

where $T_A(x)$ denotes the truth-membership function, $I_A(x)$ denotes the indeterminacy-membership function, and $F_A(x)$ denotes the falsity-membership function. $T_A(x)$,$I_A(x)$, and $F_A(x)$ are real standard or nonstandard subsets of $]^-0, 1^+[$.

There is no limitation on the sum of $T_A(x)$, $I_A(x)$, and $F_A(x)$, so $^-0 \leq T_A(x) + I_A(x) + F_A(x) \leq 3^+$.

Definition 3. *[23] Let X be a universe of discourse, with a generic element in X denoted by x. An SVNN A in X is depicted by the following:*

$$A(x) = \langle x | (T_A(x), I_A(x), F_A(x)) \rangle,$$

where $T_A(x)$ denotes the truth-membership function, $I_A(x)$ denotes the indeterminacy-membership function, and $F_A(x)$ denotes the falsity-membership function. For each point x in X, we have $T_A(x), I_A(x), F_A(x) \in [0, 1]$ and $0 \leq T_A(x) + I_A(x) + F_A(x) \leq 3$.

Based on NFN and NN, Liu (2017) defined the normal neutrosophic number (NNN).

Definition 4. *[6] Let X be a universe of discourse, with a generic element in X denoted by x, and $(a, \sigma) \in \tilde{N}$; then, an NNN A in X is expressed as:*

$$A(x) = \langle x | (a, \sigma), (T_A(x), I_A(x), F_A(x)) \rangle, x \in X,$$

where the truth-membership function $T_A(x)$ satisfies:

$$T_A(x) = T_A e^{-(\frac{x-a}{\sigma})^2}, x \in X,$$

where the indeterminacy-membership function $I_A(x)$ satisfies:

$$I_A(x) = 1 - (1 - I_A) e^{-(\frac{x-a}{\sigma})^2}, x \in X,$$

where the falsity-membership function $F_A(x)$ satisfies:

$$F_A(x) = 1 - (1 - F_A) e^{-(\frac{x-a}{\sigma})^2}, x \in X.$$

For each point x in X, we have $T_A(x), I_A(x), F_A(x) \in [0, 1]$, and $0 \leq T_A(x) + I_A(x) + F_A(x) \leq 3$. Then, we denote $\tilde{a} = \langle (a, \sigma), (T, I, F) \rangle$ as an NNN.

Some operational laws are shown in the following.

Definition 5. *[6] Let $\tilde{a}_1 = \langle (a_1, \sigma_1), (T_1, I_1, F_1) \rangle$ and $\tilde{a}_2 = \langle (a_2, \sigma_2), (T_2, I_2, F_2) \rangle$ be two NNNs; then, the operational rules are defined as follows:*

(1) $\tilde{a}_1 \oplus \tilde{a}_2 = \langle (a_1 + a_2, \sigma_1 + \sigma_2), (T_1 + T_2 - T_1 T_2, I_1 I_2, F_1 F_2) \rangle$;

(2) $\tilde{a}_1 \otimes \tilde{a}_2 = \left\langle \left(a_1 a_2, a_1 a_2 \sqrt{\frac{\sigma_1^2}{a_1^2} + \frac{\sigma_2^2}{a_2^2}} \right), (T_1 T_2, I_1 + I_2 - I_1 I_2, F_1 + F_2 - F_1 F_2) \right\rangle$;

(3) $\lambda \tilde{a}_1 = \langle (\lambda a_1, \lambda \sigma_1), (1 - (1 - T_1)^\lambda, I_1^\lambda, F_1^\lambda) \rangle \ (\lambda > 0)$;

(4) $\tilde{a}_1^\lambda = \langle (a_1^\lambda, \lambda^{\frac{1}{2}} a_1^{\lambda - 1} \sigma_1), (T_1^\lambda, 1 - (1 - I_1)^\lambda, 1 - (1 - F_1)^\lambda) \rangle (\lambda > 0)$.

Theorem 1. [6] *Let* $\tilde{a}_1 = \langle(a_1,\sigma_1),(T_1,I_1,F_1)\rangle$ *and* $\tilde{a}_2 = \langle(a_2,\sigma_2),(T_2,I_2,F_2)\rangle$ *be two NNNs, and* $\eta,\eta_1,\eta_2 > 0$*; then, we have*
(1) $\tilde{a}_1 \bigoplus \tilde{a}_2 = \tilde{a}_2 \bigoplus \tilde{a}_1$;
(2) $\tilde{a}_1 \bigotimes \tilde{a}_2 = \tilde{a}_2 \bigotimes \tilde{a}_1$;
(3) $\eta(\tilde{a}_1 \bigoplus \tilde{a}_2) = \eta\tilde{a}_1 \bigoplus \eta\tilde{a}_2$;
(4) $\eta_1\tilde{a}_1 \bigoplus \eta_2\tilde{a}_1 = (\eta_1+\eta_2)\tilde{a}_1$;
(5) $\tilde{a}_1^{\eta} \bigotimes \tilde{a}_2^{\eta} = (\tilde{a}_1 \bigotimes \tilde{a}_2)^{\eta}$;
(6) $\tilde{a}_1^{\eta_1} \bigotimes \tilde{a}_1^{\eta_2} = \tilde{a}_1^{\eta_1+\eta_2}$.

Liu (2017) proposed the score function and accuracy function for an NNN.

Definition 6. [41] *Let* $\tilde{a}_k = \langle(a_k,\sigma_k),(T_k,I_k,F_k)\rangle$ *be an NNN, then its score function is*

$$s_1(\tilde{a}_k) = a_k(2+T_k-I_k-F_k),$$
$$s_2(\tilde{a}_k) = \sigma_k(2+T_k-I_k-F_k);$$

and its accuracy function is

$$h_1(\tilde{a}_k) = a_k(2+T_k-I_k+F_k),$$
$$h_2(\tilde{a}_k) = \sigma_k(2+T_k-I_k+F_k).$$

Zhang et al. (2017) proposed the dual generalized weighted Bonferroni mean (DGWBM) operator and dual generalized weighted geometric Bonferroni mean (DGWGBM) operator.

Definition 7. [44] *Let* $a_i(i=1,2,...,n)$ *be a collection of nonnegative crisp numbers with the weight* $\omega = (\omega_1,\omega_2,...,\omega_n)^T, \omega_i \in [0,1](i=1,2,...,n)$ *and* $\sum_{i=1}^n \omega_i = 1$*. If*

$$DGWBM^R(a_1,a_2,...,a_n) = \left(\sum_{i_1,i_2,...,i_n=1}^{n}\left(\prod_{j=1}^{n}\omega_{i_j}a_{i_j}^{r_j}\right)\right)^{1/\sum_{j=1}^n r_j},$$

where $R = (r_1,r_2,...,r_n)^T$ *is the parameter vector with* $r_j \geq 0\ (i=1,2,...,n)$.

Definition 8. [44] *Let* $a_i(i=1,2,...,n)$ *be a collection of nonnegative crisp numbers with the weight* $\omega = (\omega_1,\omega_2,...,\omega_n)^T$*, where* $\omega_i \in [0,1]\ (i=1,2,...,n)$ *and* $\sum_{i=1}^n \omega_i = 1$*. if*

$$DGWBM^R(a_1,a_2,...,a_n) = \frac{1}{\sum_{j=1}^n r_j}\left(\prod_{i_1,i_2,...,i_n}^{n}\left(\sum_{j=1}^{n}\left(r_j a_{i_j}\right)\right)\right)^{\prod_{j=1}^n \omega_{i_j}},$$

where $R = (r_1,r_2,...,r_n)^T$ *is the parameter vector with* $r_i \geq 0\ (i=1,2,...,n)$.

3. Main Results

3.1. Ranking of Nonnegative Normal Neutrosophic Number

Liu and Li (2017) [6] introduced the concept of the score function s_1 and s_2, and the accuracy function h_1 and h_2, as shown in Definition 6. We found some deficiencies with the ranking of these functions, as shown below.

Let $\tilde{a}_k = \langle(a_k,\sigma_k),(T_k,I_k,F_k)\rangle\ (k=1,2)$ be any two NNNs. When $T_1 < T_2$, $I_1 > I_2$, $F_1 > F_2$, $a_1 \leq a_2$, and $\sigma_1 \geq \sigma_2$:

(1) If $a_k > 0$ or $a_k < 0$, then ranking results may be completely opposite;
(2) When s_1 can determine the ranking result of a_k, the influence of σ_k is not considered;
(3) Neither the score function nor the accuracy function satisfy the monotonicity.

We use the following example to illustrate problems (1) and (3) mentioned above.

Example 1. *Let* $\tilde{a}_1$ *and* $\tilde{a}_2$ *be two NNNs, where the specific values are as shown in Table 1. According to*

$$s_1(\tilde{a}_k) = a_k(2+T_k-I_k-F_k),$$
$$s_2(\tilde{a}_k) = \sigma_k(2+T_k-I_k-F_k),$$

$$h_1(\tilde{a}_k) = a_k(2 + T_k - I_k + F_k),$$
$$h_2(\tilde{a}_k) = \sigma_k(2 + T_k - I_k + F_k),$$

we can get its score function and accuracy function from Table 1. For number 1,

$$s_1(\tilde{a}_1) = 1 \times (2 + 0.5 - 0.2 - 0.2) = 2.1, s_1(\tilde{a}_2) = 2 \times (2 + 0.6 - 0.1 - 0.1) = 4.8,$$

by $2.1 < 4.8$*, we have* $\tilde{a}_1 < \tilde{a}_2$*. For number 2,*

$$s_1(\tilde{a}_1) = (-1) \times (2 + 0.5 - 0.2 - 0.2) = -2.1, s_1(\tilde{a}_2) = (-0.95) \times (2 + 0.6 - 0.1 - 0.1) = -2.28,$$

by $-2.1 > -2.28$*, we have* $\tilde{a}_1 > \tilde{a}_2$ *for the numerical results, which are shown in Table 2.*

From Table 2, when $T_1 < T_2$*,* $I_1 > I_2$*,* $F_1 > F_2$*,* $a_1 \leq a_2$*, and* $\sigma_1 \geq \sigma_2$ *are satisfied, we can intuitively see that the score function and accuracy function will be ranked differently if different values are taken. For example, the number 1 satisfies 0.5 < 0.6, 0.2 > 0.1, 0.2 > 0.1, 1 < 2, 0.3 > 0.1; the number 2 satisfies 0.5 < 0.6, 0.2 > 0.1, 0.2 > 0.1, −1 < −0.95, 0.2 > 0.1. However, their ranking results are completely different. The ranking results of numbers 2, 4, 6, 8 in Table 2 are counterintuitive. For example, the number 2 satisfies 0.5 < 0.6, 0.2 > 0.1, 0.2 > 0.1, −1 < −0.95, 0.2 > 0.1, and the ranking result are* $\tilde{a}_1 > \tilde{a}_2$*. However, intuitively,* $\tilde{a}_2$ *should be ranked first.*

Table 1. The numerical example.

Number	$\tilde{a}_1$	$\tilde{a}_2$
1	$\langle(1,0.3),(0.5,0.2,0.2)\rangle$	$\langle(2,0.1),(0.6,0.1,0.1)\rangle$
2	$\langle(-1,0.2),(0.5,0.2,0.2)\rangle$	$\langle(-0.95,0.1),(0.6,0.1,0.1)\rangle$
3	$\langle(2,0.4),(0.6,0.2,0.3)\rangle$	$\langle(2.5,0.2),(0.7,0.1,0.1)\rangle$
4	$\langle(2,0.4),(0.6,0.2,1)\rangle$	$\langle(2.5,0.2),(0.7,0.1,0.1)\rangle$
5	$\langle(0,1),(0.6,0.3,0.2)\rangle$	$\langle(0,0.5),(0.7,0.05,0.05)\rangle$
6	$\langle(0,1),(0.6,0.3,0.2)\rangle$	$\langle(0,0.9),(0.7,0.05,0.05)\rangle$
7	$\langle(2,2),(0.6,0.8,0.2)\rangle$	$\langle(2,1),(0.7,0,0.1)\rangle$
8	$\langle(2,2),(0.6,0.8,0.2)\rangle$	$\langle(2,1.5),(0.7,0,0.1)\rangle$

Table 2. The score function and accuracy function of the numerical example.

Number	$s_1(\tilde{a}_1)$	$s_1(\tilde{a}_2)$	$h_1(\tilde{a}_1)$	$h_1(\tilde{a}_2)$	$s_2(\tilde{a}_1)$	$s_2(\tilde{a}_2)$	$h_2(\tilde{a}_1)$	$h_2(\tilde{a}_2)$	Ranking
1	2.1	4.8	-	-	-	-	-	-	$\tilde{a}_1 < \tilde{a}_2$
2	−2.1	−2.28	-	-	-	-	-	-	$\tilde{a}_1 > \tilde{a}_2$
3	-	-	5.4	6.75	-	-	-	-	$\tilde{a}_1 < \tilde{a}_2$
4	-	-	6.8	6.75	-	-	-	-	$\tilde{a}_1 > \tilde{a}_2$
5	-	-	-	-	2.1	1.3	-	-	$\tilde{a}_1 < \tilde{a}_2$
6	-	-	-	-	2.1	2.34	-	-	$\tilde{a}_1 > \tilde{a}_2$
7	-	-	-	-	-	-	4	2.8	$\tilde{a}_1 < \tilde{a}_2$
8	-	-	-	-	-	-	4	4.2	$\tilde{a}_1 > \tilde{a}_2$

In order to avoid the disadvantages of the ranking, we propose the nonnegative normal neutrosophic number (NNNN). Additionally, we take σ into account and introduce the score function and accuracy function of the NNNN.

Definition 9. *$A(x) = \langle x|(a,\sigma),(T_A(x), I_A(x), F_A(x))\rangle$ is an NNNN if it has satisfied Definition 4 and $a \geq 0$.*

Based on the NNNN, the new score function S and accuracy function H are proposed.

Definition 10. *Suppose $\tilde{a} = \langle(a,\sigma),(T,I,F)\rangle$ is an NNNN, then its score function is*

$$S(\tilde{a}) = (a + \tfrac{1}{\sigma})(2 + T - I - F);$$

and its accuracy function is

$$H(\tilde{a}) = (a + \tfrac{1}{\sigma})(1 + T - F).$$

According to the score function and accuracy function, the following propositions are derived.

Proposition 1. *Let $\tilde{a}_k = \langle (a_k, \sigma_k), (T_k, I_k, F_k) \rangle$ $(k = 1, 2)$ be any two NNNNs, then the following conclusions are obtained.*
(1) If $a_1 \leq a_2$, $\sigma_1 \geq \sigma_2$, $T_1 < T_2$ and $I_1 > I_2$ and $F_1 > F_2$, then $S(\tilde{a}_1) < S(\tilde{a}_2)$;
(2) If $a_1 \leq a_2$, $\sigma_1 \geq \sigma_2$, $T_1 < T_2$, and $F_1 > F_2$, then $H(\tilde{a}_1) < H(\tilde{a}_2)$.

Therefore, we have the following ranking principles.

Definition 11. *Let $\tilde{a}_k = \langle (a_k, \sigma_k), (T_k, I_k, F_k) \rangle$ $(k = 1, 2)$ be any two NNNNs, then we have the following method for ranking an NNNN:*
(1) If $S(\tilde{a}_1) < S(\tilde{a}_2)$, then $\tilde{a}_1 < \tilde{a}_2$;
(2) If $S(\tilde{a}_1) = S(\tilde{a}_2)$, then
(a) If $H(\tilde{a}_1) < H(\tilde{a}_2)$, then $\tilde{a}_1 < \tilde{a}_2$;
(b) If $H(\tilde{a}_1) = H(\tilde{a}_2)$, then $\tilde{a}_1 \sim \tilde{a}_2$.

We introduce some operational laws as follows:

Definition 12. *Let $\tilde{a}_k = \langle (a_k, \sigma_k), (T_k, I_k, F_k) \rangle$ $(k = 1, 2)$ be any two NNNNs, then the operational rules are defined as follows:*
(1) $\tilde{a}_1 \hat{\otimes} \tilde{a}_2 = \langle (a_1 a_2, \sigma_1 \sigma_2), (T_1 T_2, I_1 + I_2 - I_1 I_2, F_1 + F_2 - F_1 F_2) \rangle$;
(2) $\tilde{a}_1^{*\lambda} = \langle (a_1^\lambda, \sigma_1^\lambda), (T_1^\lambda, 1 - (1 - I_1)^\lambda, 1 - (1 - F_1)^\lambda) \rangle (\lambda > 0)$.

Moreover, the relations of the operational laws are given as below, and these properties are obvious.

Proposition 2. *Let $\tilde{a}_1 = \langle (a_1, \sigma_1), (T_1, I_1, F_1) \rangle$ and $\tilde{a}_2 = \langle (a_2, \sigma_2), (T_2, I_2, F_2) \rangle$ be two NNNNs, and $\eta, \eta_1, \eta_2 > 0$; then*
(1) $\tilde{a}_1 \hat{\otimes} \tilde{a}_2 = \tilde{a}_2 \hat{\otimes} \tilde{a}_1$;
(2) $\tilde{a}_1^{*\eta} \hat{\otimes} \tilde{a}_2^{*\eta} = (\tilde{a}_1 \hat{\otimes} \tilde{a}_2)^{*\eta}$;
(3) $\tilde{a}_1^{*\eta_1} \hat{\otimes} \tilde{a}_1^{*\eta_2} = \tilde{a}_1^{*(\eta_1 + \eta_2)}$.

3.2. DGNNNWBM Operator and DGNNNWGBM Operator

This section extends the DGWBM and DGWGBM to NNNN, and proposes the dual generalized nonnegative normal neutrosophic weighted Bonferroni mean (DGNNNWBM) operator and dual generalized nonnegative normal neutrosophic weighted geometric Bonferroni mean (DGNNNWGBM) operator.

Definition 13. *Suppose $\{\tilde{a}_i | \tilde{a}_i = \langle (a_i, \sigma_i), (T_i, I_i, F_i) \rangle, i = 1, 2, ..., n\}$ is a set of NNNNs, with their weight vector being $\omega_i = (\omega_1, \omega_2, ..., \omega_n)^T$, where $\omega_i \in [0, 1]$ and $\sum_{i=1}^n \omega_i = 1$. The DGNNNWBM operator is defined as*

$$DGNNNWBM_\omega^R(\tilde{a}_1, \tilde{a}_2, ..., \tilde{a}_n) = \left(\bigoplus_{i_1, i_2, ..., i_n = 1}^{n} \left(\hat{\bigotimes}_{j=1}^{n} \omega_{i_j} \tilde{a}_{i_j}^{*r_j} \right) \right)^{*1/\sum_{j=1}^n r_j},$$

where $R = (r_1, r_2, ..., r_n)^T$ is the parameter vector with $r_i \geq 0 (i = 1, 2, ..., n)$.

The DGNNNWBM operator can consider the relationship between any elements. Here are some special cases of it.

Remark 1. *If $R = (\lambda, 0, 0, ..., 0)^T$, that is, consider the relationship of a single element, then the DGNNNWBM reduces to:*

$$DGNNNWBM_\omega^R(\tilde{a}_1, \tilde{a}_2, ..., \tilde{a}_n) = \left(\bigoplus_{i=1}^{n} \lambda \tilde{a}_i \right)^{*\frac{1}{\lambda}},$$

which is called a generalized nonnegative normal neutrosophic weighted averaging (GNNNWA) operator.

If $R = (s,t,0,0,...,0)^T$, *that is, consider the relationship between any two elements, then the DGNNNWBM reduces to:*

$$DGNNNWBM_{\omega}^{R}(\tilde{a}_1,\tilde{a}_2,...,\tilde{a}_n) = \left(\bigoplus_{i,j=1}^{n} \left(\omega_i \tilde{a}_i^{*s} \hat{\otimes} \omega_j \tilde{a}_j^{*t} \right) \right)^{*\frac{1}{s+t}},$$

which is the nonnegative normal neutrosophic weighted Bonferroni mean(NNNWBM) operator.

If $R = (s,t,r,0,0,...,0)^T$, *that is, consider the relationship between any three elements, then the DGNNNWBM reduces to:*

$$DGNNNWBM_{\omega}^{R}(\tilde{a}_1,\tilde{a}_2,...,\tilde{a}_n) = \left(\bigoplus_{i,j,k=1}^{n} \left(\omega_i \tilde{a}_i^{*s} \hat{\otimes} \omega_j \tilde{a}_j^{*t} \hat{\otimes} \omega_k \tilde{a}_k^{*r} \right) \right)^{*\frac{1}{s+t+r}},$$

which is called a generalized nonnegative normal neutrosophic weighted Bonferroni mean (GNNNWBM) operator.

Theorem 2. *Let* $\{\tilde{a}_i | \tilde{a}_i = \langle (a_i,\sigma_i),(T_i,I_i,F_i) \rangle, i = 1,2,...,n\}$ *be a set of NNNNs, then the aggregated result of the DGNNNWBM is also an NNNN and*

$$DGNNNWBM_{\omega}^{R}(\tilde{a}_1,\tilde{a}_2,...,\tilde{a}_n) = \langle (a,\sigma),(T,I,F) \rangle,$$

where

$$a = \left(\sum_{i_1,i_2,...,i_n=1}^{n} \left(\prod_{j=1}^{n} \omega_{i_j} a_{i_j}^{r_j} \right) \right)^{\frac{1}{\sum_{j=1}^{n} r_j}},$$

$$\sigma = \left(\sum_{i_1,i_2,...,i_n=1}^{n} \left(\prod_{j=1}^{n} \omega_{i_j} \sigma_{i_j}^{r_j} \right) \right)^{\frac{1}{\sum_{j=1}^{n} r_j}},$$

$$T = \left(1 - \prod_{i_1,i_2,...,i_n=1}^{n} \left(1 - \prod_{j=1}^{n} \left(1 - \left(1 - T_{i_j}^{r_j}\right)^{\omega_{i_j}}\right)\right)\right)^{\frac{1}{\sum_{j=1}^{n} r_j}},$$

$$I = 1 - \left(1 - \prod_{i_1,i_2,...,i_n=1}^{n} \left(1 - \prod_{j=1}^{n} \left(1 - \left(1 - \left(1 - I_{i_j}\right)^{r_j}\right)^{\omega_{i_j}}\right)\right)\right)^{\frac{1}{\sum_{j=1}^{n} r_j}},$$

$$F = 1 - \left(1 - \prod_{i_1,i_2,...,i_n=1}^{n} \left(1 - \prod_{j=1}^{n} \left(1 - \left(1 - \left(1 - F_{i_j}\right)^{r_j}\right)^{\omega_{i_j}}\right)\right)\right)^{\frac{1}{\sum_{j=1}^{n} r_j}}.$$

Proof. By Definition 5 and 12, we have

$$\tilde{a}_{i_j}^{*r_j} = \left\langle \left(a_{i_j}^{r_j}, \sigma_{i_j}^{r_j} \right), \left(T_{i_j}^{r_j}, 1-(1-I_{i_j})^{r_j}, 1-(1-F_{i_j})^{r_j} \right) \right\rangle,$$

and

$$\omega_{i_j}\tilde{a}_{i_j}^{*r_j} = \left\langle \left(\omega_{i_j} a_{i_j}^{r_j}, \omega_{i_j} \sigma_{i_j}^{r_j} \right), \left(1-(1-T_{i_j}^{r_j})^{\omega_{i_j}}, (1-(1-I_{i_j})^{r_j})^{\omega_{i_j}}, (1-(1-F_{i_j})^{r_j})^{\omega_{i_j}} \right) \right\rangle,$$

so

$$\hat{\bigotimes}_{j=1}^{n} \omega_{i_j} \tilde{a}_{i_j}^{*r^j} =$$

$$\left\langle \left(\prod_{j=1}^{n} \omega_{i_j} a_{i_j}^{r_j}, \prod_{j=1}^{n} \omega_{i_j} \sigma_{i_j}^{r_j} \right), \left(\prod_{j=1}^{n} \left(1 - \left(1 - T_{i_j}^{r_j}\right)^{\omega_{i_j}}\right), \right. \right.$$

$$\left. \left. \sum_{j=1}^{n} (1-(1-I_{i_j})^{r_j})^{\omega_{i_j}} - \prod_{j=1}^{n} (1-(1-I_{i_j})^{r_j})^{\omega_{i_j}}, \sum_{j=1}^{n} (1-(1-F_{i_j})^{r_j})^{\omega_{i_j}} - \prod_{j=1}^{n} (1-(1-F_{i_j})^{r_j})^{\omega_{i_j}} \right) \right\rangle$$

$$= \left\langle \left(\prod_{j=1}^{n} \omega_{i_j} a_{i_j}^{r_j}, \prod_{j=1}^{n} \omega_{i_j} \sigma_{i_j}^{r_j} \right), \left(\prod_{j=1}^{n} \left(1 - \left(1 - T_{i_j}^{r_j}\right)^{\omega_{i_j}}\right), \right. \right.$$

$$\left. \left. 1 - \prod_{j=1}^{n} \left(1 - \left(1 - \left(1 - I_{i_j}\right)^{r_j}\right)^{\omega_{i_j}}\right), 1 - \prod_{j=1}^{n} \left(1 - \left(1 - \left(1 - F_{i_j}\right)^{r_j}\right)^{\omega_{i_j}}\right) \right) \right\rangle,$$

then

$$\bigoplus_{i_1,i_2,...,i_n=1}^{n} \left(\hat{\bigotimes}_{j=1}^{n} \omega_{i_j} \tilde{a}_{i_j}^{*r_j} \right)$$

$$= \left\langle \left(\sum_{i_1,i_2,...,i_n=1}^{n} \left(\prod_{j=1}^{n} \omega_{i_j} a_{i_j}^{r_j} \right), \sum_{i_1,i_2,...,i_n=1}^{n} \left(\prod_{j=1}^{n} \omega_{i_j} \sigma_{i_j}^{r_j} \right) \right), \right.$$

$$\left(\sum_{i_1,i_2,...,i_n=1}^{n} \prod_{j=1}^{n} \left(1 - \left(1 - T_{i_j}^{r_j}\right)^{\omega_{i_j}}\right) - \prod_{i_1,i_2,...,i_n=1}^{n} \prod_{j=1}^{n} \left(1 - \left(1 - T_{i_j}^{r_j}\right)^{\omega_{i_j}}\right), \right.$$

$$\prod_{i_1,i_2,\ldots i_n=1}^{n}\left(1-\prod_{j=1}^{n}\left(1-\left(1-\left(1-I_{i_j}\right)^{r_j}\right)^{\omega_{i_j}}\right)\right), \prod_{i_1,i_2,\ldots i_n=1}^{n}\left(1-\prod_{j=1}^{n}\left(1-\left(1-\left(1-F_{i_j}\right)^{r_j}\right)^{\omega_{i_j}}\right)\right)\Bigg\rangle,$$

$$=\Bigg\langle\left(\sum_{i_1,i_2,\ldots,i_n=1}^{n}\left(\prod_{j=1}^{n}\omega_{i_j}a_{i_j}^{r_j}\right), \sum_{i_1,i_2,\ldots,i_n=1}^{n}\left(\prod_{j=1}^{n}\omega_{i_j}\sigma_{i_j}^{r_j}\right)\right), \left(1-\prod_{i_1,i_2,\ldots,i_n=1}^{n}\left(1-\prod_{j=1}^{n}\left(1-\left(1-T_{i_j}^{r_j}\right)^{\omega_{i_j}}\right)\right),\right.$$

$$\prod_{i_1,i_2,\ldots i_n=1}^{n}\left(1-\prod_{j=1}^{n}\left(1-\left(1-\left(1-I_{i_j}\right)^{r_j}\right)^{\omega_{i_j}}\right)\right), \prod_{i_1,i_2,\ldots i_n=1}^{n}\left(1-\prod_{j=1}^{n}\left(1-\left(1-\left(1-F_{i_j}\right)^{r_j}\right)^{\omega_{i_j}}\right)\right)\Bigg\rangle,$$

Let

$$a=\left(\sum_{i_1,i_2,\ldots,i_n=1}^{n}\left(\prod_{j=1}^{n}\omega_{i_j}a_{i_j}^{r_j}\right)\right)^{\frac{1}{\sum_{j=1}^{n}r_j}},$$

$$\sigma=\left(\sum_{i_1,i_2,\ldots,i_n=1}^{n}\left(\prod_{j=1}^{n}\omega_{i_j}\sigma_{i_j}^{r_j}\right)\right)^{\frac{1}{\sum_{j=1}^{n}r_j}},$$

$$T=\left(1-\prod_{i_1,i_2,\ldots,i_n=1}^{n}\left(1-\prod_{j=1}^{n}\left(1-\left(1-T_{i_j}^{r_j}\right)^{\omega_{i_j}}\right)\right)\right)^{\frac{1}{\sum_{j=1}^{n}r_j}},$$

$$I=1-\left(1-\prod_{i_1,i_2,\ldots,i_n=1}^{n}\left(1-\prod_{j=1}^{n}\left(1-\left(1-\left(1-I_{i_j}\right)^{r_j}\right)^{\omega_{i_j}}\right)\right)\right)^{\frac{1}{\sum_{j=1}^{n}r_j}},$$

$$F=1-\left(1-\prod_{i_1,i_2,\ldots,i_n=1}^{n}\left(1-\prod_{j=1}^{n}\left(1-\left(1-\left(1-F_{i_j}\right)^{r_j}\right)^{\omega_{i_j}}\right)\right)\right)^{\frac{1}{\sum_{j=1}^{n}r_j}}.$$

thus

$$\left(\bigoplus_{i_1,i_2,\ldots,i_n=1}^{n}\left(\bigotimes_{j=1}^{n}\omega_{i_j}a_{i_j}^{*r_j}\right)\right)^{*1/\sum_{j=1}^{n}r_j}=\langle(a,\sigma),(T,I,F)\rangle.$$

Thereafter

$$a\geq 0, \sigma>0, 0\leq T\leq 1, 0\leq I\leq 1, 0\leq F\leq 1.$$

Hence

$$0\leq T+I+F\leq 3.$$

which completes the proof. □

The following example is used to explain the calculation of the DGNNNWBM operator.

Example 2. *Let* $\tilde{a}_1=\langle(0.7,0.01),(0.6,0.2,0.1)\rangle$, $\tilde{a}_2=\langle(0.4,0.02),(0.8,0.1,0.3)\rangle$ *be two NNNNs. With the weighted vector* $\omega=(0.7,0.3)^T$, *and the parameter vector* $R=(2,3)^T$, *then, according to Theorem 2, we have*

$a=(\sum_{i_i,i_2=1}^{2}(\omega_{i_1}\omega_{i_2}))^{\frac{1}{2+3}}$

$=(\omega_1 a_1^{r_1\omega_1 a_1^{r_2}}+\omega_1 a_1^{r_1\omega_2 a_2^{r_2}}+\omega_2 a_2^{r_1\omega_1 a_1^{r_2}}+\omega_2 a_2^{r_1\omega_2 a_2^{r_2}})^{\frac{1}{5}}$

$=(0.7\times 0.7^2\times 0.7\times 0.7^3+0.7\times 0.7^2\times 0.3\times 0.4^3+0.3\times 0.4^2\times 0.7\times 0.7^3+0.3\times 0.4^2\times 0.3\times 0.4^3)^{\frac{1}{5}}$

= 0.6327

Similarly, we can obtain $\sigma=0.0143$.

$T=\left(1-\prod_{i_i,i_2=1}^{2}(1-(1-(1-T_{i_1}^{r_1})^{\omega_{i_1}})(1-(1-T_{i_2}^{r_2})^{\omega_{i_2}}))\right)^{\frac{1}{2+3}}$

$=(1-(1-(1-(1-T_1^{r_1})^{\omega_1})(1-(1-T_1^{r_2})^{\omega_1}))(1-(1-(1-T_1^{r_1})^{\omega_1})(1-(1-T_2^{r_2})^{\omega_2}))(1-(1-(1-T_2^{r_1})^{\omega_2})(1-(1-T_1^{r_2})^{\omega_1}))(1-(1-(1-T_2^{r_1})^{\omega_2})(1-(1-T_2^{r_2})^{\omega_2})))^{\frac{1}{5}}$

$=(1-(1-(1-(1-0.6^2)^{0.7})(1-(1-0.6^3)^{0.7}))\times(1-(1-(1-0.6^2)^{0.7})(1-(1-0.8^3)^{0.3}))\times(1-(1-(1-0.8^2)^{0.3})(1-(1-0.6^3)^{0.7}))\times(1-(1-(1-0.8^2)^{0.3})(1-(1-0.8^3)^{0.4})))^{\frac{1}{5}}$

= 0.64

$I=1-(1-\prod_{i_i,i_2=1}^{2}(1-(1-(1-(1-I_{i_1})^{r_1})^{\omega_{i_1}})(1-(1-(1-I_{i_2})^{r_2})^{\omega_{i_2}})))^{\frac{1}{2+3}}$

$=1-(1-(1-(1-(1-(1-I_1)^{r_1})^{\omega_1})(1-(1-(1-I_1)^{r_2})^{\omega_1}))(1-(1-(1-(1-I_1)^{r_1})^{\omega_1})(1-(1-(1-I_2)^{r_2})^{\omega_2}))(1-(1-(1-(1-I_2)^{r_1})^{\omega_2})(1-(1-(1-I_1)^{r_2})^{\omega_1}))(1-(1-(1-(1-I_2)^{r_1})^{\omega_2})(1-(1-(1-I_2)^{r_2})^{\omega_2})))^{\frac{1}{5}}$

$= 1-(1-(1-(1-(1-(1-0.2)^2)^{0.7})(1-(1-(1-0.2)^3)^{0.7}))(1-(1-(1-(1-0.2)^2)^{0.7})(1-(1-(1-0.1)^3)^{0.3}))(1-(1-(1-(1-0.1)^2)^{0.3})(1-(1-(1-0.2)^3)^{0.7}))(1-(1-(1-(1-0.1)^2)^{0.3})(1-(1-(1-0.1)^3)^{0.3})))^{\frac{1}{5}}$
$= 0.1265$

Similarly, we can obtain $F = 0.1195$.

So, $DGNNNWBM_{\omega}^{R}(\tilde{a}_1, \tilde{a}_2) = \langle (0.6327, 0.0143), (0.64, 0.1265, 0.1195) \rangle$

Next, we discuss some properties of the DGNNNWBM operator.

Theorem 3. *(Monotonicity) Let* $\{\tilde{a}_i | \tilde{a}_i = \langle (a_i, \sigma_i), (T_i, I_i, F_i) \rangle, i = 1, 2, ..., n\}$ *and* $\{\tilde{b}_i | \tilde{b}_i = \langle (b_i, \delta_i), (T_{b_i}, I_{b_i}, F_{b_i}) \rangle, i = 1, 2, ..., n\}$ *be two sets of NNNNs. If* $a_i \le b_i, \sigma_i \ge \delta_i$ *and* $T_{a_i} < T_{b_i}$ *and* $I_{a_i} > I_{b_i}$ *and* $F_{a_i} > F_{b_i}$ *hold for all i, then*

$$DGNNNWBM_{\omega}^{R}(\tilde{a}_1, \tilde{a}_2, ..., \tilde{a}_n) < DGNNNWBM_{\omega}^{R}(\tilde{b}_1, \tilde{b}_2, ..., \tilde{b}_n),$$

where $R = (r_1, r_2, ..., r_n)^T$ *is the parameter vector with* $r_i \ge 0 (i = 1, 2, ..., n)$.

Proof. Let

$$DGNNNWBM_{\omega}^{R}(\tilde{a}_1, \tilde{a}_2, ..., \tilde{a}_n) = \langle (a, \sigma), (T_a, I_a, F_a) \rangle,$$
$$DGNNNWBM_{\omega}^{R}(\tilde{b}_1, \tilde{b}_2, ..., \tilde{b}_n) = \langle (b, \delta), (T_b, I_b, F_b) \rangle.$$

According to the DGNNNWBM operator, we have

$$a = \Big(\sum_{i_1, i_2, ..., i_n = 1}^{n} \Big(\prod_{j=1}^{n} \omega_{i_j} a_{i_j}^{r_j} \Big) \Big)^{1/\sum_{j=1}^{n} r_j}, b = \Big(\sum_{i_1, i_2, ..., i_n = 1}^{n} \Big(\prod_{j=1}^{n} \omega_{i_j} b_{i_j}^{r_j} \Big) \Big)^{1/\sum_{j=1}^{n} r_j},$$
$$\sigma = \Big(\sum_{i_1, i_2, ..., i_n = 1}^{n} \Big(\prod_{j=1}^{n} \omega_{i_j} \sigma_{i_j}^{r_j} \Big) \Big)^{1/\sum_{j=1}^{n} r_j}, \delta = \Big(\sum_{i_1, i_2, ..., i_n = 1}^{n} \Big(\prod_{j=1}^{n} \omega_{i_j} \delta_{i_j}^{r_j} \Big) \Big)^{1/\sum_{j=1}^{n} r_j}.$$

By $a_i \le b_i, \sigma_i \ge \delta_i$ we get $a \le b, \sigma \ge \delta$.

Let

$$T_a = \left(1 - \prod_{i_1, i_2, ..., i_n = 1}^{n} \left(1 - \prod_{j=1}^{n} \left(1 - (1 - T_{a_{i_j}}^{r_j})^{\omega_{i_j}}\right)\right)\right)^{\frac{1}{\sum_{j=1}^{n} r_j}},$$

$$T_b = \left(1 - \prod_{i_1, i_2, ..., i_n = 1}^{n} \left(1 - \prod_{j=1}^{n} \left(1 - (1 - T_{b_{i_j}}^{r_j})^{\omega_{i_j}}\right)\right)\right)^{\frac{1}{\sum_{j=1}^{n} r_j}}$$

when $T_{a_i} < T_{b_i}$, we can obtain

$$(1 - T_{a_{i_j}}^{r_j})^{\omega_{i_j}} > (1 - T_{b_{i_j}}^{r_j})^{\omega_{i_j}},$$

and

$$1 - (1 - T_{a_{i_j}}^{r_j})^{\omega_{i_j}} < 1 - (1 - T_{b_{i_j}}^{r_j})^{\omega_{i_j}},$$

therefore

$$1 - \prod_{j=1}^{n} \left(1 - (1 - T_{a_{i_j}}^{r_j})^{\omega_{i_j}}\right) > 1 - \prod_{j=1}^{n} \left(1 - (1 - T_{b_{i_j}}^{r_j})^{\omega_{i_j}}\right),$$

thus

$$1 - \prod_{i_1, i_2, ..., i_n = 1}^{n} \left(1 - \prod_{j=1}^{n} \left(1 - (1 - T_{a_{i_j}}^{r_j})^{\omega_{i_j}}\right)\right) < 1 - \prod_{i_1, i_2, ..., i_n = 1}^{n} \left(1 - \prod_{j=1}^{n} \left(1 - (1 - T_{b_{i_j}}^{r_j})^{\omega_{i_j}}\right)\right),$$

then $T_a < T_b$.

Similarly, we can obtain $I_a > I_b$ and $F_a > F_b$.

According to Definition 11,

$$S(DGNNNWBM_{\omega}^{R}(\tilde{a}_1, \tilde{a}_2, ..., \tilde{a}_n)) = (a + \tfrac{1}{\sigma})(2 + T_a - I_a - F_a)$$
$$< (b + \tfrac{1}{\delta})(2 + T_b - I_b - F_b) = S(DGNNNWBM_{\omega}^{R}(\tilde{b}_1, \tilde{b}_2, ..., \tilde{b}_n)).$$

Therefore, the proof is completed. □

Remark 2. *If* $a_i \le b_i, \sigma_i \ge \delta_i, T_{a_i} \le T_{b_i}, I_{a_i} \ge I_{b_i}, F_{a_i} \ge F_{b_i}$ *and* $(T_{a_i} - T_{b_i})^2 + (I_{a_i} - I_{b_i})^2 + (F_{a_i} - F_{b_i})^2 \ne 0$ *hold for any i, Theorem 3 is still holds.*

Theorem 4. *(Boundedness) Let* $\{\tilde{a}_i|\tilde{a}_i = \langle(a_i,\sigma_i),(T_i,I_i,F_i)\rangle, i = 1,2,...,n\}$ *be a set of NNNNs. If* $a^+ = \langle(max_i(a_i), min_i(\sigma_i)),(max_i(T_i), min_i(I_i), min_i(F_i)\rangle = \langle(a_i^+,\sigma_i^+),(T_{a_i}^+, I_{a_i}^+, F_{a_i}^+)\rangle$ *and* $a^- = \langle(min_i(a_i), max_i(\sigma_i)),(min_i(T_i), max_i(I_i), max_i(F_i)\rangle = \langle(a_i^-,\sigma_i^-),(T_{a_i}^-, I_{a_i}^-, F_{a_i}^-)\rangle$*, then*
$DGNNNWBM_\omega^R(\tilde{a}^-,\tilde{a}^-,...,\tilde{a}^-) \le DGNNNWBM_\omega^R(\tilde{a}_1,\tilde{a}_2,...,\tilde{a}_n) \le DGNNNWBM_\omega^R(\tilde{a}^+,\tilde{a}^+,...,\tilde{a}^+)$.

Proof. By $a_i^- \le a_i \le a_i^+, \sigma_i^+ \le \sigma_i \le \sigma_i^-, T_i^- \le T_i \le T_i^+, I_i^+ \le I_i \le I_i^-, F_i^+ \le F_i \le F_i^-$, according to Theorem 3 and Remark 2, we get
$DGNNNWBM_\omega^R(\tilde{a}^-,\tilde{a}^-,...,\tilde{a}^-) \le DGNNNWBM_\omega^R(\tilde{a}_1,\tilde{a}_2,...,\tilde{a}_n) \le DGNNNWBM_\omega^R(\tilde{a}^+,\tilde{a}^+,...,\tilde{a}^+)$.
□

Theorem 5. *(Commutativity) Let* $\{\tilde{a}_i|\tilde{a}_i = \langle(a_i,\sigma_i),(T_i,I_i,F_i)\rangle, i = 1,2,...,n\}$ *be a set of NNNNs. If* $\tilde{a}_i'$ *is any permutation of* $\tilde{a}_i$*, then*

$$DGNNNWBM_\omega^R(\tilde{a}_1,\tilde{a}_2,...,\tilde{a}_n) = DGNNNWBM_\omega^R(\tilde{a}_1',\tilde{a}_2',...,\tilde{a}_n').$$

Unfortunately, the DGNNNWBM operator is not satisfied with idempotency, i.e., $DGNNNWBM_\omega^R(\tilde{a},\tilde{a},...,\tilde{a}) \neq a$.

Example 3. *Let* $\tilde{a} = \langle(4,0.2),(0.8,0.2,0.3)\rangle$ *be an NNNN. The weighted vector* $\omega = (0.25,0.25,0.25,0.25)^T$*, and the parameter vector* $R = (2,2,2,2)^T$*, if all* $\tilde{a}_i = \tilde{a}(i = 1,2,3,4)$*. Similar to Example 2, the following results can be obtained*

$$DGNNNWBM_\omega^R(\tilde{a},\tilde{a},\tilde{a},\tilde{a}) = \langle(4,0.2),(0.9886,0.0114,0.0886)\rangle \neq \tilde{a}.$$

Furthermore, we extend the DGWBGM to NNNNs and propose the dual generalized nonnegative normal neutrosophic weighted geometric Bonferroni mean (DGNNNWGBM) operator.

Definition 14. *Suppose* $\{\tilde{a}_i|\tilde{a}_i = \langle(a_i,\sigma_i),(T_i,I_i,F_i)\rangle, i = 1,2,...,n\}$ *is a set of NNNNs with their weight vector being* $\omega_i = (\omega_1,\omega_2,...,\omega_n)^T$*, where* $\omega_i \in [0,1]$ *and* $\sum_{i=1}^n \omega_i = 1$*. The DGNNNWGBM operator is defined as*

$$DGNNNWGBM_\omega^R(\tilde{a}_1,\tilde{a}_2,...,\tilde{a}_n) = \frac{1}{\sum_{j=1}^n r_j}\left(\bigotimes_{i_1,i_2,...i_n=1}^n\left(\bigoplus_{j=1}^n (r_j\tilde{a}_{i_j})\right)^{*\prod_{j=1}^n \omega_{i_j}}\right),$$

where $R = (r_1,r_2,...,r_n)^T$ *is the parameter vector with* $r_i \ge 0(i = 1,2,...,n)$.

The DGNNNWGBM operator can consider the relationship between any elements. Here are some special cases of it.

Remark 3. *If* $R = (\lambda,0,0,...,0)^T$*, that is, consider the relationship of a single element, then the DGNNNWGBM reduces to:*

$$DGNNNWGBM_\omega^R(\tilde{a}_1,\tilde{a}_2,...,\tilde{a}_n) = \frac{1}{\lambda}\left(\bigotimes_{i=1}^n\left(\lambda\tilde{a}_i\right)^{*\omega_i}\right)$$

which is called a generalized nonnegative normal neutrosophic weighted geometric averaging (GNNNWGA) operator.

If $R = (s,t,0,0,...,0)^T$*, that is, consider the relationship between any two elements, then the DGNNNWGBM reduces to:*

$$DGNNNWGBM_\omega^R(\tilde{a}_1,\tilde{a}_2,...,\tilde{a}_n) = \frac{1}{s+t}\bigotimes_{i,j=1}^n\left(s\tilde{a}_i \oplus t\tilde{a}_j\right)^{*\omega_i\omega_j}$$

which is called a nonnegative normal neutrosophic weighted Bonferroni geometric (NNNWBG) operator.

If $R = (s,t,r,0,0,...,0)^T$*, that is, consider the relationship between any three elements, then the DGNNNWGBM reduces to:*

$$DGNNNWGBM_\omega^R(\tilde{a}_1,\tilde{a}_2,...,\tilde{a}_n) = \frac{1}{s+t+r}\bigotimes_{i,j,k=1}^n\left(s\tilde{a}_i \oplus t\tilde{a}_j \oplus t\tilde{a}_k\right)^{*\omega_i\omega_j\omega_k}$$

which is called a generalized nonnegative normal neutrosophic weighted Bonferroni geometric (GNNNWBG) operator.

Theorem 6. *Let $\{\tilde{a}_i|\tilde{a}_i = \langle(a_i,\sigma_i),(T_i,I_i,F_i)\rangle, i=1,2,...,n\}$ be a set of NNNNs, then the aggregated result of DGNNNWGBM is also an NNNN and*

$$DGNNNWGBM_{\omega}^{R}(\tilde{a}_1,\tilde{a}_2,...,\tilde{a}_n) = \langle(\hat{a},\hat{\sigma}),(\hat{T},\hat{I},\hat{F})\rangle,$$

where

$$\hat{a} = \frac{1}{\sum_{j=1}^{n} r_j} \prod_{i_1,i_2,...i_n=1}^{n} \left(\sum_{j=1}^{n}(r_j a_{i_j})\right)^{\prod_{j=1}^{n}\omega_{i_j}},$$

$$\hat{\sigma} = \frac{1}{\sum_{j=1}^{n} r_j} \prod_{i_1,i_2,...i_n=1}^{n} \left(\sum_{j=1}^{n}(r_j \sigma_{i_j})\right)^{\prod_{j=1}^{n}\omega_{i_j}},$$

$$\hat{T} = 1-\left(1-\prod_{i_1,i_2,...,i_n=1}^{n}\left(1-\prod_{j=1}^{n}\left(1-T_{i_j}\right)^{r_j}\right)^{\prod_{j=1}^{n}\omega_{i_j}}\right)^{\frac{1}{\sum_{j=1}^{n} r_j}},$$

$$\hat{I} = \left(1-\prod_{i_1,i_2,...,i_n=1}^{n}\left(\left(1-\prod_{j=1}^{n} I_{i_j}^{r_j}\right)^{\prod_{j=1}^{n}\omega_{i_j}}\right)\right)^{\frac{1}{\sum_{j=1}^{n} r_j}},$$

$$\hat{F} = \left(1-\prod_{i_1,i_2,...,i_n=1}^{n}\left(\left(1-\prod_{j=1}^{n} F_{i_j}^{r_j}\right)^{\prod_{j=1}^{n}\omega_{i_j}}\right)\right)^{\frac{1}{\sum_{j=1}^{n} r_j}}.$$

The proof of Theorem 6 is similar to that of Theorem 2.

Likewise, an example is used to explain the calculation of the DGNNNWGBM operator.

Example 4. *Let $\tilde{a}_1 = \langle(0.5,0.03),(0.5,0.4,0.1)\rangle$, $\tilde{a}_2 = \langle(0.8,0.015),(0.9,0.2,0.2)\rangle$ be two NNNNs. The weighted vector $\omega = (0.6,0.4)^T$, and the parameter vector $R = (3,4)^T$, then, according to Theorem 6, we have*

$\hat{a} = \frac{1}{3+4}\prod_{i_1,i_2=1}^{2}(r_1 a_{i_1} + r_2 a_{i_2})^{\omega_{i_1}\omega_{i_2}}$
$= \frac{1}{7}(r_1a_1+r_2a_1)^{\omega_1\omega_1}(r_1a_1+r_2a_2)^{\omega_1\omega_2}(r_1a_2+r_2a_1)^{\omega_2\omega_1}(r_1a_2+r_2a_2)^{\omega_2\omega_2}$
$= \frac{1}{7}(3\times0.5+4\times0.5)^{0.6\times0.6}(3\times0.5+4\times0.8)^{0.6\times0.4}(3\times0.8+4\times0.5)^{0.4\times0.6}(3\times0.8+4\times0.8)^{0.4\times0.4}$
= *0.6112*

Similarly, we can obtain $\hat{\sigma} = 0.0234$.

$\hat{T} = 1-(1-\prod_{i_1,i_2=1}^{2}(1-(1-T_{i_1})^{r_1}(1-T_{i_2})^{r_2})^{\omega_{i_1}\omega_{i_2}})^{\frac{1}{3+4}}$
$= 1-(1-(1-(1-T_1)^{r_1}(1-T_1)^{r_2})^{\omega_1\omega_1}(1-(1-T_1)^{r_1}(1-T_2)^{r_2})^{\omega_1\omega_2}(1-(1-T_2)^{r_1}(1-T_1)^{r_2})^{\omega_2\omega_1}(1-(1-T_2)^{r_1}(1-T_2)^{r_2})^{\omega_2\omega_2})^{\frac{1}{7}}$
$= 1-(1-(1-(1-0.5)^3(1-0.5)^4)^{0.6\times0.6}\times(1-(1-0.5)^3(1-0.9)^4)^{0.6\times0.4}\times(1-(1-0.9)^3(1-0.5)^4)^{0.4\times0.6}(1-(1-0.9)^3(1-0.9)^4)^{0.4\times0.4})^{\frac{1}{7}}$
= *0.5674*

$\hat{I} = (1-\prod_{i_1,i_2=1}^{2}(1-I_{i_1}^{r_1}I_{i_2}^{r_2})^{\omega_{i_1}\omega_{i_2}})^{\frac{1}{3+4}}$
$= (1-(1-I_1^{r_1}I_1^{r_2})^{\omega_1\omega_1}(1-I_1^{r_1}I_2^{r_2})^{\omega_1\omega_2}(1-I_2^{r_1}I_1^{r_2})^{\omega_2\omega_1}(1-I_2^{r_1}I_2^{r_2})^{\omega_2\omega_2})^{\frac{1}{7}}$
$= (1-(1-0.4^3\times0.4^4)^{0.6\times0.6}(1-0.4^3\times0.2^4)^{0.6\times0.4}(1-0.2^3\times0.4^4)^{0.4\times0.6}(1-0.2^3\times0.2^4)^{0.4\times0.4})^{\frac{1}{7}}$
= *0.3517*

Similarly, we can obtain $\hat{I} = 0.1598$.

So, $DGNNNWGBM_{\omega}^{R}(\tilde{a}_1,\tilde{a}_2) = \langle(0.6112,0.0234),(0.5674,0.3517,0.1598)\rangle$.

The DGNNNWGBM operator has the same properties as the DGNNNWBM operator. The proof is also similar to that of the DGNNNWBM operator. Of particular note, the DGNNNWGBM operator satisfies the property of idempotency.

Theorem 7. *(Idempotency) Let $\{\tilde{a}_i|\tilde{a}_i = \langle(a_i,\sigma_i),(T_i,I_i,F_i)\rangle, i=1,2,...,n\}$ be a set of NNNNs. If all $\tilde{a}_i = \tilde{a}$, then*

$$DGNNNWGBM_{\omega}^{R}(\tilde{a}_1,\tilde{a}_2,...,\tilde{a}_n) = \tilde{a}.$$

Proof. Since $\tilde{a}_i = \tilde{a}(i = 1, 2, ..., n)$, according to operational rules,

$$DGNNNWGBM_{\omega}^{R}(\tilde{a}_1, \tilde{a}_2, ..., \tilde{a}_n)$$
$$= \frac{1}{\sum_{j=1}^{n} r_j}\left(\bigotimes_{i_1,i_2,...i_n=1}^{n}\left(\bigoplus_{j=1}^{n}(r_j\tilde{a}_{i_j})\right)^{*\prod_{j=1}^{n}\omega_{i_j}}\right)$$
$$= \frac{1}{\sum_{j=1}^{n} r_j}\left(\bigotimes_{i_1,i_2,...,i_n=1}^{n}\left(\sum_{j=1}^{n} r_j\tilde{a}\right)^{*\prod_{j=1}^{n}\omega_{i_j}}\right)$$
$$= \frac{1}{\sum_{j=1}^{n} r_j}\left(\sum_{j=1}^{n} r_j\tilde{a}\right)^{*\sum_{i_1,i_2,...,i_n=1}^{n}\prod_{j=1}^{n}\omega_{i_j}}$$

Here, $\sum_{i_1,i_2,...,i_n=1}^{n}\prod_{j=1}^{m}\omega_{i_j} = 1$ is proved by mathematical induction.

When $m = 2$, we have $\sum_{i_1,i_2=1}^{n}\omega_{i_1}\omega_{i_2} = \sum_{i_1=1}^{n}\omega_{i_1}\sum_{i_2=1}^{n}\omega_{i_2} = 1$.

Suppose $m = k - 1$, and $\sum_{i_1,i_2,...,i_{k-1}=1}^{n}\prod_{j=1}^{k-1}\omega_{i_j} = 1$,

so when $m = k$, we get

$$\sum_{i_1,i_2,...,i_k=1}^{n}\prod_{j=1}^{k}\omega_{i_j} = \sum_{i_1,i_2,...,i_k=1}^{n}\prod_{j=1}^{k-1}\omega_{i_j}\omega_{i_k} = \sum_{i_1,i_2,...,i_{k-1}=1}^{n}\prod_{j=1}^{k-1}\omega_{i_j}\sum_{i_k=1}^{k}\omega_{i_k} = 1.$$

Then

$$DGNNNWGBM_{\omega}^{R}(\tilde{a}_1, \tilde{a}_2, ..., \tilde{a}_n) = \tilde{a}.$$

That completes the proof. □

Theorem 8. *(Monotonicity) Let* $\{\tilde{a}_i|\tilde{a}_i = \langle(a_i, \sigma_i), (T_i, I_i, F_i)\rangle, i = 1, 2, ..., n\}$ *and* $\{\tilde{b}_i|\tilde{b}_i = \langle(b_i, \delta_i), (T_{b_i}, I_{b_i}, F_{b_i})\rangle, i = 1, 2, ..., n\}$ *be two sets of NNNNs. If* $a_i \leq b_i$, $\sigma_i \geq \delta_i$, $T_{a_i} \leq T_{b_i}$, $I_{a_i} \geq I_{b_i}$, $F_{a_i} \geq F_{b_i}$, *and* $(T_{a_i} - T_{b_i})^2 + (I_{a_i} - I_{b_i})^2 + (F_{a_i} - F_{b_i})^2 \neq 0$ *hold for any i, then*

$$DGNNNWGBM_{\omega}^{R}(\tilde{a}_1, \tilde{a}_2, ..., \tilde{a}_n) < DGNNNWGBM_{\omega}^{R}(\tilde{b}_1, \tilde{b}_2, ..., \tilde{b}_n).$$

Theorem 9. *(Boundedness) Let* $\{\tilde{a}_i|\tilde{a}_i = \langle(a_i, \sigma_i), (T_i, I_i, F_i)\rangle, i = 1, 2, ..., n\}$ *be a set of NNNNs. If* $\tilde{a}^+ = \langle(max_i(a_i), min_i(\sigma_i)), (max_i(T_i), min_i(I_i), min_i(F_i)\rangle$ *and* $\tilde{a}^- = \langle(min_i(a_i), max_i(\sigma_i)), (min_i(T_i), max_i(I_i), max_i(F_i))\rangle$, *then*

$$\tilde{a}^- \leq DGNNNWGBM_{\omega}^{R}(\tilde{a}_1, \tilde{a}_2, ..., \tilde{a}_n) \leq \tilde{a}^+$$

Theorem 10. *(Commutativity) Let* $\{\tilde{a}_i|\tilde{a}_i = \langle(a_i, \sigma_i), (T_i, I_i, F_i)\rangle, i = 1, 2, ..., n\}$ *be a set of NNNNs.* $\tilde{a}_i'$ *is any permutation of* $\tilde{a}_i$, *then*

$$DGNNNWGBM_{\omega}^{R}(\tilde{a}_1, \tilde{a}_2, ..., \tilde{a}_n) = DGNNNWGBM_{\omega}^{R}(\tilde{a}_1', \tilde{a}_2', ..., \tilde{a}_n').$$

4. A Multiple Attribute Decision-Making Method on the Basis of the DGNNNWBM Operator and DGNNNWGBM Operator

In this section, based on the NNNN, we utilize the DGNNNWBM operator or DGNNNWGBM operator to solve the MADM problem.

Let $A = \{A_1, A_2, ..., A_m\}$ be a set of the alternatives, and $C = \{C_1, C_2, ..., C_n\}$ be a set of the attributes; the weight vector of the attribute is $\omega = (\omega_1, \omega_2, ..., \omega_n)^T$, where $\omega_j \in [0, 1]$ and $\sum_{j=1}^{n}\omega_j = 1$. Let $D = (\tilde{a}_{ij})_{m\times n}$ be the decision matrix, and $\tilde{a}_{ij} = \langle(a_{ij}, \sigma_{ij}), (T_{ij}, I_{ij}, F_{ij})\rangle$ be the evaluation value of the alternative A_i with respect to attribute C_j, denoted by the form of NNNN.

The DGNNNWBM operator or DGNNNWGBM operator can be used to handle the MADM problem, and the steps are shown as follows:

Step 1. Standardize the decision matrix.

If all the attributes C_i are of the same type, then the attribute values do not need standardization. If there is a different type, the attributes should be converted so they are of the same type. Suppose the decision matrix $D = (\tilde{a}_{ij})_{m\times n}$ transforms to the standardized matrix $\tilde{D} = (\tilde{a}'_{ij})_{m\times n}$.

According to [6], we have the following standardization method. For the benefit attribute:

$$\tilde{a}'_{ij} = \left\langle \left(\frac{a_{ij}}{max_{1\le i\le m}(a_{ij})}, \frac{\sigma_{ij}}{max_{1\le i\le m}(\sigma_{ij})} \frac{\sigma_{ij}}{a_{ij}} \right), (T_{ij}, I_{ij}, F_{ij}) \right\rangle.$$

For the cost attribute:

$$\tilde{a}'_{ij} = \left\langle \left(\frac{min_{1\le i\le m}(a_{ij})}{a_{ij}}, \frac{\sigma_{ij}}{max_{1\le i\le m}(\sigma_{ij})} \frac{\sigma_{ij}}{a_{ij}} \right), (F_{ij}, 1 - I_{ij}, T_{ij}) \right\rangle.$$

Step 2. Utilize the DGNNNWBM operator

$$\tilde{a}_i = DGNNNWBM^R_\omega(\tilde{a}_{i1}, \tilde{a}_{i2}, ..., \tilde{a}_{in}) = \langle (a_i, \sigma_i), (T_i, I_i, F_i) \rangle$$

or the DGNNNWGBM operator

$$\tilde{a}_i = DGNNNWGBM^R_\omega(\tilde{a}_{i1}, \tilde{a}_{i2}, ..., \tilde{a}_{in}) = \langle (a_i, \sigma_i), (T_i, I_i, F_i) \rangle$$

for comprehensive evaluation.

Step 3. According to rank principles, which are shown in Definitions 10 and 11, rank the alternatives $A_1, A_2, ..., A_m$ and choose the best one.

5. Numerical Example and Comparative Analysis

In this section, the effectiveness of the proposed MADM method is illustrated, demonstrating the effect of different parameter values on the final ranking results. Finally, the advantages of the proposed method are illustrated by comparison.

5.1. The Numerical Example

In the following, the application of the proposed method is illustrated by a numerical example.

Example 5. *Patients choose a hospital according to their own needs. There are five alternatives hospitals to choose from: (1) A_1 is a people's hospital; (2) A_2 is a city hospital; (3) A_3 is a second city hospital; (4) A_4 is the first affiliated hospital; and (5) A_5 is the second affiliated hospital. There are four evaluation attributes: (1) C_1 is the hardware and software facilities; (2) C_2 is the physician team; (3) C_3 is the consumption index; and (4) C_4 is the service quality. We know the attributes C_1,C_2, and C_4 are benefit criteria, and C_3 is cost. The weight vector of the attributes is $\omega = (0.2, 0.4, 0.3, 0.1)^T$. The final evaluation outcomes are expressed by the NNNN, which is shown in Table 3.*

Table 3. The nonnegative normal neutrosophic decision matrix D.

	C_1	C_2	C_3	C_4
A_1	$\langle(4,0.3),(0.7,0.2,0.3)\rangle$	$\langle(7,0.7),(0.6,0.1,0.1)\rangle$	$\langle(5.5,0.6),(0.3,0.3,0.6)\rangle$	$\langle(6,0.4),(0.7,0.2,0.4)\rangle$
A_2	$\langle(5,0.2),(0.5,0.4,0.5)\rangle$	$\langle(8,0.5),(0.7,0.2,0.3)\rangle$	$\langle(6,0.2),(0.2,0.1,0.7)\rangle$	$\langle(7,0.6),(0.4,0.2,0.7)\rangle$
A_3	$\langle(3,0.5),(0.3,0.3,0.4)\rangle$	$\langle(6,0.2),(0.4,0.5,0.3)\rangle$	$\langle(4,0.7),(0.3,0.5,0.5)\rangle$	$\langle(5.5,0.4),(0.5,0.4,0.2)\rangle$
A_4	$\langle(4.5,0.6),(0.3,0.5,0.3)\rangle$	$\langle(5,0.4),(0.6,0.4,0.5)\rangle$	$\langle(7,0.4),(0.3,0.3,0.5)\rangle$	$\langle(4,0.5),(0.8,0.2,0.5)\rangle$
A_5	$\langle(6,0.5),(0.8,0.1,0.2)\rangle$	$\langle(6.5,0.6),(0.8,0.2,0.4)\rangle$	$\langle(5,0.3),(0.4,0.2,0.6)\rangle$	$\langle(5,0.6),(0.5,0.5,0.2)\rangle$

Step 1. Since C_1,C_2, and C_4 are benefit attributes, we have

$a'_{11} = \frac{a_{11}}{max_{1\le i\le 5} a_{i1}} = \frac{4}{6} = 0.6667, \sigma'_{11} = \frac{\sigma_{11}}{max_{1\le i\le 5}\sigma_{i1}} \frac{\sigma_{11}}{a_{11}} = \frac{0.3}{0.6}\frac{0.3}{4} = 0.0375, T'_{11} = T_{11}, I'_{11} = I_{11}, I'_{11} = I_{11},$

and C_3 is the cost attribute, so we have

$$a'_{13} = \frac{min_{1\le i\le 5}}{a_{13}} = \frac{4}{5.5} = 0.7273, \sigma_{13} = \frac{\sigma_{13}}{max_{1\le i\le 5}\sigma_{i1}} \frac{\sigma_{13}}{a_{13}} = \frac{0.6}{0.7}\frac{0.6}{5.5} = 0.0935,$$

$$T'_{13} = F_{13} = 0.6, I'_{13} = 1 - I_{13} = 0.7, F'_{13} = T_{13} = 0.3.$$

The normalized decision matrix is shown in Table 4.

Table 4. Normalized decision matrix $\tilde{D}$.

	C_1	C_2
A_1	$\langle(0.6667, 0.0375), (0.7, 0.2, 0.3)\rangle$	$\langle(0.85, 0.1), (0.6, 0.1, 0.1)\rangle$
A_2	$\langle(0.8333, 0.0133), (0.5, 0.4, 0.5)\rangle$	$\langle(1, 0.0446), (0.7, 0.2, 0.3)\rangle$
A_3	$\langle(0.5, 0.1389), (0.3, 0.3, 0.4)\rangle$	$\langle(0.75, 0.0095), (0.4, 0.5, 0.3)\rangle$
A_4	$\langle(0.75, 0.1333), (0.3, 0.5, 0.3)\rangle$	$\langle(0.625, 0.0457), (0.6, 0.4, 0.5)\rangle$
A_5	$\langle(1, 0.0694), (0.8, 0.1, 0.2)\rangle$	$\langle(0.8125, 0.0791), (0.8, 0.2, 0.4)\rangle$
	C_3	C_4
A_1	$\langle(0.7273, 0.0935), (0.6, 0.7, 0.3)\rangle$	$\langle(0.8571, 0.0444), (0.7, 0.2, 0.4)\rangle$
A_2	$\langle(0.6667, 0.0095), (0.7, 0.9, 0.2)\rangle$	$\langle(1, 0.0857), (0.4, 0.2, 0.7)\rangle$
A_3	$\langle(1, 0.1750), (0.5, 0.5, 0.3)\rangle$	$\langle(0.7857, 0.0485), (0.5, 0.4, 0.2)\rangle$
A_4	$\langle(0.5714, 0.0327), (0.5, 0.7, 0.3)\rangle$	$\langle(0.5714, 0.1042), (0.8, 0.2, 0.5)\rangle$
A_5	$\langle(0.8, 0.0257), (0.6, 0.8, 0.4)\rangle$	$\langle(0.7143, 0.12), (0.5, 0.5, 0.2)\rangle$

Step 2. Calculate the comprehensive evaluation value of each alternative by using the DGNNNWBM (DGNNNWGBM) operator (suppose $R = (1, 1, 1, 1)^T$), which is shown in Table 5. (There are 256 cases in this example, which are not listed here. MATLAB can be used for the calculations.)

Table 5. Utilization of the dual generalized nonnegative normal neutrosophic weighted Bonferroni mean (DGNNNWBM) operator and dual generalized nonnegative normal neutrosophic weighted geometric Bonferroni mean (DGNNNWGBM) operator R = (1,1,1,1).

	DGNNNWBM	DGNNNWGBM
A_1	$\langle(0.7772, 0.08), (0.8161, 0.0525, 0, 0303)\rangle$	$\langle(0.7763, 0.0788), (0.6299, 0.313, 0.2301)\rangle$
A_2	$\langle(0.8667, 0.0319), (0.8095, 0.2210, 0.1474)\rangle$	$\langle(0.8636, 0.0295), (0.6296, 0.4622, 0.3507)\rangle$
A_3	$\langle(0.7786, 0.0889), (0.5042, 0.2935, 0.1071)\rangle$	$\langle(0.7736, 0.0788), (0.4194, 0.4503, 0.31)\rangle$
A_4	$\langle(0.6286, 0.0652), (0.6902, 0.3319, 0.2185)\rangle$	$\langle(0.6277, 0.0622), (0.5287, 0.4919, 0.4004)\rangle$
A_5	$\langle(0.8364, 0.0652), (0.9198, 0.1201, 0.1236)\rangle$	$\langle(0.8353, 0.0635), (0.7098, 0.3965, 0.3402)\rangle$

Step 3. According to Definition 10, for the DGNNNWBM operator,

$$S(A_1) = (0.7772 + \frac{1}{0.08})(2 + 0.8161 - 0.0525 - 0.0303),$$
$$S(A_2) = 78.59, S(A_3) = 25.29, S(A_4) = 34.18, S(A_5) = 43.26.$$

By the ranking principle of Definition 11, we obtain $A_2 > A_5 > A_1 > A_4 > A_3$, which is shown in Table 6. The best alternative is A_2.

Table 6. The score of the alternatives.

	$S(A_1)$	$S(A_2)$	$S(A_3)$	$S(A_4)$	$S(A_5)$	Ranking
DGNNNWBM	36.29	78.59	25.29	34.18	43.26	$A_2 > A_5 > A_1 > A_4 > A_3$
DGNNNWGBM	28.10	63.07	22.34	27.33	32.73	$A_2 > A_5 > A_4 > A_1 > A_3$

5.2. Influence Analysis

To show the effects on the ranking results by altering the parameters of the DGNNNWBM and DGNNNWGBM operators, according to Definition 10 and 11, we can get the results by using MATLAB, which is shown in Tables 7 and 8.

Table 7. Ranking for different parameters of DGNNNWBM.

R	$S(A_1)$	$S(A_2)$	$S(A_3)$	$S(A_4)$	$S(A_5)$	Ranking
(1, 1, 1, 1)	36.29	78.59	25.29	34.18	43.26	$A_2 > A_5 > A_1 > A_4 > A_3$
(3, 3, 3, 3)	28.85	47.48	14.96	22.25	31.82	$A_2 > A_5 > A_1 > A_4 > A_3$
(4, 4, 4, 4)	27.84	42.48	14.06	20.61	30.08	$A_2 > A_5 > A_1 > A_4 > A_3$
(6, 6, 6, 6)	27.00	37.80	13.34	19.18	28.06	$A_2 > A_5 > A_1 > A_4 > A_3$
(10, 1, 1, 1)	29.45	41.52	15.29	21.46	29.89	$A_2 > A_5 > A_1 > A_4 > A_3$
(10, 10, 1, 1)	27.77	36.63	13.95	19.62	27.49	$A_2 > A_1 > A_5 > A_4 > A_3$
(10, 10, 10, 1)	27.04	30.94	12.34	16.30	23.70	$A_2 > A_1 > A_5 > A_4 > A_3$
(14, 15, 1, 1)	26.79	32.35	9.60	18.47	25.12	$A_2 > A_1 > A_5 > A_4 > A_3$
(16, 17, 18, 19)	26.70	31.51	9.61	18.52	24.59	$A_2 > A_1 > A_5 > A_4 > A_3$
(20, 20, 20, 20)	26.72	31.17	9.63	18.56	24.38	$A_2 > A_1 > A_5 > A_4 > A_3$

Table 8. Ranking for different parameters of DGNNNWGBM.

R	$S(A_1)$	$S(A_2)$	$S(A_3)$	$S(A_4)$	$S(A_5)$	Ranking
(1, 1, 1, 1)	28.10	63.07	22.34	27.33	32.73	$A_2 > A_5 > A_4 > A_1 > A_3$
(3, 3, 3, 3)	25.29	54.01	21.92	25.55	29.07	$A_2 > A_5 > A_4 > A_1 > A_3$
(4, 4, 4, 4)	24.47	50.6	21.74	24.80	27.89	$A_2 > A_5 > A_4 > A_1 > A_3$
(6, 6, 6, 6)	23.46	45.62	21.40	23.53	26.40	$A_2 > A_5 > A_4 > A_1 > A_3$
(10, 1, 1, 1)	24.07	46.39	21.22	23.28	27.02	$A_2 > A_5 > A_1 > A_4 > A_3$
(10, 10, 1, 1)	23.11	42.60	21.01	22.45	25.72	$A_2 > A_5 > A_1 > A_4 > A_3$
(10, 10, 10, 1)	22.81	40.93	20.92	22.10	25.14	$A_2 > A_5 > A_1 > A_4 > A_3$
(14, 15, 1, 1)	22.68	40.98	22.63	22.03	25.23	$A_2 > A_5 > A_1 > A_3 > A_4$
(16, 17, 18, 19)	31.59	53.67	31.74	28.52	37.37	$A_2 > A_5 > A_3 > A_1 > A_4$
(20, 20, 20, 20)	31.52	74.73	31.64	28.30	37.27	$A_2 > A_5 > A_3 > A_1 > A_4$

As shown in Table 7, when the parameter values are small, the ranking of the alternatives may be of little influence. When the parameter values are large, the ordering of A_1 and A_5 changes. However, the best alternative is the same, i.e., A_2. As shown in Table 8, when the parameter values are small, the ranking of the alternatives may be of little influence, but when the parameter values are large, it has a great impact on the ranking results. Although the ranking changes greatly, the best alternative is still A_2. In practical applications, we usually take $R = (1, 1, .., 1)^T$, which is not only intuitive but also takes into account the effect of multiple parameters.

5.3. Comparison Analysis

In this section, we compare the DGNNNWBM and DGNNNWGBM operators proposed in this paper with the normal neutrosophic weighted Bonferroni mean (NNWBM) operator and normal neutrosophic weighted geometric Bonferroni mean (NNWGBM) operator proposed by Liu P and Li H [6] for dealing with Example 5.1. The results are shown in Tables 8–14, where we take the first two values of the parameter R in the DGNNNWBM and DGNNNWGBM operators as the parameter values p, q in the NNWBM and NNWGBM operators.

According to the result, we conclude the following:

(1) From Tables 7 and 9, when p, q take different values and the values are small, the NNWBM operator has three different ranking results, while the DGNNNWBM operator has only one. It shows that the stability of the DGNNNWBM operator is better than that of the NNWBM operator.

(2) From Tables 8 and 10, there is only one ranking result of the NNWGBM operaotr. However, Tables 11–14 show that when the parameter values p, q are taken as $(10, 10)$ and $(14, 15)$, the result of the NNWBM operator is $T = 0, I = 1, F = 1$, and the NNWGBM operator result is $T = 1, I = 0, F = 0$. Regardless of whether the parameters p, q change, the values of a and δ in the NNWGBM operator are invariant. However, in this case, the DGNNNWBM and DGNNNWGBM

operators consider more parameters, so they can overcome these problems that arise in the NNWBM and NNWGBM operators.

From this, we know that the NNWBM and NNWGBM operators lack stability and sensitivity. Compared to the NNWBM and NNWGBM operators, the DGNNNWBM and DGNNNWGBM are not only more general, but they are also more flexible.

Table 9. Liu and Li's method [6] (ranking for different parameters of the normal neutrosophic weighted Bonferroni mean (NNWBM)).

(p,q)	$S(A_1)$	$S(A_2)$	$S(A_3)$	$S(A_4)$	$S(A_5)$	**Ranking**
(1, 1)	69.31	126.23	24.29	52.83	72.66	$A_2 > A_5 > A_1 > A_4 > A_3$
(3, 3)	78.69	209.45	57.12	77.09	108.89	$A_2 > A_5 > A_1 > A_4 > A_3$
(4, 4)	70.23	200.1	54.65	74.78	102.21	$A_2 > A_5 > A_4 > A_1 > A_3$
(6, 6)	58.8	182.3	44.67	68.65	88.42	$A_2 > A_5 > A_4 > A_1 > A_3$
(10, 1)	67.93	127.63	43.67	80.84	74.68	$A_2 > A_4 > A_5 > A_1 > A_3$

Table 10. Liu and Li's method [6] (ranking for different parameters of the normal neutrosophic weighted geometric Bonferroni mean (NNWGBM)).

(p,q)	$S(A_1)$	$S(A_2)$	$S(A_3)$	$S(A_4)$	$S(A_5)$	**Ranking**
(1, 1)	62.89	149.2	14.17	54.77	70.31	$A_2 > A_5 > A_1 > A_4 > A_3$
(3, 3)	61.89	146.55	13.78	53.1	68.49	$A_2 > A_5 > A_1 > A_4 > A_3$
(4, 4)	61.6	145.91	13.66	52.71	68.02	$A_2 > A_5 > A_1 > A_4 > A_3$
(6, 6)	61.24	145.12	13.78	52.23	67.45	$A_2 > A_5 > A_1 > A_4 > A_3$
(10, 1)	58.03	128.07	9.32	49.42	61.20	$A_2 > A_5 > A_1 > A_4 > A_3$

Table 11. The DGNNNWBM operator and DGNNWGBM operator $R = (10, 10, 10, 1)$.

	DGNNNWBM	**DGNNWGBM**
A_1	$\langle(0.8053, 0.0938), (0.6119, 0.1447, 0.1513)\rangle$	$\langle(0.776, 0.0785), (0.6137, 0.6069, 0.3126)\rangle$
A_2	$\langle(0.9374, 0.0665), (0.5108, 0.247, 0.2609)\rangle$	$\langle(0.8628, 0.0289), (0.5141, 0.7837, 0.5514)\rangle$
A_3	$\langle(0.7006, 0.1533), (0.3745, 0.3875, 0.2823)\rangle$	$\langle(0.4772, 0.0757), (0.3809, 0.4822, 0.3454)\rangle$
A_4	$\langle(0.6559, 0.1119), (0.3997, 0.3468, 0.3391)\rangle$	$\langle(0.6256, 0.0615), (0.4053, 0.6176, 0.4645)\rangle$
A_5	$\langle(0.8808, 0.0948), (0.5941, 0.1819, 0.272)\rangle$	$\langle(0.8351, 0.063), (0.5961, 0.6959, 0.3844)\rangle$

Table 12. Liu and Li's method [6] (the NNWBM operator and NNWGBM operator $p = 10, q = 10$).

	NNWBM	**NNWGBM**
A_1	$\langle(0.2491, 0.0223), (0.252, 0.6278, 0.6089)\rangle$	$\langle(0.938, 0.0457), (1, 0, 0)\rangle$
A_2	$\langle(0.2608, 0.0063), (0.3112, 0.7388, 0.6502)\rangle$	$\langle(0.962, 0.018), (1, 0.2037, 0)\rangle$
A_3	$\langle(0.159, 0.0213), (0.1708, 0.7887, 0.6887)\rangle$	$\langle(0.8029, 0.2152), (0.7628, 0.1948, 0)\rangle$
A_4	$\langle(0.1913, 0.011), (0.2205, 0.8023, 0.7488)\rangle$	$\langle(0.8898, 0.0494), (0.81, 0.2168, 0)\rangle$
A_5	$\langle(0.2574, 0.0138), (0.3343, 0.6173, 0.7303)\rangle$	$\langle(0.9555, 0.0401), (1, 0.1649, 0)\rangle$

Table 13. The DGNNNWBM operator and DGNNNWGBM operator $R = (14, 15, 1, 1)$.

	DGNNNWBM	**DGNNNWGBM**
A_1	$\langle(0.8131, 0.0944), (0.6071, 0.1287, 0.1299)\rangle$	$\langle(0.7755, 0.0778), (0.611, 0.6145, 0.3321)\rangle$
A_2	$\langle(0.9493, 0.0693), (0.4878, 0.2317, 0.2434)\rangle$	$\langle(0.8611, 0.0278), (0.4944, 0.7966, 0.5776)\rangle$
A_3	$\langle(0.7075, 0.1552), (0.3562, 0.364, 0.2669)\rangle$	$\langle(0.4607, 0.0697), (0.3683, 0.4855, 0.3546)\rangle$
A_4	$\langle(0.6684, 0.1148), (0.3741, 0.3137, 0.323)\rangle$	$\langle(0.6253, 0.0603), (0.3846, 0.6332, 0.4713)\rangle$
A_5	$\langle(0.8997, 0.0995), (0.5762, 0.1641, 0.248)\rangle$	$\langle(0.8345, 0.062), (0.5809, 0.7067, 0.3868)\rangle$

Table 14. Liu and Li's method [6] (the NNWBM operator and NNWGBM operator $p = 14, q = 15$).

	NNWBM	NNWGBM
A_1	$\langle(0.2563, 0.0275), (0, 0.6164, 0.5947)\rangle$	$\langle(0.938, 0.0457), (1, 0, 0)\rangle$
A_2	$\langle(0.2671, 0.0077), (0.3202, 1, 0.6403)\rangle$	$\langle(0.962, 0.018), (1, 0, 0)\rangle$
A_3	$\langle(0.1638, 0.0271), (0, 1, 0.6798)\rangle$	$\langle(0.8029, 0.2152), (1, 0, 0)\rangle$
A_4	$\langle(0.1955, 0.0125), (0, 1, 1)\rangle$	$\langle(0.8898, 0.0494), (1, 0, 0)\rangle$
A_5	$\langle(0.2633, 0.0168), (0.3418, 0.6064, 1)\rangle$	$\langle(0.9555, 0.0401), (1, 0, 0)\rangle$

6. Conclusions

The multiple attribute decision-making method has a wide range of applications in many domains. The nonnegative normal neutrosophic number is more suitable for dealing with uncertain information, and the dual generalized weighted Bonferroni mean operator and dual generalized weighted geometric Bonferroni mean operator take into account the relationship between arbitrary aggregation arguments. Therefore, in this paper, the definition of nonnegative normal neutrosophic number has been proposed. The score function and accuracy function have been developed to overcome the deficiency, i.e., that the original function does not satisfy the ranking principle. Considering the connections between any two or more than two aggregation arguments, the dual generalized nonnegative normal neutrosophic weighted Bonferroni mean operator and dual generalized nonnegative normal neutrosophic weighted geometric Bonferroni mean operator were discussed. Meanwhile, some properties were investigated, such as idempotency, monotonicity, boundedness, and commutativity. Based on the dual generalized nonnegative normal neutrosophic weighted Bonferroni mean operator and dual generalized nonnegative normal neutrosophic weighted geometric Bonferroni mean operator, a method was developed to deal with a multiple attribute decision-making problem with nonnegative normal neutrosophic number. Further, we used the dual generalized nonnegative normal neutrosophic weighted Bonferroni mean and dual generalized nonnegative normal neutrosophic weighted geometric Bonferroni mean operators for aggregative information. Decision making obtain the satisfactory alternative according to actual need and preference by changing the values of R, which makes our proposed multiple attribute decision-making method more flexible and reliable. Further, compared with the method in Liu [6], our method shows that when the relationship between more aggregation arguments are considered, the aggregation result is more stable; when the parameter value is larger, the aggregation result is more sensitive.

Author Contributions: J.M. conceived, wrote and revised this paper; H.-L.H. provided ideas and suggestions for the revision of the paper.

Funding: This research was funded by National Natural Science Foundation Project, grant number (11701089); Fujian Natural Science Foundation, grant number (2018J01422).

Acknowledgments: This paper is supported by Institute of Meteorological Big Data-Digital Fujian and Fujian Key Laboratory of Data Science and Statistics.

Conflicts of Interest: The authors declare no conflict of interest.

References

1. Zadeh, L.A. Fuzzy sets. *Inform. Contr.* **1965**, *8*, 338–356. [CrossRef]
2. Atanassov, K.T. Intuitionistic fuzzy sets. *Fuzzy. Set. Syst.* **1986**, *20*, 87–96. [CrossRef]
3. Smarandache, F. *A Unifying Field in Logics: Neutrosophy Logic*; American Research Press: Rehoboth, DE, USA, 1999; pp. 1–141.
4. Yang, M.S.; Ko, C.H. On a class of fuzzy c-numbers clustering procedures for fuzzy data. *Fuzzy Set. Syst.* **1996**, *84*, 49–60. [CrossRef]
5. Wang, J.Q.; Li, K.J. Multi-criteria decision-making method based on induced intuitionistic normal fuzzy related aggregation operators. *Int. J. Uncertain. Fuzziness Knowl. Base. Syst.* **2012**, *20*, 559–578. [CrossRef]
6. Liu, P.; Li, H. Multiple attribute decision-making method based on some normal neutrosophic bonferroni mean operators. *Neural Comput. Appl.* **2017**, *28*, 1–16. [CrossRef]

7. Chatterjee, P.; Mondal, S.; Boral, S.; Banerjee, A.; Chakraborty, S. A novel hybrid method for non-traditional machining process selection using factor relationship and multi-attribute border approximation method. *Facta Univ. Ser. Mech. Eng.* **2017**, *15*, 439–456. [CrossRef]
8. Petković, D.; Madić, M.; Radovanović, M.; Gečevska, V. Application of the performance selection index method for colving machining MCDM problems. *Facta Univ. Ser. Mech. Eng.* **2017**, *15*, 97–106. [CrossRef]
9. Roy, J.; Adhikary, K.; Kar, S.; Pamučar, D. A rough strength relational DEMATEL model for analysing the key success factors of hospital service quality. *Decis. Mak. Appl. Manag. Eng.* **2018**, *1*, 121–142. [CrossRef]
10. Badi, I.A.; Abdulshahed, A.M.; Shetwan, A. A case study of supplier selection for a steelmaking company in Libya by using the Combinative Distance-based ASsessment (CODAS) model. *Decis. Mak. Appl. Manag. Eng.* **2018**, *1*, 1–12. [CrossRef]
11. Lee, S.H.; Ryu, K.H.; Sohn, G. Study on entropy and similarity measure for fuzzy set. *IEICE Trans. Inform. Syst.* **2009**, *92*, 1783–1786. [CrossRef]
12. Nan, J.X.; Wang, T.; An, J.J. Intuitionistic fuzzy distance based TOPSIS method and application to MADM. *Int. J. Fuzzy Syst. Appl.* **2017**, *5*, 43–56. [CrossRef]
13. Joshi, R.; Kumar, S. A new parametric intuitionistic fuzzy entropy and its applications in multiple attribute decision making. *Int. J. Appl. Comput. Sci. Math.* **2018**, *4*, 52. [CrossRef]
14. Liu. M.; Ren, H. A new intuitionistic fuzzy entropy and application in multi-attribute decision making. *Information* **2014**, *5*, 587–601. [CrossRef]
15. Atanassov, K.; Gargov, G. Interval-valued intuitionistic fuzzy sets. *Fuzzy Set. Syst.* **1989**, *31*, 343–349. [CrossRef]
16. Huang, H.L. (T, S)-based interval-valued intuitionistic fuzzy composition matrix and its application for clustering. *Iranian J. Fuzzy Syst.* **2012**, *9*, 7–19.
17. Chen, S.M.; Huang, Z.C. Multiattribute decision making based on interval-valued intuitionistic fuzzy values and linear programming methodology. *Inform. Sci.* **2017**, *381*, 341–351. [CrossRef]
18. Biswas, A.; Kumar, S. An integrated TOPSIS approach to MADM with interval-valued intuitionistic fuzzy settings. *Adv. Comput. Commun. Paradig.* **2018**, *706*, 533–543.
19. Peng, J.J.; Wang, J.Q.; Zhang, H.Y. An outranking approach for multi-criteria decision-making problems with simplified neutrosophic sets. *Appl. Soft Comput.* **2014**, *25*, 336–346. [CrossRef]
20. Bausys, R.; Zavadskas, E.K.; Kaklauskas, A. Application of Neutrosophic Set to Multicriteria Decision Making by COPRAS. *Econ. Comput. Econ. Cybern. Stud. Res.* **2015**, *49*, 91–106.
21. Majumdarar, P.; Samant, S.K. On similarity and entropy of neutrosophic sets. *J. Intell. Fuzzy Syst.* **2014**, *26*, 1245–1252.
22. Wang, H.; Smarandache, F.; Zhang,Y.Q.; Sunderraman, R. *Interval Neutrosophic Sets and Logic: Theory and Applications in Computing*; Hexis: Corona, CA, USA, 2005.
23. Wang, H.; Smarandache, F.; Zhang, Y.Q.; Sunderraman, R. Single valued neutrosophic sets. *Multisp. Multistruct.* **2010**, *4*, 410–413.
24. Zhang, H.Y.; Wang, J.Q.; Chen, X.H. An outranking approach for multi-criteria decision-making problems with interval-valued neutrosophic sets. *Neural Comput. Appl.* **2016**, *27*, 615–627. [CrossRef]
25. Zhang, H.Y.; Ji, P.; Wang, J.Q.; Chen, X.H. An improved weighted correlation coefficient based on integrated weight for interval neutrosophic sets and its application in multi-criteria decision-making problems. *Int. J. Comput. Intell. Syst.* **2015**, *8*, 1027–1043. [CrossRef]
26. Tian, Z.P.; Zhang, H.Y.; Wang, J.; Wang, J.Q.; Chen, X.H. Multi-criteria decision-making method based on a cross-entropy with interval neutrosophic sets. *Int. J. Syst. Sci.* **2016**, *47*, 3598–3608. [CrossRef]
27. Huang, H.L. New distance measure of single-valued neutrosophic sets and its application. *Int. J. Intell. Syst.* **2016**, *31*, 1021–1032. [CrossRef]
28. Ye, J. Single valued neutrosophic cross-entropy for multicriteria decision making problems. *Appl. Math. Model.* **2014**, *38*, 1170–1175. [CrossRef]
29. Zhang, H.; Zheng, Q.; Liu, T.; Qu, Y. Mixed intuitionistic fuzzy aggregation operators decreasing results of unusual IFNs. In Proceedings of the IEEE International Conference on Fuzzy Systems (FUZZ-IEEE), Vancouver, BC, Canada, 24–29 July 2016, 896–903.
30. Sirbiladze, G.; Sikharulidze, A. Extensions of probability intuitionistic fuzzy aggregation operators in fuzzy MCDM/MADM. *Int. J. Inform. Tech. Decis. Mak.* **2018**, *17*, 621–655. [CrossRef]

31. Wang, W.; Liu, X. Intuitionistic fuzzy information aggregation using einstein operations. *IEEE Trans. Fuzzy Syst.* **2012**, *20*, 923–938. [CrossRef]
32. Wang, J.Q.; Li, K.J.; Zhang, H.Y.; Chen, X.H. A score function based on relative entropy and its application in intuitionistic normal fuzzy multiple criteria decision making. *J. Intell. Fuzzy Syst.* **2013**, *25*, 567–576.
33. Wang, J.Q.; Li, K.J. Multi-criteria decision-making method based on intuitionistic normal fuzzy aggregation operators. *Syst. Eng. Theor. Pract.* **2013**, *33*, 1501–1508.
34. Wang, J.Q.; Zhou, P.; Li, K.J.; Zhang, H.Y.; Chen, X.H. Multi-criteria decision-making method based on normal intuitionistic fuzzy induced generalized aggregation operator. *Top* **2014**, *22*, 1103–1122. [CrossRef]
35. Peng, J.J.; Wang, J.Q.; Wu, X.H. Multi-valued neutrosophic sets and power aggregation operators with their applications in multi-criteria group decision-making problems. *Int. J. Comput. Intell. Syst.* **2015**, *8*, 345–363. [CrossRef]
36. Liu, P.D.; Liu, X. The neutrosophic number generalized weighted power averaging operator and its application in multiple attribute group decision making. *Int. J. Mach. Learn. Cybern.* **2018**, *9*, 347–358. [CrossRef]
37. Ye, J. Multiple attribute decision-making method based on the possibility degree ranking method and ordered weighted aggregation operators of interval neutrosophic numbers. *J. Intell. Fuzzy Syst.* **2015**, *28*, 1307–1317.
38. Wang, R.; Li, Y.L. Generalized single-valued neutrosophic hesitant fuzzy prioritized aggregation operators and their applications to multiple criteria decision-making. *Information* **2018**, *9*, 10. [CrossRef]
39. Liu, P.D.; Teng, F. Multiple attribute decision making method based on normal neutrosophic generalized weighted power averaging operator. *Int. J. Mach. Learn. Cybern.* **2018**, *9*, 281–293. [CrossRef]
40. Liu, P.D.; Teng, F. Multiple attribute group decision making methods based on some normal neutrosophic number heronian mean operators. *J. Intell. Fuzzy Syst.* **2017**, *32*, 2375–2391. [CrossRef]
41. Liu, P.D.; Wang, P.; Liu, J.L. Normal neutrosophic frank aggregation operators and their application in multi-attribute group decision making. *Int. J. Mach. Learn. Cybern.* **2017**, *1*, 1–20. [CrossRef]
42. Şahin, R. Normal neutrosophic multiple attribute decision making based on generalized prioritized aggregation operators. *Neural. Comput. Appl.* **2017**, 1–21. [CrossRef]
43. Bonferroni, C. Sulle medie multiple di potenze. *Bollettino dell'Unione Matematica Italiana* **1950**, *5*, 267–270.
44. Zhang, R.T.; Wang, J.; Zhu, X.M.; Xia, M.; Yu, M. Some generalized pythagorean fuzzy bonferroni mean aggregation operators with their application to multiattribute group decision-making. *Complexity* **2017**, *6*, 1–16. [CrossRef]

information

MDPI

Article

Convex Aggregation Operators and Their Applications to Multi-Hesitant Fuzzy Multi-Criteria Decision-Making

Ye Mei [1], Juanjuan Peng [1] and Junjie Yang [2,*]

1 School of Electrical and Information Engineering, Hubei University of Automotive Technology, Shiyan 442002, China; meiye13635722832@163.com (Y.M.); xiaqing1981@126.com (J.P.)
2 School of Information Engineering, Lingnan Normal University, Zhanjiang 524048, China
* Correspondence: yangjunjie1998@lingnan.edu.cn

Received: 23 May 2018; Accepted: 25 July 2018; Published: 21 August 2018

Abstract: Hesitant fuzzy sets (HFSs), which were generalized from fuzzy sets, constrain the membership degree of an element to be a set of possible values between zero and one; furthermore, if two or more decision-makers select the same value, it is only counted once. However, a situation where the evaluation value is repeated several times differs from one where the value appears only once. Multi-hesitant fuzzy sets (MHFSs) can deal effectively with a case where some values are repeated more than once in a MHFS. In this paper, the novel convex combination of multi-hesitant fuzzy numbers (MHFNs) is introduced. Some aggregation operators based on convex operation, such as generalized multi-hesitant fuzzy ordered weighted average (GMHFOWA) operator, generalized multi-hesitant fuzzy hybrid weighted average (GMHFHWA) operator, generalized multi-hesitant fuzzy prioritized weighted average (GMHFPWA) operator and generalized multi-hesitant fuzzy Choquet integral weighted average (GMHFCIWA) operator, are developed and corresponding properties are discussed in detail. Then, based on the proposed aggregation operators, a novel approach for multi-criteria decision-making (MCDM) problem is proposed for ranking alternatives. Finally, an example is provided to verify the developed approach and demonstrate its validity and feasibility and the study is supported by a sensitivity analysis and a comparison analysis.

Keywords: multi-criteria decision-making; multi-hesitant fuzzy sets; aggregation operators

1. Introduction

Hesitant fuzzy sets (HFSs) and multi-hesitant fuzzy sets (MHFSs), which were originally defined by Torra [1,2], are an extension of Zadeh's fuzzy sets (FSs) [3]. They allow a membership degree to have different possible precise values between zero and one. Recently, HFSs and its extensions has been the subject of a great deal of research and have been widely applied to multi-criteria decision-making (MCDM) problems [4–20]. For example, some works on the aggregation operators of HFSs have been undertaken [9–15] and the correlation coefficient, distance and correlation measures for HFSs were developed [16–20]. For example, Zhang et al. [13] developed some induced generalized hesitant fuzzy operators and applied them to multi-criteria group decision-making (MCGDM) problems. Zhou [14] proposed hesitant fuzzy ordered accurate weighted averaging (HFOAWA) operator and hesitant fuzzy ordered accurate weighted geometric (HFOAWG) operators and applied them to project investment. Zhang [15] defined generalized hesitant fuzzy power average (GHFPA) operator and generalized hesitant fuzzy power geometric (GHFPG) operator and applied them to MCGDM problems. Yu [16] proposed some aggregation operators based on Einstein operations and applied them to MCDM problems. Wang et al. [21] proposed a wide range of hesitant multiplicative fuzzy power aggregation geometric operators on MCGDM problems for hesitant multiplicative information. Torres et al. [22]

propose a prioritized aggregation operator to combine a time sequence of hesitant fuzzy information, and applied them to the service selection problem in service-based systems. Qian and Wang [23] generalized HFSs and utilized the aggregation operators to solve MCDM problems. Meng et al. [24] developed some induced generalized hesitant fuzzy Shapley hybrid operators and applied them to MCDM problems. Zhou and Xu [25] developed an optimal discrete fitting aggregation MCDM method with HFSs. Tan et al. [26] defined some hesitant fuzzy Hamacher aggregation operators and applied them to MCDM problems. Meng and Chen [27] and Liao et al. [28] defined novel correlation coefficients between HFSs and applied them to MCDM problems. Li et al. [29] and Hu et al. [30] defined some new distance and similarity measures of HFSs and applied them to MCDM problems. Furthermore, Zhang and Wei [31] developed the E-VIKOR method to solve MCDM problems with HFSs. Zhang and Xu [32] proposed the TODIM method, which was based on measured functions with HFSs. Farhadinia [33,34] developed some information measures of HFSs and a novel method of ranking hesitant fuzzy values. Moreover, Peng et al. [35] developed an extension of ELECTRE III method to handle MCDM problems with MHFSs.

However, two main shortcomings of the existing methods of dealing with HFSs have emerged from the research to date. (1) Both distance measures, similarity measures and some comparison methods should satisfy the condition that all hesitant fuzzy numbers (HFN) must be arranged in ascending order and be of equal length. If the two HFNs being compared have different lengths, then the value of the shorter one should be increased until both are equal. However, in such cases, different methods of extension could produce different results. (2) The existing methods do not clarify: how to solve a situation where there is a repeated value in the evaluation of alternatives; and, in particular, whether decision-makers can give more than one value (possible membership degrees of an element) for each criterion or not. At the same time, the situation where the evaluation value is repeated more than once is actually different from that where a value appears only once. For example, decision-makers can determine that the possible degrees of membership by which an alternative is assessed relative to the criterion "excellence" are 0.7, 0.8, and 0.8, which is expressed in the form of an HFN as {0.7, 0.8}. However, the nature of the evaluation {0.7, 0.8} substantially differs from that expressed in the form of an MHFS as {0.7, 0.8, 0.8}, which can lead to loss of information during the data collection process. Therefore, MHFSs can overcome these shortcomings and deal with the case where some values may be repeated more than once in an HFS. In this paper, the novel MCDM approach is developed based on some convex aggregation operators of multi-hesitant fuzzy sets (MHFNs). Moreover, the proposed approach based on convex operators distinguished from other methods for MCDM method not only because the proposed approach uses MHFSs, but also due to the consideration the inter-dependent phenomena among the criteria, which makes it more consistent with the practical decision-making environment.

The remainder of this paper is organized as follows. In Section 2, the definition, as well as the comparison method, of HFSs and MHFSs is provided. In Section 3, some aggregation operators of multi-hesitant fuzzy numbers (MHFNs) based on convex operation are developed and corresponding properties are discussed. In Section 4, an MCDM method based on convex aggregation operators is proposed. In Section 5, an example to illustrate the practical application of the developed approach is provided as well as sensitivity analysis and comparison analysis. Finally, some conclusions are drawn in Section 6.

2. Hesitant Fuzzy Sets and Multi-Hesitant Fuzzy Sets

In this section, the definition of HFSs and MHFSs are reviewed. The comparison method of HFSs, which will be utilized in the latter analysis, are also presented.

Definition 1. *Let X be a reference set, and a HFS E on X be in terms of a function which will return a subset of [0, 1] in the case of it being applied to X* [1,2].

In order that it would be easily understood, Xia and Xu [9] expressed the HFS as a mathematical symbol:

$$E = \{ \langle x, h_E(x) \rangle | x \in X\} \quad (1)$$

where $h_E(x)$ is a set of values in [0, 1], denoting the possible membership degrees of the element $x \in X$ to the set E. E is called HFSs, $h_E(x)$ is called a hesitant fuzzy element (HFE) [4], and H is the set of all HFEs. In particular, if X has only one element, E is called a HFN, which can be denoted by $E = \{h_E(x)\}$. The set of all HFNs is represented by HFNS.

Torra [1,2] defined some operations on HFNs, and Xia and Xu [4] defined some new operations on HFNs as well as the score functions.

Definition 2. *Let X be a reference set, and MHFSs be defined as E_M in terms of a function H_{E_M} that returns a multi-subset of [0, 1] when applied to X* [1].

Based on Definition 1, MHFSs can be expressed by the mathematical equation:

$$E_M = \{\langle x, H_{E_M}(x)\rangle | x \in X\} \quad (2)$$

Here, $H_{E_M}(x)$ is a set of values in [0, 1] denoting the possible degrees of membership of the element $x \in X$ to the set E_M. In any $H_{E_M}(x)$, the values can be repeated multiple times. $H_{E_M}(x)$ is a multi-hesitant fuzzy element (MHFE), and H_{E_M} is the set of all MHFEs. It is noteworthy that, if X contains only a single element, E_M is called a MHFN, briefly denoted by $E_M = \{H_{E_M}(x)\}$. The set of all MHFNs is represented by MHFNS. Any HFS is a special case of an MHFS.

Moreover, the operations of HFNs between two HFNs H_A and H_B on X was defined as below [1]:

(1) $\lambda H_A = \left\{1 - (1 - \gamma_{H_A})^\lambda\right\}(\lambda > 0)$;
(2) $H_A{}^\lambda = \left\{(\gamma_{H_A})^\lambda\right\}$ $(\lambda > 0)$;
(3) $H_A \oplus H_B = \{\gamma_{H_A} + \gamma_{H_B} - \gamma_{H_A} \cdot \gamma_{H_B}\}$;
(4) $H_A \otimes H_B = \{\gamma_{H_A} \cdot \gamma_{H_B}\}$.

Apparently, the operations on HFNs presented in Definition 2 also can be suitable for MHFNs.

The ranking of two HFNs can be obtained by combining the score function and the accuracy function [9,36].

Definition 3. *Let H_A and H_B be two HFNs on X, and then the novel ranking method for MHFNs can be defined as follows* [36]:

(1) if $s(H_A) < s(H_B)$, then $H_A \prec H_B$;
(2) if $s(H_A) = s(H_B)$, then:

- *if $f(H_A) = f(H_B)$, then $H_A \sim H_B$;*
- *if $f(H_A) < f(H_B)$, then $H_A \succ H_B$;*
- *if $f(H_A) > f(H_B)$, then $H_A \prec H_B$.*

where $s(H_i) = \frac{1}{l_{H_i}} \sum_{\gamma_i \in H_i} \gamma_i$ and $f(H_i) = \frac{1}{l_{H_i} - 1} \sum_{\gamma_i \in H_i} (s(H_i) - \gamma_i)^2$ ($i = A, B$) represents the score function and accuracy function of H_i respectively [4,31], *and l_{H_i} is the number of elements in H_i. Please note that "$\prec$" means "inferior to". The score function is similar to the mean value; the greater the value of the mean, the larger the hesitant degree. The accuracy function is similar to the sample variance in statistics and can reflect the fluctuation of evaluation values of HFNs; the greater the amplitude of fluctuation, the larger the hesitant degree.*

Example 1. *Let* $h_1 = \{0.2, 0.5\}$ *and* $h_2 = \{0.3, 0.4\}$ *be two HFNs. According to Definition 3, we have* $s(h_1) = s(h_2) = 0.35$ *and* $f(h_1) = 0.045 > f(h_2) = 0.005$. *Apparently,* $H_1 \prec H_2$ *can be obtained, which is consistent with our intuitive.*

3. The Convex Combination Operation and Some Aggregation Operators of MHFNs

In this section, the convex combination operation with MHFNs is developed, and corresponding properties and aggregation operators are presented.

Definition 4. *Let* H_1 *and* H_2 *be two MHFNs. A convex combination of* H_1 *and* H_2 *is defined as*

$$C^2(w_1, H_1, w_2, H_2) = w_1 \otimes H_1 \oplus w_2 \otimes H_2 = \left\{ \left(w_1\gamma_1^\lambda + w_2\gamma_2^\lambda\right)^{1/\lambda} \middle| \gamma_1 \in H_1, \gamma_2 \in H_2 \right\}, \lambda > 0 \quad (3)$$

where $w_1 \geq 0, w_2 \geq 0$ *and* $w_1 + w_2 = 1$.

Proposition 1. *Let* H_1 *and* H_2 *be two MHFNs. For* $0 \leq w \leq 1$, *the convex combination* $C^2(w, H_1, 1-w, H_2)$ *of* H_1 *and* H_2 *is also a MHFNs.*

Proof. Based on Definition 4, we just need to prove that $0 < \left(w\gamma_1^\lambda + (1-w)\gamma_2^\lambda\right)^{1/\lambda} \leq 1$. It is obvious that $\left(w\gamma_1^\lambda + (1-w)\gamma_2^\lambda\right)^{1/\lambda} > 0$. Assume $\gamma_1' \in H_1$ and $\gamma_2' \in H_2$ do exist, so to make $\left(w\gamma_1'^\lambda + (1-w)\gamma_2'^\lambda\right)^{1/\lambda} \geq 1$ i.e., $w\gamma_1'^\lambda + (1-w)\gamma_2'^\lambda \geq 1$. If $\gamma_1' > \gamma_2'$, then $w\gamma_1'^\lambda + (1-w)\gamma_2'^\lambda \geq 1$ i.e., $w > \frac{1-\gamma_2'^\lambda}{\gamma_1'^\lambda - \gamma_2'^\lambda} > 1$ which obviously contradicts $0 \leq w \leq 1$; if $\gamma_1' = \gamma_2'$, then $w\gamma_1'^\lambda + (1-w)\gamma_1'^\lambda = \gamma_1'^\lambda > 1$ which contradicts $0 < \gamma_1'^\lambda \leq 1$; if $\gamma_1' \leq \gamma_2'$, then $w\gamma_1'^\lambda + (1-w)\gamma_2'^\lambda \geq 1$ i.e., $w < \frac{1-\gamma_2'^\lambda}{\gamma_1'^\lambda - \gamma_2'^\lambda} < 0$ which obviously contradicts $0 \leq w \leq 1$. Therefore, the hypothesis is not supported. For any $\gamma_1 \in H_1$ and $\gamma_2 \in H_2$, we have $\left(w\gamma_1^\lambda + (1-w)\gamma_2^\lambda\right)^{1/\lambda} \leq 1$. Thus, the convex combination $C^2(w, H_1, 1-w, H_2)$ of H_1 and H_2 is also a MHFNs. □

Definition 5. *Let* H_i $(i = 1, 2, \ldots, n)$ *be a collection of MHFNs. Then the generalized multi-hesitant fuzzy ordered weighted average (GMHFOWA) operator of dimension n is a mapping* $GMHFOWA : MHFN^n \to MHFN$ *that has an associated weight vector* $w = (w_1, w_2, \ldots, w_n)$ *with* $w_i \geq 0$ $(i = 1, 2, \ldots, n)$ *and* $\sum_{i=1}^{n} w_i = 1$, *and*

$$\begin{aligned} &GMHFOWA(H_1, H_2, \ldots, H_n) \\ &= C^n\left(w_k, H_{\sigma(k)}, k = 1, 2, \ldots, n\right) = w_1 \otimes H_{\sigma(1)} \oplus (1-w_1) \otimes C^{n-1}\left\{ w_i / \sum_{k=2}^{n} w_k, H_{\sigma(i)}, i = 2, 3, \ldots, n \right\}. \end{aligned} \quad (4)$$

Here $(\sigma(1), \sigma(2), \ldots, \sigma(n))$ *is a permutation of* $(1, 2, \ldots, n)$, *and such that* $H_{\sigma(1)} \leq H_{\sigma(2)} \leq \ldots \leq H_{\sigma(n)}$.

Theorem 1. *Let* H_i $(i = 1, 2, \ldots, n)$ *be a collection of MHFNs. Then their aggregated value by using the GMHFOWA operator is also a MHFN, and*

$$GMHFOWA(H_1, H_2, \ldots, H_n) = \left\{ \left(w_1\gamma_{\sigma(1)}^\lambda + w_2\gamma_{\sigma(2)}^\lambda + \ldots + w_n\gamma_{\sigma(n)}^\lambda\right)^{1/\lambda} \middle| \gamma_{\sigma(i)} \in H_{\sigma(i)}, i = 1, 2, \ldots, n \right\} \quad (5)$$

Here $(\sigma(1), \sigma(2), \ldots, \sigma(n))$ *is a permutation of* $(1, 2, \ldots, n)$, *and such that* $H_{\sigma(1)} \leq H_{\sigma(2)} \leq \ldots \leq H_{\sigma(n)}$.

Example 2. *Let* $H_1 = \{0.2, 0.2, 0.3\}, H_2 = \{0.1, 0.2\}$ *and* $H_3 = \{0.4\}$ *be three MHFNs,* $w = (0.3, 0.4, 0.3)$ *be the weight vector of them, and* $\lambda = 1$. *Based on Definition 5 and Theorem 1, if the associated vector is* $w = (0.3, 0.4, 0.3)$, *then* $H_3 > H_1 > H_2$, $H_{\sigma(1)} = H_2, H_{\sigma(2)} = H_1$ *and* $H_{\sigma(3)} = H_3$ *can be obtained. So*

$$\begin{aligned} & GMHFOWA(H_1, H_2, H_3) \\ & = \{w_1\gamma_2 + w_2\gamma_1 + w_3\gamma_3 | \gamma_1 \in H_1, \gamma_2 \in H_2, \gamma_3 \in H_3\} = \{0.23, 0.23, 0.27, 0.26, 0.26, 0.30\}. \end{aligned}$$

It can be easily proved that the GMHFOWA operator is monotonicity, commutativity and bounded, which are presented in the following.

Proposition 2. *Let* H_i $(i = 1, 2, \ldots, n)$ *be a collection of MHFNs, the following prosperities can be true.*

(1) *(Monotonicity) Let* $H_i'(i = 1, 2, \ldots, n)$ *be a collection of MHFNs. If for all i,* $H_i \leq H_i'$, *then*

$$GMHFOWA(H_1, H_2, \ldots, H_n) \leq GMHFOWA(H_1', H_2', \ldots, H_n')$$

(2) *(Commutativity) If* $H_1^*, \ldots, H_n^*$ *is a permutation of* $H_1, \ldots, H_n$, *then*

$$GMHFOWA(H_1^*, H_2^*, \ldots, H_n^*) = GMHFOWA(H_1, H_2, \ldots, H_n)$$

(3) *(Boundedness) If* $H^- = \{\gamma_1^-, \gamma_2^-, \ldots, \gamma_n^-\}$ *and* $H^+ = \{\gamma_1^+, \gamma_2^+, \ldots, \gamma_n^+\}$, *where* $\gamma_i^- = \min_{\gamma_i \in H_i} \gamma_i$ *and* $\gamma_i^+ = \max_{\gamma_i \in H_i} \gamma_i$, *then*

$$H^- \leq GMHFOWA(H_1, H_2, \ldots, H_n) \leq H^+$$

Definition 6. *Let* H_i $(i = 1, 2, \ldots, n)$ *be a collection of MHFNs. Then the generalized multi-hesitant fuzzy hybrid weighted average (GMHFHWA) operator of dimension n is a mapping* $GMHFHWA : MHFN^n \to MHFN$ *that have the weighting vector* $w = (w_1, w_2, \ldots, w_n)$ *of* $H_i (i = 1, 2, \ldots, n)$ *with* $w_i \geq 0$ $(i = 1, 2, \ldots, n)$ *and* $\sum_{i=1}^{n} w_i = 1$, *the aggregation-associated vector is* $\omega = (\omega_1, \omega_2, \ldots, \omega_n)$ *with* $\omega_i \geq 0$ $(i = 1, 2, \ldots, n)$ *and* $\sum_{i=1}^{n} \omega_i = 1$, *and*

$$\begin{aligned} & GMHFHWA(H_1, H_2, \ldots, H_n) \\ & = C^n\left(\omega_k, \dot{H}_{\sigma(k)}, k = 1, 2, \ldots, n\right) = \omega_1 \otimes \dot{H}_{\sigma(1)} \oplus (1 - \omega_1) \otimes C^{n-1}\left\{\omega_i / \sum_{k=2}^{n} \omega_k, \dot{H}_{\sigma(i)}, i = 2, 3, \ldots, n\right\}. \end{aligned} \tag{6}$$

Here $(\sigma(1), \sigma(2), \ldots, \sigma(n))$ *is a permutation of* $(1, 2, \ldots, n)$, *and* $\dot{H}_{\sigma(i)} = nw_i H_i (i = 1, 2, \ldots, n)$ *is the i-th largest of the weighted multi-hesitant fuzzy values. n is the balancing coefficient which plays a role of balance. If* $\omega = (1/n, 1/n, \ldots, 1/n)$, *then the GMHFHWA operator is reduced to the GMHFOWA operator. If* $w = (1/n, 1/n, \ldots, 1/n)$, *then the GMHFHWA operator is reduced to the GMHFWA operator.*

Theorem 2. *Let* H_i $(i = 1, 2, \ldots, n)$ *be a collection of MHFNs. Then their aggregated value by using GMHFHWA operator is also a MHFN, and*

$$GMHFHWA(H_1, H_2, \ldots, H_n) = \left\{ \left(\omega_1 \dot{\gamma}_{\sigma(1)}^{\lambda} + \omega_2 \dot{\gamma}_{\sigma(2)}^{\lambda} + \ldots + \omega_n \dot{\gamma}_{\sigma(n)}^{\lambda}\right)^{1/\lambda} \middle| \dot{\gamma}_{\sigma(i)} \in \dot{H}_{\sigma(i)}, i = 1, 2, \ldots, n \right\} \tag{7}$$

Here $(\sigma(1), \sigma(2), \ldots, \sigma(n))$ *is a permutation of* $(1, 2, \ldots, n)$, *and* $\dot{H}_{\sigma(i)} = nw_i H_i (i = 1, 2, \ldots, n)$ *is the i-th largest of the weighted multi-hesitant fuzzy values. n is the balancing coefficient which plays a role of balance.*

Example 3. *Let $H_1 = \{0.2, 0.2\}, H_2 = \{0.1, 0.3\}$ and $H_3 = \{0.4\}$ be three MHFNs. The weight vector is $w = (0.3, 0.4, 0.3)$ and aggregation-associated vector is also $\omega = (0.2, 0.4, 0.4)$, and $\lambda = 1$, then*

$$\dot{H}_1 = 3 \times 0.3 \cdot H_1 = \{0.1819, 0.1819\}; \dot{H}_2 = 3 \times 0.4 \cdot H_2 = \{0.0126, 0.0419\}; \dot{H}_3 = 3 \times 0.3 \cdot H_3 = \{0.3686\}.$$

Obviously, $s\left(\dot{H}_3\right) > s\left(\dot{H}_1\right) > s\left(\dot{H}_2\right)$. By using Theorem 3, we have

$$GMHFWA(H_1, H_2, H_3) = \left\{ \left(\omega_1 \dot{\gamma}_{\sigma(1)} + \omega_2 \dot{\gamma}_{\sigma(2)} + \ldots + \omega_n \dot{\gamma}_{\sigma(n)}\right) \middle| \dot{\gamma}_{\sigma(i)} \in \dot{H}_{\sigma(i)}, i = 1, 2, 3 \right\}$$
$$= \{0.1512, 0.1632, 0.1512, 0.1632\}.$$

Proposition 3. *Let H_i $(i = 1, 2, \ldots, n)$ be a collection of MHFNs, the following prosperities can be true.*

(1) *(Monotonicity) Let $H_i'(i = 1, 2, \ldots, n)$ be a collection of MHFNs. If for all i, $H_i \le H_i'$, then*

$$GMHFHWA(H_1, H_2, \ldots, H_n) \le GMHFHWA(H_1', H_2', \ldots, H_n')$$

(2) *(Commutativity) If $H_1^*, \ldots, H_n^*$ is a permutation of $H_1, \ldots, H_n$, then*

$$GMHFHWA(H_1^*, H_2^*, \ldots, H_n^*) = GMHFHWA(H_1, H_2, \ldots, H_n)$$

(3) *(Boundedness) If $H^- = \{\gamma_1^-, \gamma_2^-, \ldots, \gamma_n^-\}$ and $H^+ = \{\gamma_1^+, \gamma_2^+, \ldots, \gamma_n^+\}$, where $\gamma_i^- = \min_{\gamma_i \in H_i} \gamma_i$ and $\gamma_i^+ = \max_{\gamma_i \in H_i} \gamma_i$, then*

$$H^- \le GMHFHWA(H_1, H_2, \ldots, H_n) \le H^+$$

Based on the prioritization between the criteria discussed in [37], the prioritized aggregation operator can be obtained.

Definition 7. *Let H_i $(i = 1, 2, \ldots, n)$ be a collection of MHFNs. Then the generalized multi-hesitant fuzzy prioritized weighted average (GMHFPWA) operator of dimension n is a mapping $GMHFPWA : MHFN^n \to MHFN$, and*

$$\begin{aligned} &GMHFPWA(H_1, H_2, \ldots, H_n) \\ &= C^n\left(\frac{T_k}{\sum_{j=1}^n T_j}, H_k, k = 1, 2, \ldots, n\right) \\ &= \frac{T_1}{\sum_{j=1}^n T_j} \otimes H_1 \oplus \left(1 - \frac{T_1}{\sum_{j=1}^n T_j}\right) \otimes C^{n-1}\left\{\frac{T_i}{\sum_{j=1}^n T_j} / \sum_{k=2}^n \frac{T_k}{\sum_{j=1}^n T_j}, H_i, i = 2, 3, \ldots, n\right\}. \end{aligned} \tag{8}$$

Here $T_j = \prod_{k=1}^{j-1} s(H_k) (j = 2, \ldots, n)$, $T_1 = 1$ and $s(H_k)$ is the score values of $H_k (k = 1, 2, \ldots, n)$.

Theorem 3. *Let H_i $(i = 1, 2, \ldots, n)$ be a collection of MHFNs. Then their aggregated value by using the GMHFPWA operator is also a MHFN, and*

$$GMHFPWA(H_1, H_2, \ldots, H_n) = \left\{ \left(\frac{T_1}{\sum_{j=1}^n T_j} \gamma_1^\lambda + \frac{T_2}{\sum_{j=1}^n T_j} \gamma_2^\lambda + \ldots + \frac{T_n}{\sum_{j=1}^n T_j} \gamma_n^\lambda\right)^{1/\lambda} \middle| \gamma_i \in H_i, i = 1, 2, \ldots, n \right\} \tag{9}$$

Here $T_j = \prod_{k=1}^{j-1} s(H_k) (j = 2, \ldots, n)$, $T_1 = 1$ and $s(H_k)$ is the score values of $H_k (k = 1, 2, \ldots, n)$.

Example 4. *Let $H_1 = \{0.5, 0.5, 0.7\}, H_2 = \{0.4, 0.5\}, H_3 = \{0.8, 0.9\}$ and $H_4 = \{0.3, 0.4, 0.5\}$ be four MHFNs and $\lambda = 1$. Based on Definition 3, then $s(H_1) = 0.567$, $s(H_2) = 0.45$ and $s(H_3) = 0.85$ can be* *obtained. If the prioritization during four MHFNs is $H_1 \succ H_2 \succ H_3 \succ H_4$, then according to Definition 7 and Theorem 3,*

$$T_{41} = 1, T_2 = s(H_1) = 0.567, T_3 = s(H_1) \times s(H_2) = 0.2552, T_4 = 0.2169, \sum_{j=1}^{4} T_j = 2.0391.$$

So

$$GMHFPWA(H_1, H_2, H_3, H_4) = \left\{ \left(\tfrac{1}{2.0391}\gamma_1 + \tfrac{0.567}{2.0391}\gamma_2 + \tfrac{0.2552}{2.0391}\gamma_3 + \tfrac{0.2169}{2.0391}\gamma_4 \right) \middle| \gamma_i \in H_i, i = 1,2,3,4 \right\}$$
$$= \{0.4885, 0.4991, 0.5097, 0.5010, 0.5116, 0.5223, 0.5163, 0.5269, 0.5375, 0.5288, 0.5394, 0.5501, 0.4885, 0.4991, 0.5097, 0.5010, 0.5116, 0.5223, 0.5163, 0.5269, 0.5375, 0.5288, 0.5394, 0.5501, 0.5865, 0.5972, 0.6078, 0.5991, 0.6097, 0.6203, 0.6144, 0.6250, 0.6356, 0.6269, 0.6375, 0.6481\}.$$

Similarly, it can be easily proved that the GMHFPWA operator is monotonicity, commutativity and bounded, which are presented in the following.

Proposition 4. *Let H_i $(i = 1, 2, \ldots, n)$ be a collection of MHFNs, then the following properties can be true.*

(1) *(Monotonicity) Let $H_i'(i = 1, 2, \ldots, n)$ be a collection of MHFNs. If for all i, $H_i \leq H_i'$, then*

$$GMHFPWA(H_1, H_2, \ldots, H_n) \leq GMHFPWA(H_1', H_2', \ldots, H_n')$$

(2) *(Commutativity) If $H_1^*, \ldots, H_n^*$ is a permutation of $H_1, \ldots, H_n$, then*

$$GMHFPWA(H_1^*, H_2^*, \ldots, H_n^*) = GMHFPWA(H_1, H_2, \ldots, H_n)$$

(3) *(Boundedness) If $H^- = \{\gamma_1^-, \gamma_2^-, \ldots, \gamma_n^-\}$ and $H^+ = \{\gamma_1^+, \gamma_2^+, \ldots, \gamma_n^+\}$, where $\gamma_i^- = \min_{\gamma_i \in H_i} \gamma_i$ and $\gamma_i^+ = \max_{\gamma_i \in H_i} \gamma_i$, then*

$$H^- \leq GMHFPWA(H_1, H_2, \ldots, H_n) \leq H^+$$

According to the fuzzy measure (more details can be founded in [38]), the Choquet integral aggregation operator can be obtained.

Definition 8. *Let H_i $(i = 1, 2, \ldots, n)$ be a collection of MHFNs. Then the generalized multi-hesitant fuzzy Choquet integral weighted average (GMHFCIWA) operator of dimension n is a mapping $GMHFCIWA : MHFN^n \rightarrow MHFN$, and*

$$\begin{aligned} &GMHFCIWA(H_1, H_2, \ldots, H_n) \\ &= C^n\left(\left(\mu\left(A_{\sigma(k)}\right) - \mu\left(A_{\sigma(k+1)}\right)\right), H_{\sigma(k)}, k = 1, 2, \ldots, n\right) \\ &= \left(\mu\left(A_{\sigma(1)}\right) - \mu\left(A_{\sigma(2)}\right)\right) \otimes H_{\sigma(1)} \oplus \left(1 - \left(\mu\left(A_{\sigma(1)}\right) - \mu\left(A_{\sigma(2)}\right)\right)\right) \\ &\otimes C^{n-1}\left\{\left(\left(\mu\left(A_{\sigma(i)}\right) - \mu\left(A_{\sigma(i+1)}\right)\right) / \sum_{k=2}^{n}\left(\mu\left(A_{\sigma(k)}\right) - \mu\left(A_{\sigma(k+1)}\right)\right), H_{\sigma(i)}, i = 2, 3, \ldots, n\right)\right\}. \end{aligned} \tag{10}$$

Here μ is a fuzzy measure on X, $(\sigma(1), \sigma(2), \ldots, \sigma(n))$ is a permutation of $(1, 2, \ldots, n)$, and such that $H_{\sigma(1)} \leq H_{\sigma(2)} \leq \ldots \leq H_{\sigma(n)}$. $A_{\sigma(i)} = \left\{ x_{\sigma(k)} \middle| k \geq i \right\}$, and $x_{\sigma(i)}$ is the criterion corresponding to $H_{\sigma(i)}$.

Theorem 4. *Let $H_i(i = 1, 2, \ldots, n)$ be a collection of MHFNs, then their aggregated value by using the GMHFCIWA operator is also a MHFN, and*

$$\begin{aligned} &GMHFCIWA(H_1, H_2, \ldots, H_n) \\ &= \left\{ \left(\left(\mu\left(A_{\sigma(1)}\right) - \mu\left(A_{\sigma(2)}\right)\right)\gamma_{\sigma(1)}^{\lambda} + \left(\mu\left(A_{\sigma(2)}\right) - \mu\left(A_{\sigma(3)}\right)\right)\gamma_{\sigma(2)}^{\lambda} + \ldots + \left(\mu\left(A_{\sigma(n)}\right) - \mu\left(A_{\sigma(n+1)}\right)\right)\gamma_{\sigma(n)}^{\lambda} \right)^{1/\lambda} \right. \\ &\left. \middle| \gamma_{\sigma(i)} \in H_{\sigma(i)}, i = 1, 2, \ldots, n \right\}. \end{aligned} \tag{11}$$

Here $(\sigma(1), \sigma(2), \ldots, \sigma(n))$ *is a permutation of* $(1, 2, \ldots, n)$, *and such that* $H_{\sigma(1)} \leq H_{\sigma(2)} \leq \ldots \leq H_{\sigma(n)}$. $A_{\sigma(i)} = \left\{ x_{\sigma(k)} \middle| k \geq i \right\}$, *and* $x_{\sigma(i)}$ *is the criterion corresponding to* $H_{\sigma(i)}$.

Example 5. *Suppose a Venture Capital Company is going to evaluate the existed investment projects from the financial perspective. Three criteria could be considered:* c_1 *: operating capacity;* c_2*: solvency;* c_3*: profitability. Suppose three criteria are inter-dependent. The decision-makers could give the evaluation values in form of MHFNs and denoted as follows:* $H_1 = \{0.1, 0.1\}, H_2 = \{0.2, 0.4\}$ *and* $H_3 = \{0.5\}$.

Suppose that $\mu(c_1) = 0.30$, $\mu(c_2) = 0.30$, $\mu(c_3) = 0.20$, $\mu(c_1, c_2) = 0.90$, $\mu(c_1, c_3) = 0.80$, $\mu(c_2, c_3) = 0.60$, $\mu(c_1, c_2, c_3) = 1$, *then the following results can be obtained.*

$$\begin{aligned} w_{\sigma(1)} &= \mu\left(A_{\sigma(1)}\right) - \mu\left(A_{\sigma(2)}\right) \\ &= \mu\left(c_{\sigma(1)}, c_{\sigma(2)}, c_{\sigma(3)}\right) - \mu\left(c_{\sigma(2)}, c_{\sigma(3)}\right) = \mu(c_1, c_2, c_3) - \mu(c_1, c_3) = 1 - 0.80 = 0.20; \end{aligned}$$

$w_{\sigma(2)} = 0.60; w_{\sigma(3)} = 0.20$.
Then $H_{\sigma(1)} = H_2, H_{\sigma(2)} = H_1, H_{\sigma(3)} = H_3$.
Thus, the overall evaluation value can be calculated.

$$\begin{aligned} &GMHFCIWA(H_1, H_2, H_3) \\ &= C^4\left(\mu\left(A_{\sigma(k)}\right) - \mu\left(A_{\sigma(k+1)}\right), H_{\sigma(k)}, k = 1, 2, 3\right) \\ &= \left\{ w_{\sigma(1)}\gamma_{\sigma(1)} + w_{\sigma(2)}\gamma_{\sigma(2)} + w_{\sigma(3)}\gamma_{\sigma(3)} \middle| \gamma_{\sigma(i)} \in H_{\sigma(i)}, i = 1, 2, 3 \right\} \\ &= \{0.20, 0.20, 0.24, 0.24\}. \end{aligned}$$

Similarly, the GMHFCIWA operator is monotonicity, commutativity and bounded, which are presented in the following.

Proposition 5. *Let* H_i $(i = 1, 2, \ldots, n)$ *be a collection of MHFNs, and* μ *be the fuzzy measure on* X, *then the following properties can be true.*

(1) *(Monotonicity) Let* $H_i'(i = 1, 2, \ldots, n)$ *be a collection of MHFNs. If for all i,* $H_i \leq H_i'$, *then*

$$GMHFCIWA_\mu(H_1, H_2, \ldots, H_n) \leq GMHFCIWA_\mu(H_1', H_2', \ldots, H_n')$$

(2) *(Commutativity) If* $H_1^*, \ldots, H_n^*$ *is a permutation of* $H_1, \ldots, H_n$, *then*

$$GMHFCIWA_\mu(H_1^*, H_2^*, \ldots, H_n^*) = GMHFCIWA_\mu(H_1, H_2, \ldots, H_n)$$

(3) *(Boundedness) If* $H^- = \{\gamma_1^-, \gamma_2^-, \ldots, \gamma_n^-\}$ *and* $H^+ = \{\gamma_1^+, \gamma_2^+, \ldots, \gamma_n^+\}$, *where* $\gamma_i^- = \min_{\gamma_i \in H_i} \gamma_i$ *and* $\gamma_i^+ = \max_{\gamma_i \in H_i} \gamma_i$, *then*

$$H^- \leq GMHFCIWA_\mu(H_1, H_2, \ldots, H_n) \leq H^+$$

4. The MCDM Method Based on Aggregation Operators with MHFNs

The MCDM ranking/selection problems with multi-hesitant fuzzy information consists of a group of alternatives, denoted by $A = \{a_1, a_2, \ldots, a_n\}$. The alternatives are evaluated based on the criteria denoted by $C = \{c_1, c_2, \ldots, c_m\}$. a_{ij} is the value of the alternative a_i for the criterion c_j, and $a_{ij} = \left\{\gamma_{ij}^k, k = 1, 2, \ldots, l(a_{ij})\right\}$ $(i = 1, \ldots, n; j = 1, \ldots, m)$ are in the form of MHFNs, which are given by several decision-makers. Furthermore, $l(a_{ij})$ represents the number of elements in a_{ij} and the corresponding weight vector $w = (w_1, w_2, \ldots, w_m)$. This method is suitable if the number of decision-makers is small. A situation could arise where decision-makers evaluate these alternatives based on the given criteria, and one decision-maker could give several evaluation values. In particular, in the case where two or more decision-makers give the same value, it is counted repeatedly. a_{ij} is the set of evaluation values for all decision-makers.

The approach is an integration of MHFNs and aggregation operators to solve MCDM problems mentioned above. It is noted that different operators have different characteristic. The decision-makers can choose different operators according to their preference. The GMHFOWA operator mainly weights the ordered positions of the multi-hesitant fuzzy values instead of weighting the multi-hesitant fuzzy values themselves. The GMHFHWA operator reflects the importance degrees of both multi-hesitant fuzzy values and their ordered positions. Furthermore, most MCDM methods are under the assumption that the criteria are at the same priority level, and the prominent of characteristic of the GMHFPWA is that it considers prioritization among the criteria. The GMHFCIWA operator can better reflect the correlations among the elements to handle MCDM problems where the criteria are inter-dependent or interactive. Therefore, four aggregation operators can be used to deal with different relationships among the aggregated arguments, could handle MCDM problems in a flexible and objective manner under multi-hesitant fuzzy environment, and can provide more choices for decision-makers.

The procedure of this approach is shown as follows.

Step 1. Normalize the decision matrix.

For MCDM problems, the most common criteria are of maximizing and minimizing types. To unify all criteria, it is necessary to normalize the evaluation values. (Note: if all the criteria are of the maximizing type and have the same measurement unit, then there is no need to normalize them). Suppose that the matrix $R = (a_{ij})_{n\times m}$, where $a_{ij} = \left\{\gamma_{ij}^1, \gamma_{ij}^2, \ldots, \gamma_{ij}^k\right\}$ $(i = 1,2,\ldots,n; j = 1,2,\ldots,m; k = 1,2,\ldots,l(a_{ij}))$, are MHFNs, is normalized into the corresponding matrix $\widetilde{R} = (\widetilde{a}_{ij})_{n\times m}$. Where $\widetilde{a}_{ij} = \left\{\widetilde{\gamma}_{ij}^1, \widetilde{\gamma}_{ij}^2, \ldots, \widetilde{\gamma}_{ij}^k\right\}$ $(i = 1,2,\ldots,n; j = 1,2,\ldots,m; k = 1,2,\ldots,l(a_{ij}))$. $l(a_{ij})$ is the number of the elements of a_{ij}.

For the maximizing criteria, the normalization formula is

$$\widetilde{\gamma}_{ij}^k = \gamma_{ij}^k, k = 1,2,\ldots,l(a_{ij}) \tag{12}$$

for the minimizing criteria,

$$\widetilde{\gamma}_{ij}^k = 1 - \gamma_{ij}^k, k = 1,2,\ldots,l(a_{ij}) \tag{13}$$

Seemingly, the normalization values $\widetilde{a}_{ij} = \left\{\widetilde{\gamma}_{ij}^1, \widetilde{\gamma}_{ij}^2, \ldots, \widetilde{\gamma}_{ij}^k\right\}$ $\{i = 1,2,\ldots,n; j = 1,2,\ldots,m\}$ are also MHFNs.

Step 2. Aggregate the MHFNs of each decision-maker.

Utilize the GMHFOWA, GMHFHWA, GMHPWA or GMHCIWA operator to aggregate the MHFNs of each decision-maker, and the individual aggregated value y_i of the alternative a_i $(i = 1,2,\ldots,n)$ can be obtained.

Step 3. Calculate the score function value $s(y_i)$ and the accuracy function value $a(y_i)$ of y_i $(i = 1,2,\ldots,m)$ using Definition 3.

Step 4. Rank the alternatives.

5. An Illustrative Example

In this section, an example is adapted from Schmeidler [39] for further illustration of the feasibility of the proposed approach.

There is an investment company, which wants to invest in a project. There are five possible alternatives in which to invest: a_1 is a car company; a_2 is a food company; a_3 is a computer company; a_4 is an arms company; and a_5 is a TV company. The investment company must make a decision according to the following four criteria: c_1 is the environment impact; c_2 is the risk; c_3 are the growth prospects; and c_4 is the social-political impact. The environmental impact refers to the impact on the company's environment and the processes used in making the product, such as the management methods and work environment. The risk involves more than one risk factor, including product risk and development environment risk. The growth prospects include increased profitability and returns. The social-political impact refers to the government's and local residents' support for company. The four criteria are correlated with each other in the assessment process. The five possible alternatives $a_i (i = 1, 2, \ldots, 5)$ are to be evaluated using the multi-hesitant fuzzy information of two decision-makers as presented in Table 1. The evaluation values $a_{ij} (i = 1, 2, 3, 4, 5; j = 1, 2, 3, 4)$ should be in the form of MHFNs which are provided by two decision-makers based on their knowledge and experience. In the case where decision-makers give the same value, then it is counted repeatedly, and a_{ij} is the set of evaluation values for two decision-makers.

Table 1. Multi-hesitant fuzzy decision matrix.

	c_1	c_2	c_3	c_4
a_1	{0.4, 0.5, 0.7}	{0.5, 0.5, 0.8}	{0.6, 0.6, 0.9}	{0.5, 0.6}
a_2	{0.6, 0.7, 0.8}	{0.5, 0.6}	{0.6, 0.7, 0.7}	{0.4, 0.5}
a_3	{0.6, 0.8}	{0.2, 0.3, 0.5}	{0.6, 0.6}	{0.5, 0.7}
a_4	{0.5, 0.5, 0.7}	{0.4, 0.5}	{0.8, 0.9}	{0.3, 0.4, 0.5}
a_5	{0.6, 0.7}	{0.5, 0.7}	{0.7, 0.8}	{0.3, 0.3, 0.4}

5.1. An Illustration of the Proposed Approach

There are four cases that the proposed approach is used to handle the MCDM problems where the weight of criteria is known or unknown. The procedures of obtaining the optimal alternative, by using the developed approach, are shown as follows.

Case 1. If the ordered positions of the multi-hesitant fuzzy values of criteria are considered, then the GMHFOWA operator is utilized and the associated weight is $w = (0.33, 0.18, 0.37, 0.12)$. The procedures of the proposed approach can be obtained.

Step 1. Normalize the data in Table 1.

Because all the criteria are of the maximizing type and have the same measurement unit, there is no need for normalization and $\widetilde{R} = (\widetilde{a}_{ij})_{5\times 4} = (a_{ij})_{5\times 4}$.

Steps 2–3. Aggregate the MHFNs of each decision-maker and calculate the score function value and accuracy function value.

According to Definition 5 and Theorem 1, the following results can be obtained:

$$s(y_1) = 0.5810; s(y_2) = 0.5782; s(y_3) = 0.5240; s(y_4) = 0.5247; s(y_5) = 0.5485$$

Since the score function values are different, so there is no need to compute the accuracy function value.

Step 4. Rank the alternatives.

Based on Step 3, since $s(y_3) < s(y_4) < s(y_5) < s(y_2) < s(y_1)$, so the final ranking is $a_3 \prec a_4 \prec a_5 \prec a_2 \prec a_1$. The best alternative is a_1 while the worst alternative is a_3.

Case 2. If both the multi-hesitant fuzzy values of criteria and their ordered positions are considered, then the GMHFHWA operator is utilized. If the corresponding vector of criteria is $w = (0.33, 0.18, 0.37, 0.12)$ and the aggregation-associated vector is $\omega = (0.3, 0.25, 0.2, 0.25)$, according Definition 6 and Theorem 2, the following results can be obtained:

$$s(y_1) = 0.3207; s(y_2) = 0.3583; s(y_3) = 0.2940; s(y_4) = 0.3098; s(y_5) = 0.3750$$

Since $s(y_3) < s(y_4) < s(y_1) < s(y_2) < s(y_5)$, so the final ranking is $a_3 \prec a_4 \prec a_1 \prec a_2 \prec a_5$. The best alternative is a_5 while the worst alternative is a_3.

Case 3. If the prioritization among the multi-hesitant fuzzy values of criteria is taken into account, then the GMHFPWA operator is used and the prioritization relation for criteria is $c_1 \succ c_2 \succ c_3 \succ c_4$. Based on Definition 7 and Theorem 3,

$$T_{11} = 1, T_{12} = s(a_{11}) = 0.5300, T_{13} = 0.3180, T_{14} = 0.2226, \sum_{j=1}^{4} T_j = 2.0706;$$
$$T_{21} = 1, T_{22} = s(a_{21}) = 0.7000, T_{23} = 0.3850, T_{24} = 0.2568, \sum_{j=1}^{4} T_j = 2.3418;$$
$$T_{31} = 1, T_{32} = s(a_{31}) = 0.7000, T_{33} = 0.2310, T_{34} = 0.1386, \sum_{j=1}^{4} T_j = 2.0696;$$

$$T_{41} = 1, T_{42} = s(a_{41}) = 0.5670, T_{43} = s(a_{41}) \times s(a_{42}) = 0.5670 \times 0.4500 = 0.2552,$$
$$T_{44} = s(a_{41}) \times s(a_{42}) \times s(a_{44}) = 0.5670 \times 0.4500 \times 0.8500 = 0.2169, \sum_{j=1}^{4} T_j = 2.0391;$$
$$T_{51} = 1, T_{52} = s(a_{51}) = 0.6500, T_{53} = 0.3900, T_{54} = 0.2925, \sum_{j=1}^{4} T_j = 2.3325.$$

Therefore, the following results can be obtained:

$$s(y_1) = 0.5787; s(y_2) = 0.6223; s(y_3) = 0.5581; s(y_4) = 0.5520; s(y_5) = 0.6131$$

Since $s(y_4) < s(y_3) < s(y_1) < s(y_5) < s(y_2)$, so the final ranking is $a_4 \prec a_3 \prec a_1 \prec a_5 \prec a_2$. The best alternative is a_2 while the worst alternative is a_4.

Case 4. If the correlations among the multi-hesitant fuzzy values of criteria are considered, then the GMHFCIWA operator can be used. Based on Definition 8 and Theorem 4, suppose $\mu(c_1) = 0.40$, $\mu(c_2) = 0.25$, $\mu(c_3) = 0.37$, $\mu(c_4) = 0.20$, $\mu(c_1, c_2) = 0.60$, $\mu(c_1, c_3) = 0.70$, $\mu(c_1, c_4) = 0.56$, $\mu(c_2, c_3) = 0.68$, $\mu(c_2, c_4) = 0.43$, $\mu(c_3, c_4) = 0.54$, $\mu(c_1, c_2, c_3) = 0.88$, $\mu(c_1, c_2, c_4) = 0.75$, $\mu(c_2, c_3, c_4) = 0.73$, $\mu(c_1, c_3, c_4) = 0.84$, and $\mu(c_1, c_2, c_3, c_4) = 1$, then the following results can be obtained.

$$s(y_1) = 0.6200; s(y_2) = 0.6320; s(y_3) = 0.6050; s(y_4) = 0.6305; s(y_5) = 0.6400$$

Since $s(y_3) < s(y_1) < s(y_4) < s(y_2) < s(y_5)$, so the final ranking is $a_3 \prec a_1 \prec a_4 \prec a_2 \prec a_5$. The best alternative is a_5 while the worst alternative is a_3.

5.2. Sensitivity Analysis

In Step 2, four aggregation operators can be used, and the sensitivity analysis will be conducted in these cases. Since the aggregation parameter λ is a balance factor, which can be determined by decision-makers based on their preference. To investigate the influence of different λ on the ranking of alternatives, various λ are utilized. If the GMHFOWA operator, the GMHFHWA operator, the GMHPWA operator and the GMHCIWA operator are used respectively, then the ranking results are shown in Tables 2 and 3.

Table 2. Rankings obtained using the GMHFOWA operator and the GMHFHWA operator.

λ	Rankings	
	GMHFOWA	**GMHFHWA**
$\lambda = 1$	$a_3 \prec a_4 \prec a_5 \prec a_2 \prec a_1$	$a_3 \prec a_4 \prec a_1 \prec a_2 \prec a_5$
$\lambda = 2$	$a_3 \prec a_4 \prec a_5 \prec a_2 \prec a_1$	$a_3 \prec a_4 \prec a_1 \prec a_2 \prec a_5$
$\lambda = 5$	$a_3 \prec a_4 \prec a_5 \prec a_2 \prec a_1$	$a_3 \prec a_4 \prec a_1 \prec a_2 \prec a_5$
$\lambda = 10$	$a_3 \prec a_4 \prec a_5 \prec a_2 \prec a_1$	$a_3 \prec a_4 \prec a_1 \prec a_2 \prec a_5$

Table 3. Rankings obtained using the GMHFPWA and the GMHFCIWA operators.

λ	Rankings	
	GMHFPWA	GMHFCIWA
$\lambda = 1$	$a_4 \prec a_3 \prec a_1 \prec a_5 \prec a_2$	$a_3 \prec a_1 \prec a_4 \prec a_2 \prec a_5$
$\lambda = 2$	$a_4 \prec a_3 \prec a_1 \prec a_5 \prec a_2$	$a_3 \prec a_1 \prec a_4 \prec a_2 \prec a_5$
$\lambda = 5$	$a_4 \prec a_3 \prec a_1 \prec a_5 \prec a_2$	$a_3 \prec a_1 \prec a_4 \prec a_2 \prec a_5$
$\lambda = 10$	$a_4 \prec a_3 \prec a_1 \prec a_5 \prec a_2$	$a_3 \prec a_1 \prec a_4 \prec a_2 \prec a_5$

From Tables 2 and 3, it can be seen that if the GMHFCIWA operator and the GMHFHWA operator are used respectively in Step 2, then the final ranking is $a_3 \prec a_1 \prec a_4 \prec a_2 \prec a_5$ or $a_3 \prec a_4 \prec a_1 \prec a_2 \prec a_5$. The best alternative is always a_5 while the worst alternative is a_3. If the GMHFOWA operator, the GMHFPWA operator are used respectively in Step 2, then the final ranking is $a_3 \prec a_4 \prec a_5 \prec a_2 \prec a_1$ and $a_4 \prec a_3 \prec a_1 \prec a_5 \prec a_2$. The best alternative is a_1 or a_2 while the worst alternative is a_3 or a_4. However, for each operator, the rankings obtained are consistent as λ changes. Moreover, for different operator, the aggregation parameter λ also lead to different aggregation results, but the final rankings of alternatives are the same as the parameter changes. Moreover, different aggregation operators can be chosen according to the practical necessity of MCDM problems, which can represent the decision-makers' preference.

5.3. A Comparison Analysis and Discussion

In this section, to validate the feasibility of the proposed multi-hesitant fuzzy MCDM approach based on convex operators, a comparative study was conducted with other methods as shown in Xu [9,10], Zhang et al. [13], Yu [16], Zhang and Wei [31], Zhang and Xu [32], and Peng et al. [35]. Moreover, the method in Wei [11] considering the prioritization among criteria is also compared.

The method presented in Peng et al. [35] can deal with multi-hesitant fuzzy information directly. However, in other compared methods, they all do not clarify that how to solve a situation where there is a repeated value in the evaluation of alternatives. The comparison analysis was based on the same illustrative example, but the same value will be counted only once in Table 1. Suppose the weight vector of criteria is $w = (0.33, 0.18, 0.37, 0.12)$, then the compared results can be obtained as shown in Table 4.

Table 4. Comparison of different methods.

Methods		Ranking of Alternatives
Xu [9,10]		$a_3 \prec a_2 \prec a_1 \prec a_5 \prec a_4$
Wei [11]		$a_3 \prec a_1 \prec a_4 \prec a_5 \prec a_2$
Zhang [13]		$a_3 \prec a_4 \prec a_1 \prec a_2 \prec a_5$
Yu [16]		$a_3 \prec a_2 \prec a_5 \prec a_4 \prec a_1$
Zhang and Wei [31]		$a_3 \prec a_4 \prec a_2 \prec a_1 \prec a_5$
Zhang and Xu [32]		$a_4 \prec a_3 \prec a_2 \prec a_1 \prec a_5$
Peng et al. [35]		$a_4 \prec a_3 \prec a_2 \prec a_1 \prec a_5$
Proposed methods:	GMHFOWA	$a_3 \prec a_4 \prec a_5 \prec a_2 \prec a_1$
	GMHFHWA	$a_3 \prec a_4 \prec a_1 \prec a_2 \prec a_5$
	GMHFPWA	$a_4 \prec a_3 \prec a_1 \prec a_5 \prec a_2$

According to the results presented in Table 4, the following conclusions can be categorically drawn. Firstly, the repetitive values in HFSs are not taken into consideration in the existing methods. Secondly, compared with the methods relying on aggregation operators, the result of using the GMHFHWA operator is the same as that using the method of Zhang [13], and the best alternative is always a_5 while the worst alternative is always a_3; the result of using the GMHFOWA operator is the same as that using the method of Yu [16] and the best alternative is a_1 while the worst alternative is a_3; However, the results of the proposed approach are different from that using the method of

Xu [9,10]. Furthermore, the method of Wei [11] and the proposed GMHFPWA operator are all considered the prioritization among criteria. However, there exist a litter difference between the result of using GMHFPWA operator and the result of using the method of Wei [11], the final ranking is $a_4 \prec a_3 \prec a_1 \prec a_5 \prec a_2$ or $a_3 \prec a_1 \prec a_4 \prec a_5 \prec a_2$. The best alternative is always a_2 while the worst alternative is a_4 or a_3. Apparently, different operations and aggregation operators being involved in those methods can interpret the differences existing in the final rankings to some extent. Thirdly, compared with the methods relying on distance measures, the result of using the GMHFWA operator or the GMHFHWA operator is the same as that using the method of Zhang and Wei [31] and the best alternative is always a_5 while the worst alternative is always a_3. However, it is different from that using of Zhang and Xu [32] that the best alternative is a_5 while the worst alternative is a_4. Furthermore, the methods using distance measures have certain shortcomings because the condition should be satisfied that all HFNs must be arranged in ascending order and be of equal length. If two HFNs being compared have different lengths, then the value of the shorter one should be increased subjectively until both are equal. Finally, the result using the method of Peng et al. [35] is the same as that of the proposed approach. Therefore, the proposed method can effectively overcome the shortcomings of the compared methods and the computation is very simple.

From the analysis above, it can be seen that the main advantages of the approach developed in this paper over the other methods are not only due to its ability to effectively deal with the preference information expressed by MHFNs, but also due to its consideration that the weight of criteria is known or unknown. This can avoid losing and distorting the preference information provided, which makes the results better correspond with real life decision-making problems.

6. Conclusions

HFSs are considered useful in handling decision-making problems under uncertain situations where decision-makers hesitate when choosing between several values before expressing their preferences about weights and data. MHFSs can deal effectively with the case where some values are repeated more than once in an HFS. In this paper, the convex combination of MHFNs was discussed and some aggregation operators based on convex operation, such as GMHFOWA operator, GMHFHWA operator, GMHFPWA operator and GMHFCIWA operator, were developed as well. Moreover, a novel approach based on convex operators was developed to deal with MCDM problems where the data are MHFNs. Finally, an illustrative example was given to verify the proposed approach. The primary characteristic of the proposed approach is that those aggregation operators can provide more choices for decision-makers according to the actual decision-making environment. Moreover, MHFSs could overcome the shortcomings in HFSs where if two or more decision-makers set the same value, it is only counted once. Further research will investigate how to obtain the optimal values of criteria by a specified model within a multi-hesitant fuzzy environment.

Author Contributions: Conceptualization and Methodology, Y.M. and J.P.; Formal Analysis, J.P.; Validation, J.Y.; Original Draft Preparation, Y.M. and J.P.; Writing-Review & Editing, J.Y.

Funding: This work was supported by the National Natural Science Foundation, P.R. China (No. 71701065) and Educational Commission of Guangdong Province of China (2018JKZ022).

Acknowledgments: The authors thank the editors and anonymous reviewers for their helpful comments and suggestions.

Conflicts of Interest: The authors declare no conflict of interest.

References

1. Torra, V. Hesitant fuzzy sets. *Int. J. Intell. Syst.* **2010**, *25*, 529–539. [CrossRef]
2. Torra, V.; Narukawa, Y. On hesitant fuzzy sets and decision. In Proceedings of the 18th IEEE International Conference on Fuzzy Systems, Jeju Island, Korea, 20–24 August 2009; pp. 1378–1382.
3. Zadeh, L.A. Fuzzy sets. *Inf. Control* **1965**, *8*, 338–356. [CrossRef]

4. Song, C.; Xu, Z.; Zhao, H. A novel comparison of probabilistic hesitant fuzzy elements in multi-criteria decision making. *Symmetry* **2018**, *10*, 177. [CrossRef]
5. Faizi, S.; Sałabun, W.; Rashid, T.; Wątróbski, J.; Zafar, S. Group decision-making for hesitant fuzzy sets based on characteristic objects method. *Symmetry* **2017**, *9*, 136. [CrossRef]
6. Faizi, S.; Rashid, T.; Saabun, W.; Zafar, S.; Wtróbski, J. Decision making with uncertainty using hesitant fuzzy sets. *Int. J. Fuzzy Syst.* **2018**, *20*, 93–103. [CrossRef]
7. Liao, H.; Wu, D.; Huang, Y.; Ren, P.; Xu, Z.; Verma, M. Green logistic provider selection with a hesitant fuzzy linguistic thermodynamic method integrating cumulative prospect theory and PROMETHEE. *Sustainability* **2018**, *10*, 1291. [CrossRef]
8. Liu, P.; Gao, H. Multi-criteria decision making based on generalized Maclaurin symmetric means with multi-hesitant fuzzy linguistic information. *Symmetry* **2018**, *10*, 81. [CrossRef]
9. Xia, M.M.; Xu, Z.S. Hesitant fuzzy information aggregation in decision making. *Int. J. Approx. Reason.* **2011**, *52*, 395–407. [CrossRef]
10. Zhu, B.; Xu, Z.S.; Xia, M.M. Hesitant fuzzy geometric Bonferoni means. *Inf. Sci.* **2012**, *205*, 72–85. [CrossRef]
11. Wei, G.W. Hesitant fuzzy prioritized operators and their application to multiple attribute decision making. *Knowl. Based Syst.* **2012**, *31*, 176–182. [CrossRef]
12. Xia, M.M.; Xu, Z.S.; Chen, N. Some Hesitant fuzzy aggregation operators with their application in group decision making. *Group Decis. Negot.* **2013**, *22*, 259–279. [CrossRef]
13. Zhang, Z.M.; Wang, C.; Tian, D.Z.; Li, K. Induced generalized hesitant fuzzy operators and their application to multiple attribute group decision making. *Comput. Ind. Eng.* **2014**, *67*, 116–138. [CrossRef]
14. Zhou, W. An Accurate method for determining hesitant fuzzy aggregation operator weights and its application to project investment. *Int. J. Intell. Syst.* **2014**, *29*, 668–686. [CrossRef]
15. Zhang, Z.M. Hesitant fuzzy power aggregation operators and their application to multiple attribute group decision making. *Inf. Sci.* **2013**, *234*, 150–181. [CrossRef]
16. Yu, D.J. Some hesitant fuzzy information aggregation operators based on Einstein operational laws. *Int. J. Intell. Syst.* **2014**, *29*, 320–340. [CrossRef]
17. Chen, N.; Xu, Z.S.; Xia, M.M. Correlation coefficients of hesitant fuzzy sets and their applications to clustering analysis. *Appl. Math. Model.* **2013**, *37*, 2197–2211. [CrossRef]
18. Xu, Z.S.; Xia, M.M. Distance and similarity measures for hesitant fuzzy sets. *Inf. Sci.* **2011**, *181*, 2128–2138. [CrossRef]
19. Xu, Z.S.; Xia, M.M. On distance and correlation measures of hesitant fuzzy information. *Int. J. Intell. Syst.* **2011**, *26*, 410–425. [CrossRef]
20. Farhadinia, B. Distance and similarity measures for higher order hesitant fuzzy sets. *Knowl. Based Syst.* **2014**, *55*, 43–48. [CrossRef]
21. Wang, L.; Ni, M.F.; Yu, Z.K.; Zhu, L. Power geometric operators of hesitant multiplicative fuzzy numbers and their application to multiple attribute group decision making. *Math. Probl. Eng.* **2014**, *2014*, 186502. [CrossRef]
22. Torres, R.; Salas, R.; Astudillo, H. Time-based hesitant fuzzy information aggregation approach for decision-making problems. *Int. J. Intell. Syst.* **2014**, *29*, 579–595. [CrossRef]
23. Qian, G.; Wang, H.; Feng, X.Q. Generalized hesitant fuzzy sets and their application in decision support system. *Knowl. Based Syst.* **2013**, *37*, 357–365. [CrossRef]
24. Meng, F.Y.; Chen, X.H.; Zhang, Q. Induced generalized hesitant fuzzy Shapley hybrid operators and their application in multi-attribute decision making. *Appl. Soft Comput.* **2015**, *28*, 599–607. [CrossRef]
25. Zhou, W.; Xu, Z.S. Optimal discrete fitting aggregation approach with hesitant fuzzy information. *Knowl. Based Syst.* **2015**, *78*, 22–33. [CrossRef]
26. Tan, C.Q.; Yi, W.T.; Chen, X.H. Hesitant fuzzy Hamacher aggregation operators for multicriteria decision making. *Appl. Soft Comput.* **2015**, *26*, 325–349. [CrossRef]
27. Meng, F.Y.; Chen, X.H. Correlation coefficients of hesitant fuzzy sets and their application based on fuzzy measures. *Cognative Comput.* **2015**, *7*, 445–463. [CrossRef]
28. Liao, H.C.; Xu, Z.S.; Zeng, X.J. Novel correlation coefficients between hesitant fuzzy sets and their application in decision making. *Knowl. Based Syst.* **2015**, *82*, 115–127. [CrossRef]
29. Li, D.Q.; Zeng, W.Y.; Li, J.H. New distance and similarity measures on hesitant fuzzy sets and their applications in multiple criteria decision making. *Eng. Appl. Artif. Intell.* **2015**, *40*, 11–16. [CrossRef]

30. Hu, J.H.; Zhang, X.L.; Chen, X.H.; Liu, Y.M. Hesitant fuzzy information measures and their applications in multi-criteria decision making. *Int. J. Syst. Sci.* **2015**, *87*, 91–103. [CrossRef]
31. Zhang, N.; Wei, G.W. Extension of VIKOR method for decision making problem based on hesitant fuzzy set. *Appl. Math. Model.* **2013**, *37*, 4938–4947. [CrossRef]
32. Zhang, X.L.; Xu, Z.S. The TODIM analysis approach based on novel measured functions under hesitant fuzzy environment. *Knowl. Based Syst.* **2014**, *61*, 48–58. [CrossRef]
33. Farhadinia, B. A novel method of ranking hesitant fuzzy values for multiple attribute decision-making problems. *Int. J. Intell. Syst.* **2013**, *28*, 752–767. [CrossRef]
34. Farhadinia, B. Information measures for hesitant fuzzy sets and interval-valued hesitant fuzzy sets. *Inf. Sci.* **2013**, *240*, 129–144. [CrossRef]
35. Peng, J.J.; Wang, J.Q.; Wang, J.; Yang, L.J.; Chen, X.H. An extension of ELECTRE to multi-criteria decision-making problems with multi-hesitant fuzzy sets. *Inf. Sci.* **2015**, *307*, 113–126. [CrossRef]
36. Chen, N.; Xu, Z.S. Hesitant fuzzy ELECTRE II approach: A new way to handle multi-criteria decision making problems. *Inf. Sci.* **2015**, *292*, 175–197. [CrossRef]
37. Yager, R.R. Prioritized aggregation operators. *Int. J. Approx. Reason.* **2008**, *48*, 263–274. [CrossRef]
38. Wang, Z.; Klir, G.J. *Fuzzy Measure Theory*; Plenum Press: New York, NY, USA, 1992.
39. Schmeidler, D. Subjective probability and expected utility without additivity. *Econometrica* **1989**, *57*, 517–587. [CrossRef]

information

MDPI

Article

Hesitant Probabilistic Fuzzy Information Aggregation Using Einstein Operations

Jin Han Park [1,*], Yu Kyoung Park [1] and Mi Jung Son [2,*]

1 Department of Applied Mathematics, Pukyong National University, Busan 608-737, Korea; ykpark87@pknu.ac.kr
2 Department of Mathematics, Korea Maritime University, Busan 606-791, Korea
* Correspondence: jihpark@pknu.ac.kr (J.H.P.), mjson72@korea.com (M.J.S.); Tel.: +82-51-629-5530 (J.H.P.); +82-10-4542-0023 (M.J.S.)

Received: 8 August 2018; Accepted: 3 September 2018; Published: 4 September 2018

Abstract: In this paper, a hesitant probabilistic fuzzy multiple attribute group decision making is studied. First, some Einstein operations on hesitant probability fuzzy elements such as the Einstein sum, Einstein product, and Einstein scalar multiplication are presented and their properties are discussed. Then, several hesitant probabilistic fuzzy Einstein aggregation operators, including the hesitant probabilistic fuzzy Einstein weighted averaging operator and the hesitant probabilistic fuzzy Einstein weighted geometric operator and so on, are introduced. Moreover, some desirable properties and special cases are investigated. It is shown that some existing hesitant fuzzy aggregation operators and hesitant probabilistic fuzzy aggregation operators are special cases of the proposed operators. Further, based on the proposed operators, a new approach of hesitant probabilistic fuzzy multiple attribute decision making is developed. Finally, a practical example is provided to illustrate the developed approach.

Keywords: hesitant probabilistic fuzzy element (HPFE); Einstein operations; hesitant probabilistic fuzzy Einstein aggregation operators; multiple attribute decision making (MADM).

1. Introduction

Decision making problems typically consist of finding the most desirable alternative(s) out of a given set of alternatives. So far, there are applications of decision making into different disciplines, such as railroad container terminal selection, pharmaceutical supplying, hospital service quality, and so on [1–3]. Due to the increasing ambiguity and complexity of the socio-economic environment, it is difficult to obtain accurate and sufficient data for practical decision making. Therefore, uncertainty data needs to be addressed in the actual decision making process, and several other methodologies and theories have been proposed. Among them, the fuzzy set theory [4] is excellent and has been widely used in many areas of real life [5–8]. Since Zadeh [4] introduced the fuzzy set (FS) in 1965, many researchers have developed extended forms of FS, such as the intuitive fuzzy set (IFS) [9], the type-2 fuzzy set [10], the type-n fuzzy set [10], the fuzzy multiset [11] and the fuzzy hesitant set (HFS) [12]. Among these, the HFS was broadly applied to the practical decision making process. In fact, the HFS is widely used in decision making problems with the aim of resolving the difficulty of explaining hesitation in the actual assessment. The main reason is that experts may face situations in which people are hesitant to provide their preferences in the decision making process by allowing them to prefer several possible values between 0 and 1. Torra [12] introduced some basic operations of HFSs. Xia and Xu [13] defined the hesitant fuzzy element (HFE), which is the basic component of the HFS, and proposed and investigated the score function and comparison law of HFEs as the basis for its calculation and application. Li et al. [14] and Meng and Chen [15] proposed various distance measures and some correlation coefficients for HFSs. They also investigated applications based on the distance measures and correlation coefficients. Over the past decade, there many researchers [16–23] have studied the aggregation operators, one of the core issues of HFSs. Thus, many researchers have worked hard to develop the HFS theory and have helped to develop it in uncertain decision making problems [24–26].

However, there is one obvious weakness in the current approaches; namely, each of the possible values in the HFE provided by the experts has the same weight. To overcome this weakness, Xu and Zhou [27] proposed the hesitant probabilistic fuzzy set (HPFS) and hesitant probabilistic fuzzy element (HPFE) developed by introducing probabilities to HFS and HFE respectively. For example, experts evaluate a house's "comfort" using an HFE $(0.3, 0.4, 0.5)$ because they hesitate to evaluate it. However, they believe that 0.4 is appropriate and 0.3 is less appropriate than the other values in the HFE. Therefore, although the HFE $(0.3, 0.4, 0.5)$ cannot fully represent the evaluation, the HPFE $(0.3|0.2, 0.4|0.5, 0.5|0.3)$ can present this issue vividly and is more convenient than HFE. Consequently, the HPFS can overcome the defect of HFS to great extent, so it can remain the experts' evaluation information and describe their preferences better. In Ref. [27], the HPFE was combined with weighted operators to develop basic weighted operators, such as hesitant probabilistic fuzzy weighted average/geometric (HPFWA or HPFWG) operators and the hesitant probabilistic fuzzy ordered weighted averaging/geometric (HPFOWA or HPFOWG) operators. Based on the perspective of the aggregation operators, they established the consensus among decision makers in group decision making. Zhang and Wu [28] investigated some operations of HPFE and applied them to multicriteria decision making (MCDM). In another way, some scholars recently tried to solve the problem of HFSs. Bedregal et al. [29] tried to use fuzzy multisets to improve the HFSs. This method has been worked out to some extent. Wang and Li [30] proposed the picture hesitant fuzzy set to express the uncertainty and complexity of experts' opinions and applied them to solve diverse situations during MCDM processes. Interval-valued HFSs have been used in the applications of group decision making in [31]. Multiple attribute decision making (MADM) using the trapezoidal valued HFSs is discussed in [32]. Yu [33] gave the concept of triangular hesitant fuzzy sets and used it for the solution of decision making problems. Mahmood et al. [34] introduced the cubic hesitant fuzzy set and applied it to MCDM.

The study on aggregation operators to fuse hesitant probabilistic information is one of the core issues in HPFS theory. The all aggregation operators introduced previously, such as the HPFWA, HPFWG, HPFOWA, and HPFOWG operators, are based on the algebraic product and algebraic sum of HPFEs, which are a pair of the special dual t-norm and t-conorm [35]. Although the algebraic product and algebraic sum are the basic algebraic operations of HPFEs, they are not the only ones. The Einstein product and Einstein sum are good alternatives to the algebraic product and algebraic sum for structuring aggregation operators, respectively, and they have been used to aggregate the intuitionistic fuzzy values or the HFEs by many researchers [21–23,36–38]. However, it seems that in the literature, there has been little investigation on aggregation techniques using the Einstein operations to aggregate hesitant probabilistic fuzzy information. Thus, it is meaningful to research the hesitant probabilistic fuzzy information aggregation methods based on the Einstein operations. In this paper, motivated by the works of Xu and Zhou [27] and Yu [21], we propose the hesitant probabilistic fuzzy Einstein weighted aggregation operators with the help of Einstein operations, and apply them to MADM under a hesitant probabilistic fuzzy environment. To do this, the remainder of this paper is organized as follows: The following section recalls briefly some basic concepts and notions related to the HPFSs and HPFEs. In Section 3, based on the hesitant probabilistic fuzzy weighted aggregation operator and the Einstein operations, we propose the hesitant probabilistic fuzzy Einstein weighted aggregation operators including the hesitant probabilistic fuzzy Einstein weighted averaging/geometric (HPFEWA or HPFEWG) operators and the hesitant probabilistic fuzzy Einstein ordered weighted averaging/geometric (HPFEOWA or HPFEOWG) operators. Section 4 develops an approach to MADM with hesitant probabilistic fuzzy information based on the proposed operators. An example is given to demonstrate the practicality and effectiveness of the proposed approach in Section 4. Section 5 gives some concluding remarks.

2. Hesitant Fuzzy Information with Probabilities

2.1. HPFS and HPFE

The HPFS and HPFE represent hesitant fuzzy information with the following probabilities.

Definition 1. [27] *Let R be a fixed set, then an HPFS on R is expressed by a mathematical symbol:*

$$H_P = \left\{\bar{h}(\gamma_i|p_i)|\gamma_i, p_i\right\}, \tag{1}$$

where $\bar{h}(\gamma_i|p_i)$ is a set of some elements $(\gamma_i|p_i)$ denoting the hesitant fuzzy information with probabilities to the set H_P, $\gamma_i \in R$, $0 \leq \gamma_i \leq 1$, $i = 1,2,\ldots,\#\bar{h}$, where $\#\bar{h}$ is the number of possible elements in $\bar{h}(\gamma_i|p_i)$, $p_i \in [0,1]$ is the hesitant probability of γ_i, and $\sum_{i=1}^{\#\bar{h}} p_i = 1$.

For convenience, Xu and Zhou [27] called $\bar{h}(\gamma_i|p_i)$ a HPFE, and H_P the set of HPFSs. In addition, they gave the following score function, deviation function, and comparison law to compare different HPFEs.

Definition 2. [27] *Let $\bar{h}(\gamma_i|p_i)$ $(i = 1,2,,\ldots,\#\bar{h})$ be a HPFE, then*

(1) $s(\bar{h}) = \sum_{i=1}^{\#\bar{h}} \gamma_i p_i$ is called the score function of $\bar{h}(\gamma_i|p_i)$, where $\#\bar{h}$ is the number of possible elements in $\bar{h}(\gamma_i|p_i)$;

(2) $d(\bar{h}) = \sum_{i=1}^{\#\bar{h}} (\gamma_i - s(\bar{h}))^2 p_i$ is called the deviation function of $\bar{h}(\gamma_i|p_i)$, where $s(\bar{h}) = \sum_{i=1}^{\#\bar{h}} \gamma_i p_i$ is the score function of $\bar{h}(\gamma_i|p_i)$, and $\#\bar{h}$ is the number of possible elements in $\bar{h}(\gamma_i|p_i)$.

If all probabilities are equal, i.e., $p_1 = p_2 = \cdots = p_{\#\bar{h}}$, then the HPFE is reduced to the HFE. So, in this case, the score function of the HPFE is consistent with that of the HFE.

Definition 3. [27] *Let $\bar{h}_1(\gamma_i|p_i)$ $(i = 1,2,,\ldots,\#\bar{h}_1)$ and $\bar{h}_2(\gamma_j|p_j)$ $(j = 1,2,,\ldots,\#\bar{h}_2)$ be two HPFEs, $s(\bar{h}_1)$ and $s(\bar{h}_2)$ are the score functions of $\bar{h}_1$ and $\bar{h}_2$, respectively, and $d(\bar{h}_1)$ and $d(\bar{h}_2)$ are the deviation functions of $\bar{h}_1$ and $\bar{h}_2$, respectively, then*

(1) If $s(\bar{h}_1) < s(\bar{h}_2)$, then $\bar{h}_1$ is smaller than $\bar{h}_2$ which is denoted by $\bar{h}_1 < \bar{h}_2$;

(2) If $s(\bar{h}_1) = s(\bar{h}_2)$, then

(a) If $d(\bar{h}_1) > d(\bar{h}_2)$, then $\bar{h}_1$ is smaller than $\bar{h}_2$, denoted by $\bar{h}_1 < \bar{h}_2$;

(b) If $d(\bar{h}_1) = d(\bar{h}_2)$, then $\bar{h}_1$ and $\bar{h}_2$ represent the same information, denoted by $\bar{h}_1 = \bar{h}_2$.

Some operations to aggregate HPFEs based on the operations of HFEs [12,13] are defined as follows:

Definition 4. [27] *Let $\bar{h}(\gamma_i|p_i)$, $\bar{h}_1(\dot{\gamma}_j|\dot{p}_j)$ and $\bar{h}_2(\ddot{\gamma}_k|\ddot{p}_k)$ be three HPFEs, $i = 1,2,\ldots,\#\bar{h}$, $j = 1,2,\ldots,\#\bar{h}_1$, $k = 1,2,\ldots,\#\bar{h}_2$, and $\lambda > 0$, then*

(1) $(\bar{h})^c = \cup_{i=1,2,\ldots,\#\bar{h}} \left\{(1-\gamma_i)|p_i\right\}$;

(2) $\lambda\bar{h} = \cup_{i=1,2,\ldots,\#\bar{h}} \left\{1-(1-\gamma_i)^\lambda|p_i\right\}$;

(3) $\bar{h}^\lambda = \cup_{i=1,2,\ldots,\#\bar{h}} \left\{(\gamma_i)^\lambda|p_i\right\}$;

(4) $\bar{h}_1 \oplus \bar{h}_2 = \cup_{j=1,2,\ldots,\#\bar{h}_1,k=1,2,\ldots,\#\bar{h}_2} \left\{(\dot{\gamma}_j + \ddot{\gamma}_k - \dot{\gamma}_j\ddot{\gamma}_k)|\dot{p}_j\ddot{p}_k\right\}$;

(5) $\bar{h}_1 \otimes \bar{h}_2 = \cup_{j=1,2,\ldots,\#\bar{h}_1,k=1,2,\ldots,\#\bar{h}_2} \left\{\dot{\gamma}_j\ddot{\gamma}_k|\dot{p}_j\ddot{p}_k\right\}$.

Theorem 1. *Let $\bar{h}(\gamma_i|p_i)$, $\bar{h}_1(\dot{\gamma}_j|\dot{p}_j)$ and $\bar{h}_2(\ddot{\gamma}_k|\ddot{p}_k)$ be three HPFEs, $i = 1,2,\ldots,\#\bar{h}$, $j = 1,2,\ldots,\#\bar{h}_1$, $k = 1,2,\ldots,\#\bar{h}_2$, $\lambda > 0$, $\lambda_1 > 0$, and $\lambda_2 > 0$, then*

(1) $\bar{h}_1 \oplus \bar{h}_2 = \bar{h}_2 \oplus \bar{h}_1$;

(2) $\bar{h} \oplus (\bar{h}_1 \oplus \bar{h}_2) = (\bar{h} \oplus \bar{h}_1) \oplus \bar{h}_2$;

(3) $\lambda(\bar{h}_1 \oplus \bar{h}_2) = (\lambda\bar{h}_1) \oplus (\lambda\bar{h}_2)$;

(4) $\lambda_1(\lambda_2\bar{h}) = (\lambda_1\lambda_2)\bar{h}$;

(5) $\bar{h}_1 \otimes \bar{h}_2 = \bar{h}_2 \otimes \bar{h}_1$;
(6) $\bar{h} \otimes (\bar{h}_1 \otimes \bar{h}_2) = (\bar{h} \otimes \bar{h}_1) \otimes \bar{h}_2$;
(7) $(\bar{h}_1 \otimes \bar{h}_2)^\lambda = \bar{h}_1^\lambda \otimes \bar{h}_2^\lambda$;
(8) $(\bar{h}^{\lambda_1})^{\lambda_2} = \bar{h}^{(\lambda_1 \lambda_2)}$.

Proof. We only prove (3) and the other are trivial or similar to (3).

(3) Since $\bar{h}_1 \oplus \bar{h}_2 = \cup_{j=1,2,\ldots,\#\bar{h}_1,k=1,2,\ldots,\#\bar{h}_2} \{\dot{\gamma}_j + \ddot{\gamma}_k - \dot{\gamma}_j\ddot{\gamma}_k | \dot{p}_j\ddot{p}_k\}$, according to the operational law (2) in Definition 4, we have

$$\begin{aligned}\lambda(\bar{h}_1 \oplus \bar{h}_2) &= \cup_{\substack{j=1,2,\ldots,\#\bar{h}_1,\\ k=1,2,\ldots,\#\bar{h}_2}} \left\{1 - (1 - (\dot{\gamma}_j + \ddot{\gamma}_k - \dot{\gamma}_j\ddot{\gamma}_k))^\lambda \,|\dot{p}_j\ddot{p}_k\right\} \\ &= \cup_{\substack{j=1,2,\ldots,\#\bar{h}_1,\\ k=1,2,\ldots,\#\bar{h}_2}} \left\{1 - ((1 - \dot{\gamma}_j)(1 - \ddot{\gamma}_k))^\lambda \,|\dot{p}_j\ddot{p}_k\right\}.\end{aligned}$$

Since $\lambda\bar{h}_1 = \cup_{j=1,2,\ldots,\#\bar{h}_1} \{1 - (1 - \dot{\gamma}_j)^\lambda|\dot{p}_j\}$ and $\lambda\bar{h}_2 = \cup_{k=1,2,\ldots,\#\bar{h}_2} \{1 - (1 - \dot{\gamma}_k)^\lambda|\ddot{p}_k\}$, we have

$$\begin{aligned}(\lambda\bar{h}_1) \oplus (\lambda\bar{h}_2) &= \cup_{\substack{j=1,2,\ldots,\#\bar{h}_1,\\ k=1,2,\ldots,\#\bar{h}_2}} \left\{1 - (1 - \dot{\gamma}_j)^\lambda + 1 - (1 - \dot{\gamma}_k)^\lambda - (1 - (1 - \dot{\gamma}_j)^\lambda)(1 - (1 - \dot{\gamma}_k)^\lambda)|\dot{p}_j\ddot{p}_k\right\} \\ &= \cup_{\substack{j=1,2,\ldots,\#\bar{h}_1,\\ k=1,2,\ldots,\#\bar{h}_2}} \left\{1 - (1 - \dot{\gamma}_j)^\lambda(1 - \ddot{\gamma}_k)^\lambda \,|\dot{p}_j\ddot{p}_k\right\}.\end{aligned}$$

Hence, $\lambda(\bar{h}_1 \oplus \bar{h}_2) = (\lambda\bar{h}_1) \oplus (\lambda\bar{h}_2)$. □

However, for an HPFE $\bar{h}(\gamma_i|p_i)$, $i = 1,2,\ldots,\#\bar{h}$, $\lambda_1 > 0$ and $\lambda_2 > 0$, the operational laws $(\lambda_1\bar{h}) \oplus (\lambda_2\bar{h}) = (\lambda_1 + \lambda_2)\bar{h}$ and $\bar{h}^{\lambda_1} \otimes \bar{h}^{\lambda_2} = \bar{h}^{(\lambda_1+\lambda_2)}$ do not hold in general. To illustrate this case, we give the following example.

Example 1. *Let* $\bar{h}(\gamma_i|p_i) = (0.7|0.5, 0.2|0.5)$ *and* $\lambda_1 = \lambda_2 = 1$, *then*

$$\begin{aligned}(\lambda_1\bar{h}) \oplus (\lambda_2\bar{h}) = \bar{h} \oplus \bar{h} &= \cup_{i,j=1,2} \{\gamma_i + \gamma_j - \gamma_i\gamma_j|0.25\} \\ &= (0.91|0.25, 0.76|0.25, 0.76|0.25, 0.36|0.25), \\ (\lambda_1 + \lambda_2)\bar{h} = 2\bar{h} &= \cup_{i=1,2} \left\{1 - (1 - \gamma_i)^2|0.5\right\} = (0.91|0.5, 0.36|0.5)\end{aligned}$$

and $s((\lambda_1\bar{h}) \oplus (\lambda_2\bar{h})) = 0.6975 > 0.635 = s((\lambda_1 + \lambda_2)\bar{h})$ *and hence,* $(\lambda_1\bar{h}) \oplus (\lambda_2\bar{h}) > (\lambda_1 + \lambda_2)\bar{h}$. *Similarly, we have* $s(\bar{h}^{\lambda_1} \otimes \bar{h}^{\lambda_2}) = 0.2025 < 0.265 = s(\bar{h}^{(\lambda_1+\lambda_2)})$ *and thus,* $\bar{h}^{\lambda_1} \otimes \bar{h}^{\lambda_2} < \bar{h}^{(\lambda_1+\lambda_2)}$.

Based on Definition 4, in order to aggregate the HPFEs, Xu and Zhou [27] developed some hesitant probabilistic fuzzy aggregation operators, as follows:

Definition 5. [27] *Let* $\bar{h}_t$ $(t = 1,2,\ldots,T)$ *be a collection of HPFEs,* $w = (w_1, w_2, \ldots, w_T)^T$ *be the weight vector of* $\bar{h}_t$ *with* $w_t \in [0,1]$, *and* $\sum_{t=1}^T w_t = 1$, *and* p_t *be the probability of* γ_t *in the HPFE* $\bar{h}_t$, *then*

(1) the hesitant probabilistic fuzzy weighted averaging (HPFWA) operator is

$$\begin{aligned}\text{HPFWA}(\bar{h}_1, \bar{h}_2, \ldots, \bar{h}_T) &= (w_1\bar{h}_1) \oplus (w_2\bar{h}_2) \oplus \cdots \oplus (w_T\bar{h}_T) \\ &= \cup_{\gamma_1\in\bar{h}_1,\gamma_2\in\bar{h}_2,\ldots,\gamma_T\in\bar{h}_T} \left\{1 - \prod_{t=1}^{T}(1 - \gamma_t)^{w_t}\,|p_1p_2\cdots p_T\right\}.\end{aligned} \tag{2}$$

(2) the hesitant probabilistic fuzzy weighted geometric (HPFWG) operator is

$$\text{HPFWG}(\bar{h}_1, \bar{h}_2, \ldots, \bar{h}_T) = (\bar{h}_1)^{w_1} \otimes (\bar{h}_2)^{w_2} \otimes \cdots \otimes (\bar{h}_T)^{w_T}$$
$$= \cup_{\gamma_1 \in \bar{h}_1, \gamma_2 \in \bar{h}_2, \ldots, \gamma_T \in \bar{h}_T} \left\{ \prod_{t=1}^{T} (\gamma_t)^{w_t} \middle| p_1 p_2 \cdots p_T \right\}. \tag{3}$$

Definition 6. [27] *Let $\bar{h}_t$ $(t = 1, 2, \ldots, T)$ be a collection of HPFEs, $\bar{h}_{\sigma(t)}$ be the tth largest of $\bar{h}_t$ $(t = 1, 2, \ldots, T)$, and $p_{\sigma(t)}$ be the probability of $\gamma_{\sigma(t)}$ in the HPFE $\bar{h}_{\sigma(t)}$, then the following two aggregation operators, which are based on the mapping $H_P^T \to H_P$ with an associated vector $\omega = (\omega_1, \omega_2, \ldots, \omega_T)^T$ such that $\omega_t \in [0,1]$ and $\sum_{t=1}^{T} \omega_t = 1$, are given by*

(1) the hesitant probabilistic fuzzy ordered weighted averaging (HPFOWA) operator:

$$\text{HPFOWA}(\bar{h}_1, \bar{h}_2, \ldots, \bar{h}_T) = (\omega_1 \bar{h}_{\sigma(1)}) \oplus (\omega_2 \bar{h}_{\sigma(2)}) \oplus \cdots \oplus (\omega_T \bar{h}_{\sigma(T)})$$
$$= \cup_{\gamma_{\sigma(1)} \in \bar{h}_{\sigma(1)}, \gamma_{\sigma(2)} \in \bar{h}_{\sigma(2)}, \ldots, \gamma_{\sigma(T)} \in \bar{h}_{\sigma(T)}} \left\{ 1 - \prod_{t=1}^{T} (1 - \gamma_{\sigma(t)})^{w_t} \middle| p_{\sigma(1)} p_{\sigma(2)} \cdots p_{\sigma(T)} \right\}. \tag{4}$$

(2) the hesitant probabilistic fuzzy ordered weighted geometric (HPFOWG) operator:

$$\text{HPFOWG}(\bar{h}_1, \bar{h}_2, \ldots, \bar{h}_T) = (\bar{h}_{\sigma(1)})^{\omega_1} \otimes (\bar{h}_{\sigma(2)})^{\omega_2} \otimes \cdots \otimes (\bar{h}_{\sigma(T)})^{\omega_T}$$
$$= \cup_{\gamma_{\sigma(1)} \in \bar{h}_{\sigma(1)}, \gamma_{\sigma(2)} \in \bar{h}_{\sigma(2)}, \ldots, \gamma_{\sigma(T)} \in \bar{h}_{\sigma(T)}} \left\{ \prod_{t=1}^{T} (\gamma_{\sigma(t)})^{w_t} \middle| p_{\sigma(1)} p_{\sigma(2)} \cdots p_{\sigma(T)} \right\}. \tag{5}$$

2.2. Einstein Operations on HPFEs

It is well known that the t-norms and t-conorms are general concepts satisfying the requirements of the conjunction and disjunction operators. Einstein operations include the Einstein sum ($\oplus_\varepsilon$) and Einstein product ($\otimes_\varepsilon$) which are examples of t-conorms and t-norms, respectively. They were defined by Klement et al. [35] as follows:

$$x \otimes_\varepsilon y = \frac{xy}{1 + (1-x)(1-y)}, \; x \oplus_\varepsilon y = \frac{x+y}{1+xy}, \; x, y \in [0,1].$$

Based on the above Einstein operations, we give the following new operations on HPFEs:

Definition 7. *Let $\bar{h}(\gamma_i | p_i)$, $\bar{h}_1(\dot{\gamma}_j | \dot{p}_j)$ and $\bar{h}_2(\ddot{\gamma}_k | \ddot{p}_k)$ be three HPFEs, $i = 1, 2, \ldots, \#\bar{h}$, $j = 1, 2, \ldots, \#\bar{h}_1$, $k = 1, 2, \ldots, \#\bar{h}_2$, and $\lambda > 0$, then*

(1) $\bar{h}_1 \oplus_\varepsilon \bar{h}_2 = \cup_{j=1,2,\ldots,\#\bar{h}_1, k=1,2,\ldots,\#\bar{h}_2} \left\{ \frac{\dot{\gamma}_j + \ddot{\gamma}_k}{1 + \dot{\gamma}_j \ddot{\gamma}_k} \middle| \dot{p}_j \ddot{p}_k \right\}$;

(2) $\bar{h}_1 \otimes_\varepsilon \bar{h}_2 = \cup_{j=1,2,\ldots,\#\bar{h}_1, k=1,2,\ldots,\#\bar{h}_2} \left\{ \frac{\dot{\gamma}_j \ddot{\gamma}_k}{1 + (1-\dot{\gamma}_j)(1-\ddot{\gamma}_k)} \middle| \dot{p}_j \ddot{p}_k \right\}$;

(3) $\lambda \cdot_\varepsilon \bar{h} = \cup_{i=1,2,\ldots,\#\bar{h}} \left\{ \frac{(1+\gamma_i)^\lambda - (1-\gamma_i)^\lambda}{(1+\gamma_i)^\lambda + (1-\gamma_i)^\lambda} \middle| p_i \right\}$;

(4) $\bar{h}^{\wedge_\varepsilon \lambda} = \cup_{i=1,2,\ldots,\#\bar{h}} \left\{ \frac{2\gamma_i^\lambda}{(2-\gamma_i)^\lambda + \gamma_i^\lambda} \middle| p_i \right\}$.

Thus, the above four operations on the HPFEs can be suitable for the HPFSs. Moreover, some relationships are discussed for the operations on HPFEs given in Definitions 4 and 7 as follows:

Theorem 2. *Let $\bar{h}(\gamma_i | p_i)$, $\bar{h}_1(\dot{\gamma}_j | \dot{p}_j)$ and $\bar{h}_2(\ddot{\gamma}_k | \ddot{p}_k)$ be three HPFEs, $i = 1, 2, \ldots, \#\bar{h}$, $j = 1, 2, \ldots, \#\bar{h}_1$, $k = 1, 2, \ldots, \#\bar{h}_2$, and $\lambda > 0$, then*

(1) $((\bar{h})^c)^{\wedge_\varepsilon \lambda} = (\lambda \cdot_\varepsilon \bar{h})^c$;

(2) $\lambda \cdot_\varepsilon (\bar{h})^c = (\bar{h}^{\wedge_\varepsilon \lambda})^c$;

(3) $(\bar{h}_1)^c \oplus_\varepsilon (\bar{h}_2)^c = (\bar{h}_1 \otimes_\varepsilon \bar{h}_2)^c$;

(4) $(\bar{h}_1)^c \otimes_\varepsilon (\bar{h}_2)^c = (\bar{h}_1 \oplus_\varepsilon \bar{h}_2)^c$.

Proof. (1)

$$\begin{aligned}((\bar{h})^c)^{\wedge_\varepsilon\lambda} &= \cup_{i=1,2,\ldots,\#\bar{h}}\left\{\frac{2(1-\gamma_i)^\lambda}{(2-(1-\gamma_i))^\lambda+(1-\gamma_i)^\lambda}\Big|p_i\right\}\\ &= \cup_{i=1,2,\ldots,\#\bar{h}}\left\{\frac{2(1-\gamma_i)^\lambda}{(1+\gamma_i))^\lambda+(1-\gamma_i)^\lambda}\Big|p_i\right\}\\ &= \left(\cup_{i=1,2,\ldots,\#\bar{h}}\left\{\frac{(1+\gamma_i)^\lambda-(1-\gamma_i)^\lambda}{(1+\gamma_i))^\lambda+(1-\gamma_i)^\lambda}\Big|p_i\right\}\right)^c\\ &= (\lambda\cdot_\varepsilon\bar{h})^c.\end{aligned}$$

(2)

$$\begin{aligned}\lambda\cdot_\varepsilon(\bar{h})^c &= \cup_{i=1,2,\ldots,\#\bar{h}}\left\{\frac{(1+(1-\gamma_i))^\lambda-(1-(1-\gamma_i))^\lambda}{(1+(1-\gamma_i))^\lambda+(1-(1-\gamma_i))^\lambda}\Big|p_i\right\}\\ &= \cup_{i=1,2,\ldots,\#\bar{h}}\left\{1-\frac{2\gamma_i^\lambda}{(2-\gamma_i)^\lambda+\gamma_i^\lambda}\Big|p_i\right\}\\ &= \left(\cup_{i=1,2,\ldots,\#\bar{h}}\left\{\frac{2\gamma_i^\lambda}{(2-\gamma_i)^\lambda+\gamma_i^\lambda}\Big|p_i\right\}\right)^c\\ &= (\bar{h}^{\wedge_\varepsilon\lambda})^c.\end{aligned}$$

(3)

$$\begin{aligned}(\bar{h}_1)^c\oplus_\varepsilon(\bar{h}_2)^c &= \cup_{i=1,2,\ldots,\#\bar{h}_1}\{(1-\dot{\gamma}_j)|\dot{p}_j\}\oplus_\varepsilon\cup_{i=1,2,\ldots,\#\bar{h}_2}\{(1-\ddot{\gamma}_k)|\ddot{p}_k\}\\ &= \cup_{\substack{j=1,2,\ldots,\#\bar{h}_1,\\ k=1,2,\ldots,\#\bar{h}_2}}\left\{\frac{(1-\dot{\gamma}_j)+(1-\ddot{\gamma}_k)}{1+(1-\dot{\gamma}_j)(1-\ddot{\gamma}_k)}\Big|\dot{p}_j\ddot{p}_k\right\}\\ &= \left(\cup_{\substack{j=1,2,\ldots,\#\bar{h}_1,\\ k=1,2,\ldots,\#\bar{h}_2}}\left\{\frac{\dot{\gamma}_j\ddot{\gamma}_k}{1+(1-\dot{\gamma}_j)(1-\ddot{\gamma}_k)}\Big|\dot{p}_j\ddot{p}_k\right\}\right)^c\\ &= (\bar{h}_1\otimes_\varepsilon\bar{h}_2)^c.\end{aligned}$$

(4)

$$\begin{aligned}(\bar{h}_1)^c\otimes_\varepsilon(\bar{h}_2)^c &= \cup_{i=1,2,\ldots,\#\bar{h}_1}\{(1-\dot{\gamma}_j)|\dot{p}_j\}\otimes_\varepsilon\cup_{i=1,2,\ldots,\#\bar{h}_2}\{(1-\ddot{\gamma}_k)|\ddot{p}_k\}\\ &= \cup_{\substack{j=1,2,\ldots,\#\bar{h}_1,\\ k=1,2,\ldots,\#\bar{h}_2}}\left\{\frac{(1-\dot{\gamma}_j)(1-\ddot{\gamma}_k)}{1-\dot{\gamma}_j\ddot{\gamma}_k}\Big|\dot{p}_j\ddot{p}_k\right\}\\ &= \left(\cup_{\substack{j=1,2,\ldots,\#\bar{h}_1,\\ k=1,2,\ldots,\#\bar{h}_2}}\left\{\frac{\dot{\gamma}_j+\ddot{\gamma}_k}{1+\dot{\gamma}_j\ddot{\gamma}_k}\Big|\dot{p}_j\ddot{p}_k\right\}\right)^c\\ &= (\bar{h}_1\oplus_\varepsilon\bar{h}_2)^c.\end{aligned}$$

□

Theorem 3. *Let $\bar{h}(\gamma_i|p_i)$, $\bar{h}_1(\dot{\gamma}_j|\dot{p}_j)$ and $\bar{h}_2(\ddot{\gamma}_k|\ddot{p}_k)$ be three HPFEs, $i = 1,2,\ldots,\#\bar{h}$, $j = 1,2,\ldots,\#\bar{h}_1$, $k = 1,2,\ldots,\#\bar{h}_2$, $\lambda > 0$, $\lambda_1 > 0$, and $\lambda_2 > 0$, then*

(1) $\bar{h}_1\oplus_\varepsilon\bar{h}_2 = \bar{h}_2\oplus_\varepsilon\bar{h}_1$;

(2) $\bar{h}\oplus_\varepsilon(\bar{h}_1\oplus_\varepsilon\bar{h}_2) = (\bar{h}\oplus_\varepsilon\bar{h}_1)\oplus_\varepsilon\bar{h}_2$;

(3) $\lambda\cdot_\varepsilon(\bar{h}_1\oplus_\varepsilon\bar{h}_2) = (\lambda\cdot_\varepsilon\bar{h}_1)\oplus_\varepsilon(\lambda\cdot_\varepsilon\bar{h}_2)$;

(4) $\lambda_1\cdot_\varepsilon(\lambda_2\cdot_\varepsilon\bar{h}) = (\lambda_1\lambda_2)\cdot_\varepsilon\bar{h}$;

(5) $\bar{h}_1\otimes_\varepsilon\bar{h}_2 = \bar{h}_2\otimes_\varepsilon\bar{h}_1$;

(6) $\bar{h} \otimes_\varepsilon (\bar{h}_1 \otimes_\varepsilon \bar{h}_2) = (\bar{h} \otimes_\varepsilon \bar{h}_1) \otimes_\varepsilon \bar{h}_2$*;*
(7) $(\bar{h}_1 \otimes_\varepsilon \bar{h}_2)^{\wedge_\varepsilon \lambda} = \bar{h}_1^{\wedge_\varepsilon \lambda} \otimes_\varepsilon \bar{h}_2^{\wedge_\varepsilon \lambda}$*;*
(8) $(\bar{h}^{\wedge_\varepsilon \lambda_1})^{\wedge_\varepsilon \lambda_2} = \bar{h}^{\wedge_\varepsilon (\lambda_1 \lambda_2)}$*.*

Proof. Since (1), (2), (5) and (6) are trivial, and (7) and (8) are similar to (3) and (4), respectively, we only prove (3) and (4).

(3) Since $\bar{h}_1 \oplus_\varepsilon \bar{h}_2 = \cup_{j=1,2,\ldots,\#\bar{h}_1, k=1,2,\ldots,\#\bar{h}_2} \left\{ \frac{\dot{\gamma}_j + \ddot{\gamma}_k}{1+\dot{\gamma}_j \ddot{\gamma}_k} \middle| \dot{p}_j \ddot{p}_k \right\}$, by the operational law (3) in Definition 7, we have

$$\lambda \cdot_\varepsilon (\bar{h}_1 \oplus_\varepsilon \bar{h}_2) = \bigcup_{\substack{j=1,2,\ldots,\#\bar{h}_1,\\ k=1,2,\ldots,\#\bar{h}_2}} \left\{ \frac{\left(1+\frac{\dot{\gamma}_j+\ddot{\gamma}_k}{1+\dot{\gamma}_j\ddot{\gamma}_k}\right)^\lambda - \left(1-\frac{\dot{\gamma}_j+\ddot{\gamma}_k}{1+\dot{\gamma}_j\ddot{\gamma}_k}\right)^\lambda}{\left(1+\frac{\dot{\gamma}_j+\ddot{\gamma}_k}{1+\dot{\gamma}_j\ddot{\gamma}_k}\right)^\lambda + \left(1-\frac{\dot{\gamma}_j+\ddot{\gamma}_k}{1+\dot{\gamma}_j\ddot{\gamma}_k}\right)^\lambda} \middle| \dot{p}_j \ddot{p}_k \right\}$$
$$= \bigcup_{\substack{j=1,2,\ldots,\#\bar{h}_1,\\ k=1,2,\ldots,\#\bar{h}_2}} \left\{ \frac{(1+\dot{\gamma}_j)^\lambda (1+\ddot{\gamma}_k)^\lambda - (1-\dot{\gamma}_j)^\lambda (1-\ddot{\gamma}_k)^\lambda}{(1+\dot{\gamma}_j)^\lambda (1+\ddot{\gamma}_k)^\lambda + (1-\dot{\gamma}_j)^\lambda (1-\ddot{\gamma}_k)^\lambda} \middle| \dot{p}_j \ddot{p}_k \right\}.$$

Since $\lambda \cdot_\varepsilon \bar{h}_1 = \cup_{i=1,2,\ldots,\#\bar{h}_1} \left\{ \frac{(1+\dot{\gamma}_j)^\lambda - (1-\dot{\gamma}_j)^\lambda}{(1+\dot{\gamma}_j)^\lambda + (1-\dot{\gamma}_j)^\lambda} \middle| \dot{p}_j \right\}$ and $\lambda \cdot_\varepsilon \bar{h}_2 = \cup_{i=1,2,\ldots,\#\bar{h}_2} \left\{ \frac{(1+\ddot{\gamma}_k)^\lambda - (1-\ddot{\gamma}_k)^\lambda}{(1+\ddot{\gamma}_k)^\lambda + (1-\ddot{\gamma}_k)^\lambda} \middle| \ddot{p}_k \right\}$, we have

$$(\lambda \cdot_\varepsilon \bar{h}_1) \oplus_\varepsilon (\lambda \cdot_\varepsilon \bar{h}_2) = \bigcup_{\substack{j=1,2,\ldots,\#\bar{h}_1,\\ k=1,2,\ldots,\#\bar{h}_2}} \left\{ \frac{\frac{(1+\dot{\gamma}_j)^\lambda - (1-\dot{\gamma}_j)^\lambda}{(1+\dot{\gamma}_j)^\lambda + (1-\dot{\gamma}_j)^\lambda} + \frac{(1+\ddot{\gamma}_k)^\lambda - (1-\ddot{\gamma}_k)^\lambda}{(1+\ddot{\gamma}_k)^\lambda + (1-\ddot{\gamma}_k)^\lambda}}{1 + \frac{(1+\dot{\gamma}_j)^\lambda - (1-\dot{\gamma}_j)^\lambda}{(1+\dot{\gamma}_j)^\lambda + (1-\dot{\gamma}_j)^\lambda} \cdot \frac{(1+\ddot{\gamma}_k)^\lambda - (1-\ddot{\gamma}_k)^\lambda}{(1+\ddot{\gamma}_k)^\lambda + (1-\ddot{\gamma}_k)^\lambda}} \middle| \dot{p}_j \ddot{p}_k \right\}$$
$$= \bigcup_{\substack{j=1,2,\ldots,\#\bar{h}_1,\\ k=1,2,\ldots,\#\bar{h}_2}} \left\{ \frac{(1+\dot{\gamma}_j)^\lambda (1+\ddot{\gamma}_k)^\lambda - (1-\dot{\gamma}_j)^\lambda (1-\ddot{\gamma}_k)^\lambda}{(1+\dot{\gamma}_j)^\lambda (1+\ddot{\gamma}_k)^\lambda + (1-\dot{\gamma}_j)^\lambda (1-\ddot{\gamma}_k)^\lambda} \middle| \dot{p}_j \ddot{p}_k \right\}.$$

Hence $\lambda \cdot_\varepsilon (\bar{h}_1 \oplus_\varepsilon \bar{h}_2) = (\lambda \cdot_\varepsilon \bar{h}_1) \oplus_\varepsilon (\lambda \cdot_\varepsilon \bar{h}_2)$.

(4) Since $\lambda_2 \cdot_\varepsilon \bar{h} = \cup_{i=1,2,\ldots,\#\bar{h}} \left\{ \frac{(1+\gamma_i)^{\lambda_2} - (1-\gamma_i)^{\lambda_2}}{(1+\gamma_i)^{\lambda_2} + (1-\gamma_i)^{\lambda_2}} \middle| p_i \right\}$, then we have

$$\lambda_1 \cdot_\varepsilon (\lambda_2 \cdot_\varepsilon \bar{h}) = \cup_{i=1,2,\ldots,\#\bar{h}} \left\{ \frac{\left(1+\frac{(1+\gamma_i)^{\lambda_2} - (1-\gamma_i)^{\lambda_2}}{(1+\gamma_i)^{\lambda_2} + (1-\gamma_i)^{\lambda_2}}\right)^{\lambda_1} - \left(1-\frac{(1+\gamma_i)^{\lambda_2} - (1-\gamma_i)^{\lambda_2}}{(1+\gamma_i)^{\lambda_2} + (1-\gamma_i)^{\lambda_2}}\right)^{\lambda_1}}{\left(1+\frac{(1+\gamma_i)^{\lambda_2} - (1-\gamma_i)^{\lambda_2}}{(1+\gamma_i)^{\lambda_2} + (1-\gamma_i)^{\lambda_2}}\right)^{\lambda_1} + \left(1-\frac{(1+\gamma_i)^{\lambda_2} - (1-\gamma_i)^{\lambda_2}}{(1+\gamma_i)^{\lambda_2} + (1-\gamma_i)^{\lambda_2}}\right)^{\lambda_1}} \middle| p_i \right\}$$
$$= \cup_{i=1,2,\ldots,\#\bar{h}} \left\{ \frac{(1+\gamma_i)^{(\lambda_1\lambda_2)} - (1-\gamma_i)^{(\lambda_1\lambda_2)}}{(1+\gamma_i)^{(\lambda_1\lambda_2)} + (1-\gamma_i)^{(\lambda_1\lambda_2)}} \middle| p_i \right\}$$
$$= (\lambda_1\lambda_2) \cdot_\varepsilon \bar{h}.$$

□

For an HPFE, $\bar{h}(\gamma_i|p_i)$, $i = 1, 2, \ldots, \#\bar{h}$, $\lambda_1 > 0$, and $\lambda_2 > 0$, the operational laws $(\lambda_1 \cdot_\varepsilon \bar{h}) \oplus_\varepsilon (\lambda_2 \cdot_\varepsilon \bar{h}) = (\lambda_1 + \lambda_2) \cdot_\varepsilon \bar{h}$ and $\bar{h}^{\wedge_\varepsilon \lambda_1} \otimes_\varepsilon \bar{h}^{\wedge_\varepsilon \lambda_2} = \bar{h}^{\wedge_\varepsilon (\lambda_1 + \lambda_2)}$ do not hold in general. To illustrate this case, we give the following example.

Example 2. *Let* $\bar{h}(\gamma_i|p_i) = (0.3|0.5, 0.5|0.5)$ *and* $\lambda_1 = \lambda_2 = 1$*, then*

$$(\lambda_1 \cdot_\varepsilon \bar{h}) \oplus_\varepsilon (\lambda_2 \cdot_\varepsilon \bar{h}) = \bar{h} \oplus_\varepsilon \bar{h} = \cup_{i,j=1,2} \left\{ \frac{\gamma_i + \gamma_j}{1 + \gamma_i\gamma_j} \middle| 0.25 \right\}$$
$$= (0.5505|0.25, 0.6957|0.25, 0.6957|0.25, 0.8|0.25),$$

$$(\lambda_1+\lambda_2)\cdot_\varepsilon \bar{h} = 2\cdot_\varepsilon \bar{h} = \cup_{i=1,2}\left\{\frac{(1+\gamma_i)^2-(1-\gamma_i)^2}{(1+\gamma_i)^2+(1-\gamma_i)^2}\Big|0.5\right\}$$
$$= (0.5505|0.5, 0.8|0.5).$$

Clearly, $s((\lambda_1\cdot_\varepsilon \bar{h})\oplus_\varepsilon(\lambda_2\cdot_\varepsilon \bar{h})) = 0.6856 > 0.6752 = s((\lambda_1+\lambda_2)\cdot_\varepsilon \bar{h})$. *Hence,* $(\lambda_1\cdot_\varepsilon \bar{h})\oplus_\varepsilon(\lambda_2\cdot_\varepsilon \bar{h}) < (\lambda_1+\lambda_2)\cdot_\varepsilon \bar{h}$. *Similarly, we have* $s(\bar{h}^{\wedge_\varepsilon\lambda_1}\otimes_\varepsilon \bar{h}^{\wedge_\varepsilon\lambda_2}) = 0.2566 > 0.13 = s(\bar{h}^{\wedge_\varepsilon(\lambda_1+\lambda_2)})$ *and thus* $\bar{h}^{\wedge_\varepsilon\lambda_1}\otimes_\varepsilon \bar{h}^{\wedge_\varepsilon\lambda_2} < \bar{h}^{\wedge_\varepsilon(\lambda_1+\lambda_2)}$.

3. Some HPFE Weighted Aggregation Operators Based on Einstein Operation

One important issue is the question of how to extend Einstein operations to aggregate the HPFE information provided by the decision makers. The optimal approach is weighted aggregation operators, in which the widely used technologies are the weighted averaging (WA) operator, the ordered weighted averaging (OWA) operator, and their extended forms [39,40]. Yu [21] proposed the hesitant fuzzy Einstein weighted averaging (HFEWA) operator, the hesitant fuzzy Einstein ordered weighted averaging (HFEOWA) operator, the hesitant fuzzy Einstein weighted geometric (HFEWG) operator, and the hesitant fuzzy Einstein ordered weighted geometric (HFEOWG) operator based on those operators. Similar to these hesitant fuzzy information aggregation operators, we propose the corresponding hesitant probabilistic fuzzy Einstein weighted and ordered operators to aggregate the HPFEs.

Definition 8. *Let* $\bar{h}_t$ $(t=1,2,\ldots,T)$ *be a collection of HPFEs; then, a hesitant probabilistic fuzzy Einstein weighted averaging (HPFEWA) operator is a mapping* $H_P^T \to H_P$ *such that*

$$\mathrm{HPFEWA}(\bar{h}_1,\bar{h}_2,\ldots,\bar{h}_T) = (w_1\cdot_\varepsilon \bar{h}_1)\oplus_\varepsilon(w_2\cdot_\varepsilon \bar{h}_2)\oplus_\varepsilon\cdots\oplus_\varepsilon(w_T\cdot_\varepsilon \bar{h}_T), \tag{6}$$

where $w=(w_1,w_2,\ldots,w_T)^T$ *is the weight vector of* $\bar{h}_t$ $(t=1,2,\ldots,T)$ *with* $w_t\in[0,1]$ *and* $\sum_{t=1}^T w_t = 1$, *and* p_t *is the probability of* γ_t *in HPFE* $\bar{h}_t$. *In particular, if* $w=\left(\frac{1}{T},\frac{1}{T},\ldots,\frac{1}{T}\right)^T$, *then the HPFEWA operator is reduced to the hesitant probabilistic fuzzy Einstein averaging (HPFEA) operator:*

$$\mathrm{HPFEA}(\bar{h}_1,\bar{h}_2,\ldots,\bar{h}_T) = (\frac{1}{T}\cdot_\varepsilon \bar{h}_1)\oplus_\varepsilon(\frac{1}{T}\cdot_\varepsilon \bar{h}_2)\oplus_\varepsilon\cdots\oplus_\varepsilon(\frac{1}{T}\cdot_\varepsilon \bar{h}_T). \tag{7}$$

From Definitions 7 and 8, we can get the following result by using mathematical induction.

Theorem 4. *Let* $\bar{h}_t$ $(t=1,2,\ldots,T)$ *be a collection of HPFEs; then, their aggregated value obtained using the* HPFEWA *operator is also a HPFE, and*

$$\mathrm{HPFEWA}(\bar{h}_1,\bar{h}_2,\ldots,\bar{h}_T) = \bigcup_{\gamma_1\in\bar{h}_1,\gamma_2\in\bar{h}_2,\ldots,\gamma_T\in\bar{h}_T}\left\{\frac{\prod_{t=1}^T(1+\gamma_t)^{w_t}-\prod_{t=1}^T(1-\gamma_t)^{w_t}}{\prod_{t=1}^T(1+\gamma_t)^{w_t}+\prod_{t=1}^T(1-\gamma_t)^{w_t}}\Big|p_1p_2\cdots p_T\right\}, \tag{8}$$

where $w=(w_1,w_2,\ldots,w_T)^T$ *is the weight vector of* $\bar{h}_t$ $(t=1,2,\ldots,T)$ *with* $w_t\in[0,1]$ *and* $\sum_{t=1}^T w_t = 1$, *and* p_t *is the probability of* γ_t *in HPFE* $\bar{h}_t$.

Proof. We prove Equation (8) by mathematical induction. For $T=2$, since $w_1\cdot_\varepsilon \bar{h}_1 = \cup_{\gamma_1\in\bar{h}_1}\left\{\frac{(1+\gamma_1)^{w_1}-(1-\gamma_1)^{w_1}}{(1+\gamma_1)^{w_1}+(1-\gamma_1)^{w_1}}\Big|p_1\right\}$ and $w_2\cdot_\varepsilon \bar{h}_2 = \cup_{\gamma_2\in\bar{h}_2}\left\{\frac{(1+\gamma_2)^{w_2}-(1-\gamma_2)^{w_2}}{(1+\gamma_2)^{w_2}+(1-\gamma_2)^{w_2}}\Big|p_2\right\}$, then

$$(w_1\cdot_\varepsilon \bar{h}_1)\oplus_\varepsilon(w_2\cdot_\varepsilon \bar{h}_2) = \bigcup_{\gamma_1\in\bar{h}_1,\gamma_2\in\bar{h}_2}\left\{\frac{\frac{(1+\gamma_1)^{w_1}-(1-\gamma_1)^{w_1}}{(1+\gamma_1)^{w_1}+(1-\gamma_1)^{w_1}}+\frac{(1+\gamma_2)^{w_2}-(1-\gamma_2)^{w_2}}{(1+\gamma_2)^{w_2}+(1-\gamma_2)^{w_2}}}{1+\frac{(1+\gamma_1)^{w_1}-(1-\gamma_1)^{w_1}}{(1+\gamma_1)^{w_1}+(1-\gamma_1)^{w_1}}\cdot\frac{(1+\gamma_2)^{w_2}-(1-\gamma_2)^{w_2}}{(1+\gamma_2)^{w_2}+(1-\gamma_2)^{w_2}}}\Big|p_1p_2\right\}$$
$$= \bigcup_{\gamma_1\in\bar{h}_1,\gamma_2\in\bar{h}_2}\left\{\frac{\prod_{t=1}^2(1+\gamma_t)^{w_t}-\prod_{t=1}^2(1-\gamma_t)^{w_t}}{\prod_{t=1}^2(1+\gamma_t)^{w_t}+\prod_{t=1}^2(1-\gamma_t)^{w_t}}\Big|p_1p_2\right\}.$$

If Equation (8) holds for $T = k$, that is

$$(w_1 \cdot_\varepsilon \bar{h}_1) \oplus_\varepsilon (w_2 \cdot_\varepsilon \bar{h}_2) \oplus_\varepsilon \cdots \oplus_\varepsilon (w_k \cdot_\varepsilon \bar{h}_k) = \bigcup_{\gamma_1 \in \bar{h}_1, \gamma_2 \in \bar{h}_2, \ldots, \gamma_k \in \bar{h}_k} \left\{ \frac{\prod_{t=1}^{k}(1+\gamma_t)^{w_t} - \prod_{t=1}^{k}(1-\gamma_t)^{w_t}}{\prod_{t=1}^{k}(1+\gamma_t)^{w_t} + \prod_{t=1}^{k}(1-\gamma_t)^{w_t}} \middle| p_1 p_2 \cdots p_k \right\},$$

then, when $T = k+1$, according to the Einstein operations of HPFEs, we have

$$\begin{aligned}
&(w_1 \cdot_\varepsilon \bar{h}_1) \oplus_\varepsilon (w_2 \cdot_\varepsilon \bar{h}_2) \oplus_\varepsilon \cdots \oplus_\varepsilon (w_{k+1} \cdot_\varepsilon \bar{h}_{k+1}) \\
&= ((w_1 \cdot_\varepsilon \bar{h}_1) \oplus_\varepsilon (w_2 \cdot_\varepsilon \bar{h}_2) \oplus_\varepsilon \cdots \oplus_\varepsilon (w_k \cdot_\varepsilon \bar{h}_k)) \oplus_\varepsilon (w_{k+1} \cdot_\varepsilon \bar{h}_{k+1}) \\
&= \bigcup_{\gamma_1 \in \bar{h}_1, \gamma_2 \in \bar{h}_2, \ldots, \gamma_k \in \bar{h}_k} \left\{ \frac{\prod_{t=1}^{k}(1+\gamma_t)^{w_t} - \prod_{t=1}^{k}(1-\gamma_t)^{w_t}}{\prod_{t=1}^{k}(1+\gamma_t)^{w_t} + \prod_{t=1}^{k}(1-\gamma_t)^{w_t}} \middle| p_1 p_2 \cdots p_k \right\} \\
&\quad \oplus_\varepsilon \bigcup_{\gamma_{k+1} \in \bar{h}_{k+1}} \left\{ \frac{(1+\gamma_{k+1})^{w_{k+1}} - (1-\gamma_{k+1})^{w_{k+1}}}{(1+\gamma_{k+1})^{w_{k+1}} + (1-\gamma_{k+1})^{w_{k+1}}} \middle| p_{k+1} \right\} \\
&= \bigcup_{\gamma_1 \in \bar{h}_1, \gamma_2 \in \bar{h}_2, \ldots, \gamma_k \in \bar{h}_k, \gamma_{k+1} \in \bar{h}_{k+1}} \left\{ \frac{\prod_{t=1}^{k+1}(1+\gamma_t)^{w_t} - \prod_{t=1}^{k+1}(1-\gamma_t)^{w_t}}{\prod_{t=1}^{k+1}(1+\gamma_t)^{w_t} + \prod_{t=1}^{k+1}(1-\gamma_t)^{w_t}} \middle| p_1 p_2 \cdots p_k p_{k+1} \right\},
\end{aligned}$$

i.e., Equation (8) holds for $T = k+1$. Hence, Equation (8) holds for all T. Thus,

$$\text{HPFEWA}(\bar{h}_1, \bar{h}_2, \ldots, \bar{h}_T) = \bigcup_{\gamma_1 \in \bar{h}_1, \gamma_2 \in \bar{h}_2, \ldots, \gamma_T \in \bar{h}_T} \left\{ \frac{\prod_{t=1}^{T}(1+\gamma_t)^{w_t} - \prod_{t=1}^{T}(1-\gamma_t)^{w_t}}{\prod_{t=1}^{T}(1+\gamma_t)^{w_t} + \prod_{t=1}^{T}(1-\gamma_t)^{w_t}} \middle| p_1 p_2 \cdots p_T \right\},$$

which completes the proof of theorem. □

Based on Theorem 4, we have basic properties of the HPFEWA operator, as follows:

Theorem 5. *Let $\bar{h}_t(\gamma_i^{(t)}|p_t)$ $(t = 1, 2, \ldots, T)$ be a collection of HPFEs, $w = (w_1, w_2, \ldots, w_T)^T$ be the weight vector of $\bar{h}_t$ $(t = 1, 2, \ldots, T)$ such that $w_t \in [0, 1]$ and $\sum_{t=1}^{T} w_t = 1$, and p_t be the corresponding probability of $\gamma_i^{(t)}$ in HPFE $\bar{h}_t$; then, we have the following:*

(1) (Boundary):

$$\bar{h}^- \leq \text{HPFEWA}(\bar{h}_1, \bar{h}_2, \ldots, \bar{h}_T) \leq \bar{h}^+, \tag{9}$$

where $\bar{h}^- = (\min_{1 \leq t \leq T} \min_{\gamma_t \in \bar{h}_t} \gamma_t | p_1 p_2 \cdots p_T)$ and $\bar{h}^+ = (\max_{1 \leq t \leq T} \max_{\gamma_t \in \bar{h}_t} \gamma_t | p_1 p_2 \cdots p_T)$.

(2) (Monotonicity): Let $\bar{h}_t^*(\dot{\gamma}_i^{(t)}|p_t)$ $(t = 1, 2, \ldots, T)$ be a collection of HPFEs with $\#_t = \#\bar{h}_t = \#\bar{h}_t^*$ for $t = 1, 2, \ldots, T$, $w = (w_1, w_2, \ldots, w_T)^T$ be the weight vector of $\bar{h}_t^*$ $(t = 1, 2, \ldots, T)$, such that $w_t \in [0, 1]$ and $\sum_{t=1}^{T} w_t = 1$, and p_t is the probability of $\dot{\gamma}_i^{(t)}$ in HPFE $\bar{h}_t^*$. If $\gamma_i^{(t)} \leq \dot{\gamma}_i^{(t)}$ for each $i = 1, 2, \ldots, \#_t$, $t = 1, 2, \ldots, T$; then,

$$\text{HPFEWA}(\bar{h}_1, \bar{h}_2, \ldots, \bar{h}_T) \leq \text{HPFEWA}(\bar{h}_1^*, \bar{h}_2^*, \ldots, \bar{h}_T^*). \tag{10}$$

Proof. (1) Let $f(x) = \frac{1-x}{1+x}$, $x \in [0, 1]$, then $f'(x) = \frac{-2}{(1+x)^2} < 0$, i.e., $f(x)$ is a decreasing function. Let $\max \gamma_t = \max_{1 \leq t \leq T} \max_{\gamma_t \in \bar{h}_t} \gamma_t$ and $\min \gamma_t = \min_{1 \leq t \leq T} \min_{\gamma_t \in \bar{h}_t} \gamma_t$. For any $\gamma_t \in \bar{h}_t$ $(t = 1, 2, \ldots, T)$, since $\min_{\gamma_t \in \bar{h}_t} \gamma_t \leq \gamma_t \leq \max_{\gamma_t \in \bar{h}_t} \gamma_t$, then $f(\max_{\gamma_t \in \bar{h}_t} \gamma_t) \leq f(\gamma_t) \leq f(\min_{\gamma_t \in \bar{h}_t} \gamma_t)$, and so

$$\frac{1 - \max \gamma_t}{1 + \max \gamma_t} \leq \frac{1 - \max_{\gamma_t \in \bar{h}_t} \gamma_t}{1 + \max_{\gamma_t \in \bar{h}_t} \gamma_t} \leq \frac{1 - \gamma_t}{1 + \gamma_t} \leq \frac{1 - \min_{\gamma_t \in \bar{h}_t} \gamma_t}{1 + \min_{\gamma_t \in \bar{h}_t} \gamma_t} \leq \frac{1 - \min \gamma_t}{1 + \min \gamma_t}.$$

Since $w = (w_1, w_2, \ldots, w_T)^T$ is the weight vector of $\bar{h}_t$ $(t = 1,2,\ldots,T)$ with $w_t \in [0,1]$ and $\sum_{t=1}^{T} w_t = 1$, we have

$$\prod_{t=1}^{T}\left(\frac{1-\max\gamma_t}{1+\max\gamma_t}\right)^{w_t} \leq \prod_{t=1}^{T}\left(\frac{1-\gamma_t}{1+\gamma_t}\right)^{w_t} \leq \prod_{t=1}^{T}\left(\frac{1-\min\gamma_t}{1+\min\gamma_t}\right)^{w_t}.$$

Since $\prod_{t=1}^{T}\left(\frac{1-\max\gamma_t}{1+\max\gamma_t}\right)^{w_t} = \left(\frac{1-\max\gamma_t}{1+\max\gamma_t}\right)^{\sum_{t=1}^{T} w_t} = \frac{1-\max\gamma_t}{1+\max\gamma_t}$ and $\prod_{t=1}^{T}\left(\frac{1-\min\gamma_t}{1+\min\gamma_t}\right)^{w_t} = \left(\frac{1-\min\gamma_t}{1+\min\gamma_t}\right)^{\sum_{t=1}^{T} w_t} = \frac{1-\min\gamma_t}{1+\min\gamma_t}$, we get

$$\begin{aligned}
&\frac{1-\max\gamma_t}{1+\max\gamma_t} \leq \prod_{t=1}^{T}\left(\frac{1-\gamma_t}{1+\gamma_t}\right)^{w_t} \leq \frac{1-\min\gamma_t}{1+\min\gamma_t} \\
&\Leftrightarrow \frac{2}{1+\max\gamma_t} \leq 1+\prod_{t=1}^{T}\left(\frac{1-\gamma_t}{1+\gamma_t}\right)^{w_t} \leq \frac{2}{1+\min\gamma_t} \\
&\Leftrightarrow \frac{1+\min\gamma_t}{2} \leq \frac{1}{1+\prod_{t=1}^{T}\left(\frac{1-\gamma_t}{1+\gamma_t}\right)^{w_t}} \leq \frac{1+\max\gamma_t}{2} \\
&\Leftrightarrow \min\gamma_t \leq \frac{2}{1+\prod_{t=1}^{T}\left(\frac{1-\gamma_t}{1+\gamma_t}\right)^{w_t}} - 1 \leq \max\gamma_t,
\end{aligned}$$

i.e.,

$$\min\gamma_t \leq \frac{\prod_{t=1}^{T}(1+\gamma_t)^{w_t} - \prod_{t=1}^{T}(1-\gamma_t)^{w_t}}{\prod_{t=1}^{T}(1+\gamma_t)^{w_t} + \prod_{t=1}^{T}(1-\gamma_t)^{w_t}} \leq \max\gamma_t. \tag{11}$$

Let $\text{HPFEWA}(\bar{h}_1, \bar{h}_2, \ldots, \bar{h}_T) = \bar{h}(\gamma_i|p_1p_2\cdots p_T)$, $i = 1,2,\ldots,\#\bar{h}$, where $\#\bar{h} = \#\bar{h}_1 \times \#\bar{h}_2 \times \cdots \times \#\bar{h}_T$, $\bar{h}^- = (\min\gamma_t|p_1p_2\cdots p_T)$ and $\bar{h}^+ = (\max\gamma_t|p_1p_2\cdots p_T)$; then, Equation (11) is transformed into the following form: $\min\gamma_t \leq \gamma_i \leq \max\gamma_t$ for all $i = 1,2,\ldots,\#\bar{h}$. Thus, $s(\bar{h}^-) = \min\gamma_t p_1p_2\cdots p_T \leq \sum_{i=1}^{\#\bar{h}} \gamma_i p_1p_2\cdots p_T = s(\bar{h})$ and $s(\bar{h}) = \sum_{i=1}^{\#\bar{h}} \gamma_i p_1p_2\cdots p_T \leq \max\gamma_t p_1p_2\cdots p_T = s(\bar{h}^+)$.

If $s(\bar{h}^-) < s(\bar{h})$ and $s(\bar{h}) < s(\bar{h}^+)$, then by Definition 3, we have $\bar{h}^- < \text{HPFEWA}(\bar{h}_1, \bar{h}_2, \ldots, \bar{h}_T) < \bar{h}^+$. If $s(\bar{h}) = s(\bar{h}^+)$, i.e., $\max\gamma_t = \sum_{i=1}^{\#\bar{h}} \gamma_i$, then $d(\bar{h}) = \sum_{i=1}^{\#\bar{h}}(\gamma_i - s(\bar{h}))^2 p_1p_2\cdots p_T = (\max\gamma_t - s(\bar{h}))^2 p_1p_2\cdots p_T = d(\bar{h}^+)$. In this case, in accordance with Definition 3, it follows that $\text{HPFEWA}(\bar{h}_1, \bar{h}_2, \ldots, \bar{h}_T) = \bar{h}^+$. If $s(\bar{h}) = s(\bar{h}^-)$, then similarly, we have $\text{HPFEWA}(\bar{h}_1, \bar{h}_2, \ldots, \bar{h}_T) = \bar{h}^-$.

(2) Let $f(x) = \frac{1-x}{1+x}$, $x \in [0,1]$; then, $f(x)$ is a decreasing function. If $\gamma_i^{(t)} \leq \dot{\gamma}_i^{(t)}$ for each $i = 1,2,\ldots,\#_t$, $t = 1,2,\ldots,T$; then, $f(\gamma_i^{(t)}) \geq f(\dot{\gamma}_i^{(t)})$, for each $i = 1,2,\ldots,\#_t$, $t = 1,2,\ldots,T$, i.e., $\frac{1-\gamma_i^{(t)}}{1+\gamma_i^{(t)}} \geq \frac{1-\dot{\gamma}_i^{(t)}}{1+\dot{\gamma}_i^{(t)}}$, for each $i = 1,2,\ldots,\#_t$, $t = 1,2,\ldots,T$. For any $\gamma_i^{(t)} \in \bar{h}_t$ $(t = 1,2,\ldots,T)$, since $w = (w_1, w_2, \ldots, w_T)^T$ is the weight vector of $\bar{h}_t$ $(t = 1,2,\ldots,T)$ such that $w_t \in [0,1]$, $t = 1,2,\ldots,T$ and $\sum_{t=1}^{T} w_t = 1$, we have

$$\left(\frac{1-\gamma_i^{(t)}}{1+\gamma_i^{(t)}}\right)^{w_t} \geq \left(\frac{1-\dot{\gamma}_i^{(t)}}{1+\dot{\gamma}_i^{(t)}}\right)^{w_t}, \; t = 1,2,\ldots,T.$$

Then,

$$\prod_{t=1}^{T}\left(\frac{1-\gamma_i^{(t)}}{1+\gamma_i^{(t)}}\right)^{w_t} \geq \prod_{t=1}^{T}\left(\frac{1-\dot{\gamma}_i^{(t)}}{1+\dot{\gamma}_i^{(t)}}\right)^{w_t} \Leftrightarrow 1+\prod_{t=1}^{T}\left(\frac{1-\gamma_i^{(t)}}{1+\gamma_i^{(t)}}\right)^{w_t} \geq 1+\prod_{t=1}^{T}\left(\frac{1-\dot{\gamma}_i^{(t)}}{1+\dot{\gamma}_i^{(t)}}\right)^{w_t}$$

$$\Leftrightarrow \frac{1}{1+\prod_{t=1}^{T}\left(\frac{1-\gamma_i^{(t)}}{1+\gamma_i^{(t)}}\right)^{w_t}} \leq \frac{1}{1+\prod_{t=1}^{T}\left(\frac{1-\dot{\gamma}_i^{(t)}}{1+\dot{\gamma}_i^{(t)}}\right)^{w_t}}$$

$$\Leftrightarrow \frac{2}{1+\prod_{t=1}^{T}\left(\frac{1-\gamma_i^{(t)}}{1+\gamma_i^{(t)}}\right)^{w_t}} - 1 \leq \frac{2}{1+\prod_{t=1}^{T}\left(\frac{1-\dot{\gamma}_i^{(t)}}{1+\dot{\gamma}_i^{(t)}}\right)^{w_t}} - 1,$$

i.e.,

$$\frac{\prod_{t=1}^{T}(1+\gamma_i^{(t)})^{w_t} - \prod_{t=1}^{T}(1-\gamma_i^{(t)})^{w_t}}{\prod_{t=1}^{T}(1+\gamma_i^{(t)})^{w_t} + \prod_{t=1}^{T}(1-\gamma_i^{(t)})^{w_t}} \leq \frac{\prod_{t=1}^{T}(1+\dot{\gamma}_i^{(t)})^{w_t} - \prod_{t=1}^{T}(1-\dot{\gamma}_i^{(t)})^{w_t}}{\prod_{t=1}^{T}(1+\dot{\gamma}_i^{(t)})^{w_t} + \prod_{t=1}^{T}(1-\dot{\gamma}_i^{(t)})^{w_t}}. \tag{12}$$

Let $\text{HPFEWA}(\bar{h}_1, \bar{h}_2, \ldots, \bar{h}_T) = \bar{h}(\gamma_i | p_1 p_2 \cdots p_T)$ and $\text{HPFEWA}(\bar{h}_1^*, \bar{h}_2^*, \ldots, \bar{h}_T^*) = \bar{h}^*(\dot{\gamma}_i | p_1 p_2 \cdots p_T)$, where $i = 1, 2, \ldots, \#$, and $\# = \#_1 \times \#_2 \times \cdots \times \#_T$ is the number of possible elements in $\bar{h}(\gamma_i | p_1 p_2 \cdots p_T)$ and $\bar{h}^*(\dot{\gamma}_i | p_1 p_2 \cdots p_T)$, respectively, then the Equation (12) is transformed into the form $\gamma_i \leq \dot{\gamma}_i$ $(i = 1, 2, \ldots, \#)$. Thus, $s(\bar{h}) = \sum_{i=1}^{\#} \gamma_i p_1 p_2 \cdots p_T \leq \sum_{i=1}^{\#} \dot{\gamma}_i p_1 p_2 \cdots p_T = s(\bar{h}^*)$.

If $s(\bar{h}) < s(\bar{h}^*)$, then, according to Definition 3, we have $\text{HPFEWA}(\bar{h}_1, \bar{h}_2, \ldots, \bar{h}_T) < \text{HPFEWA}(\bar{h}_1^*, \bar{h}_2^*, \ldots, \bar{h}_T^*)$. If $s(\bar{h}) = s(\bar{h}^*)$, i.e., $\sum_{i=1}^{\#} \gamma_i = \sum_{i=1}^{\#} \dot{\gamma}_i$, then $d(\bar{h}) = \sum_{i=1}^{\#}(\gamma_i - s(\bar{h}))^2 p_1 p_2 \cdots p_T = \sum_{i=1}^{\#}(\dot{\gamma}_i - s(\bar{h}^*))^2 p_1 p_2 \cdots p_T = d(\bar{h}^*)$. In this case, based on Definition 3, it follows that $\text{HPFEWA}(\bar{h}_1, \bar{h}_2, \ldots, \bar{h}_T) = \text{HPFEWA}(\bar{h}_1^*, \bar{h}_2^*, \ldots, \bar{h}_T^*)$. □

However, the HPFEWA operator does not satisfy the idempotency. To illustrate this, we give the following example.

Example 3. *Let* $\bar{h}_1 = \bar{h}_2 = (0.3|0.5, 0.7|0.5)$, *and* $w = (0.2, 0.8)^T$ *is the weight vector* $\bar{h}_t$ $(t = 1, 2)$*; then,*

$$\begin{aligned}\text{HPFEWA}(\bar{h}_1, \bar{h}_2) &= \cup_{\gamma_1 \in \bar{h}_1, \gamma_2 \in \bar{h}_2}\left\{\frac{\prod_{t=1}^{2}(1+\gamma_t)^{w_t} - \prod_{t=1}^{2}(1-\gamma_t)^{w_t}}{\prod_{t=1}^{2}(1+\gamma_t)^{w_t} + \prod_{t=1}^{2}(1-\gamma_t)^{w_t}}|p_1 p_2\right\} \\ &= (0.3|0.25, 0.398|0.25, 0.639|0.25, 0.7|0.25)\end{aligned}$$

and thus $\text{HPFEWA}(\bar{h}_1, \bar{h}_2) \neq (0.3|0.5, 0.7|0.5)$.

Based on the HPFWG operator and Einstein operation, we developed the hesitant probabilistic fuzzy Einstein weighted geometric operator as follows:

Definition 9. *Let* $\bar{h}_t$ $(t = 1, 2, \ldots, T)$ *be a collection of HPFEs; then, the hesitant probabilistic fuzzy Einstein weighted geometric (HPFEWG) operator is a mapping* $(H_P^T \rightarrow H_P)$ *such that*

$$\text{HPFEWG}(\bar{h}_1, \bar{h}_2, \ldots, \bar{h}_T) = \bar{h}_1^{\wedge_\varepsilon w_1} \otimes_\varepsilon \bar{h}_2^{\wedge_\varepsilon w_2} \otimes_\varepsilon \cdots \otimes_\varepsilon \bar{h}_T^{\wedge_\varepsilon w_T}, \tag{13}$$

where $w = (w_1, w_2, \ldots, w_T)^T$ *is the weight vector of* $\bar{h}_t$ $(t = 1, 2, \ldots, T)$ *with* $w_t \in [0, 1]$ *and* $\sum_{t=1}^{T} w_t = 1$, *and* p_t *is the probability of* γ_t *in HPFE* $\bar{h}_t$. *In particular, if* $w = \left(\frac{1}{T}, \frac{1}{T}, \ldots, \frac{1}{T}\right)^T$, *then the HPFEWG operator is reduced to the hesitant probabilistic fuzzy Einstein geometric (HPFEG) operator:*

$$\mathrm{HPFEG}(\bar{h}_1, \bar{h}_2, \ldots, \bar{h}_T) = \bar{h}_1^{\wedge_\varepsilon \frac{1}{T}} \otimes_\varepsilon \bar{h}_2^{\wedge_\varepsilon \frac{1}{T}} \otimes_\varepsilon \cdots \otimes_\varepsilon \bar{h}_T^{\wedge_\varepsilon \frac{1}{T}}. \tag{14}$$

Theorem 6. *Let* $\bar{h}_t$ $(t = 1, 2, \ldots, T)$ *be a collection of HPFEs; then, their aggregated value obtained using the* HPFEWG *operator is also a HPFE and*

$$\mathrm{HPFEWG}(\bar{h}_1, \bar{h}_2, \ldots, \bar{h}_T) = \bigcup_{\gamma_1 \in \bar{h}_1, \gamma_2 \in \bar{h}_2, \ldots, \gamma_T \in \bar{h}_T} \left\{ \frac{2\prod_{t=1}^{T} \gamma_t^{w_t}}{\prod_{t=1}^{T}(2-\gamma_t)^{w_t} + \prod_{t=1}^{T} \gamma_t^{w_t}} \Big| p_1 p_2 \cdots p_T \right\}, \tag{15}$$

where $w = (w_1, w_2, \ldots, w_T)^T$ *is the weight vector of* $\bar{h}_t$ $(t = 1, 2, \ldots, T)$ *with* $w_t \in [0, 1]$ *and* $\sum_{t=1}^{T} w_t = 1$, *and* p_t *is the probability of* γ_t *in HPFE* $\bar{h}_t$.

Proof. We prove Equation (15) by mathematical induction on T. When $T = 2$, since $\bar{h}_1^{\wedge_\varepsilon w_1} = \cup_{\gamma_1 \in \bar{h}_1} \left\{ \frac{2\gamma_1^{w_1}}{(2-\gamma_1)^{w_1} + \gamma_1^{w_1}} \Big| p_1 \right\}$ and $\bar{h}_2^{\wedge_\varepsilon w_2} = \cup_{\gamma_2 \in \bar{h}_2} \left\{ \frac{2\gamma_2^{w_2}}{(2-\gamma_2)^{w_2} + \gamma_2^{w_2}} \Big| p_2 \right\}$, we have

$$\begin{aligned}
\bar{h}_1^{\wedge_\varepsilon w_1} \otimes_\varepsilon \bar{h}_2^{\wedge_\varepsilon w_2} &= \bigcup_{\gamma_1 \in \bar{h}_1, \gamma_2 \in \bar{h}_2} \left\{ \frac{\frac{2\gamma_1^{w_1}}{(2-\gamma_1)^{w_1} + \gamma_1^{w_1}} \cdot \frac{2\gamma_2^{w_2}}{(2-\gamma_2)^{w_2} + \gamma_2^{w_2}}}{1 + \left(1 - \frac{2\gamma_1^{w_1}}{(2-\gamma_1)^{w_1} + \gamma_1^{w_1}}\right)\left(1 - \frac{2\gamma_2^{w_2}}{(2-\gamma_2)^{w_2} + \gamma_2^{w_2}}\right)} \Big| p_1 p_2 \right\} \\
&= \bigcup_{\gamma_1 \in \bar{h}_1, \gamma_2 \in \bar{h}_2} \left\{ \frac{2\prod_{t=1}^{2} \gamma_t^{w_t}}{\prod_{t=1}^{2}(2-\gamma_t)^{w_t} + \prod_{t=1}^{2} \gamma_t^{w_t}} \Big| p_1 p_2 \right\}.
\end{aligned}$$

Assume that Equation (15) holds for $T = k$, i.e.,

$$\bar{h}_1^{\wedge_\varepsilon w_1} \otimes_\varepsilon \bar{h}_2^{\wedge_\varepsilon w_2} \otimes_\varepsilon \cdots \otimes_\varepsilon \bar{h}_k^{\wedge_\varepsilon w_k} = \bigcup_{\gamma_1 \in \bar{h}_1, \gamma_2 \in \bar{h}_2, \cdots, \gamma_k \in \bar{h}_k} \left\{ \frac{2\prod_{t=1}^{k} \gamma_t^{w_t}}{\prod_{t=1}^{k}(2-\gamma_t)^{w_t} + \prod_{t=1}^{k} \gamma_t^{w_t}} \Big| p_1 p_2 \cdots p_k \right\}.$$

In accordance with the Einstein operational laws of HPFEs for $T = k + 1$, we have

$$\begin{aligned}
\bar{h}_1^{\wedge_\varepsilon w_1} \otimes_\varepsilon \bar{h}_2^{\wedge_\varepsilon w_2} \otimes_\varepsilon \cdots \otimes_\varepsilon \bar{h}_{k+1}^{\wedge_\varepsilon w_{k+1}} &= \left(\bar{h}_1^{\wedge_\varepsilon w_1} \otimes_\varepsilon \bar{h}_2^{\wedge_\varepsilon w_2} \otimes_\varepsilon \cdots \otimes_\varepsilon \bar{h}_k^{\wedge_\varepsilon w_k}\right) \otimes_\varepsilon \bar{h}_{k+1}^{\wedge_\varepsilon w_{k+1}} \\
&= \bigcup_{\gamma_1 \in \bar{h}_1, \gamma_2 \in \bar{h}_2, \ldots, \gamma_k \in \bar{h}_k} \left\{ \frac{2\prod_{t=1}^{k} \gamma_t^{w_t}}{\prod_{t=1}^{k}(2-\gamma_t)^{w_t} + \prod_{t=1}^{k} \gamma_t^{w_t}} \Big| p_1 p_2 \cdots p_k \right\} \\
&\quad \otimes_\varepsilon \bigcup_{\gamma_{k+1} \in \bar{h}_{k+1}} \left\{ \frac{2\gamma_{k+1}^{w_{k+1}}}{(2-\gamma_{k+1})^{w_{k+1}} + \gamma_{k+1}^{w_{k+1}}} \Big| p_{k+1} \right\} \\
&= \bigcup_{\gamma_1 \in \bar{h}_1, \gamma_2 \in \bar{h}_2, \ldots, \gamma_k \in \bar{h}_k, \gamma_{k+1} \in \bar{h}_{k+1}} \left\{ \frac{2\prod_{t=1}^{k+1} \gamma_t^{w_t}}{\prod_{t=1}^{k+1}(2-\gamma_t)^{w_t} + \prod_{t=1}^{k+1} \gamma_t^{w_t}} \Big| p_1 p_2 \cdots p_k p_{k+1} \right\},
\end{aligned}$$

i.e., Equation (15) holds for $T = k + 1$. Then, Equation (15) holds for all T. Hence, we complete the proof of the theorem. □

Based on Theorem 6, we have basic properties of the HPFEWG operator, as follows:

Theorem 7. *Let $\bar{h}_t(\gamma_i^{(t)}|p_t)$ $(t = 1,2,\ldots,T)$ be a collection of HPFEs, $w = (w_1, w_2, \ldots, w_T)^T$ be the weight vector of $\bar{h}_t$ $(t = 1,2,\ldots,T)$ such that $w_t \in [0,1]$ and $\sum_{t=1}^{T} w_t = 1$, and p_t be the corresponding probability of $\gamma_i^{(t)}$ in HPFE $\bar{h}_t$. Then, we have the following.*

(1) (Boundary):

$$\bar{h}^- \leq \mathrm{HPFEWG}(\bar{h}_1, \bar{h}_2, \ldots, \bar{h}_T) \leq \bar{h}^+, \tag{16}$$

where $\bar{h}^- = (\min_{1\leq t\leq T} \min_{\gamma_t\in\bar{h}_t} \gamma_t | p_1 p_2 \cdots p_T)$ and $\bar{h}^+ = (\max_{1\leq t\leq T} \max_{\gamma_t\in\bar{h}_t} \gamma_t | p_1 p_2 \cdots p_T)$.

(2) (Monotonicity): Let $\bar{h}_t^*(\dot{\gamma}_i^{(t)}|p_t)$ $(t = 1,2,\ldots,T)$ be a collection of HPFEs with $\#_t = \#\bar{h}_t = \#\bar{h}_t^*$ for $t = 1,2,\ldots,T$, $w = (w_1, w_2, \ldots, w_T)^T$ be the weight vector of $\bar{h}_t^*$ $(t = 1,2,\ldots,T)$ such that $w_t \in [0,1]$ and $\sum_{t=1}^{T} w_t = 1$, and p_t be the probability of $\dot{\gamma}_i^{(t)}$ in HPFE $\bar{h}_t^*$. If $\gamma_i^{(t)} \leq \dot{\gamma}_i^{(t)}$ for each $i = 1,2,\ldots,\#_t$, $t = 1,2,\ldots,T$, then

$$\mathrm{HPFEWG}(\bar{h}_1, \bar{h}_2, \ldots, \bar{h}_T) \leq \mathrm{HPFEWG}(\bar{h}_1^*, \bar{h}_2^*, \ldots, \bar{h}_T^*). \tag{17}$$

Proof. (1) Let $g(x) = \frac{2-x}{x}$, $x \in (0,1]$; then, $g'(x) = \frac{-2}{x^2} < 0$, i.e., $g(x)$ is a decreasing function. Let $\max \gamma_t = \max_{1\leq t\leq T} \max_{\gamma_t\in\bar{h}_t} \gamma_t$ and $\min \gamma_t = \min_{1\leq t\leq T} \min_{\gamma_t\in\bar{h}_t} \gamma_t$. For any $\gamma_t \in \bar{h}_t$ $(t = 1,2,\ldots,T)$, since $\min_{\gamma_t\in\bar{h}_t} \gamma_t \leq \gamma_t \leq \max_{\gamma_t\in\bar{h}_t} \gamma_t$; then, $g(\max_{\gamma_t\in\bar{h}_t} \gamma_t) \leq g(\gamma_t) \leq g(\min_{\gamma_t\in\bar{h}_t} \gamma_t)$, and so

$$\frac{2-\max\gamma_t}{\max\gamma_t} \leq \frac{2-\max_{\gamma_t\in\bar{h}_t}\gamma_t}{\max_{\gamma_t\in\bar{h}_t}\gamma_t} \leq \frac{2-\gamma_t}{\gamma_t} \leq \frac{2-\min_{\gamma_t\in\bar{h}_t}\gamma_t}{\min_{\gamma_t\in\bar{h}_t}\gamma_t} \leq \frac{2-\min\gamma_t}{\min\gamma_t}.$$

Since $w = (w_1, w_2, \ldots, w_T)^T$ is the weight vector of $\bar{h}_t$ $(t = 1,2,\ldots,T)$ with $w_t \in [0,1]$ and $\sum_{t=1}^{T} w_t = 1$, we have

$$\prod_{t=1}^{T}\left(\frac{2-\max\gamma_t}{\max\gamma_t}\right)^{w_t} \leq \prod_{t=1}^{T}\left(\frac{2-\gamma_t}{\gamma_t}\right)^{w_t} \leq \prod_{t=1}^{T}\left(\frac{2-\min\gamma_t}{\min\gamma_t}\right)^{w_t}.$$

Since $\prod_{t=1}^{T}\left(\frac{2-\max\gamma_t}{\max\gamma_t}\right)^{w_t} = \left(\frac{2-\max\gamma_t}{\max\gamma_t}\right)^{\sum_{t=1}^{T} w_t} = \frac{2-\max\gamma_t}{\max\gamma_t}$ and $\prod_{t=1}^{T}\left(\frac{2-\min\gamma_t}{\min\gamma_t}\right)^{w_t} = \left(\frac{2-\min\gamma_t}{\min\gamma_t}\right)^{\sum_{t=1}^{T} w_t} = \frac{2-\min\gamma_t}{\min\gamma_t}$, we obtain

$$\begin{aligned}
\frac{2-\max\gamma_t}{\max\gamma_t} \leq \prod_{t=1}^{T}\left(\frac{2-\gamma_t}{\gamma_t}\right)^{w_t} \leq \frac{2-\min\gamma_t}{\min\gamma_t} &\Leftrightarrow \frac{2}{\max\gamma_t} \leq 1+\prod_{t=1}^{T}\left(\frac{2-\gamma_t}{\gamma_t}\right)^{w_t} \leq \frac{2}{\min\gamma_t} \\
&\Leftrightarrow \frac{\min\gamma_t}{2} \leq \frac{1}{1+\prod_{t=1}^{T}\left(\frac{2-\gamma_t}{\gamma_t}\right)^{w_t}} \leq \frac{\max\gamma_t}{2} \\
&\Leftrightarrow \min\gamma_t \leq \frac{2}{1+\prod_{t=1}^{T}\left(\frac{2-\gamma_t}{\gamma_t}\right)^{w_t}} \leq \max\gamma_t,
\end{aligned}$$

i.e.,

$$\min\gamma_t \leq \frac{2\prod_{t=1}^{T}\gamma_t^{w_t}}{\prod_{t=1}^{T}(2-\gamma_t)^{w_t}+\prod_{t=1}^{T}\gamma_t^{w_t}} \leq \max\gamma_t. \tag{18}$$

Let $\mathrm{HPFEWG}(\bar{h}_1, \bar{h}_2, \ldots, \bar{h}_T) = \bar{h}(\gamma_i|p_1 p_2 \cdots p_T)$, $i = 1,2,\ldots,\#\bar{h}$, where $\#\bar{h} = \#\bar{h}_1 \times \#\bar{h}_2 \times \cdots \times \#\bar{h}_T$, $\bar{h}^- = (\min\gamma_t|p_1 p_2 \cdots p_T)$ and $\bar{h}^+ = (\max\gamma_t|p_1 p_2 \cdots p_T)$. Then, Equation (18) is transformed into the following forms: $\min\gamma_t \leq \gamma_i \leq \max\gamma_t$ for all $i = 1,2,\ldots,\#\bar{h}$. Thus, $s(\bar{h}^-) = \min\gamma_t p_1 p_2 \cdots p_T \leq \sum_{i=1}^{\#\bar{h}} \gamma_i p_1 p_2 \cdots p_T = s(\bar{h})$ and $s(\bar{h}) = \sum_{i=1}^{\#\bar{h}} \gamma_i p_1 p_2 \cdots p_T \leq \max\gamma_t p_1 p_2 \cdots p_T = s(\bar{h}^+)$. If $s(\bar{h}^-) < s(\bar{h})$ and $s(\bar{h}) < s(\bar{h}^+)$. Then, based on Definition 3, we have $\bar{h}^- < \mathrm{HPFEWG}(\bar{h}_1, \bar{h}_2, \ldots, \bar{h}_T) < \bar{h}^+$. If $s(\bar{h}) = s(\bar{h}^+)$, i.e., $\max\gamma_t = \sum_{i=1}^{\#\bar{h}} \gamma_i$, then

$d(\bar{h}) = \sum_{i=1}^{\#\bar{h}} (\gamma_i - s(\bar{h}))^2 p_1 p_2 \cdots p_T = (\max \gamma_t - s(\bar{h}))^2 p_1 p_2 \cdots p_T = d(\bar{h}^+)$. In this case, based on Definition 3, it follows that $\text{HPFEWG}(\bar{h}_1, \bar{h}_2, \ldots, \bar{h}_T) = \bar{h}^+$. If $s(\bar{h}) = s(\bar{h}^-)$. Then, similarly, we have $\text{HPFEWG}(\bar{h}_1, \bar{h}_2, \ldots, \bar{h}_T) = \bar{h}^-$.

(2) Let $g(x) = \frac{2-x}{x}$, $x \in (0,1]$; then, $g(x)$ is a decreasing function. If $\gamma_i^{(t)} \leq \dot{\gamma}_i^{(t)}$ for each $i = 1,2,\ldots,\#_t$, $t = 1,2,\ldots,T$, then $g(\gamma_i^{(t)}) \geq g(\dot{\gamma}_i^{(t)})$, for each $i = 1,2,\ldots,\#_t$, $t = 1,2,\ldots,T$, i.e., $\frac{2-\gamma_i^{(t)}}{\gamma_i^{(t)}} \geq \frac{2-\dot{\gamma}_i^{(t)}}{\dot{\gamma}_i^{(t)}}$, for each $i = 1,2,\ldots,\#_t$, $t = 1,2,\ldots,T$. For any $\gamma_i^{(t)} \in \bar{h}_t$ $(t = 1,2,\ldots,T)$, since $w = (w_1, w_2, \ldots, w_T)^T$ is the weight vector of $\bar{h}_t$ $(t = 1,2,\ldots,T)$ such that $w_t \in [0,1]$, $t = 1,2,\ldots,T$ and $\sum_{t=1}^T w_t = 1$, we have

$$\left(\frac{2-\gamma_i^{(t)}}{\gamma_i^{(t)}}\right)^{w_t} \geq \left(\frac{2-\dot{\gamma}_i^{(t)}}{\dot{\gamma}_i^{(t)}}\right)^{w_t}, \ i = 1,2,\ldots,\#_t, \ t = 1,2,\ldots,T.$$

Then,

$$\begin{aligned}
\prod_{t=1}^{T}\left(\frac{2-\gamma_i^{(t)}}{\gamma_i^{(t)}}\right)^{w_t} \geq \prod_{t=1}^{T}\left(\frac{2-\dot{\gamma}_i^{(t)}}{\dot{\gamma}_i^{(t)}}\right)^{w_t} &\Leftrightarrow 1+\prod_{t=1}^{T}\left(\frac{2-\gamma_i^{(t)}}{\gamma_i^{(t)}}\right)^{w_t} \geq 1+\prod_{t=1}^{T}\left(\frac{2-\dot{\gamma}_i^{(t)}}{\dot{\gamma}_i^{(t)}}\right)^{w_t} \\
&\Leftrightarrow \frac{1}{1+\prod_{t=1}^{T}\left(\frac{2-\gamma_i^{(t)}}{\gamma_i^{(t)}}\right)^{w_t}} \leq \frac{1}{1+\prod_{t=1}^{T}\left(\frac{2-\dot{\gamma}_i^{(t)}}{\dot{\gamma}_i^{(t)}}\right)^{w_t}} \\
&\Leftrightarrow \frac{2}{1+\prod_{t=1}^{T}\left(\frac{2-\gamma_i^{(t)}}{\gamma_i^{(t)}}\right)^{w_t}} - 1 \leq \frac{2}{1+\prod_{t=1}^{T}\left(\frac{2-\dot{\gamma}_i^{(t)}}{\dot{\gamma}_i^{(t)}}\right)^{w_t}} - 1,
\end{aligned}$$

i.e.,

$$\frac{2\prod_{t=1}^{T}(\gamma_i^{(t)})^{w_t}}{\prod_{t=1}^{T}(2-\gamma_i^{(t)})^{w_t} + \prod_{t=1}^{T}(\gamma_i^{(t)})^{w_t}} \leq \frac{2\prod_{t=1}^{T}(\gamma_i^{(t)})^{w_t}}{\prod_{t=1}^{T}(2-\gamma_i^{(t)})^{w_t} + \prod_{t=1}^{T}(\gamma_i^{(t)})^{w_t}}. \tag{19}$$

Let $\text{HPFEWG}(\bar{h}_1, \bar{h}_2, \ldots, \bar{h}_T) = \bar{h}(\gamma_i | p_1 p_2 \cdots p_T)$ and $\text{HPFEWG}(\bar{h}_1^*, \bar{h}_2^*, \ldots, \bar{h}_T^*) = \bar{h}^*(\dot{\gamma}_i | p_1 p_2 \cdots p_T)$, where $i = 1,2,\ldots,\#$, and $\# = \#_1 \times \#_2 \times \cdots \times \#_T$ is the number of possible elements in $\bar{h}(\gamma_i | p_1 p_2 \cdots p_T)$ and $\bar{h}^*(\dot{\gamma}_i | p_1 p_2 \cdots p_T)$, respectively. Then, the Equation (19) is transformed into the form $\gamma_i \leq \dot{\gamma}_i$ $(i = 1,2,\ldots,\#)$. Thus, $s(\bar{h}) = \sum_{i=1}^{\#} \gamma_i p_1 p_2 \cdots p_T \leq \sum_{i=1}^{\#} \dot{\gamma}_i p_1 p_2 \cdots p_T = s(\bar{h}^*)$. If $s(\bar{h}) < s(\bar{h}^*)$, then based on Definition 3, $\text{HPFEWG}(\bar{h}_1, \bar{h}_2, \ldots, \bar{h}_T) < \text{HPFEWG}(\bar{h}_1^*, \bar{h}_2^*, \ldots, \bar{h}_T^*)$. If $s(\bar{h}) = s(\bar{h}^*)$, i.e., $\sum_{i=1}^{\#} \gamma_i = \sum_{i=1}^{\#} \dot{\gamma}_i$, then $d(\bar{h}) = \sum_{i=1}^{\#} (\gamma_i - s(\bar{h}))^2 p_1 p_2 \cdots p_T = \sum_{i=1}^{\#} (\dot{\gamma}_i - s(\bar{h}^*))^2 p_1 p_2 \cdots p_T = d(\bar{h}^*)$. In this case, based on Definition 3, it follows that $\text{HPFEWG}(\bar{h}_1, \bar{h}_2, \ldots, \bar{h}_T) = \text{HPFEWG}(\bar{h}_1^*, \bar{h}_2^*, \ldots, \bar{h}_T^*)$. □

If all probabilities of values in each HPFE are equal, i.e., $p_1 = p_2 = \cdots = p_{\#\bar{h}_t}$ $(t = 1,2,\ldots,T)$, then the HPFE is reduced to the HFE. In this case, the score function of the HPFEWA (resp. HPFEWG) operator is consistent with that of the HFEWA (resp. HFEWG) operator [21]. So, we can conclude that the HPFEWA (resp. HPFEWG) operator is reduced to the HFEWA (resp. HFEWG) operator [21]. In order to analyze the relationship between the HPFEWA (resp. HPFEWG) operator and the HPFWA (resp. HPFWG) operator [27], we introduce the following lemma.

Lemma 1. [41,42] *Let $x_i > 0$, $w_i > 0$, $i = 1,2,\ldots,N$, and $\sum_{i=1}^N w_i = 1$, then $\prod_{i=1}^N x_i^{w_i} \leq \sum_{i=1}^N w_i x_i$, with equality if and only if $x_1 = x_2 = \cdots = x_N$.*

Theorem 8. *If $\bar{h}_t$ $(t = 1,2,\ldots,T)$ are a collection of HPFEs and $w = (w_1, w_2, \ldots, w_T)^T$ is the weight vector of $\bar{h}_t$, with $w_t \in [0,1]$ and $\sum_{t=1}^T w_t = 1$, and p_t is the probability of γ_t in HPFE $\bar{h}_t$, then*

(1) $\text{HPFEWA}(\bar{h}_1, \bar{h}_2, \ldots, \bar{h}_T) \leq \text{HPFWA}(\bar{h}_1, \bar{h}_2, \ldots, \bar{h}_T)$;

(2) $\text{HPFEWG}(\bar{h}_1, \bar{h}_2, \ldots, \bar{h}_T) \geq \text{HPFWG}(\bar{h}_1, \bar{h}_2, \ldots, \bar{h}_T)$.

Proof. (1) For any $\gamma_t \in \bar{h}_t$ $(t = 1, 2, \ldots, T)$, based on Lemma 1, we obtain the inequality $\prod_{t=1}^{T}(1+\gamma_t)^{w_t} + \prod_{t=1}^{T}(1-\gamma_t)^{w_t} \leq \sum_{t=1}^{T} w_t(1+\gamma_t) + \sum_{t=1}^{T} w_t(1-\gamma_t) = 2$, and then

$$\frac{\prod_{t=1}^{T}(1+\gamma_t)^{w_t} - \prod_{t=1}^{T}(1-\gamma_t)^{w_t}}{\prod_{t=1}^{T}(1+\gamma_t)^{w_t} + \prod_{t=1}^{T}(1-\gamma_t)^{w_t}} = 1 - \frac{2\prod_{t=1}^{T}(1-\gamma_t)^{w_t}}{\prod_{t=1}^{T}(1+\gamma_t)^{w_t} + \prod_{t=1}^{T}(1-\gamma_t)^{w_t}} \leq 1 - \prod_{t=1}^{T}(1-\gamma_t)^{w_t}.$$

Hence, we can obtain the inequality

$$\bigcup_{\gamma_1 \in \bar{h}_1, \gamma_2 \in \bar{h}_2, \ldots, \gamma_T \in \bar{h}_T} \left\{ \frac{\prod_{t=1}^{T}(1+\gamma_t)^{w_t} - \prod_{t=1}^{T}(1-\gamma_t)^{w_t}}{\prod_{t=1}^{T}(1+\gamma_t)^{w_t} + \prod_{t=1}^{T}(1-\gamma_t)^{w_t}} \right\} \leq \bigcup_{\gamma_1 \in \bar{h}_1, \gamma_2 \in \bar{h}_2, \ldots, \gamma_T \in \bar{h}_T} \left\{ 1 - \prod_{t=1}^{T}(1-\gamma_t)^{w_t} \right\}. \tag{20}$$

Let $\text{HPFEWA}(\bar{h}_1, \bar{h}_2, \ldots, \bar{h}_T) = \bar{h}(\gamma_i|p_i)$ and $\text{HPFWA}(\bar{h}_1, \bar{h}_2, \ldots, \bar{h}_T) = \bar{h}^*(\gamma_i^*|p_i)$, $i = 1, 2, \ldots \#$, where $\# = \#\bar{h} = \#\bar{h}^*$ is the number of possible elements in $\bar{h}(\gamma_i|p_i)$ and $\bar{h}^*(\dot{\gamma}_i|p_i)$, respectively. Then, Equation (21) is transformed into the form $\gamma_i \leq \gamma_i^*$ $(i = 1, 2, \ldots, \#)$. According to $s(\bar{h}) = \sum_{i=1}^{\#\bar{h}} \gamma_i p_i$, we have $\text{HPFEWA}(\bar{h}_1, \bar{h}_2, \ldots, \bar{h}_T) \leq \text{HPFWA}(\bar{h}_1, \bar{h}_2, \ldots, \bar{h}_T)$.

(2) For any $\gamma_t \in \bar{h}_t$ $(t = 1, 2, \ldots, T)$, bsed on Lemma 1, we have $\prod_{t=1}^{T}(2-\gamma_t)^{w_t} + \prod_{t=1}^{T}\gamma_t^{w_t} \leq \sum_{t=1}^{T} w_t(2-\gamma_t) + \sum_{t=1}^{T} w_t\gamma_t = 2$, and then

$$\frac{2\prod_{t=1}^{T}\gamma_t^{w_t}}{\prod_{t=1}^{T}(2-\gamma_t)^{w_t} + \prod_{t=1}^{T}\gamma_t^{w_t}} \geq \prod_{t=1}^{T}\gamma_t^{w_t}.$$

Hence, similarly to (1), we have $\text{HPFEWG}(\bar{h}_1, \bar{h}_2, \ldots, \bar{h}_T) \geq \text{HPFWG}(\bar{h}_1, \bar{h}_2, \ldots, \bar{h}_T)$. □

Example 4. *Let* $\bar{h}_1 = (0.5|0.5, 0.6|0.5)$ *and* $\bar{h}_2 = (0.1|0.2, 0.3|0.3, 0.4|0.5)$ *be two HPFEs and* $w = (0.6, 0.4)^T$ *be the weight vector of them. Then, based on Equation (8), the aggregated value from the HPFEWA operator is*

$$\begin{aligned}
\text{HPFEWA}(\bar{h}_1, \bar{h}_2) &= (w_1 \cdot_\varepsilon \bar{h}_1) \oplus_\varepsilon (w_2 \cdot_\varepsilon \bar{h}_2) \\
&= \bigcup_{\gamma_1 \in \bar{h}_1, \gamma_2 \in \bar{h}_2} \left\{ \frac{\prod_{t=1}^{2}(1+\gamma_t)^{w_t} - \prod_{t=1}^{2}(1-\gamma_t)^{w_t}}{\prod_{t=1}^{2}(1+\gamma_t)^{w_t} + \prod_{t=1}^{2}(1-\gamma_t)^{w_t}} \middle| p_1 p_2 \right\} \\
&= \{0.3537|0.1, 0.4247|0.15, 0.4614|0.25, 0.4268|0.1, 0.4928|0.15, 0.5265|0.25\}.
\end{aligned}$$

If we use the HPFWA operator (Equation (2)) to aggregate two HPFEs, then we have

$$\begin{aligned}
\text{HPFWA}(\bar{h}_1, \bar{h}_2) &= (w_1\bar{h}_1) \oplus (w_2\bar{h}_2) \\
&= \bigcup_{\gamma_1 \in \bar{h}_1, \gamma_2 \in \bar{h}_2} \left\{ 1 - \prod_{t=1}^{2}(1-\gamma_t)^{w_t} \middle| p_1 p_2 \right\} \\
&= \{0.3675|0.1, 0.4280|0.15, 0.4622|0.25, 0.4467|0.1, 0.4996|0.15, 0.5296|0.25\}.
\end{aligned}$$

Then, $s(\text{HPFEWA}(\bar{h}_1, \bar{h}_2)) = 0.4627$ *and* $s(\text{HPFWA}(\bar{h}_1, \bar{h}_2)) = 0.4685$, *and thus,* $\text{HPFEWA}(\bar{h}_1, \bar{h}_2) < \text{HPFWA}(\bar{h}_1, \bar{h}_2)$.

On the other hand, based on Equation (15), the aggregated value by HPFEWG operator is

$$\begin{aligned}\mathrm{HPFEWG}(\bar{h}_1,\bar{h}_2) &= \bar{h}_1^{\wedge_\varepsilon w_1} \otimes_\varepsilon \bar{h}_2^{\wedge_\varepsilon w_2}\\ &= \bigcup_{\gamma_1\in\bar{h}_1,\gamma_2\in\bar{h}_2}\left\{\frac{2\prod_{t=1}^{2}\gamma_t^{w_t}}{\prod_{t=1}^{2}(2-\gamma_t)^{w_t}+\prod_{t=1}^{2}\gamma_t^{w_t}}\middle|p_1p_2\right\}\\ &= \{0.2748|0.1, 0.4108|0.15, 0.4581|0.25, 0.3126|0.1, 0.4622|0.15, 0.5135|0.25\}.\end{aligned}$$

If we use the HPFWG operator (Equation (3)) to aggregate two HPFEs, then we get

$$\begin{aligned}\mathrm{HPFWG}(\bar{h}_1,\bar{h}_2) &= (\bar{h}_1)^{w_1}\otimes(\bar{h}_2)^{w_2}\\ &= \bigcup_{\gamma_1\in\bar{h}_1,\gamma_2\in\bar{h}_2}\left\{\prod_{t=1}^{2}(\gamma_t)^{w_t}|p_1p_2\right\}\\ &= \{0.2627|0.1, 0.4076|0.15, 0.4573|0.25, 0.2930|0.1, 0.4547|0.15, 0.5102|0.25\}.\end{aligned}$$

It is clear that $\mathrm{HPFEWG}(\bar{h}_1,\bar{h}_2) > \mathrm{HPFWG}(\bar{h}_1,\bar{h}_2)$.

Theorem 9. *If $\bar{h}_t$ $(t = 1,2,\ldots,T)$ are a collection of HPFEs, $w = (w_1, w_2, \ldots, w_T)^T$ is the weight vector of $\bar{h}_t$ with $w_t \in [0,1]$ and $\sum_{t=1}^{T} w_t = 1$, and p_t is the probability of γ_t in HPFE $\bar{h}_t$. Then,*
(1) $\mathrm{HPFEWA}((\bar{h}_1)^c, (\bar{h}_2)^c, \ldots, (\bar{h}_T)^c) = (\mathrm{HPFEWG}(\bar{h}_1, \bar{h}_2, \ldots, \bar{h}_T))^c$;
(2) $\mathrm{HPFEWG}((\bar{h}_1)^c, (\bar{h}_2)^c, \ldots, (\bar{h}_T)^c) = (\mathrm{HPFEWA}(\bar{h}_1, \bar{h}_2, \ldots, \bar{h}_T))^c$.

Proof. Since (2) is similar (1), we only prove (1).

$$\begin{aligned}&\mathrm{HPFEWA}((\bar{h}_1)^c, (\bar{h}_2)^c, \ldots, (\bar{h}_T)^c)\\ &= \bigcup_{\gamma_1\in\bar{h}_1,\gamma_2\in\bar{h}_2,\ldots,\gamma_T\in\bar{h}_T}\left\{\frac{\prod_{t=1}^{T}(1+(1-\gamma_t))^{w_t}-\prod_{t=1}^{T}(1-(1-\gamma_t))^{w_t}}{\prod_{t=1}^{T}(1+(1-\gamma_t))^{w_t}+\prod_{t=1}^{T}(1-(1-\gamma_t))^{w_t}}\middle|p_1p_2\cdots p_T\right\}\\ &= \bigcup_{\gamma_1\in\bar{h}_1,\gamma_2\in\bar{h}_2,\ldots,\gamma_T\in\bar{h}_T}\left\{1-\frac{2\prod_{t=1}^{T}\gamma_t^{w_t}}{\prod_{t=1}^{T}(2-\gamma_t)^{w_t}+\prod_{t=1}^{T}\gamma_t^{w_t}}\middle|p_1p_2\cdots p_T\right\}\\ &= (\mathrm{HPFEWG}(\bar{h}_1,\bar{h}_2,\ldots,\bar{h}_T))^c.\end{aligned}$$

□

Theorem 8 shows that (1) the values aggregated by the HPFEWA operator are not larger than those obtained by the HPFWA operator. That is to say, the HPFEWA operator reflects the decision maker's pessimistic attitude rather than the HPFWA operator in the aggregation process; and (2) the values aggregated by the HPFWG operator are not larger than those obtained by the HPFEWG operator. Thus, the HPFEWG operator reflects the decision maker's optimistic attitude rather than the HPFWG operator in the aggregation process. Moreover, we developed the following ordered weighted operators based on the HPFOWA operator [27] and the HPFOWG operator [27] to aggregate the HPFEs.

Let $\bar{h}_t$ $(t = 1,2,\ldots,T)$ be a collection of HPFEs, $\bar{h}_{\sigma(t)}$ be the tth largest of $\bar{h}_t$ $(t = 1,2,\ldots,T)$, and $p_{\sigma(t)}$ be the probability of $\gamma_{\sigma(t)}$ in the HPFE $\bar{h}_{\sigma(t)}$; then, we have the following two aggregation operators, which are based on the mapping $H_P^T \to H_P$ with an associated vector $\omega = (\omega_1, \omega_2, \ldots, \omega_T)^T$, such that $\omega_t \in [0,1]$ and $\sum_{t=1}^{T}\omega_t = 1$:

(1) The hesitant probabilistic fuzzy Einstein ordered weighted averaging (HPFEOWA) operator is

$$\begin{aligned}&\mathrm{HPFEOWA}(\bar{h}_1,\bar{h}_2,\ldots,\bar{h}_T) = (\omega_1\cdot_\varepsilon\bar{h}_{\sigma(1)})\oplus_\varepsilon(\omega_2\cdot_\varepsilon\bar{h}_{\sigma(2)})\oplus_\varepsilon\cdots\oplus_\varepsilon(\omega_T\cdot_\varepsilon\bar{h}_{\sigma(T)})\\ &= \bigcup_{\gamma_{\sigma(1)}\in\bar{h}_{\sigma(1)},\gamma_{\sigma(2)}\in\bar{h}_{\sigma(2)},\ldots,\gamma_{\sigma(T)}\in\bar{h}_{\sigma(T)}}\left\{\frac{\prod_{t=1}^{T}(1+\gamma_{\sigma(t)})^{\omega_t}-\prod_{t=1}^{T}(1-\gamma_{\sigma(t)})^{\omega_t}}{\prod_{t=1}^{T}(1+\gamma_{\sigma(t)})^{\omega_t}+\prod_{t=1}^{T}(1-\gamma_{\sigma(t)})^{\omega_t}}\middle|p_{\sigma(1)}p_{\sigma(2)}\cdots p_{\sigma(T)}\right\}.\end{aligned} \tag{21}$$

(2) The hesitant probabilistic fuzzy Einstein ordered weighted geometric (HPFEOWG) operator is

$$\begin{aligned} &\mathrm{HPFEOWG}(\bar{h}_1, \bar{h}_2, \ldots, \bar{h}_T) = (\bar{h}_{\sigma(1)}^{\wedge_\varepsilon \omega_1}) \otimes_\varepsilon (\bar{h}_{\sigma(2)}^{\wedge_\varepsilon \omega_2}) \otimes_\varepsilon \cdots \otimes_\varepsilon (\bar{h}_{\sigma(T)}^{\wedge_\varepsilon \omega_T}) \\ &= \bigcup_{\gamma_{\sigma(1)} \in \bar{h}_{\sigma(1)}, \gamma_{\sigma(2)} \in \bar{h}_{\sigma(2)}, \ldots, \gamma_{\sigma(T)} \in \bar{h}_{\sigma(T)}} \left\{ \frac{2\prod_{t=1}^{T} \gamma_{\sigma(t)}^{\omega_t}}{\prod_{t=1}^{T} (2-\gamma_{\sigma(t)})^{\omega_t} + \prod_{t=1}^{T} \gamma_{\sigma(t)}^{\omega_t}} \middle| p_{\sigma(1)} p_{\sigma(2)} \cdots p_{\sigma(T)} \right\}. \end{aligned} \tag{22}$$

Example 5. *Let* $\bar{h}_1 = (0.5|0.5, 0.6|0.5)$ *and* $\bar{h}_2 = (0.1|0.2, 0.3|0.3, 0.4|0.5)$ *be two HPFEs, and suppose that the associated aggregated vector is* $\omega = (0.55, 0.45)^T$*. Based on Definition 3, the score values of* $\bar{h}_1$ *and* $\bar{h}_2$ *are* $s(\bar{h}_1) = 0.55$ *and* $s(\bar{h}_2) = 0.31$*. Since* $s(\bar{h}_1) > s(\bar{h}_2)$*; then,*

$$\bar{h}_{\sigma(1)} = \bar{h}_1 = (0.5|0.5, 0.6|0.5),\ \bar{h}_{\sigma(2)} = \bar{h}_2 = (0.1|0.2, 0.3|0.3, 0.4|0.5).$$

Based on Equation (21), the aggregated values by the HPFEOWA operator are

$$\begin{aligned} \mathrm{HPFEOWA}(\bar{h}_1, \bar{h}_2) &= (\omega_1 \cdot_\varepsilon \bar{h}_{\sigma(1)}) \oplus_\varepsilon (\omega_2 \cdot_\varepsilon \bar{h}_{\sigma(2)}) \\ &= \{0.3340|0.1, 0.4023|0.1, 0.4148|0.15, 0.4564|0.25, 0.4781|0.15, 0.5167|0.25\}. \end{aligned}$$

On the other hand, based on Equation (22), the aggregated values by the HPFEOWG operator are

$$\begin{aligned} \mathrm{HPFEOWG}(\bar{h}_1, \bar{h}_2) &= (\bar{h}_{\sigma(1)}^{\wedge_\varepsilon \omega_1}) \otimes_\varepsilon (\bar{h}_{\sigma(2)}^{\wedge_\varepsilon \omega_2}) \\ &= \{0.2937|0.1, 0.2859|0.1, 0.4005|0.15, 0.4466|0.15, 0.4530|0.25, 0.5033|0.25\}. \end{aligned}$$

In the following section, we look at the HPFEOWA and HPFEOWG operators for some special cases of the associated vector ω.

(1) If $\omega = (1, 0, \ldots, 0)^T$, then

$$\begin{aligned} \mathrm{HPFEOWA}(\bar{h}_1, \bar{h}_2, \ldots, \bar{h}_T) &= \bar{h}_{\sigma(1)} = \max\{\bar{h}_i\}, \\ \mathrm{HPFEOWG}(\bar{h}_1, \bar{h}_2, \ldots, \bar{h}_n) &= \bar{h}_{\sigma(1)} = \max\{\bar{h}_t\}. \end{aligned}$$

(2) If $\omega = (0, 0, \ldots, 1)^T$, then

$$\begin{aligned} \mathrm{HPFEOWA}(\bar{h}_1, \bar{h}_2, \ldots, \bar{h}_T) &= \bar{h}_{\sigma(T)} = \min\{\bar{h}_t\}, \\ \mathrm{HPFEOWG}(\bar{h}_1, \bar{h}_2, \ldots, \bar{h}_T) &= \bar{h}_{\sigma(T)} = \min\{\bar{h}_t\}. \end{aligned}$$

(3) If $\omega_s = 1,\ w_t = 0,\ s \neq t$, then

$$\begin{aligned} \bar{h}_{\sigma(T)} \leq \mathrm{HPFEOWA}(\bar{h}_1, \bar{h}_2, \ldots, \bar{h}_T) = \bar{h}_{\sigma(s)} \leq \bar{h}_{\sigma(1)}, \\ \bar{h}_{\sigma(T)} \leq \mathrm{HPFEOWG}(\bar{h}_1, \bar{h}_2, \ldots, \bar{h}_T) = \bar{h}_{\sigma(s)} \leq \bar{h}_{\sigma(1)}, \end{aligned}$$

where $\bar{h}_{\sigma(s)}$ is the sth largest $\bar{h}_t$ ($t = 1, 2, \ldots, T$).

(4) If $\omega=(\frac{1}{T},\frac{1}{T},\ldots,\frac{1}{T})^T$, then

$$\begin{aligned}
&\mathrm{HPFEOWA}(\bar{h}_1,\bar{h}_2,\ldots,\bar{h}_T)\\
&=\bigcup_{\gamma_{\sigma(1)}\in\bar{h}_{\sigma(1)},\gamma_{\sigma(2)}\in\bar{h}_{\sigma(2)},\ldots,\gamma_{\sigma(T)}\in\bar{h}_{\sigma(T)}}\left\{\frac{\prod_{t=1}^T(1+\gamma_{\sigma(t)})^{\frac{1}{T}}-\prod_{t=1}^T(1-\gamma_{\sigma(t)})^{\frac{1}{T}}}{\prod_{t=1}^T(1+\gamma_{\sigma(t)})^{\frac{1}{T}}+\prod_{t=1}^T(1-\gamma_{\sigma(t)})^{\frac{1}{T}}}\middle|p_{\sigma(1)}p_{\sigma(2)}\cdots p_{\sigma(T)}\right\}\\
&=\bigcup_{\gamma_1\in\bar{h}_1,\gamma_2\in\bar{h}_2,\ldots,\gamma_T\in\bar{h}_T}\left\{\frac{\prod_{t=1}^T(1+\gamma_t)^{\frac{1}{T}}-\prod_{t=1}^T(1-\gamma_t)^{\frac{1}{T}}}{\prod_{t=1}^T(1+\gamma_t)^{\frac{1}{T}}+\prod_{t=1}^T(1-\gamma_t)^{\frac{1}{T}}}\middle|p_1p_2\cdots p_T\right\}\\
&=\mathrm{HPFEA}(\bar{h}_1,\bar{h}_2,\ldots,\bar{h}_T),
\end{aligned}$$

$$\begin{aligned}
&\mathrm{HPFEOWA}(\bar{h}_1,\bar{h}_2,\ldots,\bar{h}_T)\\
&=\bigcup_{\gamma_{\sigma(1)}\in\bar{h}_{\sigma(1)},\gamma_{\sigma(2)}\in\bar{h}_{\sigma(2)},\ldots,\gamma_{\sigma(T)}\in\bar{h}_{\sigma(T)}}\left\{\frac{2\prod_{t=1}^T\gamma_{\sigma(t)}^{\frac{1}{T}}}{\prod_{t=1}^T(2-\gamma_{\sigma(t)})^{\frac{1}{T}}+\prod_{t=1}^T\gamma_{\sigma(t)}^{\frac{1}{T}}}\middle|p_{\sigma(1)}p_{\sigma(2)}\cdots p_{\sigma(T)}\right\}\\
&=\bigcup_{\gamma_1\in\bar{h}_1,\gamma_2\in\bar{h}_2,\ldots,\gamma_T\in\bar{h}_T}\left\{\frac{2\prod_{t=1}^T\gamma_t^{\frac{1}{T}}}{\prod_{t=1}^T(2-\gamma_t)^{\frac{1}{T}}+\prod_{t=1}^T\gamma_t^{\frac{1}{T}}}\middle|p_1p_2\cdots p_T\right\}\\
&=\mathrm{HPFEG}(\bar{h}_1,\bar{h}_2,\ldots,\bar{h}_T),
\end{aligned}$$

i.e., the HPFEOWA (resp. HPFEOWG) operator is reduced to HPFEA (resp. HPFEG) operator.

Similar to Theorems 8 and 9, the above ordered weighted operators have the relationship below.

Theorem 10. *If $\bar{h}_t$ ($t=1,2,\ldots,T$) is a collection of HPFEs, $\omega=(\omega_1,\omega_2,\ldots,\omega_T)^T$ is the associated vector of the aggregation operator such that $\omega_t\in[0,1]$ and $\sum_{t=1}^T\omega_t=1$. Then,*
(1) $\mathrm{HPFEOWA}(\bar{h}_1,\bar{h}_2,\ldots,\bar{h}_T)\le\mathrm{HPFOWA}(\bar{h}_1,\bar{h}_2,\ldots,\bar{h}_T)$;
(2) $\mathrm{HPFEOWG}(\bar{h}_1,\bar{h}_2,\ldots,\bar{h}_T)\ge\mathrm{HPFOWG}(\bar{h}_1,\bar{h}_2,\ldots,\bar{h}_T)$.

Theorem 11. *If $\bar{h}_t$ ($t=1,2,\ldots,T$) is a collection of HPFEs, $\omega=(\omega_1,\omega_2,\ldots,\omega_T)^T$ is the associated vector of the aggregation operator, such that $\omega_t\in[0,1]$ and $\sum_{t=1}^T\omega_t=1$. Then,*
(1) $\mathrm{HPFEOWA}((\bar{h}_1)^c,(\bar{h}_2)^c,\ldots,(\bar{h}_T)^c)=(\mathrm{HPFEOWG}(\bar{h}_1,\bar{h}_2,\ldots,\bar{h}_T))^c$;
(2) $\mathrm{HPFEOWG}((\bar{h}_1)^c,(\bar{h}_2)^c,\ldots,(\bar{h}_T)^c)=(\mathrm{HPFEOWA}(\bar{h}_1,\bar{h}_2,\ldots,\bar{h}_T))^c$.

Clearly, the fundamental characteristic of the HPFEWA and HPFEWG operators is that they consider the importance of each given HPFE, whereas the fundamental characteristic of the HPFEOWA and HPFEOWG operators is the weighting of the ordered positions of the HPFEs instead of weighting the given HPFEs themselves. By combining the advantages of the HPFEWA (resp. HPFEWG) and HPFEOWA (resp. HPFEOWG) operators, in the following text, we develop some hesitant probabilistic fuzzy hybrid aggregation operators that weight both the given HPFEs and their ordered positions.

Let $\bar{h}_t$ ($t=1,2,\ldots,T$) be a collection of HPFEs, $w=(w_1,w_2,\ldots,w_T)^T$ be the weight vector of $\bar{h}_t$ with $w_t\in[0,1]$ and $\sum_{t=1}^T w_t=1$, and p_t be the probability of γ_t in the HPFE $\bar{h}_t$. Then, we have the following two aggregation operators which are based on the mapping $H_P^T\to H_P$ with an associated vector $\omega=(\omega_1,\omega_2,\ldots,\omega_T)^T$, such that $\omega_t\in[0,1]$ and $\sum_{t=1}^T\omega_t=1$:

(1) The hesitant probabilistic fuzzy Einstein hybrid averaging (HPFEHA) operator is

$$\begin{aligned}
&\mathrm{HPFEHA}(\bar{h}_1,\bar{h}_2,\ldots,\bar{h}_T)=(\omega_1\cdot_\varepsilon\dot{h}_{\sigma(1)})\oplus_\varepsilon(\omega_2\cdot_\varepsilon\dot{h}_{\sigma(2)})\oplus_\varepsilon\cdots\oplus_\varepsilon(\omega_T\cdot_\varepsilon\dot{h}_{\sigma(T)})\\
&=\bigcup_{\dot{\gamma}_{\sigma(1)}\in\dot{h}_{\sigma(1)},\dot{\gamma}_{\sigma(2)}\in\dot{h}_{\sigma(2)},\ldots,\dot{\gamma}_{\sigma(T)}\in\dot{h}_{\sigma(T)}}\left\{\frac{\prod_{t=1}^T(1+\dot{\gamma}_{\sigma(t)})^{\omega_t}-\prod_{t=1}^T(1-\dot{\gamma}_{\sigma(t)})^{\omega_t}}{\prod_{t=1}^T(1+\dot{\gamma}_{\sigma(t)})^{\omega_t}+\prod_{t=1}^T(1-\dot{\gamma}_{\sigma(t)})^{\omega_t}}\middle|\dot{p}_{\sigma(1)}\dot{p}_{\sigma(2)}\cdots\dot{p}_{\sigma(T)}\right\},
\end{aligned}\tag{23}$$

where $\dot{h}_{\sigma(t)}$ is the tth largest of the weighted HPFEs $\dot{h}_t = Tw_t \cdot_\varepsilon \bar{h}_t$ $(t = 1,2,\ldots,T)$, T is the balancing coefficient, and $\dot{p}_{\sigma(t)}$ be the probability of $\dot{\gamma}_{\sigma(t)}$ in the HPFE $\dot{h}_{\sigma(t)}$.

(2) The hesitant probabilistic fuzzy Einstein hybrid geometric (HPFEHG) operator is

$$\begin{aligned} \text{HPFEHG}(\bar{h}_1, \bar{h}_2, \ldots, \bar{h}_T) &= (\ddot{h}_{\sigma(1)}^{\wedge_\varepsilon \omega_1}) \otimes_\varepsilon (\ddot{h}_{\sigma(2)}^{\wedge_\varepsilon \omega_2}) \otimes_\varepsilon \cdots \otimes_\varepsilon (\ddot{h}_{\sigma(T)}^{\wedge_\varepsilon \omega_T}) \\ &= \cup_{\ddot{\gamma}_{\sigma(1)} \in \ddot{h}_{\sigma(1)}, \ddot{\gamma}_{\sigma(2)} \in \ddot{h}_{\sigma(2)}, \ldots, \ddot{\gamma}_{\sigma(T)} \in \ddot{h}_{\sigma(T)}} \left\{ \frac{2\prod_{t=1}^{T} \ddot{\gamma}_{\sigma(t)}^{\omega_t}}{\prod_{t=1}^{T}(2-\ddot{\gamma}_{\sigma(t)})^{\omega_t} + \prod_{t=1}^{T} \ddot{\gamma}_{\sigma(t)}^{\omega_t}} \middle| \ddot{p}_{\sigma(1)} \ddot{p}_{\sigma(2)} \cdots \ddot{p}_{\sigma(T)} \right\}, \end{aligned} \tag{24}$$

where $\ddot{h}_{\sigma(t)}$ is the tth largest of the weighted HPFEs $\ddot{h}_t = \bar{h}_t^{\wedge_\varepsilon Tw_t}$ $(t = 1,2,\ldots,T)$, T is the balancing coefficient, and $\ddot{p}_{\sigma(t)}$ is the probability of $\ddot{\gamma}_{\sigma(t)}$ in the HPFE $\ddot{h}_{\sigma(t)}$.

Especially, if $w = (\frac{1}{T}, \frac{1}{T}, \ldots, \frac{1}{T})^T$, then $\dot{h}_t = \ddot{h}_t = \bar{h}_t$ $(t = 1,2,\ldots,T)$. In this case, the HPFEHA (resp. HPFEHG) operator is reduced to the HPFEOWA (resp. HPFEOWG) operator. If $\omega = (\frac{1}{T}, \frac{1}{T}, \ldots, \frac{1}{T})^T$, then since $\frac{1}{T} \cdot_\varepsilon \dot{h}_t = \frac{1}{T} \cdot_\varepsilon (Tw_t \cdot_\varepsilon \bar{h}_t) = \cup_{\gamma_t \in \bar{h}_t} \left\{ \frac{(1+\gamma_t)^{w_t} - (1-\gamma_t)^{w_t}}{(1+\gamma_t)^{w_t} + (1-\gamma_t)^{w_t}} \middle| p_t \right\}$ and $\ddot{h}_t^{\wedge_\varepsilon \frac{1}{T}} = (\bar{h}_t^{\wedge_\varepsilon Tw_t})^{\wedge_\varepsilon \frac{1}{T}} = \cup_{\gamma_t \in \bar{h}_t} \left\{ \frac{2\gamma_t^{w_t}}{(2-\gamma_t)^{w_t} + \gamma_t^{w_t}} \middle| p_t \right\}$, we have

$$\begin{aligned} \text{HPFEHA}(\bar{h}_1, \bar{h}_2, \ldots, \bar{h}_T) &= (\frac{1}{T} \cdot_\varepsilon \dot{h}_{\sigma(1)}) \oplus_\varepsilon (\frac{1}{T} \cdot_\varepsilon \dot{h}_{\sigma(2)}) \oplus_\varepsilon \cdots \oplus_\varepsilon (\frac{1}{T} \cdot_\varepsilon \dot{h}_{\sigma(T)}) \\ &= \bigcup_{\gamma_1 \in \bar{h}_1, \gamma_2 \in \bar{h}_2, \ldots, \gamma_T \in \bar{h}_T} \left\{ \frac{\prod_{t=1}^{T}(1+\gamma_t)^{w_t} - \prod_{t=1}^{T}(1-\gamma_t)^{w_t}}{\prod_{t=1}^{T}(1+\gamma_t)^{w_t} + \prod_{t=1}^{T}(1-\gamma_t)^{w_t}} \middle| p_1 p_2 \cdots p_T \right\} \\ &= \text{HPFEWA}(\bar{h}_1, \bar{h}_2, \ldots, \bar{h}_T), \end{aligned}$$

$$\begin{aligned} \text{HPFEHG}(\bar{h}_1, \bar{h}_2, \ldots, \bar{h}_T) &= (\ddot{h}_{\sigma(1)}^{\wedge_\varepsilon \frac{1}{T}}) \otimes_\varepsilon (\ddot{h}_{\sigma(2)}^{\wedge_\varepsilon \frac{1}{T}}) \otimes_\varepsilon \cdots \otimes_\varepsilon (\ddot{h}_{\sigma(T)}^{\wedge_\varepsilon \frac{1}{T}}) \\ &= \bigcup_{\gamma_1 \in \bar{h}_1, \gamma_2 \in \bar{h}_2, \ldots, \gamma_T \in \bar{h}_T} \left\{ \frac{2\prod_{t=1}^{T} \gamma_t^{w_t}}{\prod_{t=1}^{T}(2-\gamma_t)^{w_t} + \prod_{t=1}^{T} \gamma_t^{w_t}} \middle| p_1 p_2 \cdots p_T \right\} \\ &= \text{HPFEWG}(\bar{h}_1, \bar{h}_2, \ldots, \bar{h}_T), \end{aligned}$$

i.e., the HPFEHA (resp. HPFEHG) operator is reduced to the HPFEWA (resp. HPFEWG) operator.

Example 6. *Let* $\bar{h}_1 = (0.5|0.5, 0.6|0.5)$ *and* $\bar{h}_2 = (0.1|0.2, 0.3|0.3, 0.5|0.5)$ *be two HPFEs. Suppose that the weight vector of them is* $w = (0.63, 0.37)^T$*, and the aggregation associated vector is* $\omega = (0.3, 0.7)^T$*. Then,*

$$\begin{aligned} \dot{h}_1 &= \left(\frac{(1+0.5)^{2\times0.63} - (1-0.5)^{2\times0.63}}{(1+0.5)^{2\times0.63} + (1-0.5)^{2\times0.63}} \middle| 0.5, \frac{(1+0.6)^{2\times0.63} - (1-0.6)^{2\times0.63}}{(1+0.6)^{2\times0.63} + (1-0.6)^{2\times0.63}} \middle| 0.5 \right) \\ &= (0.5993|0.5, 0.7031|0.5), \\ \dot{h}_2 &= \left(\frac{(1+0.1)^{2\times0.37} - (1-0.1)^{2\times0.37}}{(1+0.1)^{2\times0.37} + (1-0.1)^{2\times0.37}} \middle| 0.2, \frac{(1+0.3)^{2\times0.37} - (1-0.3)^{2\times0.37}}{(1+0.3)^{2\times0.37} + (1-0.3)^{2\times0.37}} \middle| 0.3, \right. \\ &\quad \left. \frac{(1+0.5)^{2\times0.37} - (1-0.5)^{2\times0.37}}{(1+0.5)^{2\times0.37} + (1-0.5)^{2\times0.37}} \middle| 0.2 \right) \\ &= (0.7411|0.2, 0.2251|0.3, 0.3851|0.5) \end{aligned}$$

and $s(\dot{h}_1) = 0.6512$ *and* $s(\dot{h}_2) = 0.4083$*. Since* $s(\dot{h}_1) > s(\dot{h}_2)$*, we have*

$$\dot{h}_{\sigma(1)} = \dot{h}_1 = (0.5993|0.5, 0.7031|0.5),\ \dot{h}_{\sigma(2)} = \dot{h}_2 = (0.7411|0.2, 0.2251|0.3, 0.3851|0.5).$$

From Equation (23), we have

$$\begin{aligned}\text{HPFEHA}(\bar{h}_1,\bar{h}_2) &= (\omega_1 \cdot_\varepsilon \dot{h}_{\sigma(1)}) \oplus_\varepsilon (\omega_2 \cdot_\varepsilon \dot{h}_{\sigma(2)}) \\ &= \bigcup_{\dot{\gamma}_{\sigma(1)}\in\dot{h}_{\sigma(1)},\dot{\gamma}_{\sigma(2)}\in\dot{h}_{\sigma(2)}} \left\{ \frac{\prod_{t=1}^{2}(1+\dot{\gamma}_{\sigma(t)})^{\omega_t} - \prod_{t=1}^{2}(1-\dot{\gamma}_{\sigma(t)})^{\omega_t}}{\prod_{t=1}^{2}(1+\dot{\gamma}_{\sigma(t)})^{\omega_t} + \prod_{t=1}^{2}(1-\dot{\gamma}_{\sigma(t)})^{\omega_t}} \middle| \dot{p}_{\sigma(1)}\dot{p}_{\sigma(2)} \right\} \\ &= \{0.3715|0.15, 0.4175|0.15, 0.4557|0.25, 0.4977|0.25, 0.7037|0.1, 0.7302|0.1\}.\end{aligned}$$

On the other hand,

$$\begin{aligned}\ddot{h}_1 &= \left(\frac{2\times 0.5^{2\times 0.63}}{(2-0.5)^{2\times 0.63} + 0.5^{2\times 0.63}}|0.5, \frac{2\times 0.6^{2\times 0.63}}{(2-0.6)^{2\times 0.63} + 0.6^{2\times 0.63}}|0.5 \right) \\ &= (0.4007|0.5, 0.5117|0.5), \\ \ddot{h}_2 &= \left(\frac{2\times 0.1^{2\times 0.37}}{(2-0.1)^{2\times 0.37} + 0.1^{2\times 0.37}}|0.2, \frac{2\times 0.3^{2\times 0.37}}{(2-0.3)^{2\times 0.37} + 0.3^{2\times 0.37}}|0.3, \frac{2\times 0.5^{2\times 0.37}}{(2-0.5)^{2\times 0.37} + 0.5^{2\times 0.37}}|0.5 \right) \\ &= (0.2033|0.2, 0.4339|0.3, 0.6145|0.5)\end{aligned}$$

and since $s(\ddot{h}_1) = 0.4562 > 0.4465 = s(\ddot{h}_2)$, we have $\ddot{h}_{\sigma(1)} = \ddot{h}_1$ and $\ddot{h}_{\sigma(2)} = \ddot{h}_2$. From Equation (24), we have

$$\begin{aligned}\text{HPFEHG}(\bar{h}_1,\bar{h}_2) &= (\ddot{h}_{\sigma(1)}^{\wedge_\varepsilon \omega_1}) \otimes_\varepsilon (\ddot{h}_{\sigma(2)}^{\wedge_\varepsilon \omega_2}) \\ &= \bigcup_{\ddot{\gamma}_{\sigma(1)}\in\ddot{h}_{\sigma(1)},\ddot{\gamma}_{\sigma(2)}\in\ddot{h}_{\sigma(2)}} \left\{ \frac{2\prod_{t=1}^{2}\gamma_{\sigma(t)}^{\omega_t}}{\prod_{t=1}^{2}(2-\gamma_{\sigma(t)})^{\omega_t} + \prod_{t=1}^{2}\gamma_{\sigma(t)}^{\omega_t}} \middle| \ddot{p}_{\sigma(1)}\ddot{p}_{\sigma(2)} \right\} \\ &= \{0.2512|0.1, 0.2728|0.1, 0.4237|0.15, 0.4563|0.25, 0.5441|0.15, 0.5825|0.25\}.\end{aligned}$$

4. An Approach to MADM with Hesitant Probabilistic Fuzzy Information

In this section, we utilize the proposed aggregation operators to develop an approach for MADM with hesitant probabilistic fuzzy information.

Let $X = \{x_1, x_2, \ldots, x_n\}$ be a set of n alternatives and $G = \{g_1, g_2, \ldots, g_m\}$ be a set of m attributes whose weight vector is $w = (w_1, w_2, \ldots, w_m)^T$, satisfying $w_i > 0$ $(i = 1, 2, \ldots, m)$ and $\sum_{i=1}^{m} w_i = 1$, where w_i denotes the importance degree of attribute g_i. Suppose the decision makers provide the evaluating values that the alternatives x_j $(i = 1, 2, \ldots, n)$ satisfy the attributes g_i $(j = 1, 2, \ldots, m)$ represented by the HPFEs $\bar{h}_{ij}(\gamma_{ij}|p_{ij})$ $(i = 1, 2, \ldots, m; j = 1, 2, \ldots, n)$. All of these HPFEs are contained in the hesitant probabilistic fuzzy decision matrix $D = \left(\bar{h}_{ij}(\gamma_{ij}|p_{ij})\right)_{m\times n}$ (see Table 1).

Table 1. Hesitant probabilistic fuzzy decision matrix (D).

	x_1	x_2	$\cdots$	x_n
g_1	$\bar{h}_{11}(\gamma_{11}\|p_{11})$	$\bar{h}_{12}(\gamma_{12}\|p_{12})$	$\cdots$	$\bar{h}_{1n}(\gamma_{1n}\|p_{1n})$
g_2	$\bar{h}_{21}(\gamma_{21}\|p_{21})$	$\bar{h}_{22}(\gamma_{22}\|p_{22})$	$\cdots$	$\bar{h}_{2n}(\gamma_{11}\|p_{2n})$
$\vdots$	$\vdots$	$\vdots$	$\ddots$	$\vdots$
g_m	$\bar{h}_{m1}(\gamma_{m1}\|p_{m1})$	$\bar{h}_{m2}(\gamma_{m2}\|p_{m2})$	$\cdots$	$\bar{h}_{mn}(\gamma_{mn}\|p_{mn})$

The following steps can be used to solve the MADM problem under the hesitant probabilistic fuzzy environment and obtain an optimal alternative.

Step 1: Obtain the normalized hesitant probabilistic fuzzy decision matrix. In general, the attribute set (G) can be divided two subsets, G_1 and G_2, where G_1 and G_2 are the set of benefit attributes and cost attributes, respectively. If all of the attributes are of the same type, then the evaluation values do not need normalization, whereas if there are benefit attributes and cost attributes in MADM, in such cases, we may transform the evaluation values of cost type into the evaluation values of the benefit type by the following normalization formula:

$$\bar{r}_{ij}(\beta_{ij}|p_{ij}) = \begin{cases} \bar{h}_{ij}, & i \in G_1 \\ \bar{h}_{ij}^c, & i \in G_2, \end{cases} \tag{25}$$

where $\bar{h}_{ij}^c = \cup_{\gamma_{ij} \in \bar{h}_{ij}} \{(1-\gamma_{ij})|p_{ij}\}$ is the complement of $\bar{h}_{ij}$. Then, we obtain the normalized hesitant probabilistic fuzzy decision matrix $H = \left(\bar{r}_{ij}(\beta_{ij}|p_{ij})\right)_{m \times n}$ (see Table 2).

Table 2. Normalized hesitant probabilistic fuzzy decision matrix (H).

	x_1	x_2	$\cdots$	x_n
g_1	$\bar{r}_{11}(\beta_{11}\|p_{11})$	$\bar{r}_{12}(\beta_{12}\|p_{12})$	$\cdots$	$\bar{r}_{1n}(\beta_{1n}\|p_{1n})$
g_2	$\bar{r}_{21}(\beta_{21}\|p_{21})$	$\bar{r}_{22}(\beta_{22}\|p_{22})$	$\cdots$	$\bar{r}_{2n}(\beta_{11}\|p_{2n})$
$\vdots$	$\vdots$	$\vdots$	$\ddots$	$\vdots$
g_m	$\bar{r}_{m1}(\beta_{m1}\|p_{m1})$	$\bar{r}_{m2}(\beta_{m2}\|p_{m2})$	$\cdots$	$\bar{r}_{mn}(\beta_{mn}\|p_{mn})$

Step 2: Compute the overall assessment of alternatives. Utilize the HPFEWA operator

$$\begin{aligned} \bar{r}_j &= \text{HPFEWA}(\bar{r}_{1j}, \bar{r}_{2j}, \ldots, \bar{r}_{mj}) \\ &= \bigcup_{\beta_{1j} \in \bar{r}_{1j}, \beta_{2j} \in \bar{r}_{2j}, \ldots, \beta_{mj} \in \bar{r}_{mj}} \left\{ \frac{\prod_{i=1}^m (1+\beta_{ij})^{w_i} - \prod_{i=1}^m (1-\beta_{ij})^{w_i}}{\prod_{i=1}^m (1+\beta_{ij})^{w_i} + \prod_{i=1}^m (1-\beta_{ij})^{w_i}} \middle| p_{1j} p_{2j} \cdots p_{mj} \right\} \end{aligned} \tag{26}$$

or the HPFEWG operator

$$\begin{aligned} \bar{r}_j &= \text{HPFEWG}(\bar{r}_{1j}, \bar{r}_{2j}, \ldots, \bar{r}_{mj}) \\ &= \bigcup_{\beta_{1j} \in \bar{r}_{1j}, \beta_{2j} \in \bar{r}_{2j}, \ldots, \beta_{mj} \in \bar{r}_{mj}} \left\{ \frac{2\prod_{i=1}^m (\beta_{ij})^{w_i}}{\prod_{i=1}^m (2-\beta_{ij})^{w_i} + \prod_{i=1}^m (\beta_{ij})^{w_i}} \middle| p_{1j} p_{2j} \cdots p_{mj} \right\} \end{aligned} \tag{27}$$

to aggregate all the evaluating values $\bar{r}_{ij}$ $(1 = 1, 2, \ldots, m)$ of the jth column and get the overall rating value $\bar{r}_j$ corresponding to the alternative $(x_j$ $(j = 1, 2, \ldots, n))$.

Step 3: Rank the order of all alternatives. Utilize the method in Definition 3 to rank the overall rating values $\bar{r}_j$ $(j = 1, 2, \ldots, n)$. Rank all the alternatives(x_j $(j = 1, 2, \ldots, n)$) in accordance with $\bar{r}_j$ $(j = 1, 2, \ldots, n)$ in descending order, and finally, select the most desirable alternative(s) with the largest overall evaluation value(s).

Step 4: End.

In the above-mentioned procedure, the HPFEWA (or HPFEWG) operator is utilized to aggregate the evaluating values of each alternative with respect to a collection of the attributes to rank and select the alternative(s). So we give a detail illustration of the decision making procedure with a propulsion/manoeuvring system selection problem.

Example 7. The propulsion/manoeuvring system selection is based on a study that was conducted for the selection of propulsion/manoeuvring system of a double ended passenger ferry to operate across the Bosphorus in Istanbul with the aim of reducing the journey time in highly congested seaway traffic (adopted from Ölçer and Odabaşi [43] and Wang and Liu [37]).

The propulsion/manoeuvring system alternatives are given as the set of alternatives $X = \{x_1, x_2, x_3\}$. (1) x_1 is the conventional propeller and high lift rudder; (2) x_2 is the Z drive; and (3) x_3 is the cycloidal propeller. The selection decision is made on the basis of one objective and seven subjective attributes, which are the following: (1) g_1 is the investment cost; (2) g_2 is the operating cost;

(3) g_3 is the manoeuvrability; (4) g_4 is the propulsive power requirement; (5) g_5 is the reliability.; (6) g_6 is the propulsive power requirement; and (7) g_7 is the propulsive arrangement requirement. Note that the attributes are cost attributes, except for attributes g_3 and g_5, and the corresponding weight vector is $w = (0.15, 0.2, 0.3, 0.2, 0.15)^T$.

Assume that the decision makers use the linguistic terms shown in Table 3 to represent the evaluating values of the alternatives with respect to different attributes, respectively, and they provide their linguistic decision matrices (D) as listed in Tables 4.

Table 3. Linguistic terms and their corresponding hesitant probabilistic fuzzy elements (HPFEs).

Linguistic Terms	HPFEs
Very low (VL)	$(0\|0.7, 0.1\|0.3)$
Low (L)	$(0.15\|0.6, 0.25\|0.4)$
Medium low (ML)	$(0.3\|0.6, 0.4\|0.4)$
Medium (M)	$(0.45\|0.5, 0.55\|0.5)$
Medium high (MH)	$(0.6\|0.45, 0.7\|0.55)$
High (H)	$(0.75\|0.4, 0.85\|0.6)$
Very high (VH)	$(0.9\|0.4, 1\|0.6)$

Table 4. Linguistic decision matrix (D).

	x_1	x_2	x_3
g_1	ML	M	H
g_2	M	ML	H
g_3	MH	M	MH
g_4	H	H	L
g_5	MH	MH	M
g_6	H	M	M
g_7	L	MH	MH

Step 1: Based on Tables 3 and 4, we can get the hesitant probabilistic fuzzy decision matrix $D = (\bar{h}_{ij})_{7\times 3}$ (see Table 5).

Table 5. Hesitant probabilistic fuzzy decision matrix (D).

	x_1	x_2	x_3
g_1	$(0.3\|0.6, 0.4\|0.4)$	$(0.45\|0.5, 0.55\|0.5)$	$(0.75\|0.4, 0.85\|0.6)$
g_2	$(0.45\|0.5, 0.55\|0.5)$	$(0.3\|0.6, 0.4\|0.4)$	$(0.75\|0.4, 0.85\|0.6)$
g_3	$(0.6\|0.45, 0.7\|0.55)$	$(0.45\|0.5, 0.55\|0.5)$	$(0.6\|0.45, 0.7\|0.55)$
g_4	$(0.75\|0.4, 0.85\|0.6)$	$(0.75\|0.4, 0.85\|0.6)$	$(0.15\|0.6, 0.25\|0.4)$
g_5	$(0.6\|0.45, 0.7\|0.55)$	$(0.6\|0.45, 0.7\|0.55)$	$(0.45\|0.5, 0.55\|0.5)$
g_6	$(0.75\|0.4, 0.85\|0.6)$	$(0.45\|0.5, 0.55\|0.5)$	$(0.45\|0.5, 0.55\|0.5)$
g_7	$(0.15\|0.6, 0.25\|0.4)$	$(0.6\|0.45, 0.7\|0.55)$	$(0.6\|0.45, 0.7\|0.55)$

Then, considering that the attributes are cost attributes, except for attributes g_3 and g_5, based on Equation (25), the hesitant probabilistic fuzzy decision matrix (D) can be transformed into the following normalized hesitant probabilistic fuzzy decision matrix: $H = (\bar{r}_{ij})_{7\times 3}$ (see Table 6).

Table 6. Normalized hesitant probabilistic fuzzy decision matrix (H).

	$\mathbf{x}_1$	$\mathbf{x}_2$	$\mathbf{x}_3$
g_1	$(0.6\|0.4, 0.7\|0.6)$	$(0.45\|0.5, 0.55\|0.5)$	$(0.15\|0.6, 0.25\|0.4)$
g_2	$(0.45\|0.5, 0.55\|0.5)$	$(0.6\|0.4, 0.7\|0.6)$	$(0.15\|0.6, 0.25\|0.4)$
g_3	$(0.6\|0.45, 0.7\|0.55)$	$(0.45\|0.5, 0.55\|0.5)$	$(0.6\|0.45, 0.7\|0.55)$
g_4	$(0.15\|0.6, 0.25\|0.4)$	$(0.15\|0.6, 0.25\|0.4)$	$(0.75\|0.4, 0.85\|0.6)$
g_5	$(0.6\|0.45, 0.7\|0.55)$	$(0.6\|0.45, 0.7\|0.55)$	$(0.45\|0.5, 0.55\|0.5)$
g_6	$(0.15\|0.6, 0.25\|0.4)$	$(0.45\|0.5, 0.55\|0.5)$	$(0.45\|0.5, 0.55\|0.5)$
g_7	$(0.75\|0.4, 0.85\|0.6)$	$(0.3\|0.55, 0.4\|0.45)$	$(0.3\|0.55, 0.4\|0.45)$

Step 2: Utilize the decision information given in matrix H and the HPFEWA operator (26) to derive the overall rating values ($\bar{r}_j$) of the alternative x_j ($j = 1, 2, 3$):

$$\bar{r}_1 = \{0.4953|0.0243, 0.5148|0.0297, 0.5109|0.0162, 0.5299|0.0198, 0.5337|0.0297, 0.5521|0.0363, 0.5484|0.0198, 0.5664|0.0242, 0.5152|0.0243, 0.5341|0.0297, 0.5304|0.0162, 0.5489|0.0198, 0.5525|0.0297, 0.5704|0.0363, 0.5669|0.0198, 0.5843|0.0242, 0.5148|0.0365, 0.5337|0.0446, 0.5299|0.0243, 0.5484|0.0297, 0.5521|0.0446, 0.5700|0.0545, 0.5664|0.0297, 0.5839|0.0363, 0.5341|0.0365, 0.5525|0.0446, 0.5489|0.0243, 0.5669|0.0297, 0.5704|0.0446, 0.5878|0.0545, 0.5843|0.0297, 0.6013|0.0363\},$$

$$\bar{r}_2 = \{0.4550|0.0270, 0.4754|0.0330, 0.4713|0.0180, 0.4914|0.0220, 0.4862|0.0270, 0.5059|0.0330, 0.5019|0.0180, 0.5212|0.0220, 0.4821|0.0405, 0.5019|0.0495, 0.4980|0.0270, 0.5174|0.0330, 0.5123|0.0405, 0.5313|0.0495, 0.5275|0.0270, 0.5461|0.0330, 0.4707|0.0270, 0.4908|0.0330, 0.4868|0.0180, 0.5065|0.0220, 0.5013|0.0270, 0.5206|0.0330, 0.5168|0.0180, 0.5357|0.0220, 0.4974|0.0405, 0.5168|0.0495, 0.5129|0.0270, 0.5319|0.0330, 0.5270|0.0405, 0.5456|0.0495, 0.5419|0.0270, 0.5601|0.0330\},$$

$$\bar{r}_3 = \{0.4840|0.0324, 0.4992|0.0324, 0.5261|0.0486, 0.5404|0.0486, 0.5230|0.0396, 0.5374|0.0396, 0.5629|0.0594, 0.5764|0.0594, 0.4997|0.0216, 0.5146|0.0216, 0.5410|0.0324, 0.5550|0.0324, 0.5379|0.0264, 0.5520|0.0264, 0.5769|0.0396, 0.5902|0.0396, 0.4958|0.0216, 0.5108|0.0216, 0.5373|0.0324, 0.5514|0.0324, 0.5342|0.0264, 0.5484|0.0264, 0.5734|0.0396, 0.5867|0.0396, 0.5114|0.0144, 0.5260|0.0144, 0.5519|0.0216, 0.5657|0.0216, 0.5489|0.0176, 0.5628|0.0176, 0.5873|0.0264, 0.6002|0.0264\}.$$

Step 3: Calculate the score values of the overall rating values ($\bar{r}_j$) of the alternatives (x_j ($j = 1, 2, 3$)):

$$s(\bar{r}_1) = 0.5533,\ s(\bar{r}_2) = 0.5110,\ s(\bar{r}_3) = 0.5473.$$

Since $s(\bar{r}_1) > s(\bar{r}_3) > s(\bar{r}_2)$, the ranking order of the alternatives x_j ($j = 1, 2, 3$) is

$$x_1 \succ x_3 \succ x_2.$$

Therefore, the best alternative is x_1.

If we utilize the HPFEWG operator (27) in Step 2 to get the overall rating values ($\bar{r}_j$) of the alternatives (x_j ($j = 1, 2, 3$)), we obtain

$$\bar{r}_1 = \{0.4426|0.0243, 0.4545|0.0297, 0.4828|0.0162, 0.4955|0.0198, 0.4666|0.0297, 0.4790|0.0363, 0.5083|0.0198, 0.5215|0.0242, 0.4613|0.0243, 0.4736|0.0297, 0.5027|0.0162, 0.5127|0.0198, 0.4860|0.0297, 0.4987|0.0363, 0.5289|0.0198, 0.5423|0.0242, 0.4545|0.0365, 0.4667|0.0446, 0.4955|0.0243, 0.5084|0.0297, 0.4790|0.0446, 0.4916|0.0545, 0.5215|0.0297, 0.5348|0.0363, 0.4726|0.0365, 0.4861|0.0446, 0.5157|0.0243, 0.5290|0.0297, 0.4987|0.0446, 0.5117|0.0545, 0.5423|0.0297, 0.5560|0.0363\},$$

$$\bar{r}_2 = \{0.4100|0.0270, 0.4213|0.0330, 0.4481|0.0180, 0.4602|0.0220, 0.4367|0.0270, 0.4485|0.0330, 0.4766|0.0180, 0.4892|0.0220, 0.4250|0.0405, 0.4366|0.0495, 0.4641|0.0270, 0.4765|0.0330, 0.4525|0.0405, 0.4646|0.0495, 0.4933|0.0270, 0.5062|0.0330, 0.4232|0.0270, 0.4347|0.0330, 0.4622|0.0180, 0.4745|0.0220, 0.4505|0.0270, 0.4626|0.0330, 0.4913|0.0180, 0.5041|0.0220, 0.4386|0.0405, 0.4505|0.0495, 0.4786|0.0270, 0.4912|0.0330, 0.4666|0.0405, 0.4790|0.0495, 0.5084|0.0270, 0.5215|0.0330\},$$

$$\bar{r}_3 = \{0.3890|0.0324, 0.4017|0.0324, 0.4022|0.0486, 0.4152|0.0486, 0.4109|0.0396, 0.4241|0.0396, 0.4246|0.0594, 0.4382|0.0594, 0.4257|0.0216, 0.4393|0.0216, 0.4398|0.0324, 0.4537|0.0324, 0.4491|0.0264, 0.4632|0.0264, 0.4638|0.0396, 0.4782|0.0396, 0.4163|0.0216, 0.4297|0.0216, 0.4302|0.0324, 0.4439|0.0324, 0.4393|0.0264, 0.4532|0.0264, 0.4538|0.0396, 0.4680|0.0396, 0.4549|0.0144, 0.4691|0.0144, 0.4697|0.0216, 0.4843|0.0216, 0.4794|0.0176, 0.49421|0.0176, 0.4948|0.0264, 0.5098|0.0264\}.$$

Then, we calculate the scores of the overall rating values $\bar{r}_j$ of the alternatives:

$$s(\bar{r}_1) = 0.4968,\ s(\bar{r}_2) = 0.4621,\ s(\bar{r}_3) = 0.4429.$$

Since $s(\bar{r}_1) > s(\bar{r}_2) > s(\bar{r}_3)$, the ranking order of the alternatives x_j ($j = 1, 2, 3$) is

$$x_1 \succ x_2 \succ x_3.$$

Then, the best alternative is also x_1.

In order to compare the performance with the existing operators, in the following text, the HPFWA operator (2) and HPFWG operator (3) proposed by Xu and Zhou [27] are used to computing the overall rating values. If we first utilize the HPFWA operator (2) presented in Step 2, then we get the overall rating values $\bar{r}_j$ of the alternatives (x_j ($j = 1, 2, 3$)):

$$\bar{r}_1 = \{0.5043|0.0243, 0.5252|0.0297, 0.5165|0.0162, 0.5369|0.0198, 0.5453|0.0297, 0.5645|0.0363, 0.5566|0.0198, 0.5753|0.0242, 0.5238|0.0243, 0.5439|0.0297, 0.5356|0.0162, 0.5552|0.0198, 0.5632|0.0297, 0.5817|0.0363, 0.5740|0.0198, 0.5920|0.0242, 0.5252|0.0365, 0.5453|0.0446, 0.5369|0.0243, 0.5565|0.0297, 0.5645|0.0446, 0.5829|0.0545, 0.5753|0.0297, 0.5932|0.0363, 0.5439|0.0365, 0.5632|0.0446, 0.5552|0.0243, 0.5739|0.0297, 0.5817|0.0446, 0.5993|0.0545, 0.5920|0.0297, 0.6092|0.0363\},$$

$$\bar{r}_2 = \{0.4632|0.0270, 0.4859|0.0330, 0.4765|0.0180, 0.4986|0.0220, 0.4946|0.0270, 0.5159|0.0330, 0.5071|0.0180, 0.5279|0.0220, 0.4933|0.0405, 0.5147|0.0495, 0.5058|0.0270, 0.5267|0.0330, 0.5229|0.0405, 0.5430|0.0495, 0.5347|0.0270, 0.5543|0.0330, 0.4792|0.0270, 0.5012|0.0330, 0.4920|0.0180, 0.5135|0.0220, 0.5096|0.0270, 0.5303|0.0330, 0.5217|0.0180, 0.5419|0.0220, 0.5083|0.0405, 0.5291|0.0495, 0.5205|0.0270, 0.5407|0.0330, 0.5370|0.0405, 0.5566|0.0495, 0.5485|0.0270, 0.5675|0.0330\},$$

$$\bar{r}_3 = \{0.5027|0.0324, 0.5175|0.0324, 0.5510|0.0486, 0.5643|0.0486, 0.5439|0.0396, 0.5574|0.0396, 0.5882|0.0594, 0.6004|0.0594, 0.5150|0.0216, 0.5294|0.0216, 0.5621|0.0324, 0.5751|0.0324, 0.5552|0.0264, 0.5684|0.0264, 0.5984|0.0396, 0.6103|0.0396, 0.5120|0.0216, 0.5264|0.0216, 0.5594|0.0324, 0.5724|0.0324, 0.5524|0.0264, 0.5656|0.0264, 0.5958|0.0396, 0.6078|0.0396, 0.5240|0.0144, 0.5381|0.0144, 0.5702|0.0216, 0.5810|0.0216, 0.5634|0.0176, 0.5764|0.0176, 0.6058|0.0264, 0.6175|0.0264\}.$$

Then, the scores of the overall rating values ($\bar{r}_j$ $(j = 1, 2, 3)$) are $s(\bar{r}_1) = 0.5630$, $s(\bar{r}_2) = 0.5202$, and $s(\bar{r}_3) = 0.5672$, and so, the ranking order of the alternatives (x_j $(j = 1, 2, 3)$) is $x_3 \succ x_1 \succ x_2$. Thus, the best alternative is x_3.

Next, if we utilize the HPFWG operator (3) presented in Step 2, we get the overall rating values ($\bar{r}_j$) of the alternatives x_j $(j = 1, 2, 3)$:

$$\bar{r}_1 = \{0.4293|0.0243, 0.4393|0.0297, 0.4754|0.0162, 0.4866|0.0198, 0.4496|0.0297, 0.4601|0.0363, 0.4979|0.0198, 0.5096|0.0242, 0.4469|0.0243, 0.4573|0.0297, 0.4949|0.0162, 0.5065|0.0198, 0.4680|0.0297, 0.4790|0.0363, 0.5183|0.0198, 0.5305|0.0242, 0.4393|0.0365, 0.4496|0.0446, 0.4866|0.0243, 0.4980|0.0297, 0.4601|0.0446, 0.4709|0.0545, 0.5096|0.0297, 0.5216|0.0363, 0.4573|0.0365, 0.4680|0.0446, 0.5065|0.0243, 0.5184|0.0297, 0.4790|0.0446, 0.4902|0.0545, 0.5305|0.0297, 0.5429|0.0363\},$$

$$\bar{r}_2 = \{0.3995|0.0270, 0.4089|0.0330, 0.4425|0.0180, 0.4529|0.0220, 0.4243|0.0270, 0.4342|0.0330, 0.4699|0.0180, 0.4809|0.0220, 0.4120|0.0405, 0.4217|0.0495, 0.4563|0.0270, 0.4670|0.0330, 0.4375|0.0405, 0.4478|0.0495, 0.4846|0.0270, 0.4960|0.0330, 0.4117|0.0270, 0.4214|0.0330, 0.4560|0.0180, 0.4667|0.0220, 0.4373|0.0270, 0.4475|0.0330, 0.4843|0.0180, 0.4956|0.0220, 0.4246|0.0405, 0.4345|0.0495, 0.4703|0.0270, 0.4813|0.0330, 0.4509|0.0405, 0.4615|0.0495, 0.4994|0.0270, 0.5111|0.0330\},$$

$$\bar{r}_3 = \{0.3699|0.0324, 0.3812|0.0324, 0.3792|0.0486, 0.3908|0.0486, 0.3874|0.0396, 0.3992|0.0396, 0.3972|0.0594, 0.4093|0.0594, 0.4097|0.0216, 0.4222|0.0216, 0.4200|0.0324, 0.4329|0.0324, 0.4291|0.0264, 0.4422|0.0264, 0.4399|0.0396, 0.4534|0.0396, 0.3994|0.0216, 0.4116|0.0216, 0.4095|0.0324, 0.4220|0.0324, 0.4183|0.0264, 0.4311|0.0264, 0.4289|0.0396, 0.4420|0.0396, 0.4423|0.0144, 0.4559|0.0144, 0.4535|0.0216, 0.4674|0.0216, 0.4633|0.0176, 0.4774|0.0176, 0.4750|0.0264, 0.4895|0.0264\}.$$

Then, the scores of the overall rating values ($\bar{r}_j$ $(j = 1, 2, 3)$) are $s(\bar{r}_1) = 0.4817$, $s(\bar{r}_2) = 0.4501$, and $s(\bar{r}_3) = 0.4210$, and so the ranking order of the alternatives (x_j $(j = 1, 2, 3)$) is $x_1 \succ x_2 \succ x_3$. Thus, the best alternative is x_1.

The relative comparison of the methods using different operators proposed by Xu and Zhou [27] is shown in Table 7. From Table 7, we can see that the obtained overall rating values of the alternatives are different with each of the four operators, respectively, and then, the ranking orders of the alternatives also are different. Each of the methods using different hesitant probabilistic fuzzy operators has its advantages and disadvantages, and none of them always perform better than the others in any situation. It depends on how we look at things, and not on how they are themselves.

Table 7. Comparison of overall rating values and ranking orders of alternatives.

Aggregation Operator	Overall Rating Values	Ranking Orders
HPFWA operator [27]	$s(\bar{r}_1) = 0.5630$, $s(\bar{r}_2) = 0.5202$, $s(\bar{r}_3) = 0.5672$	$x_3 \succ x_1 \succ x_2$
HPFWG operator [27]	$s(\bar{r}_1) = 0.4817$, $s(\bar{r}_2) = 0.4501$, $s(\bar{r}_3) = 0.4210$	$x_1 \succ x_2 \succ x_3$
HPFEWA operator	$s(\bar{r}_1) = 0.5533$, $s(\bar{r}_2) = 0.5110$, $s(\bar{r}_3) = 0.5473$	$x_1 \succ x_3 \succ x_2$
HPFEWG operator	$s(\bar{r}_1) = 0.4968$, $s(\bar{r}_2) = 0.4621$, $s(\bar{r}_3) = 0.4429$	$x_1 \succ x_2 \succ x_3$

Consequently, the use of different hesitant probabilistic fuzzy aggregation operators reflects the decision maker's pessimistic (or optimistic) attribute. For example, the proposed HPFEWA operator shows that the decision maker has a more pessimistic attribute than the HPFWA operator [27], and the proposed HPFEWG operator shows that the decision maker has a more optimistic attribute than the HPFWG operator [27] in the aggregation process.

5. Conclusions

The hesitant probabilistic fuzzy MADM is an important research topic in HPFS theory and decision science with uncertain information. Information aggregation is one of the core issues. Based on the Einstein operational rules of HPFEs, in this paper, we developed a series of hesitant probabilistic fuzzy Einstein aggregation operators, including the HPFEWA, HPFEWG, HPFEOWA, HPFEOWG, HPFEHA, and HPFEHG operators. Some basic properties of the proposed aggregation operators, such as boundedness and monotonicity, and the relationships between them were investigated. We compared the proposed operators with the existing hesitant probabilistic fuzzy aggregation operators proposed by Xu and Zhou [27] and presented corresponding relations. These proposed hesitant probabilistic Einstein aggregation operators provide a fine supplement to the existing work on HPFSs. Based on the HPFEWA and HPFEWG operators, a new method for MADM was developed in hesitant probabilistic fuzzy environments. A practical example was provided to illustrate the hesitant probabilistic fuzzy MADM process. Through a comparison between the proposed method with the previously proposed hesitant probabilistic fuzzy MADM method [27], we showed some advantages of the proposed hesitant probabilistic fuzzy MADM method.

This paper only considered decision makers with equl weights in the decision making process, but further studies on unequal weights are needed. Moreover, research using other operations, such as Hamacher and Frank t-conoms and t-norms instead of the Einstein t-conorm and t-norm, should be discussed in future studies.

Author Contributions: J.H.P. drafted the initial manuscript and conceived the MADM framework. M.J.S. provided the relevant literature review and illustrated example. J.H.P. and Y.K.P. revised the manuscript and analyzed the data.

Funding: This work was supported by a Research Grant from Pukyong National University (2017).

Conflicts of Interest: The authors declare no conflict of interest.

References

1. Milosavljevića, M.; Bursaća, M.; Tričkovića, G. Selection of the railroad container terminal in Serbia based on multi criteria decision-making methods. *Decis. Mak. Appl. Manag. Eng.* **2018**, *1*. [CrossRef]
2. Badi, I.; Ballem, M. Supplier selection using the rough BWM-MAIRCA model: A case study in pharmaceutical supplying in Libya. *Decis. Maki. Appl. Manag. Eng.* **2018**, *1*. [CrossRef]
3. Jagannath Roy, J.; Krishnedu Adhikary, K.; Samarjit Kar, S.; Dragan Pamucar, D. A rough strength relational DEMATEL model for analysing the key success factors of hospital service quality. *Decis. Mak. Appl. Manag. Eng.* **2018**, *1*, 121–142.
4. Zadeh, L.A. Information and control. *Fuzzy Sets* **1965**, *8*, 338–353.
5. Zhou, W. An accurate method for determining hesitant fuzzy aggregation operator weights and its application to project investment. *Int. J. Intell. Syst.* **2014**, *29*, 668–686. [CrossRef]
6. Zhou, W.; Xu, Z.S. Generalized asymmetric linguistic term set and its application to qualitative decision making involving risk appetities. *Eur. J. Oper. Res.* **2016**, *254*, 610–621. [CrossRef]
7. Dong, Y.C.; Chen, X.; Herrera, F. Minimizing adjusted simple terms in the consensus reaching process with hesitant linguistic assessments in group decision making. *Inf. Sci.* **2015**, *297*, 95–117. [CrossRef]
8. Gou, X.J.; Xu, Z.S. Exponential operations for intuitionistic fuzzy numbers and interval numbers in multi-attribute decision making. *Fuzzy Optim. Decis. Making* **2016**. [CrossRef]
9. Atanassov, K. Intuitionistic fuzzy sets. *Fuzzy Sets Syst.* **1986**, *20*, 87–96. [CrossRef]
10. Dubois, D.; Prade, H. *Fuzzy Sets and Systems: Theory and Applications*; Academic Press: Cambridge, MA, USA, 1980.
11. Miyamoto, S. Fuzzy multisets and their generalizations. In Proceedings of the International Conference on Membrane Computing, Curtea de Arges, Romania, 21–25 August 2000, pp. 225–236.
12. Torra, V. Hesitant fuzzy sets. *Int. J. Intell. Syst.* **2010**, *25*, 529-539. [CrossRef]
13. Xia, M.M.; Xu, Z.S. Hesitant fuzzy information aggregation in decision making. *Int. J. Approx. Reason.* **2011**, *52*, 395–407. [CrossRef]
14. Li, D.Q.; Zeng, W.Y.; Zhao, Y.B. Note on distance measure of hesitant fuzzy sets. *Inf. Sci.* **2015**, *321*, 103–115. [CrossRef]
15. Meng, F.Y.; Chen, X.H. Correlation coefficients of hesitant fuzzy sets and their application based on fuzzy measures. *Cognit. Comput.* **2015**, *7*, 445–463. [CrossRef]
16. Bedregal, B.; Reiser, R.; Bustince, H.; Lopez-Molina, C.; Torra, V. Aggregation functions for typical hesitant fuzzy elements and the cation of automorphisms. *Inf. Sci.* **2014**, *255*, 82–99. [CrossRef]
17. Xu, Z.S.; Cai, X.Q. Recent advances in intuitionistic fuzzy information aggregation. *Fuzzy Optim. Decis. Mak.* **2010**, *9*, 359–381. [CrossRef]
18. Zhu, B.; Xu, Z.S.; Xia, M.M. Hesitant fuzzy geometric Bonferroni means. *Inf. Sci.* **2012**, *205*, 72–85. [CrossRef]
19. Xia, M.M.; Xu, Z.S.; Chen, N. Some hesitant fuzzy aggregation operators with their application in group decision making. *Group Dec. Negoit.* **2013**, *22*, 259–279. [CrossRef]
20. Zhang, Z.M. Hesitant fuzzy power aggregation operators and their application to multiple attribute group decision making. *Inf. Sci.* **2013**, *234*, 150–181. [CrossRef]
21. Yu, D.J. Some hesitant fuzzy infromation aggregation operators based on Einstein operational laws. *Int. J. Intell. Syst.* **2014**, *29*, 320-340. [CrossRef]
22. Zhou, X.; Li, Q. Multiple attribute decision making based on hesitant fuzzy Einstein geometric aggregation operators. *J. App. Math.* **2014**, *2014*. [CrossRef]
23. Zhou, X.; Li, Q. Generalized hesitant fuzzy prioritized Einstein aggregation operators and their application in group decision making. *Int. J. Fuzzy. Syst.* **2014**, *16*, 303-316.
24. Liao, H.C.; Xu, Z.S. A VIKOR-based method for hesitant fuzzy multi-criteria decision making. *Fuzz. Optim. Decis. Mak.* **2016**, *12*, 372–392. [CrossRef]
25. Rodriguez, R.M.; Martínez, L.; Torra, V.; Xu, Z.S.; Herrera, F. Hesitant fuzzy sets: State of the art and future directions. *Int. J. Intell. Syst.* **2014**, *29*, 495–524. [CrossRef]
26. Zhou, W.; Xu, Z.S. Asmmetric hesitant fuzzy sigmoid preference relations in analytic hierarchy process. *Inf. Sci.* **2016**, *358*, 191–207. [CrossRef]
27. Xu, Z.S.; Zhou, W. Consensus building with a group of decision makers under the hesitant probabilistic fuzzy environment. *Fuzz. Optim. Decis. Mak.* **2017**, *16*, 481–503. [CrossRef]

28. Zhang, Z.; Wu, C. Weighted hesitant fuzzy sets and their application to multi-criteria decision making. *Br. J. Math. Comput. Sci.* **2014**, *4*, 1091–1123. [CrossRef]
29. Bedregal, B.; Beliakov, G.; Bustine, H.; Calvo, T.; Mesiar, R.; Patermain, D. A class of fuzzy multisets with a fixed number of memberships. *Inf. Sci.* **2012**, *189*, 1–17. [CrossRef]
30. Wang, R.; Li, Y.. Picture hesitant fuzzy set and its application to multiple criteria decision-making. *Symmetry* **2018**, *10*, 295–324. [CrossRef]
31. Chen, N.; Xu, Z.S.; Xia, M.M. Interval-valued hesitant preference relations and their applications to group decision making. *Knowl. Based Syst.* **2013**, *37*, 528–540. [CrossRef]
32. Pathinathan, T.; Savarimuthu, J.S. Trapezoidal hesitant fuzzy multi-attribute decision making based on TOPSIS. *Int. Arch. App. Sci. Technol.* **2015**, *6*, 39–49.
33. Yu, D. Triangular hesitant fuzzy set and its application to teaching quality evaluation. *J. Inform. Comput. Sci.* **2013**, *10*, 1925–1934. [CrossRef]
34. Mahmood, T.; Mehmood F.; Khan, Q. Cubic hesitant fuzzy sets and their applications to multi criteria decision making. *Int. J. Algebra Stat.* **2016**, *5*, 19–51. [CrossRef]
35. Klement, E.; Mesiar, R.; Pap, E. Book Review: "Triangular Norms". *Int. J. Uncertain. Fuzz. Knowl. Based Syst.* **2003**, *11*, 257–259. [CrossRef]
36. Wang, W.Z.; Liu, X.W. Intuitionistic fuzzy geometric aggregation operators based on Einstein operations. *Int. J. Intell. Syst.* **2011**, *26*, 1049–1075. [CrossRef]
37. Wang, W.Z.; Liu, X.W. Intuitionistic fuzzy infromation aggregation using Einstein operations. *IEEE Trans. Fuzzy Syst.* **2012**, *20*, 923–938. [CrossRef]
38. Zhao, X.; Wei, G. Some intuitionistic fuzzy Einstein hybrid aggregation operators and their application to multiple attribute decision making. *Knowl. Based Syst.* **2013**, *37*, 472–479. [CrossRef]
39. Yager, R.R. On ordered weighted averaging aggregation operators in multi-criteria decision making. *IEEE Trans. Syst. Man Cybern.* **1988**, *18*, 183–190. [CrossRef]
40. Xu, Z.S. An overview of methods for determining OWA weights. *Int. J. Intell. Syst.* **2005**, *20*, 843–865. [CrossRef]
41. Xu, Z.S. On consistency of the weighted geometric mean complex judgement matrix in AHP. *Eur. J. Oper. Res.* **2000**, *126*, 683–687. [CrossRef]
42. Torra, V.; Narukawa, Y. *Modeling Decisions: Information Fusion and Aggregation Operators*; Springer: Berlin/Heidelberg, Germany, 2007.
43. Ölçer, A.I.; Odabaşi, A.Y. A new fuzzy multiple attribute group decision making methodology and its application to propulsion/manoeuvring system selectio problem. *Eur. J. Oper. Res.* **2005**, *166*, 93–114. [CrossRef]

information

MDPI

Article

Development of an ANFIS Model for the Optimization of a Queuing System in Warehouses

Mirko Stojčić [1], Dragan Pamučar [2,*], Eldina Mahmutagić [3] and Željko Stević [1]

[1] Faculty of Transport and Traffic Engineering, University of East Sarajevo, Vojvode Mišića 52, 74000 Doboj, Bosnia and Herzegovina; mirkostojcic1@hotmail.com (M.S.); zeljkostevic88@yahoo.com (Ž.S.)

[2] Department of Logistics, Military academy, University of Defence in Belgrade, Pavla Jurisica Sturma 33, 11000 Belgrade, Serbia

[3] NERI d.o.o. Ljesnica bb 74250 Maglaj, Bosnia and Herzegovina; mahmutagiceldina24@gmail.com

* Correspondence: dpamucar@gmail.com; Tel: +381113603932

Received: 3 September 2018; Accepted: 21 September 2018; Published: 22 September 2018

Abstract: Queuing systems (QS) represent everyday life in all business and economic systems. On the one hand, and there is a tendency for their time and cost optimization, but on the other hand, they have not been sufficiently explored. This especially applies to logistics systems, where a large number of transportation and storage units appear. Therefore, the aim of this paper is to develop an ANFIS (Adaptive neuro-fuzzy inference system) model in a warehouse system with two servers for defining QS optimization parameters. The research was conducted in a company for the manufacturing of brown paper located in the territory of Bosnia and Herzegovina, which represents a significant share of the total export production of the country. In this paper, the optimization criterion is the time spent in the system, which is important both from the aspect of all customers of the system, and from that of the owner of the company. The time criterion directly affects the efficiency of the system, but also the overall costs that this system causes. The developed ANFIS model was compared with a mathematical model through a sensitivity analysis. The mathematical model showed outstanding results, which justifies its development and application.

Keywords: ANFIS; warehouse; queuing systems; logistics

1. Introduction

In the daily performance of various activities and processes, logistics, as an integral and indispensable part of every business system, plays a very important role. It is necessary to rationalize the activities and processes that can significantly affect a competitive position of a company. A warehouse, as an individual logistics subsystem, together with transportation, represent the biggest causes of logistics costs, and there is a constant search for potential places of savings in these subsystems. Long ago, a warehouse was just a place used to separate surplus products, while its function today is completely different. Compared to the former static function, today's warehouses represent a dynamic system in which the movement of goods is dominant. Therefore, in this paper, the emphasis is on the storage system of Natron-Hayat company, which is one of the largest companies in Bosnia and Herzegovina; this is sufficiently proved by the fact that it is one of the top five exporters of Bosnia and Herzegovina [1]. The current storage system of the company is decentralized, whereby each manufacturing facility has its own warehouse. Under such circumstances, there is the accumulation of demands for loading goods into vehicles and queuing, which again causes certain costs. In order to be successful in conditions of great competition, one of the most important segments is to satisfy the needs of customers, which is an integral part of a supply chain. Thus, it is necessary, according to Stević et al. [2], to optimize from the perspective of all participants in the complete supply chain. Today, customers pay more attention to the time they spend queuing; this time affects their decision about whether they will use the service again. This paper considers and analyzes the storage system of Natron-Hayat, a working group-warehouse paper machine (PM4), where arrivals of transportation means, queues, and service time depend on a number of factors. Throughout the research carried out in this paper, data on the arrivals of transportation means, which are registered at the weighing scale for loading in the PM4 warehouse, and the loading time for each vehicle, have been collected. Taking into account the capacities of all manufacturing machines in the company, the calculation of the basic parameters of the queuing system was conducted only for the PM4 warehouse, while the warehouses of the other manufacturing facilities were not taken into consideration at the moment. The PM4 storage system was into consideration, since it is a manufacturing machine with the largest capacity in the company; the company's operations largely depends on its work, as can be seen in more detail in [3].

This paper has several goals. The first is to determine the state of the queuing systems of the company i.e., warehouse paper machine (PM4), which is the object of the research, by observing the system on a monthly basis. The second goal relates to the calculation of the system's indicators by using a mathematical model. The third and most important goal of this research, which also represents a research contribution, is the development of an ANFIS model with three input variables: the inter-arrival time of trucks, the cumulative arrival time, and the service time (transloading-manipulative operations). The developed model provides meaningful information to all participants in the complete queuing system about the time in the system and the possibilities of its deviation, which can play an important role in planning and modeling the most important processes and business activities.

The proposed ANFIS model implies the union of all the advantages that the two artificial intelligence areas possess; the most important is the possibility of adaptation or learning from the example, and an approximate reasoning. Previous works provide an insight into the use of ANFIS in the field of traffic and transport, but in this paper, it is used in combination with the principles of queuing theory. Also, queuing theory is mainly based on analytical optimization models whose resolution can be complex. Numerous studies show the low sensitivity of ANFIS to inaccurate and incomplete input data, and its good ability to model non-linear dependencies, which is a characteristic of queuing systems.

In addition to introductory considerations where the basic reasons for the research are presented, the work is structured in five other sections. The second section provides a review of the literature referring to queuing systems and their optimization in different areas, as well as the application of ANFIS. Also, this section provides a literature review referring to multi-criteria decision-making methods (MCDM). The third section presents the methods in which a complete research algorithm is shown, by recognizing a need to perform it to the final goal. In addition, the basic settings of the QS and ANFIS models have been given. The fourth section provides a case study which consists of data collection and creation, and training of the model. Also, in this section the statistical inference of the distributions of input flow into the system and service time was determined. The fifth section contains the results and discussion. The sixth section provides a sensitivity analysis related to the QS mathematical model. The paper ends with conclusions in which the directions for future research are given.

2. Literature Review

This section is divided into three parts. The first is related to the application of the queuing systems theory and associated models in traffic and transportation. The second is oriented towards ANFIS models in this specific field, while the third refers to the overview of the application of multi-criteria analysis methods in traffic and transport.

2.1. Models of Queuing System Theory in Traffic And Transportation

Queues can occur wherever there is a need for the service of a large number of customers and the number of servers is limited. We encounter such situations in our lives every day, and taking into account that traffic and transportation affect the lives of every individual, queuing systems theory has a significant potential for application in this field. Every day, the number of transportation means is increasing on the streets, while the existing infrastructure, i.e., roads, does not undergo such a rapid expansion; consequently, traffic jams, waiting times on roads, travel time, costs increase, etc. In addition to these negative phenomena, from the ecological aspect, environmental pollution is also increasing. In order to optimize traffic flows, according to Guerrouahane et al. [4] and Raheja [5], exponential distribution of the inter-arrival time/general distribution of the service time/number of servers/system capacity systems are used for modeling. In [6], it is stated that highway flows are modeled as exponential distribution of the inter-arrival time/general distribution of the service time/one server, when there is no congestion, and as general distribution of the inter-arrival time/general distribution of the service time/number of servers in cases of congestion. The study [7] describes several different models for a traffic flow analysis, including exponential distribution of the inter-arrival time/exponential distribution of the service time/one server, in addition to those previously mentioned. Although queuing system theory is mainly used to model traffic flow on highways, intersections in large cities are bottlenecks, and can be modeled as exponential distributions of the inter-arrival time/exponential distribution of the service time/one server systems [8–10]. No matter which model is used, they all include a certain mathematical apparatus, i.e., mathematical models that are selected based on the system functioning (input flow, service flow). Therefore, not all are equally adequate for a particular system. Another example of applying the queuing system theory in traffic and transportation relates to the routing of transportation means in real-time. According to Chen [11], two strategies are given: FCFS (First Come First Served) and Median Repositioning, where the second one shows better results. The modeling of queues is also applicable to supply chains, where each sub-process is a queuing system. It is particularly important to optimize green supply chains by reducing fuel consumption, transportation time, and waiting time [12,13]. Warehouses can also be modeled using queuing theory. Some of the requirements that are imposed on optimization are shorter time of the implementation of operations, work with larger number of units of goods, the provision of the required quality of service, minimum costs. In addition, the application of queuing systems theory can be used to determine the size and capacity of a warehouse, and therefore, the necessary equipment within it [14,15]. Since the automation of warehouses is a trend nowadays, the AS/RS system (Automated Storage and Retrieval System) sets complex synchronization requirements that are resolved by queuing theory [16].

2.2. ANFIS Models in Traffic And Transportation

Intelligent transportation systems involve the application of various technologies, including artificial neural networks and fuzzy logic, i.e., neuro-fuzzy systems. Some of the basic objectives are to increase passenger safety, optimize routes, optimize the choice of means of transportation, reduce travel time, reduce costs, reduce traffic jams, waiting, etc. Using ANFIS models, a prediction of traffic flows in intelligent transportation systems can be performed. According to Bao-ping and Zeng-Qiang [17], the prediction model consists of 104 parameters that are adjusted during a training process. In order to increase safety, in research [18], ANFIS is used to assess the impact of intelligent transportation system technologies—such as video surveillance and drowsiness warnings—on the number of fatalities due to traffic accidents. Similarly, in [19], critical points on a road in rural areas are identified on the basis of collected data on traffic accidents. Traffic control at intersections with traffic lights can also be carried out using ANFIS models [20,21]. This involves reading out external data on the current state of the intersection, and forwarding them to the model that processes them and reacts in accordance with the learned rules [22]. From an ecological point of view, it is possible to estimate the noise level, as indicated in [23]. Traffic flow density, vehicle speed, and the noise level of horns can be taken as input independent variables. The selection of an optimum transportation route using the ANFIS model, based on the criteria specified by the dispatcher, is considered in [24], while [25] deals with a choice of an optimum mode of transportation. Warehousing is an important part of the entire supply chain. It can be said that the efficiency of a warehouse affects the overall efficiency of the chain, and therefore, requires intelligent optimization solutions. [26] deals with determining the number of forklifts in a warehouse which are required for loading goods using an ANFIS model, based on a given number of pallets and time available. Another example of the implementation of expert knowledge in an ANFIS model for using forklifts is presented in [27].

2.3. Methods of Multi-Criteria Decision-Making in Traffic And Transportation

Multi-criteria decision-making can be used as an adequate tool for making valid decisions. In [28], the EMDS (Ecosystem Management Decision Support System) and SADfLOR (Web-Based Forest and Natural Resources Decision Support System) imply the integration of multiple approaches to determine optimal bundles of ecosystem services. In traffic and transport, the role of MCDM (Multi-Criteria Decision Making) is of paramount importance. For the determination of optimal locations, spatial information obtained by the GIS (Geographic Information System) is often used with multi-criteria decision making [29]. Karczmarczyk et al. [30] represents the application of the novel method, COMET (Characteristic Objects Method) to determine the best model of the electric car for sustainable city transport with respect to increasing pollution in cities. The same method was used in [31] for the selection of the best scenarios for the transport of dangerous goods. With the aforementioned COMET method, the theory of the fuzzy sets for the modeling of imprecise data is used. The supply chain involves a large number of participants, and with the correct choice of suppliers, at the initial stage, good conditions for optimizing the entire process are created. Stevic et al. [2] developed a new approach: Rough EDAS (Evaluation based on Distance from Average Solution), Rough COPRAS (Complex Proportional Assessment) and Rough MULTIMOORA (Multi-Objective Optimization by Ratio Analysis Plus the Full Multiplicative Form) to solve the problem of correct choice of suppliers. In [32], a model of dynamic or temporal choice of a supplier is proposed using multi-criteria decision making. The observed neuro-fuzzy system, ANFIS, finds its application in the field of decision-making on the basis of several criteria. Khalili-Damghani et al. [33] divides the process of selecting suppliers in two phases, the first of which involves the application of ANFIS to determine the overall usefulness of the supplier based on expert knowledge. According to Torquaybade [34], in the supply chain optimization, ANFIS has the role of assessing the performance of each of the Pull Control Policies based on the input variables. A multi-criteria approach was also applied in [35] for the identification of priority black spots in order to increase the safety in traffic. The applicability in the field of traffic engineering of the MCDM methods is also confirmed in paper [36], where it is used for the evaluation and selection of roundabouts in an urban area.

3. Methods

The first step of the research is data collection, i.e., the values of selected variables that will be used first for statistical analyses, and then for the creation and training of the model based on fuzzy logic and artificial neural networks. The basic method used for the realization of the research is modeling. In addition, a statistical method is used, as well as theoretical analysis explaining basic concepts and principles of queuing systems and the ANFIS model. Used software packages, such as MATLAB, Minitab, and EasyFit, make the application of these methods much easier.

Figure 1 shows the proposed model in the study. It consists of a total of three phases and 11 steps. The first phase includes four steps: the first relates to the recognition of the need to conduct research that will help both customers (transportation companies) and the company's management increase the efficiency of their business. The second step is the formation of a team and the distribution of tasks, as well as counseling with staff at a tactical level on how to interact with the system, i.e. collect data, which is the third step of this phase. In the final step of the first phase, the sorting of data collected, and their processing, are performed. The second phase consists of three steps. The first two relate to determining the distribution of the input flow and the flow of service, respectively, while the third involves the formation of a mathematical model in the Minitab software for the calculation of the basic parameters of the system. The third phase includes the development of the ANFIS model, discussion of the obtained results, and a comparison with a mathematical model.

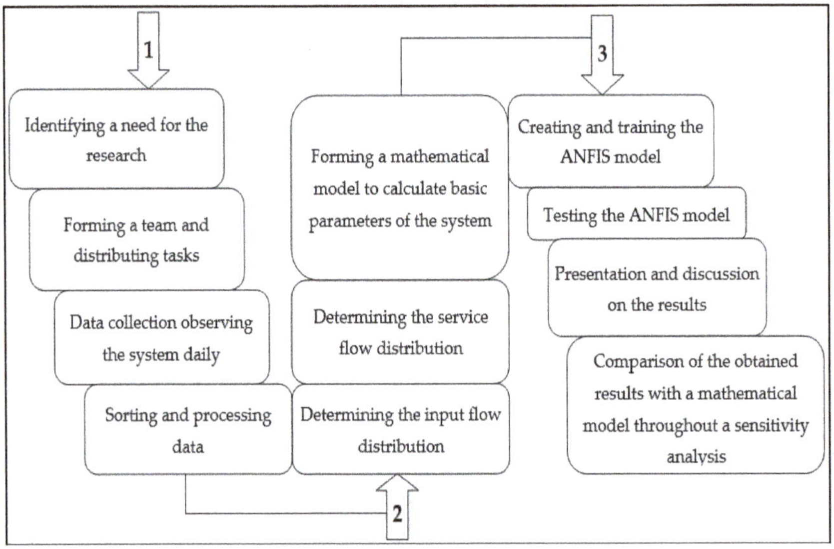

Figure 1. Diagram of the research flow.

3.1. Basic Principles of Queuing Systems Theory

Today, queuing theory has a very wide range of applications in many branches of human activity where customers come into a system, by some mathematical distribution, due to a particular service, and in the case of occupied servers, form one or more queues. Upon the completion of service, where its time also corresponds to some distribution, the customer leaves the system. Such examples can be seen daily in traffic, logistics systems, banks, post offices, in telecommunications traffic, at a gas station, etc. Naturally, situations that are more complex are possible when a customer passes throughout a network of interconnected queuing systems. The task of queuing systems theory is to explain and model the behavior of such systems using a mathematical apparatus. In addition to other methods, it is widely applicable to operational research [37]. Modeling a queuing system enables the analysis and optimization of its performance.

As the basic features of the queuing systems model, the following can be identified: the input process, service mechanism, and queue discipline [38]. According to Maragatha and Srinivasan [39], every model can be described using the following features:

- The distribution of inter-arrival time; this most often corresponds to a Poisson, exponential or general distribution. Arrivals can be individual or in groups [40].
- The distribution of service time: exponential, hyper-exponential, hypo-exponential, constant, general.
- The number of servers can be one or more.
- The length of queue can be precisely defined or infinite. In case of arrival when the queue capacity is maximally filled, the costumer is denied, which is known as 'balking'.
- System capacity implies the maximum number of customers in the system, being served or in the queue.

System disciplines:

- FIFO (First in, First out)—in the order of arrival,
- LIFO (Last in, First out)—a customer that comes last will be served first,
- Random Service—customers are served in random order,
- Round Robin—a customer gets a time slot within which he/she will be served. If the service is not completed, the customer returns to the beginning of the queue,
- Priority Disciplines—the order of customer service is determined according to the priority that each one receives [37].

The Kendall notation uses these six features to describe the queuing system:

$$A/B/m/K/n/D$$

where

- A—the distribution of the inter-arrival time,
- B—the distribution of the service time. Positions A and B can be replaced by M (Markov processes, exponential distribution); D (deterministic distribution); E (Erlang distribution); H (hyper-exponential distribution); G (general distribution),
- m—the number of servers,
- K—system capacity,
- n—population size,
- D—queuing discipline [40]. Unless stated, it is assumed that it is the FIFO [41].

The most common and simplest queuing systems are of M/M/m type.

3.2. Adaptive Neuro-Fuzzy Inference Model

Unlike biological neural networks, artificial neural networks represent an attempt to model the human brain through modern computing. They consist of a number of process elements, or artificial neurons, which are analogous to the brain, in which the basic elements are nerve cells. Artificial neurons, as well as nerve cells, are characterized by parallel work in the processing of various types of information [42]. Their basic feature is the ability to learn, which means that it is necessary to first train the network to efficiently perform tasks such as recognizing shapes, images, speech, function approximation, prediction, optimization, data clustering, processing inaccurate and incomplete data, etc. Accordingly, the basic task of an artificial neural network is to combine different inputs, and to process and forward signals to one or more outputs. There are various types of artificial neural networks depending on the number of neurons, i.e., layers, network training methods, the way to transmit signals throughout the network, etc.

Fuzzy technologies allow the computer to work with uncertainties, thereby achieving a similarity with the human way of thinking. Fuzzy logic is an extension of classical logic in which variables can have only two values: correct (1) and incorrect (0). In this way, variables can occupy any real value between 0 and 1. Fuzzy sets are basic elements for presenting and processing unclear things and uncertainties in fuzzy logic, and they are mathematically presented by membership functions. The inference system in fuzzy logic implies defined membership functions of individual variables and inference rules, which connect input variables with an inference; they are called IF-THEN rules.

The systems that integrate the principles of artificial neural networks and fuzzy logic are called neuro-fuzzy systems. They use a learning ability of artificial neural networks based on training data in order to adapt the forms of membership fuzzy functions and inference fuzzy rules. In this way, in one system, the advantages of logical inference and learning are combined. One of the most commonly used neuro-fuzzy systems is ANFIS (Adaptive Neuro-Fuzzy Inference System). ANFIS is a multilayer neural network that, based on data (input-output vector) for training, provides a certain value of an output variable for certain inputs. An important feature is that ANFIS can effectively model nonlinear connections of inputs and outputs [43]. ANFIS training is based on the application of an algorithm of error propagation backward, either alone or in combination with the method of least squared error, i.e., hybrid algorithm [44]. ANFIS uses the Takagi-Sugeno method of inference, and a typical fuzzy rule, assuming two inputs (x and y) and a logical AND operation, can be written as follows:

IF x is A **AND** y is B, **THEN** z = f(x,y)

A and B denote fuzzy sets of input variables x and y, while z is an output function.

The ANFIS structure consists of five layers, as shown in Figure 2. The nodes of the first layer define fuzzy sets, i.e., membership fuzzy functions corresponding to input variables. This layer is often called a fuzzification layer, because it determines the membership degree of the value of a variable to a particular fuzzy set [45]. The nodes of this layer are adaptive, which means that their parameters are adjusted during a training period [44]. The first-layer nodes that represent the membership functions of the input variable X can be defined as $\mu_{A_j}(x)$, where j (j = 1 ,..., 2) denotes the number of membership functions [46].

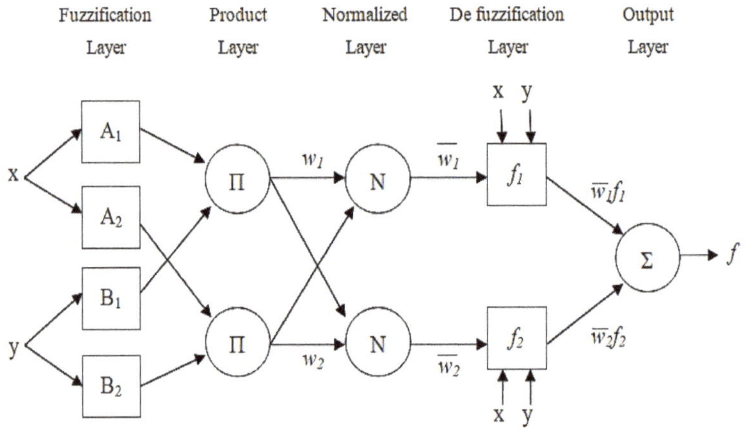

Figure 2. The ANFIS structure with two inputs.

The second-layer nodes are fixed and perform an operation of multiplying the input signals (operation AND). The *ith* neuron has the output of the form: $w_i = \mu_{A_i}(x) \times \mu_{B_i}(y)$. The output of the second layer is equal to the minimum value of the two inputs [47].

The third layer normalizes the values obtained at the output of the nodes of the second hidden layer. In the case shown in Figure 2, with two nodes in the second layer, the normalized value at the output of the *ith* node of the third hidden layer has the following mathematical form:

$$\overline{w}_i = \frac{w_i}{w_1 + w_2} \quad (1)$$

Each node of the fourth layer is an adaptive node with the function it completes, which can be written as follows [48]:

$$\overline{w}_i f_i = \overline{w}_i (p_i x + q_i y + r_i) \quad (2)$$

where p_i, q_i and r_i are inference parameters. The fifth layer calculates the output as a sum of all input signals:

$$f = \sum_i \overline{w}_i f_i = \frac{\sum_i w_i f_i}{\sum_i w_i} \quad (3)$$

The set of fuzzy inference rules that apply to the structure given in Figure 2 consists of two rules:

$$\text{IF } x \text{ is } A_1 \text{ AND } y \text{ is } B_1, \text{ THEN } f_1 = p_1x + q_1y + r_1$$

$$\text{IF } x \text{ is } A_2 \text{ AND } y \text{ is } B_2, \text{ THEN } f_2 = p_2x + q_2y + r_2$$

ANFIS is most often trained with a hybrid algorithm. It requires two passes through the network in each epoch. In the forward pass, a method of least squares is used to modify the parameters of the linear functions of rule inferences (layer 4) [34]. When going backward, the parameters of fuzzy membership functions of input variables (layer 1) are modified by the algorithm of error back-propagation.

4. Case Study

Figure 3 shows the basic processes of the QS at Natron-Hayat for the PM4 storage system. The road freight transportation means of different companies represent customers of the system. They enter the system with a certain intensity. Upon entering the company's property, they form a queue depending on the current service intensity. There are a total of two transloading fronts (servers) where forklifts are engaged in the loading of goods. The queue formed and servers represent the service system, while loaded vehicles are serviced customers.

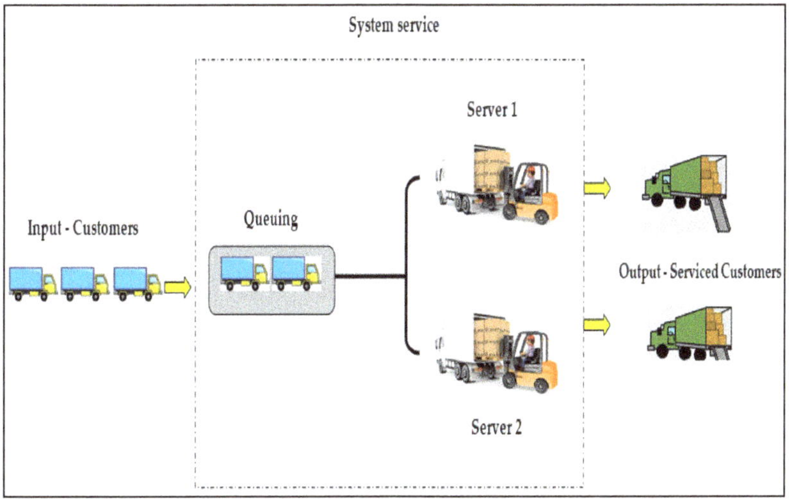

Figure 3. The basic process of the queuing system for the PM4 warehouse4.1. Data collection.

4.1. Data Collection

In order to model the queuing system of the PM4 working group of Natron-Hayat, it was necessary to collect data on the basic features of the system. The values of the following variables were monitored:

- the inter-arrival time of trucks,
- the cumulative arrival time since an initial time (for each day),
- the service time (transloading-manipulative operations) and
- the time in the system.

The first three listed variables represent the input variables of the ANFIS model that predicts the time spent in the system. Therefore, it is clear that time represents the output variable, taking into account the well-known fact that the time in queuing systems is one of the most important criteria for optimizing and modeling them.

The data collection period lasted 11 working days in two shifts of 8 h, based on which a monthly report was received for 22 working days, which means a total of 352 h. A total of 237 trucks entered the system. Out of the total set of data, the values of the seventh day of monitoring are excluded, after 2 p.m. to the end of working hours. The reason is the emergence of unusual and extremely high values of time spent in the system during the arrival of trucks at the end of the first and the beginning of the second shift. Such values adversely affect the performance of the ANFIS model. Figure 4 graphically shows the time spent in the system and the cumulative arrival time for each truck, as well as the given deviations occurring from the arrivals of the 65th to the 76th truck.

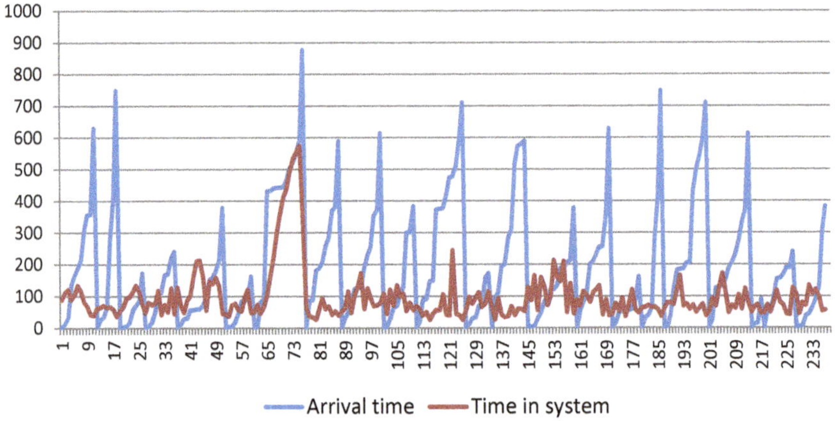

Figure 4. Time in the system and cumulative arrival time for 237 trucks.

In addition to the extreme deviations, the values of variables for the 122nd truck that entered the system due to the extreme value of the service time of 240 min were neglected. The time of its arrival in the system is in the 8th hour on the last day of data collection, so that period was also omitted. The final set of data, which is statistically analyzed and used to create the ANFIS model, is reduced to the time of 352 h, during which 224 trucks entered the system.

The examination of the input flow, i.e. the arrivals of trucks in the system, is essential for determining the distribution of the probabilities of inter-arrival time and the distribution of probabilities of the arrival of certain number of trucks at a given interval. Table 1 gives frequencies related to the number of trucks that arrived in a period of one hour. The biggest frequency is 214, when no truck entered the system during one hour. For a larger number of trucks that arrived in one hour, frequencies of hours were reduced, so in the end, the largest number of trucks that arrived during one hour was five, and with a frequency of two.

Based on Table 1, a statistical procedure is used to determine the distribution of the input flow. The EasyFit software provides graphical and tabular results for the procedure to determine the best fitting of data with a particular distribution. Figure 5 shows the Poisson distribution that best fits the input flow. The distribution parameter is $\lambda = 0.6747$, which represents the arrival intensity of the number of trucks within one hour.

Table 1. Frequencies of the input flow for a period of one hour.

Number of Trucks	Frequency
0	214
1	48
2	46
3	14
4	8
5	2

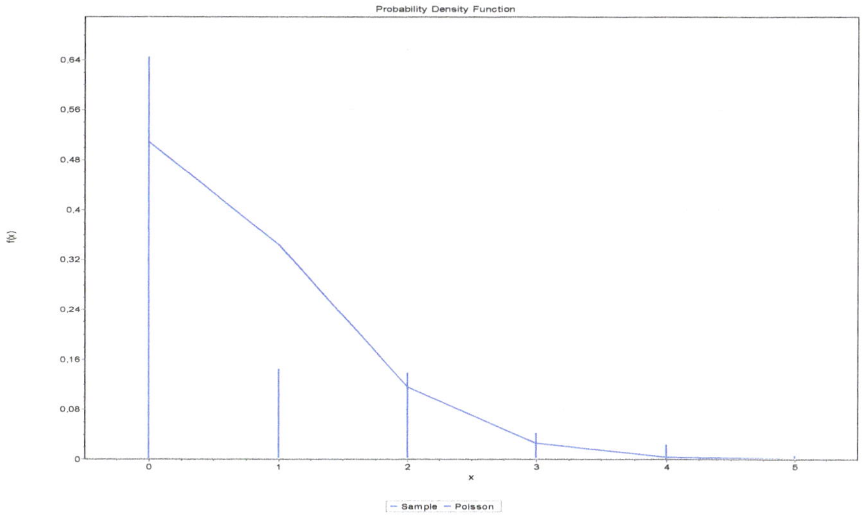

Figure 5. Poisson distribution of the input flow.

Table 2 gives an overview of the completed statistics of the Anderson Darling (AD) test for determining data fitting to a particular distribution. The Poisson distribution is ranked as the best in the Anderson Darling test with the statistic of 88.26. A total of 8 distributions are given, but it is possible to determine the fitting only for Poisson, Geometric, and D. Uniform.

Table 2. The results of statistical tests for determining the fitting of the input flow to a certain distribution.

Distribution	Anderson Darling	
	Statistic	Rank
Poisson	88.26	1
Geometric	112.79	2
D. Uniform	197.01	3
Bernoulli	No fit (data max > 1)	
Binomial	No fit	
Hyper-geometric	No fit	
Logarithmic	No fit (data min < 1)	
Neg. Binomial	No fit	

In order to determine the distribution of service time, it is necessary to divide the data into classes. Taking into account that the maximum value of service time is 95 min, and the minimum is 15 min, a division into 8 classes per 11 min of duration is performed. Table 3 shows the number of trucks that is served within a certain interval (class). The largest number of trucks is served within the class with the limits between 26 and 36 min, while no truck is served from 70 to 80 min.

Table 3. Frequencies of the service time.

Class Limits	Arithmetic Mean of Class-Interval	Frequency (Number of Trucks)
15–25	20	46
26–36	31	72
37–47	42	68
48–58	53	22
59–69	64	6
70–80	75	0
81–91	83	6
92–102	97	4

Figure 6 shows the Levy distribution curve that best fits the service time frequencies presented in Table 3. The probability density function of the shown distribution can be expressed as follows:

$$f(x) = \sqrt{\frac{\sigma}{\gamma}} \frac{\exp(-0,5\sigma/(x-\gamma))}{(x-\gamma)^{3/2}} \tag{4}$$

where the parameters, in this specific case, are $\sigma = 32.634$ and $\gamma = 0$. The average service time is 38.62 min, which means that the service intensity is $\mu = 1.55$ trucks/hour.

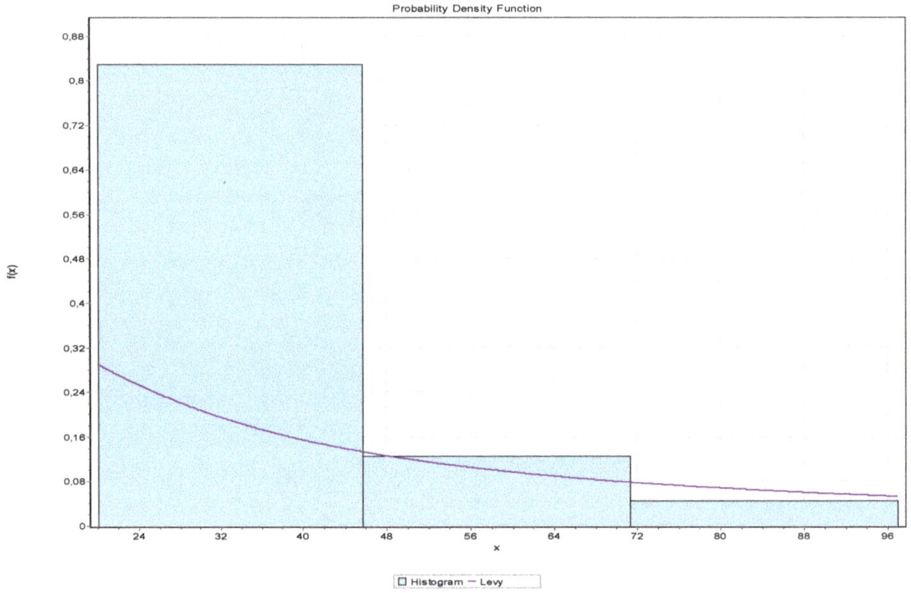

Figure 6. Distribution of the service time.

Table 4 presents the values of ranges and the statistics of the Anderson Darling test for testing the fitting of service times to certain distributions. Results are given for 8 distributions, and Levy is ranked as first, with a statistic of 1.5611.

Table 4. Results of statistical tests for determining the fitting of service times to a certain distribution.

Distribution	Anderson Darling	
	Statistic	**Rank**
Levy	1.5611	1
Levy (2P)	1.6041	2
Pareto 2	2.3671	3
Exponential	2.9128	4
Rayleigh	3.6386	5
Reciprocal	3.8271	6
Log-Logistic (3P)	4.2251	7
Fatigue Life (3P)	4.4674	8

As mentioned, we used the software Easyfit to obtain the statistical distributions. This software supports Kolmogorov-Smirnov (KS) and AD tests. We used the AD test, because it is better, and according to Engmann and Cousineau [49], it has two extra advantages over the KS test. First, it is especially sensitive to differences at the tails of distributions. Second, there is evidence that the AD test is more capable of detecting very small differences, even between large sample sizes. This is one of its main advantages in the field of engineering. Also, the KS test is less able to detect changes in asymmetry, requiring almost twice as many data compared to the AD test. Finally, this test, according to Stephens [50], is a good all-purpose test. The AD test is also used in [51,52].

4.2. Creation and Training of the Model

Creation, training and testing of the ANFIS model is performed in the MATLAB software package, which, thanks to the graphical user interface of the ANFIS editor, allows easy manipulation of the model's parameters and variables. As a result, a large number of graphic displays of parameters and performance are obtained.

The total set of data on the inter-arrival time, cumulative arrival time, service time, and the time spent in the system for each truck that enters the system, is divided into three parts:

- Training data, consisting of 73.21% or 164 input-output vectors, providing the so-called "Learning with a teacher", where the outputs from the network are known in advance for appropriate inputs.
- Checking data, which is primarily aimed at preventing the occurrence of training data overfitting. The ANFIS model monitors the value of the checking error in each training epoch and retains learned parameters at its minimum value. Checking data consists of 13.39% or 30 input-output vectors.
- Testing data enables us to perform an evaluation of the abilities of the ANFIS model to perform a prediction of the time spent in the system as accurately as possible. The outputs of the ANFIS model are compared with known values, and the goal is to select a model that makes a minimum error. As well as checking data, testing data consists of 13.39% of the total set of data.

The process, from creation to model testing, can be summarized by the algorithmic steps given in Figure 7.

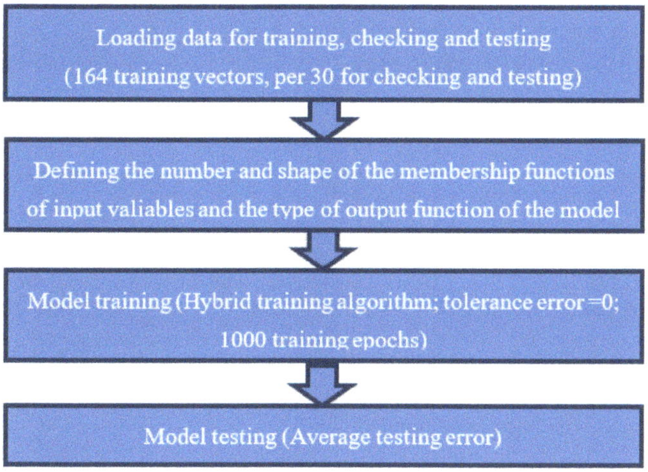

Figure 7. Steps of the process from creation to testing of the ANFIS model.

5. Results and discussion

The ANFIS model performance is estimated based on an average testing error, which in fact is an average square error-*RMSE* (*Root Mean Square Error*), and is calculated as:

$$RMSE = \sqrt{\frac{1}{N}\sum_{k=1}^{N}[n(k) - \hat{n}(k)]^2} \quad (5)$$

where N is a number of testing vectors, $n(k)$ is expected (measured) value, and $\hat{n}(k)$ is the value obtained by the model. Table 5 gives an overview of RMSE values depending on the shape and number of fuzzy membership functions for each of the three input variables for the constant shape of the output function. The values of the average testing error for different ANFIS models in the case of a linear shape of the output function are given in Table 6. The model training was carried out in 1000 epochs. With a larger number of membership functions, the average testing error increases, so that a maximum of three are considered here. Table 5 gives an overview of RMSE values depending on the shape and number of the membership fuzzy functions for each of the three input variables for the constant shape of the output function.

Table 5. The values of average testing errors of different ANFIS models with constant output.

Shape of Fuzzy Membership Functions	Number of Fuzzy Membership Functions for Each of Three Input Variables							
	2 2 2	3 3 3	2 2 3	2 3 2	2 3 3	3 3 2	3 2 3	3 2 2
Trimf	18.66	64.64	21.53	22.94	30.46	66.73	23.30	20.72
Trapmf	14.23	27.72	16.16	14.33	16.64	16.46	20.55	18.11
Gbellmf	19.46	47.00	19.66	21.42	58.65	17.34	46.14	16.29
Gaussmf	16.91	307.84	20.26	17.81	26.91	31.49	22.64	17.79
Gauss2mf	14.30	20.50	63.56	15.70	22.40	26.89	19.16	17.83
Pimf	14.78	23.21	17.10	14.05	17.19	15.07	21.25	17.42
Dsigmf	13.67	64.43	24.38	17.09	60.50	28.00	19.29	17.46
Psigmf	13.67	64.43	24.38	17.09	60.50	28.00	19.29	17.46

Table 6. The values of average testing errors of different ANFIS models with linear output.

Shape of Fuzzy Membership Functions	Number of Fuzzy Membership Functions for Each of Three Input Variables							
	2 2 2	3 3 3	2 2 3	2 3 2	2 3 3	3 3 2	3 2 3	3 2 2
Trimf	217.63	13557.47	577.56	836.02	1726.74	35925.48	19459.66	1219.73
Trapmf	116.57	318.80	219.14	38.82	25.42	60.76	294.55	44.87
Gbellmf	22.19	34006.56	187.06	5305.95	23183.85	14176.49	50418.31	6937.78
Gaussmf	33.02	24328.30	257.99	3293.97	11566.52	5466.29	51114.92	108.12
Gauss2mf	253.78	3727.55	319.52	9477.00	34019.82	3765.83	14012.94	2537.84
Pimf	841.49	597.19	2104.52	222.63	23.77	44.86	415.37	190.11
Dsigmf	588.50	4961.53	2289.36	774.68	65680.67	2563.84	19890.09	449.44
Psigmf	427.85	5027.61	1118.50	1120.38	62986.63	7788.29	15398.09	4170.70

By comparing the values in Tables 5 and 6, it is concluded that the linear output model gives drastically higher error values than the constant output shape. The least average testing error from Table 6 is 25.42 min, and from Table 5, 13.67 min. Therefore, a model with a lower error is selected from Table 5, which has two fuzzy membership functions for each input variable. The functions are in the shape of dsig, and represent the difference of two sigmoid functions, which can be written as follows:

$$f(x;a,c) = \frac{1}{1+e^{-a(x-c)}} \tag{6}$$

Since it relates to the difference, the dsig function has four parameters: a_1, a_2, c_1, and c_2. Figure 8 shows the certain membership functions for the first input variable-inter-arrival time. The learned parameters of the first function marked by red color in Figure 8 are 0.0578, −172, 0.0572, 174.

The values of the prediction of time spent in the system, based on the input values of checking data of the selected model, are given in Table 7.

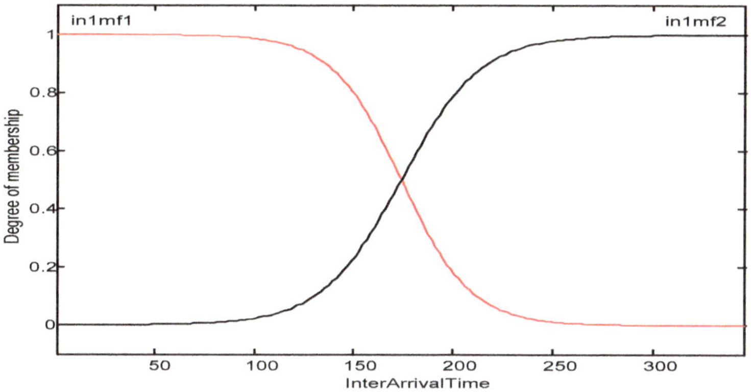

Figure 8. Fuzzy functions of membership for the input variable of inter-arrival time.

Table 7. The values of the prediction of the time spent in the system and checking data.

Ordinal Number	Checking Data				ANFIS Output
	Inter-Arrival Interval	Arrival Time	Service Time	Time Spent in the System	
1st	2	357	30	42	58.36
2nd	272	629	25	40	37.84
3rd	1	1	40	63	83.49
4th	24	25	45	64	85.68
5th	8	33	30	71	82.53
6th	1	1	30	50	82.54
7th	11	12	30	66	82.54
8th	2	14	25	73	82.47
9th	77	91	30	46	82.34
10th	40	109	40	99	83.34
11th	20	129	65	109	118.59
12th	170	299	30	52	66.57
13th	1	300	45	79	74.73
14th	83	383	30	55	52.54
15th	1	1	35	48	82.77
16th	54	55	35	73	82.70
17th	3	58	30	50	82.51
18th	35	93	40	78	83.39
19th	60	153	15	118	82.03
20th	3	156	55	80	101.73
21st	12	168	50	73	90.92
22nd	21	189	15	44	81.54
23rd	1	190	30	42	81.62
24th	50	240	95	125	118.06
25th	1	1	40	107	83.49
26th	2	3	35	45	82.77
27th	1	4	45	79	85.69
28th	34	38	45	71	85.66
29th	7	45	85	134	121.85
30th	20	65	65	109	118.79

In addition to the tabular overview, the accuracy of the prediction can also be shown graphically, as in Figure 9. Red asterisks denote the outputs of the ANFIS model, while blue points denote measured checking data. The RMSE for such a set of data is 22.06. Although while testing the model it showed the least error of 13.67 over the testing data, the set of checking data is different from it, and that is the reason why, in this case, the RMSE has the given value.

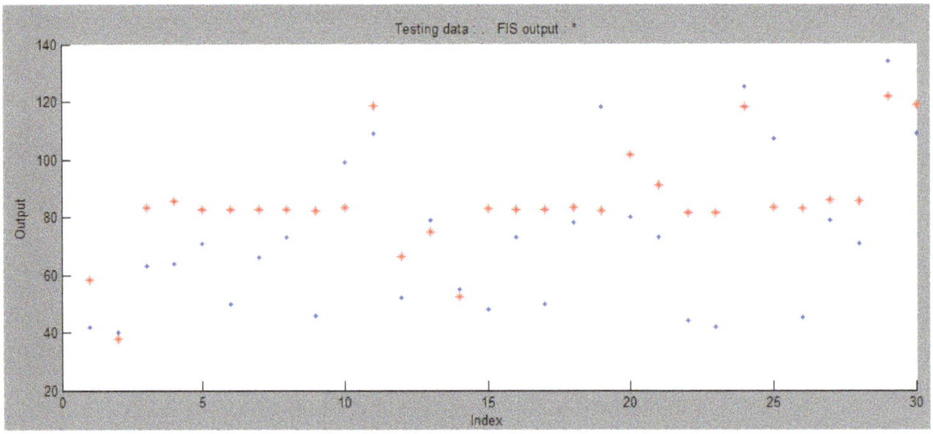

Figure 9. Deviations of the ANFIS model output from the measured testing data.

The structure of the selected model is shown in Figure 10, where the number of nodes in each layer of the neural network can be seen. It is obvious that in all fuzzy inference rules, of which there are 8 (the number of nodes in the third layer), the logical operator (AND) figures.

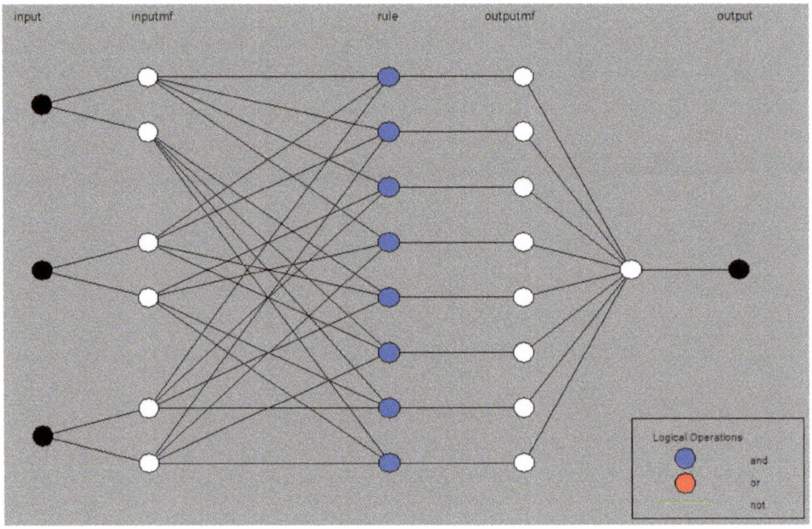

Figure 10. The structure of the selected ANFIS model.

Figure 11 shows the surface of dependence, i.e. a portable function of the selected model. Taking into account that there are three input variables, the dependence of the output from the input is given for all three combinations. It is evident that the time spent in the system has a greater value in reducing inter-arrival times and the cumulative arrival time of trucks. Regarding the influence of the service time on the observed output, it is concluded that the time spent in the system increases with the increase of the specified variable and the decrease in inter-arrival times. An increase in the value of the output variable is also caused by an increase in the service time and simultaneous reduction in the cumulative arrival time.

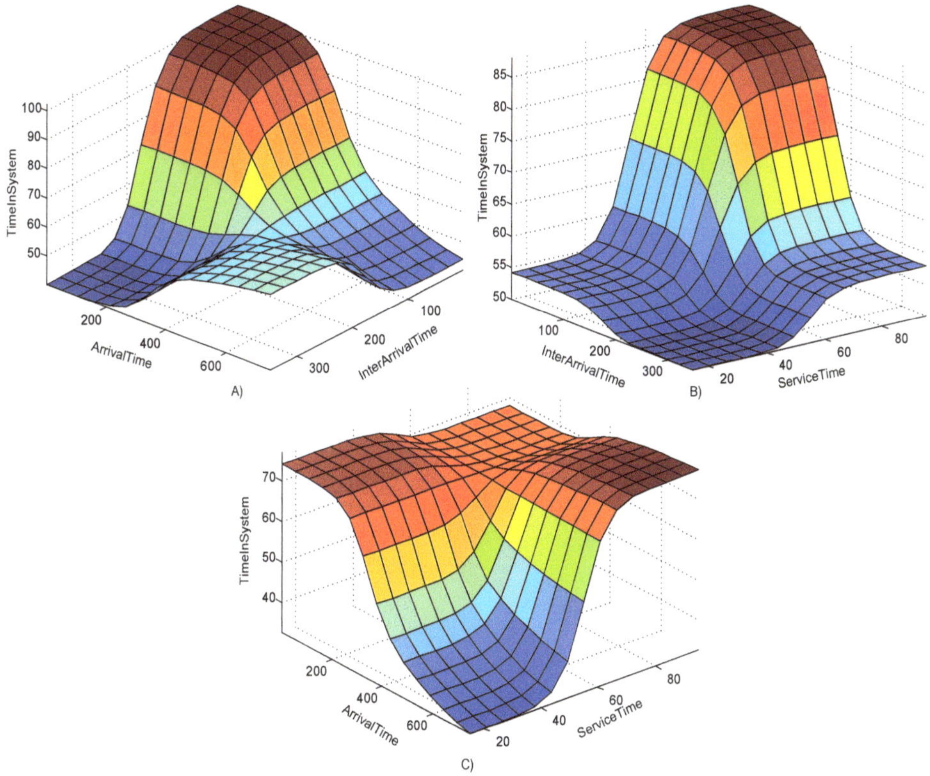

Figure 11. The surface of the dependence of the time spent in the system from: (**a**) inter-arrival time and arrival time; (**b**) inter-arrival time and service time; (**c**) arrival time and service time.

6. Sensitivity Analysis

In order to validate the developed ANFIS model, it was compared with a mathematical model developed by regression analysis of training data. A polynomial mathematical model with the highest correlation index $R^2 = 15.58$ was selected. The model is of the second degree and has the following form:

$$\begin{aligned} Time\ in\ system = \ & 61.52 - 0.094\ Inter\ Arrival\ time + 0.683\ Service\ time - 7.834 \\ & *10^{-5} Arrival\ time^2 \end{aligned} \tag{7}$$

Table 8 gives an overview of the measured values of the time spent in the system and predicted values determined by the mathematical model and ANFIS for the same input data set. The RMSE value, in the case of the mathematical model, is 22.96, which means that the ANFIS model shows better performance.

Figure 12 gives a graphic display of the values shown in Table 8. The red squares represent the real or measured values of the time spent in the system. Blue rhombuses represent the predicted values obtained by the mathematical model of second degree, which is given by the expression (7). Predicted values obtained by the ANFIS model are marked with green triangles. The figure provides a visual performance analysis of the two models compared to the real values of the time spent in the system.

Table 8. Measured values of time in the system and predicted values determined by the mathematical and ANFIS model.

Measured, Real Values	Mathematical Model	ANFIS Model
42	71.84	58.36
40	22.03	37.84
63	88.75	83.49
64	89.95	85.68
71	81.17	82.53
50	81.92	82.54
66	80.96	82.54
73	78.39	82.47
46	74.12	82.34
99	84.15	83.34
109	102.73	118.59
52	59.03	66.57
79	85.11	74.73
55	62.72	52.54
48	85.33	82.77
73	80.11	82.70
50	81.46	82.51
78	84.87	83.39
118	64.29	82.03
80	96.90	101.73
73	92.33	90.92
44	66.99	81.54
42	79.09	81.62
125	117.19	118.06
107	88.75	83.49
45	85.24	82.77
79	92.16	85.69
71	88.95	85.66
134	118.76	121.85
109	103.70	118.79

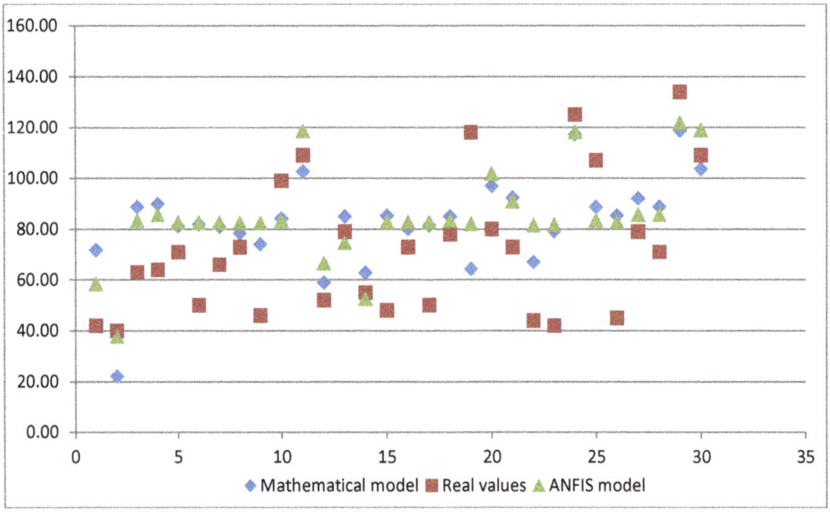

Figure 12. Values obtained by the mathematical and ANFIS model and the actual measured ones.

7. Conclusion

In this paper, a study on the modeling of a queuing system in a logistics company for the manufacturing of brown paper was conducted. An ANFIS model for modeling a time component of the system was developed as a criterion for optimization. The contributions of this research can be described in the following ways. The proposed neuro-fuzzy model extends the theoretical framework of knowledge in the field of QS. The QS problem is considered by a new methodology, and thus, a basis for further theoretical and practical upgrading is formed. In addition, the presented model emphasizes the unique practical parameters (the inter-arrival time of trucks, the cumulative time of arrival since an initial time, the service time), which, in former MSS models, have not been considered as unified, despite being of importance for this logistics company and its customers.

The proposed neuro-fuzzy model has four main advantages over other methods. First, in comparison with classical QS, it has an adaptability feature, which is reflected in its ability to adjust a fuzzy rule base. Fuzzy rules are very important for managing queuing systems, especially for a descriptive approach that prefers intuitive, heuristic searches of solutions in a queuing system process. This flexibility allows us to overcome the limitations of conventional QS models that perform a prediction of the flow throughout statistical consideration of parameters without intuitiveness. Second, the neuro-fuzzy model is effective under conditions of uncertainty, and can provide support to decision-makers when there is uncertainty in logistics processes. Third, it can be implemented as a computer system that supports a dynamic decision-making process in the QS. And fourth, the proposed model allows relatively fast and objective estimates to be made of serving the vehicles in a transportation company, under conditions of a changing environment. The continuation of this research may also include the optimization of other, smaller queuing systems for the decentralized storage system of this company.

Author Contributions: Each author has participated and contributed sufficiently to take public responsibility for appropriate portions of the content.

Conflicts of Interest: The authors declare no conflicts of interest.

References

1. Stević, Ž.; Pamučar, D.; Kazimieras Zavadskas, E.; Ćirović, G.; Prentkovskis, O. The selection of wagons for the internal transport of a logistics company: A novel approach based on rough BWM and rough SAW methods. *Symmetry* **2017**, *9*, 264. [CrossRef]
2. Stević, Ž.; Pamučar, D.; Vasiljević, M.; Stojić, G.; Korica, S. Novel Integrated multi-criteria model for supplier selection: Case study construction company. *Symmetry* **2017**, *9*, 279. [CrossRef]
3. Stević, Ž.; Mulalić, E.; Božičković, Z.; Vesković, S.; Đalić, I. Economic analysis of the project of warehouse centralization in the paper production company. *Serbian J. Manag.* **2018**, *13*, 47–62. [CrossRef]
4. Guerrouahane, N.; Aissani, D.; Bouallouche-Medjkoune, L.; Farhi, N. M/g/c/c state dependent queueing model for road traffic simulation. *arXiv*, 2016; arXiv:1612.09532.
5. Raheja, T. Modelling traffic congestion using queuing networks. *Sadhana* **2010**, *35*, 427–431. [CrossRef]
6. Van Woensel, T.; Vandaele, N. Queueing models for uninterrupted traffic flows. In Proceedings of the 13th Mini-EURO Conference Handling Uncertainty in the Analysis of Traffic and Transportation Systems, Bari, Italy, 2002; pp. 636–640.
7. Vandaele, N.; Van Woensel, T.; Verbruggen, A. A queueing based traffic flow model. *Transport. Res. D-Tr. E.* **2000**, *5*, 121–135. [CrossRef]
8. Osorio, C.; Bierlaire, M. Network performance optimization using a queueing network model. In Proceedings of the European Transport Conference, Langelaan, The Netherlands, 6–8 October 2008.
9. Anokye, M.; Abdul-Aziz, A.R.; Annin, K.; Oduro, F.T. Application of queuing theory to vehicular traffic at signalized intersection in Kumasi-Ashanti region, Ghana. *Am. Int. J. Cont. Res.* **2013**, *3*, 23–29.
10. Wang, F.; Ye, C.; Zhang, Y.; Li, Y. Simulation analysis and improvement of the vehicle queuing system on intersections based on MATLAB. *Open Cybernet. Syst. J.* **2014**, *8*, 217–223. [CrossRef]

11. Chen, W.N. Application of queuing theory to dynamic vehicle routing problem. *Glob. J. Bus. Res.* **2009**, *3*, 85–91.
12. Azizi, A.; Yarmohammadi, Y.; Yasini, A.; Sadeghifard, A. A queuing model to reduce energy consumption and pollutants production through transportation vehicles in green supply chain management. *J. Sci. Res. Rep.* **2015**, *5*, 571–581. [CrossRef]
13. Aziziankohan, A.; Jolai, F.; Khalilzadeh, M.; Soltani, R.; Tavakkoli-Moghaddam, R. Green supply chain management using the queuing theory to handle congestion and reduce energy consumption and emissions from supply chain transportation fleet. *J. Ind. Eng. Manag.* **2017**, *10*, 213–236. [CrossRef]
14. Gong, Y.; De Koster, R.B. A review on stochastic models and analysis of warehouse operations. *Log. Res.* **2011**, *3*, 191–205. [CrossRef]
15. Masek, J.; Camaj, J.; Nedeliakova, E. Application the queuing theory in the warehouse optimization. *Int. J. Soc. Behav. Educ. Econ. Bus. Ind. Eng.* **2015**, *9*, 3744–3748.
16. Cai, X.; Heragu, S.S.; Liu, Y. Modeling automated warehouses using semi-open queueing networks. In *Handbook of Stochastic Models and Analysis of Manufacturing System Operations*; Smith, J.M., Tan, B., Eds.; Springer-Verlag: New York, NY, USA, 2013; pp. 29–71.
17. Bao-ping, C.; Zeng-Qiang, M. Short-term traffic flow prediction based on ANFIS. In Proceedings of the International Conference on Communication Software and Networks, Sichuan, China, 27–28 February 2009; pp. 791–793.
18. Rahimi, A.M. Neuro-fuzzy system modelling for the effects of intelligent transportation on road accident fatalities. *Tehnički Vjesn.* **2017**, *24*, 1165–1171.
19. Hosseinlou, M.H.; Sohrabi, M. Predicting and identifying traffic hot spots applying neuro-fuzzy systems in intercity roads. *Int. J. Environ. Sci. Technol.* **2009**, *6*, 309–314. [CrossRef]
20. Suraj, S.; Jagrut, G. Smart traffic control using adaptive neuro-fuzzy Inference system (ANFIS). *Int. J. Adv. Eng. Res. Dev.* **2015**, *2*, 295–302.
21. Araghi, S.; Khosravi, A.; Creighton, D. ANFIS traffic signal controller for an isolated intersection. In Proceedings of the International Conference on Fuzzy Computation Theory and Applications, Rome, Italy, 22–24 October 2014; pp. 175–180.
22. Udofia, K.M.; Emagbetere, J.O.; Edeko, F.O. Dynamic traffic signal phase sequencing for an isolated intersection using ANFIS. *Auto. Control Intell. Syst.* **2014**, *2*, 21–26.
23. Sharma, A.; Vijay, R.; Bodhe, G.L.; Malik, L.G. Adoptive neuro-fuzzy inference system for traffic noise prediction. *Int. J. Comput. Appl.* **2014**, *98*, 14–19. [CrossRef]
24. Pamučar, D.; Ćirović, G. Vehicle route selection with an adaptive neuro fuzzy inference system in uncertainty conditions. *Decis. Mak. Appl. Manag. Eng.* **2018**, *1*, 13–37. [CrossRef]
25. Andrade, K.; Uchida, K.; Kagaya, S. Development of transport mode choice model by using adaptive neuro-fuzzy inference system. *Transport. Res. Rec.-J. Transport. Res. Board* **2006**, *1977*, 8–16. [CrossRef]
26. Mircetic, D.; Lalwani, C.; Lirn, T.; Maslaric, M.; Nikolicic, S. ANFIS expert system for cargo loading as part of decision support system in warehouse. In Proceedings of the 19th International Symposium on Logistics (ISL 2014), Ho Chi Minh City, Vietnam, 6–9 July 2014; pp. 10–20.
27. Mirčetić, D.; Ralević, N.; Nikoličić, S.; Maslarić, M.; Stojanović, Đ. Expert system models for forecasting forklifts engagement in a warehouse loading operation: A case study. *PROMET-Zagreb.* **2016**, *28*, 393–401. [CrossRef]
28. Marto, M.; Reynolds, K.; Borges, J.; Bushenkov, V.; Marques, S. Combining Decision Support Approaches for Optimizing the Selection of Bundles of Ecosystem Services. *Forests* **2018**, *9*, 438. [CrossRef]
29. Al-Anbari, M.A.; Thameer, M.Y.; Al-Ansari, N. Landfill Site Selection by Weighted Overlay Technique: Case Study of Al-Kufa, Iraq. *Sustainability* **2018**, *10*, 999. [CrossRef]
30. Sałabun, W.; Karczmarczyk, A. Using the COMET Method in the Sustainable City Transport Problem: An Empirical Study of the Electric Powered Cars. *Procedia Comput. Sci.* **2018**, *126*, 2248–2260. [CrossRef]
31. Wątróbski, J.; Sałabun, W.; Karczmarczyk, A.; Wolski, W. Sustainable decision-making using the COMET method: An empirical study of the ammonium nitrate transport management. In Proceedings of the 2017 Federated Conference on Computer Science and Information Systems, Prague, Czech Republic, 3–6 September 2017; pp. 949–958.

32. Wątróbski, J.; Sałabun, W.; Ladorucki, G. The temporal supplier evaluation model based on multicriteria decision analysis methods. In Proceedings of the Asian Conference on Intelligent Information and Database Systems, Kanazawa, Japan, 3–5 April 2017; pp. 432–442.
33. Khalili-Damghani, K.; Dideh-Khani, H.; Sadi-Nezhad, S. A two-stage approach based on ANFIS and fuzzy goal programming for supplier selection. *Int. J. Appl. Decis. Sci.* **2013**, *6*, 1–14.
34. Torkabadi, A.M.; Mayorga, R.V. Optimization of Supply Chain based on JIT Pull Control Policies: An Integrated Fuzzy AHP and ANFIS Approach. *WSEAS Trans. Comput.* **2017**, *16*, 366–377.
35. Pirdavani, A.; Brijs, T.; Wets, G. A Multiple Criteria Decision-Making Approach for Prioritizing Accident Hotspots in the Absence of Crash Data. *Transp. Rev.* **2010**, *30*, 97–113. [CrossRef]
36. Stević, Ž.; Pamučar, D.; Subotić, M.; Antuchevičiene, J.; Zavadskas, E. The Location Selection for Roundabout Construction Using Rough BWM-Rough WASPAS Approach Based on a New Rough Hamy Aggregator. *Sustainability* **2018**, *10*, 2817. [CrossRef]
37. Stidham, S., Jr. Analysis, design, and control of queueing systems. *Oper. Res.* **2002**, *50*, 197–216. [CrossRef]
38. Cooper, R.B. *Introduction to queueing theory*, 2nd ed.; Fineman, J., Schreiber, L.C., Eds.; North Holland: New York, NY, USA, 1981.
39. Maragatha, S.; Srinivasan, S. Analysis of M/M/I queueing model for ATM facility. *Glob. J. Theor.Appl. Mathematics Sci.* **2012**, *2*, 41–46.
40. Defraeye, M.; Van Nieuwenhuyse, I. Staffing and scheduling under nonstationary demand for service: A literature review. *Omega* **2016**, *58*, 4–25. [CrossRef]
41. Stević, Ž. Calculation of the basic parameters of queuing systems using winqsb software. In Proceedings of the XI International May Conference on Strategic Management, Bor, Serbia, 29–31 May 2015; pp. 91–100.
42. Sremac, S.; Tanackov, I.; Kopić, M.; Radović, D. ANFIS model for determining the economic order quantity. *Decis. Mak. Appl. Manag. Eng.* **2018**, *1*, 1–12. [CrossRef]
43. Tiwari, S.; Babbar, R.; Kaur, G. Performance evaluation of two ANFIS models for predicting water quality Index of River Satluj (India). *Adv. in Civ. Eng.* **2018**, *2018*, 1–10. [CrossRef]
44. Billah, M.; Waheed, S.; Ahmed, K.; Hanifa, A. Real time traffic sign detection and recognition using adaptive neuro fuzzy inference system. *Commun. Appl. Electron.* **2015**, *3*, 1–5. [CrossRef]
45. Qasem, S.N.; Ebtehaj, I.; Riahi Madavar, H. Optimizing ANFIS for sediment transport in open channels using different evolutionary algorithms. *J. Appl. Res. Water Wastewater* **2017**, *4*, 290–298.
46. Lukovac, V.; Pamučar, D.; Popović, M.; Đorović, B. Portfolio model for analyzing human resources: An approach based on neuro-fuzzy modeling and the simulated annealing algorithm. *Expert Syst. Appl.* **2017**, *90*, 318–331. [CrossRef]
47. Pamučar, D.; Ljubojević, S.; Kostadinović, D.; Đorović, B. Cost and risk aggregation in multi-objective route planning for hazardous materials transportation—A neuro-fuzzy and artificial bee colony approach. *Expert Syst. Appl.* **2016**, *65*, 1–15. [CrossRef]
48. Das, R.D.; Winter, S. Detecting urban transport modes using a hybrid knowledge driven framework from GPS trajectory. *ISPRS Int. J. Geo-Inf.* **2016**, *5*, 207. [CrossRef]
49. Engmann, S.; Cousineau, D. Comparing distributions: The two-sample Anderson-Darling test as an alternative to the Kolmogorov-Smirnoff test. *J. Appl. Quant. Methods* **2011**, *6*, 1–17.
50. Stephens, M.A. Tests based on EDF statistics. In *Goodness of Fit. Techniques* Chapter 4; D'Agostino, R.B., Stephens, M.A., Eds.; Routledge: New York, NY, USA, 1986.
51. Barford, P.; Crovella, M. Generating representative web workloads for network and server performance evaluation. In *ACM SIGMETRICS Performance Evaluation Review*; ACM: New York, NY, USA, 1998; pp. 151–160.
52. Jovanović, B.; Grbić, T.; Bojović, N.; Kujačić, M.; Šarac, D. Application of ANFIS for the Estimation of Queuing in a Postal Network Unit: A Case Study. *Acta Polytech. Hung.* **2015**, *12*, 25–40.

information

MDPI

Article

Multiple Criteria Decision-Making in Heterogeneous Groups of Management Experts

Virgilio López-Morales

Information Technology and Systems Research Center (CITIS), Universidad Autónoma del Estado de Hidalgo, 42184 Pachuca, Hidalgo, Mexico; virgilio@uaeh.edu.mx; Tel.: +52-771-71-72000 (ext. 6734)

Received: 31 October 2018; Accepted: 21 November 2018; Published: 27 November 2018

Abstract: In commercial organizations operations, frequently some dynamic events occur which involve operational, managerial, and valuable information aspects. Then, in order to make a sound decision, the business professional could be supported by a Multi Criteria Decision-Making (MCDM) system for taking an external course of action, as, for instance, forecasting a new market or product, up to an inner decision concerning for instance, the volume of manufacture. Thus, managers need, in a collective manner, to analyze the actual problems, to evaluate various options according to diverse criteria, and finally choose the best solution from a set of various alternatives. Throughout these processes, uncertainty and hesitancy easily arise, when it comes to define and judge criteria or alternatives. Several approaches have been introduced to allow Decision Makers (DMs) to deal with. The Interval Multiplicative Preference Relations (IMPRs) approach is a useful technique and the basis of our proposed methodology to provide reliable consistent and in consensus IMPRs. In this manner, DMs' choices are implicitly including their uncertainty while maintaining both an acceptable individual consistency, as well as group consensus levels. The present method is based on some recent results and an optimization algorithm to derive reliable consistent and in consensus IMPRs. In order to illustrate our results and compare them with other methodologies, a few examples are addressed and solved.

Keywords: uncertain group decision-making support systems; multiple criteria decision-making; reliable group decision-making; interval multiplicative preference relations

---

1. Introduction

Among the vast world of commercial operations organizations and despite their differences, they have common business operations or activities such as acquiring inventory, hiring employees and cashing from customers. Nowadays, inside each modern organization, we can frequently find an information system working in synchrony with these business operations. Furthermore, several important information systems are fed by operating departments (work centers of the organization), and, as a result, these systems' outcomes can be used to manage these operations. As a consequence, managers analyze their corresponding information system in light of the work that the organization performs. For instance, when a marketing manager is required to advise management and to have some reports for management decision-making, she/he must understand the organization's product cycles.

There are various events which occur while organizations engage their business operations as, for instance, diverse trends in purchases and sales. This dynamic data coming from these events are frequently recorded and kept up in a database to mirror and supervise business operations. These records include operational, managerial, and valuable information details. Thus, in order to design and use a group decision-making system, the business professional must previously consider what kind of event data is needed and the necessary process to extract the useful information.

In essence, the type of decision under consideration rules what kind, amount and quality of the information must be used to make a sound decision. Furthermore, information is more valuable when it recognizes and doesn't disregard the personal management styles, the main choices of the the decision maker, the weighting of each decision maker with respect to the entire decision-making process, and managers' uncertainty and hesitancy when a heterogenous group tries to achieve an integral decision.

Multi criteria decision-making then could support human-centered management in taking a course of action for a sound decision-making to elucidate several management problems. For instance, what products to sell and the suitable targeted market to sell them, or the structure of the organization better suited for this process or even the direction and motivation of employees based on some known standards. Another type of management problem can be an inner decision as for instance, human resources information, volume of manufacture and the available delivering chains, which is useful to develop alternative methods for manufacturing and delivering a new product.

On the other hand, the essence of the information provided to managers must be according to the management level. For example, strategic level managers need information allowing them to assess the environment to forecast future events and conditions, where s/he may not be as concerned with the timeliness or accuracy of the information as her/his interest is in the trends. In this case, most of the information is exogenous to the organization, or tactical management needs information mainly coming from pertinent operational units. Some exogenous information is needed, as well as more detailed and accurate information than the information required at the strategic level. Finally, operational management unit usually requires exogenous narrower information but more detailed and accurate information. It comes largely from within the organization where frequent decisions are made, with shorter lead times to respond in a timely manner to current variations as, for instance, in sales patterns.

As we noted in this section, managers support the entire firms' making decision to achieve diverse kind of goals. They have basically different kinds of perspectives and local goals to reach a high efficiency level at their respective departments. However, the main goal of the firm needs, at certain moments, to combine local goals and perspectives of the work centers, with particular global targets.

In this scenario of heterogeneous managers group, an MCDM system for supporting the management for group decision-making becomes paramount.

Fortunately, there are a vast variety of MCDM techniques ranging from: analytic hierarchy process (AHP) [1], multi attribute utility theory (MAUT) [2], simple multi attribute rating technique (SMART) [3], fuzzy set theory (FST) [4], data envelopment analysis (DEA) [5], case-based reasoning (CBR) [6], simple additive weighting (SAW) [7], elimination et choice translating reality (ELECTRE) [8], technique for order of preference by similarity to ideal solution (TOPSIS) [9], preference ranking and organization method for enrichment evaluation (PROMETHEE) [10], and goal programming (GP) [11]. Another interesting method for addressing uncertainty is the Hesitant Fuzzy Sets (HFS) [12] where some recent contributions for heterogeneous information are found [13–18]. These methods are appropriate for uncertainty problems, since fuzzy logic aims to represent human preferences based on individual opinions expressed through a linguistic setting. Its major drawback is that membership functions (which can be seen as intervals) are fixed and also, until now, there does not exist detailed work related to an appropriate analytical tool (norms, aggregation operators, etc.) for a particular study case. Thus, it results in a greater uncertainty and vagueness to the problem.

Recently, several frameworks have been employed and successfully applied to solve decision problems in many areas, including international politics and laws [19], transportation [20–23], business intelligence [24], information and communication technologies [25], water resources management [26], environmental risk analysis [27], flood risk management [28], environmental impact assessment and environmental sciences [14,29], solid waste management [30], climate change [31], remote sensing [32], energy [33], health technology assessment [34] and nanotechnology research [35]. Furthermore, MCDM techniques have been integrated with known systems such as genetic algorithms, geographic information systems, fuzzy logic and intelligent systems, automatic control systems and neural networks which recently are being applied.

Group Decision Making (GDM) is a main MCDM issue, where multiple DMs (managers in our case) act collaboratively and collectively, analyze decision-making problems, evaluate goals according to a set of criteria, and finally choose the best solution from a set of alternatives [36]. As noted in [37], when organizations gather specialized groups or larger groups, and the number of alternatives increase, unanimity may be difficult to attain. For this reason, flexible or milder benchmarks (definitions) of Group Consensus and Individual Consistency have been employed.

Group Consensus (GC) has to do with group cooperation and agreement since the alternative, option, or goal to be achieved is the best course of action for the whole organization. On the other hand, Individual Consistency (IC) concerns each DM to have her/his information, and, consequently, her/his judgments, free of contradictions.

Derived of the blend of heterogenous DM group, a common problem in big organizations is that managers often can accurately state definitions and assessments on the priority rate of the set of criteria or alternatives, when it concerns their own operating departments. However, when it comes to define and judge any other set of criteria or alternatives, they face uncertainty and hesitation problems.

In this paper, we address and solve these problems by allowing the DM to utilize blended assessments. For example, DMs can use crisp values for her/his ratio judgments of criteria or alternatives when s/he is confident; and then they can use intervals where these are used to express her/his uncertainty and hesitant assessments.

The aim of this paper is to synthesize a novel approach in order to provide a reliable measure of both the IC and the GC of a set of these blended assessments, which we called Interval-Multiplicative Preference Relations (I-MPRs). Then, in the next step, the approach is verified by a constrained optimization algorithm where the improved I-MPRs will finally fulfill the Individual Consistency and Group Consensus Indices since both restrictions are involved in the constraints.

In order to verify the requirements of acceptable IC and GC levels, we use the Hadamard's operator. As soon as IC and then GC are validated, an Interval Priority Vector can be obtained from this set of ordered judgment I-MPRs, in order to rank the alternatives as a final result of the DM analysis. Various techniques are revised and the prioritization method which indicates the entire order of the intervals and the preference degrees, is addressed according to our results.

The main advantages of our approach follow:

- It is provided through a couple of algorithms and a nonlinear optimization approach concurrently applied.
- Through the Hadamard's operator and some easy algebraic manipulations, objective functionals are synthesized (as it will be detailed further on), to be used in the optimization algorithm.
- When the I-MPRs improved by the methodology are reduced into an MPR (defined in the I-MPR), our approach can still give reliable results. For example, for this MPR, we can verify the results of IC or GC, with an alternative method.
- The IC or the GC accepted indices (threshold values) have been previously investigated and fixed. Nevertheless, the project designer could assign a different value depending on the project requirements.
- Obtained results are independent of the method of prioritization utilized in the consensus operation.

This paper is organized as follows: In Section 2, some preliminaries (Definitions, Theorems and Lemmata) are given to support a basis of the main methodologies and techniques previously described above and the approach introduced here. In Section 3, an extension of analysis and results derived in the former section is used in the I-MPR framework for obtaining our main results. Then, in Section 4, a slight modification of a prioritization method is given in light of the results. In Section 5, some numerical examples are solved and compared with other methodologies. Finally, in Section 6, concluding remarks are given about the main advantages, drawbacks and future research of our methodology.

2. Preliminaries

Let us consider a Group Decision-Making problem and let $D = \{d_1, d_2, \cdots, d_m\}$ be the set of DMs, and $C = \{c_1, c_2, \cdots, c_n\}$ be a finite set of criteria (or alternatives), where c_i denotes the ith criteria.

In the AHP framework [1], a pairwise comparison matrix or an MPR is given by a DM where s/he provides judgments through a ratio (c_i/c_j) for every pair of criteria (or alternatives) (c_i and c_j) to represent the preference degree of the first criteria (or alternative) over the second. A Saaty's Scale is frequently used to pick up a value for this preference ratio where $SS = [1/9 \;\; 9]$.

Thus, an MPR for instance A = $(a_{ij})_{n\times n}$, is a positive reciprocal $n \times n$ matrix, $a_{ij} > 0$, such that $a_{ji} = 1/a_{ij}$, $a_{ii} = 1$, $\forall i, j \in N$. Note that $a_{ij} \in [1/9 \;\; 9]$.

Let $\lambda = \{\lambda_1, \cdots, \lambda_m\}$ be the weight vector of the $m - th$ DM, where $\lambda_s > 0$, $s \in M$, $\sum_{s=1}^{m} \lambda_s = 1$, which can be derived with several techniques (see for instance [38] and the references cited therein).

An MPR $n \times n$ matrix is called a *completely consistent* MPR (cf. [1]) if

$$a_{ij} = a_{il}a_{lj}, \;\; \forall i, j, l \in N. \tag{1}$$

Thus, their corresponding completely consistent MPR $K = (k_{ij})_{n\times n}$ can be constructed from each MPR $n \times n$ matrix as follows:

$$k_{ij} \;\; = \;\; \prod_{l=1}^{n} \left(a_{il}a_{lj}\right)^{1/n} \;\; = a_{ij}^{2/n} \prod^{n}_{\substack{l=1\\ i\neq l,\; j\neq l}} (a_{il}a_{lj})^{1/n}, \tag{2}$$

where $i = 1, 2, \cdots, n-1$, $j = i+1, \cdots, n$.

Furthermore, let A^c be the group MPR (cf. [39]) which represents the group opinion utilizing the geometric average operator:

$$A^c = (a^c_{ij})_{n\times n} = \prod_{t=1}^{m} \left(a_{ij}^{(t)}\right)^{\lambda_t}, \quad i, j \in N; \quad t \in M. \tag{3}$$

Let us denote (cf. [40]), an I-MPR A_t given by the $t - th$ expert as

$$A_t \;\; = \;\; (a_{ij}^{(t)})_{n\times n} = \begin{pmatrix} 1 & \left[\overset{-(t)}{a}_{12} \; \overset{+(t)}{a}_{12}\right] & \cdots & \left[\overset{-(t)}{a}_{1n} \; \overset{+(t)}{a}_{1n}\right] \\ \left[\overset{-(t)}{a}_{21} \; \overset{+(t)}{a}_{21}\right] & 1 & \cdots & \left[\overset{-(t)}{a}_{2n} \; \overset{+(t)}{a}_{2n}\right] \\ \vdots & \cdots & \ddots & \vdots \\ \left[\overset{-(t)}{a}_{n1} \; \overset{+(t)}{a}_{n1}\right] & \left[\overset{-(t)}{a}_{n2} \; \overset{+(t)}{a}_{n2}\right] & \cdots & 1 \end{pmatrix}, \tag{4}$$

where $\overset{-(t)}{a}_{ij}, \overset{+(t)}{a}_{ij} > 0$, $\overset{-(t)}{a}_{ij} \leq \overset{+(t)}{a}_{ij}$, $\overset{-(t)}{a}_{ij} = 1/\overset{+(t)}{a}_{ji}$ and $\overset{+(t)}{a}_{ij} = 1/\overset{-(t)}{a}_{ji}$. Furthermore, $A_t = [\overset{-}{A_t} \;\; \overset{+}{A_t}]$. For example,

$$\overset{-}{A_t} = (\overset{-(t)}{a}_{ij})_{n\times n} = \begin{cases} \overset{-(t)}{a}_{ij}, & i < j, \\ 1, & i = j, \\ \overset{+(t)}{a}_{ij}, & i > j, \end{cases} \quad \overset{+}{A_t} = (\overset{+(t)}{a}_{ij})_{n\times n} = \begin{cases} \overset{+(t)}{a}_{ij}, & i < j, \\ 1, & i = j, \\ \overset{-(t)}{a}_{ij}, & i > j. \end{cases} \tag{5}$$

2.1. Measuring the Dissimilarity between Matrices

A useful operator to measure the degree of dissimilarity between two MPRs is the Hadamard Product (HP). The HP of $A = (a_{ij})_{n\times n}$ and $B = (b_{ij})_{n\times n}$ is defined by

$$C = (c_{ij})_{n\times n} = A \circ B = a_{ij}b_{ij}. \tag{6}$$

Consequently, the degree of dissimilarity between A and B is defined as $d(A,B) = \frac{1}{n^2}e^T A \circ B^T e$ or:

$$\frac{1}{n^2}\Sigma_{i=1}^{n}\Sigma_{j=1}^{n}a_{ij}b_{ji} = \frac{1}{n}\left[\frac{1}{n}\Sigma_{i=1}^{n-1}\Sigma_{j=i+1}^{n}\left(a_{ij}b_{ji} + a_{ji}b_{ij}\right) + 1\right], \tag{7}$$

where $e = (1,1,\cdots,1)^T_{n\times 1}$. Note that $d(A,B) \geq 1$, $d(A,B) = d(B,A)$ and $d(A,B) = 1$ if and only if $A = B$.

In order to have an assessment of Individual Consistency (CI), one can measure the compatibility of A_l with respect to (w.r.t.) its own completely consistent matrix K given by Equation (2). Thus,

$$CI_K(A_l) = d(A_l, K) \leq \overline{CI}, \tag{8}$$

where $\overline{CI} = 1.1$, $A_l, l = 1,2,\cdots,m$ is an individual MPR.

In a similar manner, the group consensus index of each MPR, i.e., $GCI_{A^c}(A_l), l = 1,2,\cdots,m$, is based on the compatibility of A_l w.r.t. the group opinion given by A^c in Equation (3). Thus, the assessment of group consensus for each MPR A_l given by $GCI_{A^c}(A_l)$ read as:

$$GCI_{A^c}(A_l) = d(A_l, A^c) \leq \overline{GCI}, \tag{9}$$

where the index is usually set at $\overline{GCI} = 1.1$ and A_l is an individual MPR.

From Equations (7) and (8), for CI_K, it follows

$$\begin{array}{l} d(A_l,K) \leq \overline{CI} \Rightarrow \\ 1.0 \leq \frac{1}{n}\left[\frac{1}{n}\Sigma_{i=1}^{n-1}\Sigma_{j=i+1}^{n}\left(a_{ij}\prod_{k=1}^{n}(a_{ik}a_{kj}) + a_{ji}\prod_{k=1}^{n}(a_{jk}a_{ki})\right) + 1\right] \leq \overline{CI}. \end{array} \tag{10}$$

Respectively from Equations (7) and (9), for GCI_{A^c} evaluated for an A_l follows:

$$d(A_l, A^c) \leq \overline{GCI} \Rightarrow 1.0 \leq \frac{1}{n}\left[\frac{1}{n}\Sigma_{i=1}^{n-1}\Sigma_{j=i+1}^{n}\left(a_{ij,l}a^c_{ji} + a_{ji,l}a^c_{ij}\right) + 1\right] \leq \overline{GCI}. \tag{11}$$

An MPR A_t is completely consistent if and only if $CI_K(A_t) = 1$. Thus, a threshold useful to measure the similarity of two MPRs up to an acceptable level of consistency was suggested by [41,42] as $\overline{CI} = 1.1$. Similarly, an MPR A_t is completely in consensus if and only if $GCI_{A^c}(A_t) = 1$ and an acceptable level of consensus is $\overline{GCI} = 1.1$.

In the following up to the end of the section, we utilize Definitions and Theorems recently introduced to measure the Individual Consistency Index and Group Consensus Index of a set of I-MPRs (cf. [43,44]). For details and proofs, please refer to these works.

Definition 1. *Let* $\{\overset{o}{A}_t\}_{n\times n} = \{a_{ij}^{(t)}\}_{n\times n}$ *be the set of MPRs generated by the combinations of* $\overset{-(t)}{a}_{ij}$ *and* $\overset{+(t)}{a}_{ij}$ *entries of* A_t *given by Equation (5), where* $o = 1,2,\cdots,\mu$ *and* $\mu = 2^{\frac{n(n-1)}{2}}$.

Definition 2. *The Individual Consistency Index of the I-MPRs* $(A_t)_{n\times n}$ *given by Equation (4) when one has generated the set of MPRs* $\{\overset{o}{A}_t\}_{n\times n}$ *given by Definition 1, is defined by*

$$CI_{\underset{K}{*}}(\overset{*}{A}_t) \equiv max\{CI_{\underset{K}{1}}(\overset{1}{A}_t), CI_{\underset{K}{2}}(\overset{2}{A}_t), \cdots, CI_{\underset{K}{\mu-1}}(\overset{\mu-1}{A}_t), CI_{\underset{K}{\mu}}(\overset{\mu}{A}_t)\}. \tag{12}$$

Definition 3. *The smallest Individual Consistency Index of the I-MPRs $(A_t)_{n\times n}$ given by Equation (4) when one has generated the set of MPR $\{\overset{o}{A_t}\}_{n\times n}$ given by Definition 1, is defined by*

$$CI_{\tilde{K}}(\tilde{A}_t) \equiv min\{CI_{\underset{K}{1}}(\overset{1}{A_t}), CI_{\underset{K}{2}}(\overset{2}{A_t}), \cdots, CI_{\underset{K}{\mu-1}}(\overset{\mu-1}{A}_t), CI_{\underset{K}{\mu}}(\overset{\mu}{A_t})\}. \tag{13}$$

From Definition 2, the next theorem is set forth.

Theorem 1. *Let $(A_t)_{n\times n} = (a_{ij}^{(t)})_{n\times n}$ be an I-MPR given by Equation (4) and generate the set of μ MPRs $\{\overset{o}{A_t}\}_{n\times n}$ by using the Definition 1. If one has an MPR, $(A_x)_{n\times n} = (a_{ij}^{(x)})_{n\times n}$ within the intervals $\left[\overset{-}{A_t}\ \overset{+}{A_t}\right]$ given by the I-MPRs $(A_t)_{n\times n}$, then*

$$CI_K(A_x) \leq CI_{\underset{K}{*}}(\overset{*}{A_t}). \tag{14}$$

In order to illustrate how useful are these Definitions and Theorem, we apply them on two I-MPRs given by two experts to assessing three criteria.

Example 1. *Let us consider two experts evaluating a set of three criteria through:*

$$A_1 = \begin{pmatrix} 1 & [1\ 2] & [3\ 4] \\ [1/2\ 1] & 1 & [5\ 6] \\ [1/4\ 1/3] & [1/6\ 1/5] & 1 \end{pmatrix}, \quad A_2 = \begin{pmatrix} 1 & [1.1\ 1.2] & [1.3\ 1.4] \\ [1/1.2\ 1/1.1] & 1 & [1.5\ 1.6] \\ [1/1.4\ 1/1.3] & [1/1.6\ 1/1.5] & 1 \end{pmatrix}, \tag{15}$$

where for A_1 it implies $w_1 \succ w_2 \succ w_3$, and for A_2 one gets $w_1 \succ w_2 \succ w_3$.

Note: Experts coincide in their rankings of the three criteria. Nevertheless, the assessment ratios are different.

In the following, the Definitions 1–3 are figured out for these two I-MPRs.

From Definition 1 applied to A_1:

$$\begin{aligned} \overset{1}{A_1} &= \begin{pmatrix} 1 & 1 & 3 \\ 1 & 1 & 5 \\ 1/3 & 1/5 & 1 \end{pmatrix}, \overset{2}{A_1} = \begin{pmatrix} 1 & 1 & 3 \\ 1 & 1 & 6 \\ 1/3 & 1/6 & 1 \end{pmatrix}, \overset{3}{A_1} = \begin{pmatrix} 1 & 1 & 4 \\ 1 & 1 & 5 \\ 1/4 & 1/5 & 1 \end{pmatrix}, \cdots, \\ \overset{8}{A_1} &= \begin{pmatrix} 1 & 2 & 4 \\ 1/2 & 1 & 6 \\ 1/4 & 1/6 & 1 \end{pmatrix}. \end{aligned} \tag{16}$$

In a similar manner, from Definition 1 applied to A_2:

$$\begin{aligned} \overset{1}{A_2} &= \begin{pmatrix} 1 & 1.1 & 1.3 \\ * & 1 & 1.5 \\ * & * & 1 \end{pmatrix}, \overset{2}{A_2} = \begin{pmatrix} 1 & 1.1 & 1.3 \\ * & 1 & 1.6 \\ * & * & 1 \end{pmatrix}, \overset{3}{A_2} = \begin{pmatrix} 1 & 1.1 & 1.4 \\ * & 1 & 1.5 \\ * & * & 1 \end{pmatrix}, \cdots, \\ \overset{8}{A_2} &= \begin{pmatrix} 1 & 1.2 & 1.4 \\ * & 1 & 1.6 \\ * & * & 1 \end{pmatrix}, \end{aligned} \tag{17}$$

where, from now on, symbol $*$ corresponds to the inverted entries, respectively.

From Definition 2 applied to A_1 and A_2, it read as:

$$CI_{\underset{K}{*}}(\overset{*}{A_1}) = 1.072454,\ CI_{\tilde{K}}(\tilde{A}_1) = 1.0018;\ CI_{\underset{K}{*}}(\overset{*}{A_2}) = 1.005640,\ CI_{\tilde{K}}(\tilde{A}_2) = 1.0010. \tag{18}$$

The first expert has the higher Individual Consistency Index which means that s/he has a weaker consistency in her/his judgments. It coincides with our logical expectations by analyzing and comparing both I-MPRs.

By defining for A_1 and A_2 an MPR within their own intervals, for instance

$$A_{1x} = \begin{pmatrix} 1 & 1.5 & 3.7 \\ * & 1 & 5.6 \\ * & * & 1 \end{pmatrix}, \; A_{2x} = \begin{pmatrix} 1 & 1.15 & 1.37 \\ * & 1 & 1.56 \\ * & * & 1 \end{pmatrix}, \tag{19}$$

and by applying Theorem 1 to both MPR, one has:

$$CI_{K_{1x}}(A_{1x}) = 1.0251 \leq CI_{\overset{*}{K}}(\overset{*}{A_1}); \; CI_{K_{2x}}(A_{2x}) = 1.0027 \leq CI_{\overset{*}{K}}(\overset{*}{A_2}). \tag{20}$$

In order to reach a Group Consensus solution, the experts are supposed to participate in rounds of discussion to meet a common goal. Then, definitions are next introduced to compute the dissimilarity amongst DM′ assessments for finally obtaining the GCI level.

In a similar manner as Definition 1, one can compute the whole combination of the values given by the set of I-MPRs through:

Definition 4. *From the m I-MPRs given as in Equation (4), $(A_t)_{n\times n}$, $t = 1, 2, \cdots, m$, their set of MPRs $\{\breve{A}\}_{n\times n}$ which is the whole combination of the interval values of the set of $(A_t)_{n\times n}$ I-MPRs with $\overset{-(t)}{a_{ij}}$ and $\overset{+(t)}{a_{ij}}$, where $r = 1, 2, \cdots, v$ and $v = 2^{m\cdot\mu} = 2^{\frac{m(n)(n-1)}{2}}$, is given by*

$$\{\breve{A}\}_{n\times n} = \{\overset{1}{\breve{A}}, \overset{2}{\breve{A}}, \cdots, \overset{v}{\breve{A}}\}. \tag{21}$$

Furthermore, a set of $\{\breve{A}^c\}_{n\times n}$ which represents each group opinion utilizing the geometric average operator corresponding to each element of the set $\{\breve{A}\}_{n\times n}$, can be defined.

Definition 5. *Based on Definition 4, one can compute for the set of I-MPRs $(A_t)_{n\times n}$ and $\{\breve{A}\}_{n\times n}$, $t = 1, 2, \cdots, m$, its set of MPR $\{\breve{A}^c\}_{n\times n}$. Then, the corresponding set of $\{\breve{A}^c\}_{n\times n}$ which is the Group I-MPRs Opinion utilizing the geometric operator is given by*

$$\{\breve{A}^c\}_{n\times n} = \{\breve{a}^c_{ij}\}_{n\times n} = \{\prod_{t=1}^{m} (\breve{A})^{\lambda_t}\}, \; t \in M, \tag{22}$$

where $\{\breve{a}^c_{ij}\}_{n\times n}$ is the combination of the whole set of values ($2^{m\cdot\mu} = 2^{\frac{m\cdot n\cdot(n-1)}{2}}$).

Definition 6. *The Group Consensus Index of a set of I-MPRs $(A_t)_{n\times n}$, $t = 1, 2, \cdots, m$, given by Equation (4) when one has generated the set of MPRs $\{\breve{A}^c\}_{n\times n} = \{\overset{1}{\breve{A}^c}, \overset{2}{\breve{A}^c}, \cdots, \overset{v}{\breve{A}^c}\}$, given by Definition 5, is defined by*

$$GCI_{\breve{H}_t}(A_t)_{n\times n} \equiv \max\{GCI_{\overset{1}{\breve{A}^c}}(\overset{1}{\breve{A}}), GCI_{\overset{2}{\breve{A}^c}}(\overset{2}{\breve{A}}), \cdots, GCI_{\overset{v-1}{\breve{A}^c}}(\overset{v-1}{\breve{A}}), GCI_{\overset{v}{\breve{A}^c}}(\overset{v}{\breve{A}})\}, \tag{23}$$

where $v = 2^{\frac{m\cdot n\cdot(n-1)}{2}}$ and $\breve{H}_t$ is associated with the corresponding $\overset{p}{\breve{A}^c}$ matrix and $p \in \{1, 2, \cdots, v\}$.

Definition 7. *The smallest Group Consensus Index of a set of I-MPRs $(A_t)_{n\times n}$, $t = 1, 2, \cdots, m$, given by Equation (4) when one has generated the set of MPRs $\{\breve{A}^c\}_{n\times n} = \{\overset{1}{\breve{A}^c}, \overset{2}{\breve{A}^c}, \cdots, \overset{v}{\breve{A}^c}\}$, given by Definition 5, is defined by*

$$GCI_{\bar{H}_t}\left(\bar{A}_t\right)_{n\times n} \equiv \min\{GCI_{\overset{1}{\breve{A}^c}}(\overset{1}{\breve{A}}), GCI_{\overset{2}{\breve{A}^c}}(\overset{2}{\breve{A}}), \cdots, GCI_{\overset{\nu-1}{\breve{A}^c}}(\overset{\nu-1}{\breve{A}}), GCI_{\overset{\nu}{\breve{A}^c}}(\overset{\nu}{\breve{A}})\}, \tag{24}$$

where $\nu = 2^{\frac{m\cdot n\cdot(n-1)}{2}}$ *and* $\bar{H}_t$ *is associated with the corresponding* $\overset{p}{\breve{A}^c}_{n\times n}$ *matrix and* $p \in \{1, 2, \cdots, \nu\}$.

Example 2. *Let us consider the confidence weight of the two experts through the following weight vector* $\lambda = [2/7\ \ 5/7]$. *Then, from Definition 4, Equations (16) and (17), it follows:*

$$\overset{1}{\breve{A}} = \{\overset{1}{A_1}, \overset{1}{A_2}\}; \overset{2}{\breve{A}} = \{\overset{1}{A_1}, \overset{2}{A_2}\}; \overset{3}{\breve{A}} = \{\overset{1}{A_1}, \overset{3}{A_2}\}; \cdots; \overset{64}{\breve{A}} = \{\overset{8}{A_1}, \overset{8}{A_2}\}. \tag{25}$$

From Definition 5 applied to Equation (25), one has:

$$\{\breve{A}^c\} = \{\overset{1}{\breve{A}^c}, \overset{2}{\breve{A}^c}, \cdots, \overset{64}{\breve{A}^c}\}, \tag{26}$$

where, from Equation (3), it implies:

$$\begin{aligned} \overset{1}{\breve{A}^c} &= (\overset{1}{\breve{a}^c}_{ij}) = \left(\overset{1}{a}{}^{(1)}_{ij}\right)^{2/7} * \left(\overset{1}{a}{}^{(2)}_{ij}\right)^{5/7}; \overset{2}{\breve{A}^c} = (\overset{2}{\breve{a}^c}_{ij}) = \left(\overset{1}{a}{}^{(1)}_{ij}\right)^{2/7} * \left(\overset{2}{a}{}^{(2)}_{ij}\right)^{5/7}; \\ \overset{3}{\breve{A}^c} &= (\overset{3}{\breve{a}^c}_{ij}) = \left(\overset{1}{a}{}^{(1)}_{ij}\right)^{2/7} * \left(\overset{3}{a}{}^{(2)}_{ij}\right)^{5/7}; \cdots; \overset{64}{\breve{A}^c} = (\overset{64}{\breve{a}^c}_{ij}) = \left(\overset{8}{a}{}^{(1)}_{ij}\right)^{2/7} * \left(\overset{8}{a}{}^{(2)}_{ij}\right)^{5/7}. \end{aligned} \tag{27}$$

From Definition 6, it results:

$$GCI_{\breve{H}_1}(\breve{A}_1) = 1.2143;\ \ GCI_{\breve{H}_2}(\breve{A}_2) = 1.0325. \tag{28}$$

These results indicate that the first expert A_1, *which has the lowest weight* ($\lambda_1 = 2/7$), *is not in consensus. In addition, from Definition 7, it follows:*

$$GCI_{\bar{H}_1}\left(\bar{A}_1\right) = 1.1120;\ \ GCI_{\bar{H}_2}\left(\bar{A}_2\right) = 1.0173. \tag{29}$$

As soon as we obtain ICI and GCI of a set of I-MPRs, the accepted threshold values ($\overline{ICI}$ and $\overline{GCI}$ respectively) give how acceptable in Consistency and Consensus the assessments of the DMs are. Whenever some I-MPRs are over their index values, we can apply the following strategy to improve their Individual Consistency and/or their Group Consensus Indices.

3. Reliable Intervals for Individual Consistency and Group Consensus

Once the ICI has been calculated, it may happen that one or more I-MPRs are not consistent. For example, r I-MPRs were not consistent. In that case, we need to improve the Individual Consistency of those I-MPRs and then let us define this set of MPRs as:

$$A_I^o \equiv \begin{bmatrix} \overset{-}{a}{}^{(o)}_{ij} & \overset{+}{a}{}^{(o)}_{ij} \end{bmatrix}_{n\times n},\quad i, j \in N,\ \ o = 1, 2, \cdots, r. \tag{30}$$

Naturally, each one of the A_I^o, $I = 1, 2, \cdots, r$ doesn't verify for a special combination of its values (given by Equation (12)), the inequality given in Equation (10).

In a similar manner, once the GCI has been calculated, it may happen that one or more I-MPRs are not in consensus. For example, for instance y I-MPRs were not in consensus. In that case, we need to improve the Group Consensus of those I-MPRs and then let us define this set of MPRs as

$$A_J^p \equiv \begin{bmatrix} \overline{a}_{ij}^{(p)} & \overline{a}_{ij}^{(p)} \end{bmatrix}_{n \times n}, \quad i,j \in N, \ p = 1,2,\cdots,y. \tag{31}$$

Once again, each one of the $A_J^P, J = 1,2,\cdots,y$ doesn't verify, for a special combination of values (given by Equation (23)), the inequality given in Equation (11).

3.1. Sequential Quadratic Programming Methodology

When one has a constrained nonlinear optimization problem (NLP) and wants a numerical solution, the Sequential Quadratic Programming (SQP) [45] is a useful approach. Let us consider the implementation of the sequential quadratic programming methodology to NLP of the form:

$$\begin{aligned} \text{minimize} \quad & f(x), \\ \text{over } x \quad & \in \quad \mathbb{R}^n, \\ \text{subject to } h(x) \quad & = \quad 0, \\ g(x) \quad & \leq \quad 0, \end{aligned} \tag{32}$$

where $f : \mathbb{R}^n \longrightarrow \mathbb{R}$ is the objective functional, the functions $h : \mathbb{R}^n \longrightarrow \mathbb{R}^m$ and $g : \mathbb{R}^n \longrightarrow \mathbb{R}^p$ describe the equality and inequality constraints. The sequential quadratic programming is an iterative method for modelling the NLP for a given iterate x^k, $k \in \mathbb{N}_0$, by a Quadratic Programming (QP) subproblem. As soon as the method solves that QP subproblem, the solution is used to build a new iterate x^{k+1}. This is figured out in such a way that the sequence $(x^k)_{k \in \mathbb{N}_0}$ converges to a local minimum x^* of the NLP Equation (32) as $k \to \infty$.

Then, as we can observe, the problem of minimizing an objective functional (given in our case by Equation (7)), where one has some inequalities constraints from Equation (10), (related to $g(x)$) with the additional observation of constraints imposed by Equation (30) (related to $h(x)$), defines an NLP to improve the Individual Consistency Index of an I-MPR.

In a similar manner, the problem of minimizing an objective functional (given by Equation (7)) providing some inequalities constraints from Equation (11), (related to $g(x)$) with the additional observation of constraints imposed by Equation (31) (related to $h(x)$), defines an NLP to improve the Group Consensus Index of a set of I-MPRs.

An additional observation is that since we have interval judgments in I-MPRs, we should apply the SQP to minimize the objective functional to found the minimal values verifying the required constraints. After that, we should then apply the SQP to maximize the objective functional to found the maximal values. In this manner, we will finally find reliable intervals where are fulfilled inequalities and equalities constraints and conditions of Equations (10) and (11), to obtain acceptable Individual Consistency and Group Consensus, respectively.

3.2. Matching the Problem with the SQP for Improving I-MPRs

From Equation (30) (Equation (31) resp.), the following inequalities can be stated in terms of optimization variables x_i, as follows:

$$\begin{aligned} \overline{a}_{12}^{\delta} \quad & \leq \quad x_1 \quad \leq \quad \overset{+}{a}{}_{12}^{\delta}, \\ & \vdots \\ \overline{a}_{1n}^{\delta} \quad & \leq \quad x_{n-1} \quad \leq \quad \overset{+}{a}{}_{1n}^{\delta}, \\ & \vdots \\ \overline{a}_{(n-1)n}^{\delta} \quad & \leq \quad x_{\frac{n^2-n}{2}} \quad \leq \quad \overset{+}{a}{}_{(n-1)n}^{\delta}, \end{aligned} \tag{33}$$

where δ stands for o in the Individual Consistency assessments and for p in the Group Consensus assessments. For the Individual Consistency, the inequalities given from Equation (10) are

$1.0 \leq d(\overset{*}{A}_I, \overset{*}{K}) \leq 1.1$ which impose some inequality constraints to be fulfilled. For Group Consensus improvement, they are given from Equation (11) as $1.0 \leq d(\check{A}_J, \check{H}_J) \leq 1.1$.

The relationship between x_k, $k = 1, 2, \cdots, \frac{n^2-n}{2}$ and the set of I-MPRs $(A_t)_{n \times n}$, $t = 1, 2, \cdots, m$ under analysis is given by:

$$
\begin{aligned}
A_t &= \begin{pmatrix}
1 & \overset{-+}{a}{}^{(t)}_{12} & \overset{-+}{a}{}^{(t)}_{13} & \cdots & \overset{-+}{a}{}^{(t)}_{1(n-1)} & \overset{-+}{a}{}^{(t)}_{1n} \\
* & 1 & \overset{-+}{a}{}^{(t)}_{23} & \cdots & \overset{-+}{a}{}^{(t)}_{2(n-1)} & \overset{-+}{a}{}^{(t)}_{2n} \\
* & * & 1 & \cdots & \overset{-+}{a}{}^{(l)}_{3(n-1)} & \overset{-+}{a}{}^{(t)}_{3n} \\
\vdots & \cdots & \cdots & \ddots & \vdots & \vdots \\
* & * & \cdots & \cdots & 1 & \overset{-+}{a}{}^{(t)}_{(n-1)n} \\
* & * & \cdots & \cdots & * & 1
\end{pmatrix} \equiv \\
X_t &= \begin{pmatrix}
1 & x_{1+\frac{n(n-1)}{2}(t-1)} & x_{2+\frac{n(n-1)}{2}(t-1)} & \cdots & x_{n-1+\frac{n(n-1)}{2}(t-1)} \\
* & 1 & x_{n+\frac{n(n-1)}{2}(t-1)} & \cdots & x_{2(n-1)-1+\frac{n(n-1)}{2}(t-1)} \\
* & * & 1 & \cdots & x_{3(n-1)-3+\frac{n(n-1)}{2}(t-1)} \\
\vdots & \cdots & \cdots & \ddots & \vdots \\
* & * & \cdots & \cdots & x_{\frac{n(n-1)}{2}t} \\
* & * & \cdots & \cdots & 1
\end{pmatrix},
\end{aligned} \tag{34}
$$

where $\overset{-+}{a}{}^{(t)}_{ij}$ stands for the corresponding crisp or interval value of the I-MPR.

From the definition of constrained nonlinear optimization problem, we note that when the I-MPR is in fact just an MPR (when the respective expert has a high confidence in her/his assessments), the SQP could not give any different solution. This is because the optimization variables will remain unchanged since the variables x_i, $i = 1, 2, 3, \cdots, (n^2 - n)/2$ will be defined as a crisp values (cf. Equation (33)).

Thus, in order to provide a general benchmark to address and solve any possible case of assessments given through I-MPRs, we will modify slightly the SQP. We introduce a design parameter ϵ which will be used as an additive or subtractive element, for high and low bounds, respectively, of the I-MPRs to be improved.

For example, we modify Equation (33) as follows:

$$
\begin{array}{rcccl}
\overline{a}^{\delta}_{12} - \epsilon & \leq & x_1 & \leq & \overset{+}{a}{}^{\delta}_{12} + \epsilon, \\
& & \vdots & & \\
\overline{a}^{\delta}_{1n} - \epsilon & \leq & x_{n-1} & \leq & \overset{+}{a}{}^{\delta}_{1n} + \epsilon, \\
& & \vdots & & \\
\overline{a}^{\delta}_{(n-1)n} - \epsilon & \leq & x_{\frac{n^2-n}{2}} & \leq & \overset{+}{a}{}^{\delta}_{(n-1)n} + \epsilon,
\end{array} \tag{35}
$$

where $\epsilon > 0$. For the sake of compactness, let us integrate in the same terms the parameter ϵ. For example, $\overline{a}^{(o)}_{ij} \equiv \overline{a}^{o}_{ij} - \epsilon$ and $\overset{+}{a}{}^{(o)}_{ij} \equiv \overset{+}{a}{}^{o}_{ij} + \epsilon$ for Individual Consistency analysis, and $\overline{a}^{(p)}_{ij} \equiv \overline{a}^{p}_{ij} - \epsilon$ and $\overset{+}{a}{}^{(p)}_{ij} \equiv \overset{+}{a}{}^{p}_{ij} + \epsilon$, for Group Consensus analysis.

In that manner, when the SQP does not provide a feasible solution, we will again iterate with an incremental ϵ-value, until a solution is found.

It is worth mentioning that ϵ must be initialized, with a small value, to keep the maximum of the information provided by the expert.

3.2.1. Individual Consistency Objective Functional

Thus, when an I-MPR is inconsistent, we need to synthesize the objective functional. For example, from Equations (7) and (10), one obtains:

$$1.0 \leq \frac{1}{n}\left[\frac{1}{n}\Sigma_{i=1}^{n-1}\Sigma_{j=i+1}^{n}\left(a_{ij}\prod_{k=1}^{n}(a_{ik}a_{kj}) + a_{ji}\prod_{k=1}^{n}(a_{jk}a_{ki})\right) + 1\right] \leq \overline{CI}. \tag{36}$$

After some algebraic manipulations, it implies:

$$1.0 \leq \frac{1}{n^2}\left[\Sigma_{i=1}^{n-1}\Sigma_{j=i+1}^{n}\left(a_{ij}^{\frac{n-2}{n}}\prod^{n}_{\substack{l=1\\ i\neq l, l\neq j}}(a_{jl}a_{li}) + a_{ji}^{\frac{n-2}{n}}\prod^{n}_{\substack{l=1\\ i\neq l, l\neq j}}(a_{il}a_{lj})\right)\right] + \frac{1}{n} \leq \overline{CI}. \tag{37}$$

3.2.2. Group Consensus Objective Functional

In a similar manner, when an I-MPR $(A_J^o)_{n\times n}$ is not in consensus, we need to synthesize the objective functional. For example, from Equation (11), one obtains:

$$1.0 \leq \frac{1}{n^2}\left[\Sigma_{i=1}^{n-1}\Sigma_{j=i+1}^{n}\left(a_{ij}^{(J)}\prod_{t=1}^{m}(a_{ji}^{(t)})^{\lambda_t} + a_{ji}^{(J)}\prod_{t=1}^{m}(a_{ij}^{(t)})^{\lambda_t}\right)\right] + \frac{1}{n} \leq \overline{GCI}. \tag{38}$$

After some algebraic manipulations, it implies for $(A_J)_{n\times n}$:

$$\begin{aligned} 1.0 \leq \frac{1}{n^2}\Bigg(& \underbrace{\frac{a_{12}^{(J)(1-\lambda_J)}}{a_{12}^{(1)(\lambda_1)} \cdot a_{12}^{(2)(\lambda_2)} \cdots a_{12}^{(m)(\lambda_m)}}} + \frac{1}{\alpha_1} + \underbrace{\frac{a_{13}^{(J)(1-\lambda_J)}}{a_{13}^{(1)(\lambda_1)} \cdot a_{13}^{(2)(\lambda_2)} \cdots a_{13}^{(m)(\lambda_m)}}} + \frac{1}{\alpha_2} + \\ & \cdots + \underbrace{\frac{a_{(n-1)n}^{(J)(1-\lambda_J)}}{a_{(n-1)n}^{(1)(\lambda_1)} \cdot a_{(n-1)n}^{(2)(\lambda_2)} \cdots a_{(n-1)n}^{(m)(\lambda_m)}}} + \frac{1}{\alpha_m}\Bigg) + \frac{1}{n} \leq \overline{GCI}, \end{aligned} \tag{39}$$

where α_1 equals to the term over the first brace, α_2 equals to the term over the second brace, and so on. Furthermore, note that the term $a_{12}^{(1)(\lambda_1)} \cdot a_{12}^{(2)(\lambda_2)} \cdots a_{12}^{(m)(\lambda_m)}$ excepts the term $a_{12}^{(J)(\lambda_J)}$, the term $a_{13}^{(1)(\lambda_1)} \cdot a_{13}^{(2)(\lambda_2)} \cdots a_{13}^{(m)(\lambda_m)}$ excepts the term $a_{13}^{(J)(\lambda_J)}$, and so on.

Finally, the initialization point x_0 used in the NLP can be set at any point within the corresponding interval given by Equation (33).

In the following, an algorithm based on an NLP is used to obtain reliable intervals for the assessment of based decision models such as those given by the set of I-MPRs for both indices ($\overline{CI}$ and $\overline{GCI}$). Two similar methods are blended to obtain both reliable I-MPR Consistency and Consensus levels. For example, for Individual Consistency improvement, we can find an I-MPR verifying Individual Consistency. For the other case, for Group Consensus improvement, a reliable Consensus Index for I-MPRs is obtained. In order to use an NLP, an SQP algorithm can be found in [45]. In the following, our algorithm is described in detail.

3.3. *Improving the Individual Consistency of an I-MPR*

A scheme of the Algorithm 1 implementation is depicted in Figure 1. Once the set of I-MPRs are all acceptably consistent, we can improve the Group Consensus level through the following algorithm.

Algorithm 1: Algorithm IC-I-MPR

Input: $A_I^o = (\overset{-(o)}{a_{ij}} \; \overset{+(o)}{a_{ij}})_{n\times n}$: the initial interval I-MPR; x_0: the initialization value for the nonlinear optimization, which is to be defined within the corresponding interval or as the corresponding crisp value; the threshold value of $\overline{CI}$ for the Individual Consistency assessment; ϵ_c : Design parameter value allowing the enlargement pace of the searching space of the algorithm.

Output: $\overline{A}_I^o$: the consistency interval matrix computed and verifying interval conditions given by Equation (10).

Step 1: Get the function for the assessment of Individual Consistency given by Equation (37).

Step 2: Define for the nonlinear optimization algorithm, the allowed intervals:

$$\overset{-(o)}{a_{ij}} \leq x_i \leq \overset{+(o)}{a_{ij}} ; j > i, \; i,j = 1,2,\cdots,n. \tag{40}$$

Thus, assign the former linear inequality constraints as follows:

$$\begin{aligned} c(2k+1) &= x_i - \overset{+(o)}{a_{ij}} ; \; c(2(k+1)) = \overset{-(o)}{a_{ij}} - x_i, \\ j > i, \; i,j &= 1,2,\cdots,n, \quad k = 0,1,2,\cdots,\tfrac{n^2-n}{2}. \end{aligned} \tag{41}$$

Step 3: Obtain the acceptable index of Individual Consistency, $1.0 \leq d(\overset{*}{A}_I, \overset{*}{K}) \leq \overline{CI}$. Thus, based on matrix K given by Equation (2), the nonlinear inequality constraints imposed are given by Equation (10).

Step 4: Solve the former nonlinear optimization problem using the SQP algorithm to minimize it.

Step 5: If an unfeasible solution is obtained, assign $\epsilon_c = \beta * \epsilon_c$, where $\beta = 1,2,3,\cdots$, increments at each iteration and return to **Step 4**. Otherwise, continue to the next step.

Step 6: Obtain $\overline{A}_{Imin}^o = (\overset{-(o)}{a_{ij}})_{n\times n}$. Solve again the same nonlinear optimization problem but this time in order to maximize it. In order to do so, assign the objective functional as $-f(x)$. Obtain $\overline{A}_{Imax}^o = (\overset{+(o)}{a_{ij}})_{n\times n}$.

Step 7: Compose the Consistency Interval Matrix $\overline{A}_I^o$ as follows:

$$\overline{A}_I^o = \begin{bmatrix} \overset{-(o)}{a_{ij}} & \overset{+(o)}{a_{ij}} \end{bmatrix}_{n\times n}, \tag{42}$$

where $\begin{bmatrix} \overset{-(o)}{a_{ij}} & \overset{+(o)}{a_{ij}} \end{bmatrix}$, stands for the interval or crisp value obtained.

Step 8: end.

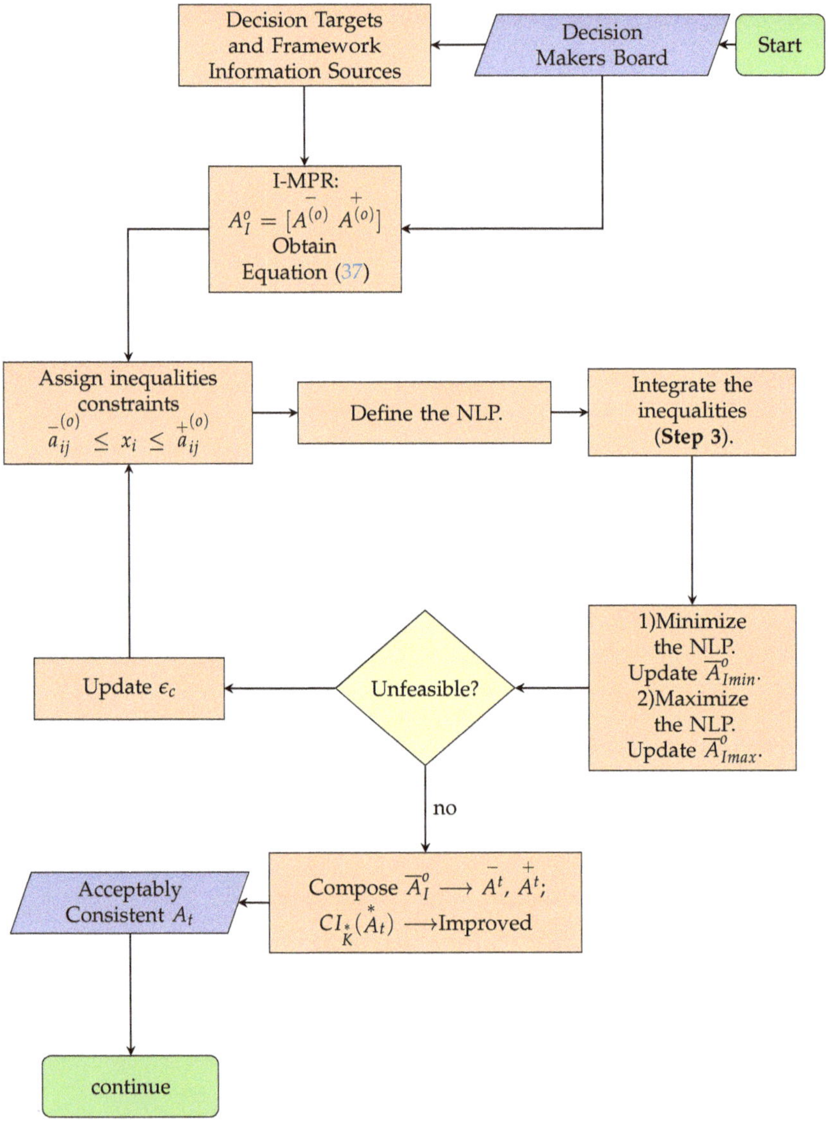

Figure 1. Process flowchart for improving consistency of an I-MPR.

3.4. Improving the Group Consensus of a Set of I-MPRs

A scheme of the Algorithm 2 implementation is depicted in Figure 2.

Algorithm 2: Algorithm GC-I-MPR

Input: $A_J^P = (\overset{-(p)}{a_{ij}}\ \overset{+(p)}{a_{ij}})_{n\times n}$: the initial interval I-MPRs; x_0: the initialization value for the nonlinear optimization, which is to be defined within the corresponding interval or as the corresponding crisp value; $\overline{GCI}$ for the Group consensus assessment; ϵ_g : design parameter allowing the enlargement of the searching space of the algorithm.

Output: $\overline{A}_J^P$: the I-MPRs computed and verifying interval conditions given by Equation (11).

Step 1: Get the function for the assessment of Group Consensus given by Equation (39).

Step 2: Define for the nonlinear optimization algorithm, the allowed intervals:

$$\overset{-(p)}{a_{ij}} \leq x_i \leq \overset{+(p)}{a_{ij}}\ ; j > i,\ \ i,j = 1,2,\cdots,n. \tag{43}$$

Thus, assign the former linear inequality constraints as follows:

$$\begin{aligned} c(2k+1) &= x_i - \overset{+(p)}{a_{ij}}\ ;\ c(2(k+1)) = \overset{-(p)}{a_{ij}} - x_i; \\ j > i,\ i,j &= 1,2,\cdots,n,\ \ k = 0,1,2,\cdots,\tfrac{n^2-n}{2}. \end{aligned} \tag{44}$$

Step 3: Obtain the acceptable Group Consensus Index $d(\check{A}_J, \check{H}_J)$. Thus, based on the set of $\{\check{A}^c\}_{n\times n}$ given by Equation (22), the nonlinear inequality constraints imposed is given by $1.0 \leq d(\check{A}_J, \check{H}_J) \leq 1.1$.

Step 4: If $GCI(A_t) \leq \overline{GCI}$, $t = 1,2,\cdots,m$, then goto **Step 9**. Otherwise, continue with the next step.

Step 5: Solve the former nonlinear optimization problem (NLP) using the SQP algorithm to minimize it.

Step 6: If an unfeasible solution is obtained, assign $\epsilon_g = \theta * \epsilon$, where $\theta = 1,2,3,\cdots$, increments at each iteration and return to **Step 4**. Otherwise, continue to the next step.

Step 7: Obtain the matrix $\overline{A}_{Jmin}^P = (\overset{-(p)}{a_{ij}})_{n\times n}$. Solve again the same nonlinear optimization problem but this time in order to maximize it. Obtain $\overline{A}_{Jmax}^P = (\overset{+(p)}{a_{ij}})_{n\times n}$.

Step 8: Goto to **Step 2**.

Step 9: Compose the $J-th$ Group Consensus Interval Matrix $\overline{A}_J^P$ as follows:

$$\overline{A}_J^P = \begin{bmatrix} \overset{-(p)}{a_{ij}} & \overset{+(p)}{a_{ij}} \end{bmatrix}_{n\times n}, \tag{45}$$

where $\begin{bmatrix} \overset{-(p)}{a_{ij}} & \overset{+(p)}{a_{ij}} \end{bmatrix}$, stands for the interval or crisp value obtained.

Step 10: end.

Section Remarks:

- In the case that an expert has provided a crisp value(s) in her/his judgement(s), this value(s) drives the process of nonlinear optimization since they will slightly change with the pace of ϵ. It is very useful since precisely in that value(s), the expert has shown her/his highest confidence level.
- At the end of both algorithms, one gets reliable I-MPRs, i.e., where the consistency and consensus constraints are fulfilled.

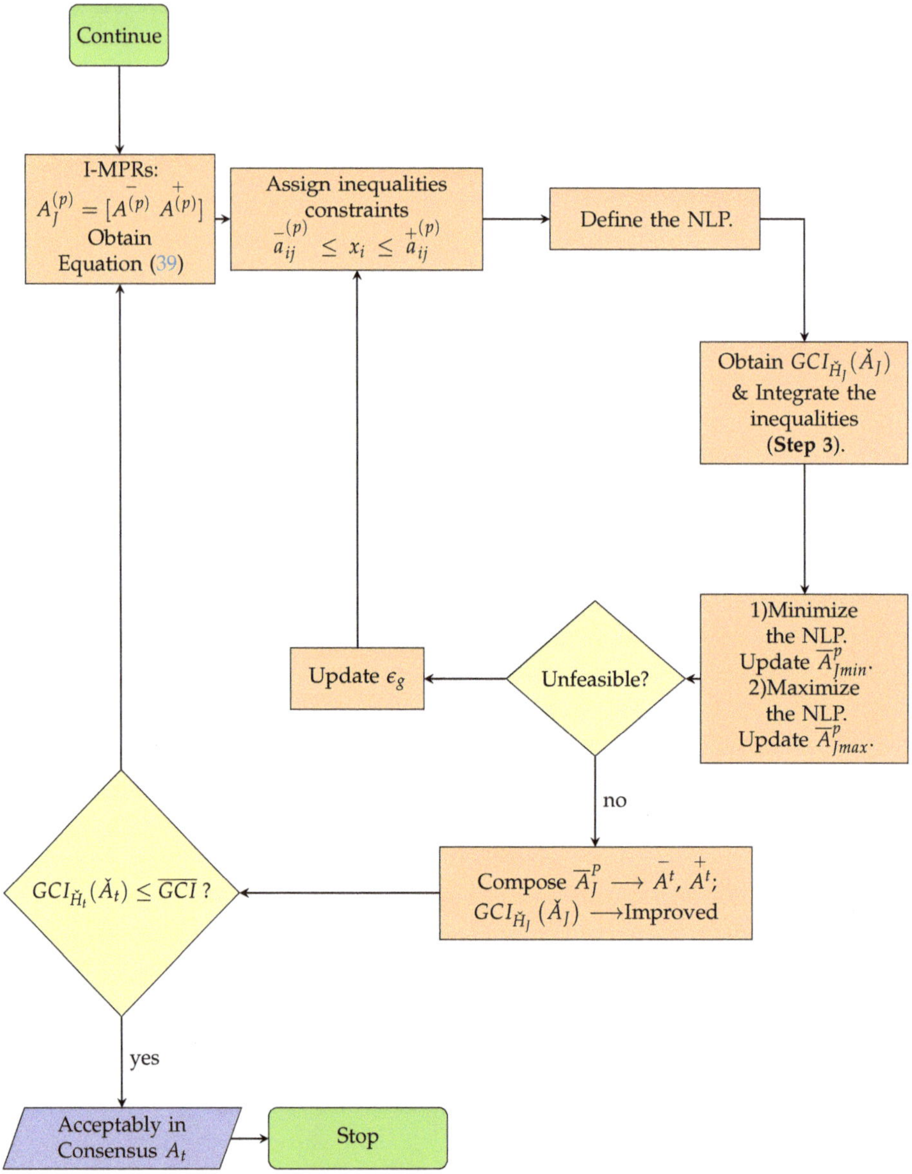

Figure 2. Process flowchart for improving consensus of I-MPRs.

4. Prioritization Method and Methodology Application

The process of deriving a priority weight vector of the alternatives, $w_i = (w_1, w_2, \cdots, w_n)^T$ from an I-MPR, for instance an A_t, is called a prioritization method, where $w_l \geq 0$ and $\sum_{l=1}^{n} w_l = 1$. Then, when one has the evaluation for $\check{A}$ which is associated to the maximum value (cf. Equation (23)) of the Group Consensus Index of I-MPRs A_t, it follows from Equation (9):

$$\check{w}_i = \left(\left(\prod_{j=1}^{n} \overset{*}{a}_{ij}\right)^{1/n}\right) \cdot \left(\left(\prod_{j=1}^{n} \check{H}_t\right)^{1/n}\right) = w_i(\check{A})w_i(\check{H}_t), \tag{46}$$

where $w_i(\check{A})$ and $w_i(\check{H}_t)$ are the weights of $\check{A}$ and $\check{H}_t$, respectively.

In a similar manner, for $\overline{A}$ which is associated to the minimum value (cf. Equation (13)) of the Group Consensus Index of I-MPRs A_t, one has:

$$\overline{w}_i(\theta) = \left(\left(\prod_{j=1}^{n} \overline{a}_{ij}\right)^{1/n}\right) \cdot \left(\left(\prod_{j=1}^{n} \overline{H_t}\right)^{1/n}\right) = w_i(\overline{A})w_i(\overline{H_t}), \tag{47}$$

where $w_i(\overline{A})$ and $w_i(\overline{H_t})$ are the weights of $\overline{A}$ and $\overline{H_t}$, respectively.

Consequently:

$$w_i = [\overline{w}_i,\ \check{w}_i]\,. \tag{48}$$

As soon as the set of I-MPRs is acceptably consistent and in consensus, the interval priority vector to rank alternatives is obtained. To do so, an interval ranking is used.

Interval Priority Vector Synthesis

An interval priority vector should reflect different expert's risk preferences for her/his interval judgments. There are mainly two prioritization methods: (among others cf. [46] and the references therein) Eigenvalue-based Methods (EM), (cf. [1,47]) and the Row Geometric Mean Method (RGMM), (cf. [48]) that are utilized to derive a priority weight vector from an ordered judgment matrix which is the method here utilized.

Based on the results given in [49], a slight modification of their method is addressed below.

Let us consider for instance two intervals $a = \left[\overset{-}{a}, \overset{+}{a}\right]$ and $b = \left[\overset{-}{b}, \overset{+}{b}\right]$ where $\overset{-}{a}, \overset{-}{b} > 0$ and a, b are positioned on the x and y axis, respectively. A uniform probability distribution is assumed on the constrained area composed of a and b. For upper left points on $y = x$, the y values are larger than x values and viceversa for the lower right points (cf. Figure 3 where one possible case is shown).

Then, the preference degree of $P(a > b)$ is equal to $\frac{S_1}{d(a)d(b)}$, where $d(a) = \overset{+}{a} - \overset{-}{a}$ and $d(b) = \overset{+}{b} - \overset{-}{b}$ and the following ranking interval method can be stated.

Definition 8.

$$P(a > b) = \begin{cases} 1 & \overset{-}{a} \geq \overset{+}{b}, \\ 1 - \frac{(\overset{+}{b} - \overset{-}{a})^2}{2d(a)d(b)} & \overset{-}{b} \leq \overset{-}{a} < \overset{+}{b} \leq \overset{+}{a}, \\ \frac{2\overset{+}{a} - (\overset{+}{b} + \overset{-}{b})}{2d(a)} & \overset{-}{a} < \overset{-}{b} < \overset{+}{b} \leq \overset{+}{a}, \end{cases} \tag{49}$$

where if $P(a > b) > 0.5$, then $a > b$; if $P(a > b) = 0.5$, then $a = b$ and if $P(a > b) < 0.5$, then $a < b$; and in this manner Equation (49) indicates the total order of intervals and their preference degrees [50].

In order to provide access to the public as on-line security and resource conditions allow, these complete methodology (definitions, applied theorems, algorithms and interval priority vector synthesis) will be soon available in a site. Then, this benchmark will be useful to test a submitted set of I-MPRs for assessment on their Individual Consistency (ICI) and the Group Consensus Indices (GCI). This site will test different data on various based decision models.

In the following, our main methodology is applied to numerical examples under diverse considerations in order to test different situations when the system is working in a real scenario.

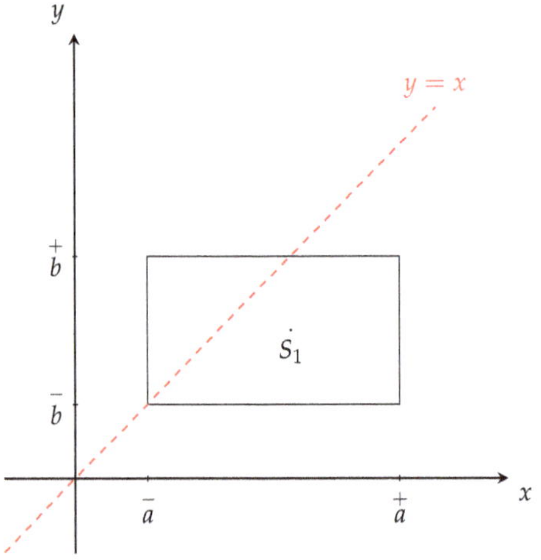

Figure 3. One case for the interval analysis.

5. Illustration of the Methodology through Numerical Examples

In the following example, let us suppose that three managers from different operating departments are participating in a group decision about the evaluation of a new market to have the best return of investment of a set of products. There are four decision criteria which involve the details of judgments, which are: development, legal restrictions for economic activities, society and infrastructure denoted as C_1, C_2, C_3, and C_4, respectively. The three mangers state their preferences over the four criteria and, since they are confident on the analysis carried out by their operating department, they have a crisp value for criteria judgments related to their department. Nevertheless, they provide an interval for others criteria assessments. Their associated weight vector is $\lambda = [1/3,\ 1/3,\ 1/3]$.

Example 3. *The three managers provide their I-MPRs through*

$$\begin{aligned} A_1 &= \begin{pmatrix} 1 & [1\ 2] & 1/5 & [1/3\ 1/2] \\ * & 1 & 1/7 & [5\ 6] \\ * & * & 1 & 1/2 \\ * & * & * & 1 \end{pmatrix}, A_2 = \begin{pmatrix} 1 & 5/2 & 1/3 & 3/4 \\ * & 1 & [1/2\ 3/2] & [2\ 3] \\ * & * & 1 & [1\ 2] \\ * & * & * & 1 \end{pmatrix}, \\ A_3 &= \begin{pmatrix} 1 & 2 & [1/2\ 3/2] & [1\ 3/2] \\ * & 1 & 3/4 & 3 \\ * & * & 1 & [2\ 3] \\ * & * & * & 1 \end{pmatrix}, \end{aligned} \tag{50}$$

where the initial ranking given by expert 1 is $w_3 \overset{1}{>} w_2 \overset{0.3724}{>} w_4 \overset{1}{>} w_1$, *by expert 2 is:* $w_3 \overset{1}{>} w_2 \overset{0.967}{>} w_1 \overset{1}{>} w_4$, *and by expert 3 is:* $w_3 \overset{0.8179}{>} w_1 \overset{1}{>} w_2 \overset{1}{>} w_4$.

Note that the first manager has more information and expertise on the third criteria. On the other hand, the second manager is focused on the first criteria and finally the third manager is focused on the second criteria.

Let us now apply Definition 1 to A_1, A_2 *and* A_3 *as follows:*

$$\{\overset{o}{A_1}\} = \{\overset{1}{A_1}, \overset{2}{A_1}, \cdots, \overset{63}{A_1}, \overset{64}{A_1}\};\ \{\overset{o}{A_2}\} = \{\overset{1}{A_2}, \overset{2}{A_2}, \cdots, \overset{63}{A_2}, \overset{64}{A_2}\};\ \{\overset{o}{A_3}\} = \{\overset{1}{A_3}, \overset{2}{A_3}, \cdots, \overset{63}{A_3}, \overset{64}{A_3}\}, \tag{51}$$

where $\{\overset{1}{A}_r, \overset{2}{A}_r, \cdots, \overset{64}{A}_r\}$, $r = 1,2,3$ *are obtained in a similar manner as in Example 1.*

Then, $CI_{\overset{*}{K}}(\overset{*}{A}_i), i = 1,2,3$ *is calculated through Equation (2) where, for the first expert, one obtains:*

$$\overset{*}{A}_1 = \begin{pmatrix} 1 & 2 & 1/5 & 1/3 \\ * & 1 & 1/7 & 6 \\ * & * & 1 & 1/2 \\ * & * & * & 1 \end{pmatrix}, \tag{52}$$

and its corresponding consistent MPR (calculated by Equation (2)) is

$$\overset{*}{K}_1 = \begin{pmatrix} 1 & 0.7468 & 0.2954 & 0.6043 \\ * & 1 & 0.3956 & 0.8091 \\ * & * & 1 & 2.0453 \\ * & * & * & 1 \end{pmatrix}. \tag{53}$$

Similar calculations are carried out for A_2 *and* A_3.

Then, from Definition 3, $CI_{\tilde{K}}(\tilde{A}_i), i = 1,2,3$ *is calculated through Equation (13) where for the first expert one obtains:*

$$\tilde{A}_1 = \begin{pmatrix} 1 & 1 & 1/5 & 1/2 \\ * & 1 & 1/7 & 5 \\ * & * & 1 & 1/2 \\ * & * & * & 1 \end{pmatrix}, \tag{54}$$

and its corresponding consistent MPR (calculated by Equation (2)) is

$$\tilde{K}_1 = \begin{pmatrix} 1 & 0.6117 & 0.2749 & 0.5946 \\ * & 1 & 0.4495 & 0.9721 \\ * & * & 1 & 2.1627 \\ * & * & * & 1 \end{pmatrix}. \tag{55}$$

Similar calculations are carried out for A_2 *and* A_3.

From Equation (10), one obtains: $CI_{\overset{*}{K}_1}(\overset{*}{A}_1) = 1.66161$, $CI_{\overset{*}{K}_2}(\overset{*}{A}_2) = 1.207088$ *and* $CI_{\overset{*}{K}_3}(\overset{*}{A}_3) = 1.09$. *As a consequence,* A_1 *and* A_2 *are not consistent and* A_3 *is acceptably consistent.*

We apply Algorithm IC-I-MPR to $A_1^{(1)} = A_1$ *and* $A_2^{(2)} = A_2$ *with initialization points* $x_0^{(1)} = [3/2, 1/5, 1/3, 1/7, 11/2, 1/2]$ *and* $x_0^{(2)} = [5/2, 1/3, 3/4, 1, 5/2, 3/2]$ *which are the midpoints of* A_1 *and* A_2*, respectively.*

From Equation (30), $\overset{o}{A}_1 = A_1$ *and* $\overset{o}{A}_2 = A_2$.

Let us provide the solution for A_1 *which has the highest CI. Then, from Equation (33) and the first expert, the following inequalities will be utilized in the optimization process:*

$$\begin{array}{rcccl} 1 - \epsilon_c & \leq & x_1 & \leq & 2 + \epsilon_c, \\ 1/5 - \epsilon_c & \leq & x_2 & \leq & 1/5 + \epsilon_c, \\ 1/3 - \epsilon_c & \leq & x_3 & \leq & 1/2 + \epsilon_c, \\ 1/7 - \epsilon_c & \leq & x_4 & \leq & 1/7 + \epsilon_c, \\ 5 - \epsilon_c & \leq & x_5 & \leq & 6 + \epsilon_c, \\ 1/2 - \epsilon_c & \leq & x_6 & \leq & 1/2 + \epsilon_c, \end{array} \tag{56}$$

where $\epsilon_c = 0.001$*, but for the first iteration* $\epsilon_c = 0.0$.

From Equation (37) and variable's definition given by Equation (34), the objective functional is obtained by replacing $a_{12} = x_1$, $a_{13} = x_2$, $a_{14} = x_3$, $a_{23} = x_4$, $a_{24} = x_5$ and $a_{34} = x_6$. Thus, one obtains:

$$
\begin{aligned}
f(x) \;=\; sign * \Bigg[\frac{1}{16}\Bigg[& \underbrace{\left(\frac{x_2 x_3}{x_1^2 x_4 x_5}\right)^{1/4}}+\frac{1}{\beta_1}+\underbrace{\left(\frac{x_2^2 x_6}{x_1 x_3 x_4}\right)^{1/4}}+\frac{1}{\beta_2}+\underbrace{\left(\frac{x_3^2}{x_1 x_2 x_5 x_6}\right)^{1/4}}+\frac{1}{\beta_3}+\underbrace{\left(\frac{x_4^2 x_1 x_6}{x_2 x_5}\right)^{1/4}}+\frac{1}{\beta_4}+ \\
& \underbrace{\left(\frac{x_5^2 x_1}{x_3 x_4 x_6}\right)^{1/4}}+\frac{1}{\beta_5}+\underbrace{\left(\frac{x_6^2 x_2 x_4}{x_3 x_5}\right)^{1/4}}+\frac{1}{\beta_6}\Bigg]+\frac{1}{4}\Bigg];
\end{aligned} \tag{57}
$$

where β_1 equals the term over the first brace, β_2 equals the term over the second brace, and so on. In addition, $sign = 1$ is defined to obtain the minimization and $sign = -1$ for the maximization.

The same procedure is used for the second expert (A_2). After applying the SQP, the results follow:

$$
\overline{A}_1 = \begin{pmatrix} 1 & 0.5610 & [0.5713\ 0.5890] & 0.9390 \\ * & 1 & 0.5819 & 4.5610 \\ * & * & 1 & 0.9390 \\ * & * & * & 1 \end{pmatrix}, \quad \overline{A}_2 = \begin{pmatrix} 1 & 2.5 & 1/3 & 3/4 \\ * & 1 & 1/2 & [2\ 2.0303] \\ * & * & 1 & [1.7628\ 2] \\ * & * & * & 1 \end{pmatrix}. \tag{58}
$$

Now that we have the set of I-MPRs in acceptable Individual Consistency, we can proceed with solving the Group Consensus for $\overline{A}_1$, $\overline{A}_2$ and A_3.

Let us now apply Definition 4 to these I-MPRs. The set $\{\overset{1}{\check{A}}\}$ is calculated as in Equation (25). For example,

$$
\overset{1}{\check{A}} = \{\overset{1}{A_1}, \overset{1}{A_2}, \overset{1}{A_3}\}; \overset{2}{\check{A}} = \{\overset{1}{A_1}, \overset{1}{A_2}, \overset{2}{A_3}\}; \overset{3}{\check{A}} = \{\overset{1}{A_1}, \overset{2}{A_2}, \overset{1}{A_3}\}; \cdots ; \overset{262143}{\check{A}} = \{\overset{64}{A_1}, \overset{64}{A_2}, \overset{63}{A_3}\}, \overset{262144}{\check{A}} = \{\overset{64}{A_1}, \overset{64}{A_2}, \overset{64}{A_3}\}. \tag{59}
$$

From Definition 5, let us apply it to $\overline{A}_1$, $\overline{A}_2$ and A_3 where, by Equation (3), one obtains:

$$
\overset{1}{\check{A}^c} = \left[\left(a_{ij}^{(1)}\right)^{\lambda_1} \cdot \left(a_{ij}^{(2)}\right)^{\lambda_2} \cdot \left(a_{ij}^{(3)}\right)^{\lambda_3}\right], \quad i,j = 1,2,3,4, \tag{60}
$$

where $(a_{ij}^{(x)})^{\lambda_x}$, $x = 1,2,3$ comes from $\overset{1}{\check{A}_1}$ (cf. Equation (59)), for $\overset{2}{\check{A}^c}$ it comes from $\overset{2}{\check{A}_1}$, and so on. Note that, in this example, $\lambda_1 = \lambda_2 = \lambda_3 = 1/3$.

From Definition 6, we note that the Group Consensus of each DM gives the following result. Since $GCI_{\check{H}_t}(\check{A}_1) = 1.0985$, $GCI_{\check{H}_t}(\check{A}_2) = 1.0473$ and $GCI_{\check{H}_t}(\check{A}_3) = 1.0381$, all the experts are in acceptable consensus with a global ranking order $w_3 \overset{0.9524}{\succ} w_2 \overset{0.8081}{\succ} w_1 \overset{1}{\succ} w_4$.

Since we have detailed the calculations of the Algorithm 1, in the next example, we provide details of the Algorithm 2.

In the next example, we address and solve a problem of partner selection that is determined for the formation of a virtual enterprise already considered in [51,52].

A virtual enterprise is a dynamic association of enterprises, working together to benefit from a market opportunity by giving a solution that could not be delivered individually. Naturally, it comes with a market opportunity and then it is stopped after that market particular event. Thus, a virtual enterprise is, by definition, a non-permanent alliance of diverse, autonomous, and in some cases geographically distributed organizations sharing resources and skills having common objectives and profiting from a benefit window in the market opportunities [51].

Example 4. *In this example, the main enterprise needs to select a partner to grasp a new market opportunity where four candidates (alternatives) must be analyzed by four DMs. The CEO's management staff (DMs) is* *involved in the partner selection process which should select the most suitable partner. Then, each DM compares each pair of alternatives c_i and c_j, and gives her/his preference assessment through the next I-MPRs:*

$$A_1 = \begin{pmatrix} 1 & [2/3\ 1] & [3/2\ 2] & [5/3\ \ 2] \\ * & 1 & [1\ 3] & [1\ \ 2] \\ * & * & 1 & [1/2\ 1] \\ * & * & * & 1 \end{pmatrix}, \quad A_2 = \begin{pmatrix} 1 & [2\ 3] & [3\ 4] & [2\ \ 6] \\ * & 1 & [3/2\ 3] & [3\ \ 4] \\ * & * & 1 & [9/10\ 1] \\ * & * & * & 1 \end{pmatrix},$$
$$A_3 = \begin{pmatrix} 1 & [1/2\ 1] & [2\ 3] & [1\ \ 3] \\ * & 1 & [3\ 4] & [2\ \ 4] \\ * & * & 1 & [1/4\ 1/2] \\ * & * & * & 1 \end{pmatrix}, \quad A_4 = \begin{pmatrix} 1 & [1\ 2] & [2/3\ 2] & [2\ \ 3] \\ * & 1 & [2/3\ 6/5] & [2\ \ 5] \\ * & * & 1 & [3\ 7/2] \\ * & * & * & 1 \end{pmatrix}, \tag{61}$$

where the initial individual order of intervals for each DM is given by
DM1:$x_2 \overset{1}{\succ} x_1 \overset{1}{\succ} x_4 \overset{1}{\succ} x_3$; *DM2:*$x_1 \overset{1}{\succ} x_2 \overset{1}{\succ} x_3 \overset{.6698}{\succ} x_4$; *DM3:*$x_2 \overset{1}{\succ} x_1 \overset{1}{\succ} x_4 \overset{1}{\succ} x_3$ *& DM4:*$x_3 \overset{.6382}{\succ} x_1 \overset{.9904}{\succ} x_2 \overset{1}{\succ} x_4$.

In [52], it is found that the third expert is more consistent and then it is assigned a higher weight. They obtained the final ranking of the alternatives as: $x_2 \succ x_1 \succ x_4 \succ x_3$. On the other hand, when all experts are considered equally weighted, they obtained the final ranking of the alternatives as $x_1 \succ x_2 \succ x_3 \succ x_4$.

Let us apply our method to check for consistency and consensus indices.

By applying the first part of our Definitions, the Individual Consistency of each DM (by using Equation (12)) is given by:

$$CI_{A_1} = 1.060286,\ CI_{A_2} = 1.076585,\ CI_{A_3} = 1.101763,\ CI_{A_4} = 1.078917, \tag{62}$$

where it is noted that the third DM (A_3) has provided a slightly inconsistent I-MPR.

Since the third expert's judgments A_3 are not consistent, let us apply the ***Algorithm IC-I-MPR****.*

Take an initialization point $x_0^{(1)} = [3/4, 5/2, 2, 7/2, 3, 3/4]$ and $\epsilon_c = 0.001$ and the result follows:

$$\overline{A}_3 = \begin{pmatrix} 1 & [0.7413\ 1] & [2\ 2.9656] & [1.4825\ 2.9962] \\ * & 1 & [3\ 4] & [2\ 3.8970] \\ * & * & 1 & [1/4\ 1/2] \\ * & * & * & 1 \end{pmatrix}. \tag{63}$$

Now that we have the whole set of I-MPRs (A_1, A_2, $\overline{A}_3$ and A_4) in acceptable consistency, we can proceed to address the group consensus analysis.

First case: Although the first expert has the most consistent I-MPR (cf. Equation (62)), let us consider that the third expert A_3 is the more relevant, by assigning the next following experts' weighting $\lambda = [1/5, 1/5, 2/5, 1/5]$.

From Definition 6 applied to the first DM (A_1), it follows:

$$GCI_{\check{H}_1}(\check{A}_1) = \max\{GCI_{\check{A}^c}^{1}(\check{A}), GCI_{\check{A}^c}^{2}(\check{A}), \cdots, GCI_{\check{A}^c}^{\nu-1}(\check{A}), GCI_{\check{A}^c}^{\nu}(\check{A})\}, \tag{64}$$

where in this case $\nu = 16777216$. Similar calculations are carried out for A_2, $\overline{A}_3$ and A_4.

Then, we obtain $GCI_{\check{H}_1}(\check{A}_1) = 1.0657$, $GCI_{\check{H}_2}(\check{A}_2) = 1.0646$, $GCI_{\check{H}_3}(\check{A}_3) = 1.0908$, and finally $GCI_{\check{H}_4}(\check{A}_4) = 1.5364$. Thus, the fourth expert is not in acceptable consensus.

From Equation (31) $A_1^p = A_4$ and from Equation (33) the following inequalities will be utilized in the optimization process:

$$\begin{array}{rcccl} 1-\epsilon_g & \leq & x_1 & \leq & 2+\epsilon_g, \\ 2/3-\epsilon_g & \leq & x_2 & \leq & 2+\epsilon_g, \\ 2-\epsilon_g & \leq & x_3 & \leq & 3+\epsilon_g, \\ 2/3-\epsilon_g & \leq & x_4 & \leq & 6/5+\epsilon_g, \\ 2-\epsilon_g & \leq & x_5 & \leq & 5+\epsilon_g, \\ 3-\epsilon_g & \leq & x_6 & \leq & 7/2+\epsilon_g, \end{array} \tag{65}$$

where $\epsilon_g = 0.001$, but for the first iteration $\epsilon_g = 0.0$.

From Equation (39) and variable's definition given by Equation (34), the objective functional is obtained by replacing in it $a_{12}^{(4)}, a_{13}^{(4)}, a_{14}^{(4)}, a_{23}^{(4)}, a_{24}^{(4)}$ and $a_{34}^{(4)}$. Then, one obtains:

$$\begin{aligned} f(x) \;=\; sign * \Bigg[\frac{1}{16} \Bigg[& \Bigg(\underbrace{\frac{x_{19}^{(1-\lambda_4)}}{a_{12}^{(1)\lambda_1} a_{12}^{(2)\lambda_2} a_{12}^{(3)\lambda_3}}} \Bigg) + \frac{1}{\alpha_1} + \Bigg(\underbrace{\frac{x_{20}^{(1-\lambda_4)}}{a_{13}^{(1)\lambda_1} a_{13}^{(2)\lambda_2} a_{13}^{(3)\lambda_3}}} \Bigg) + \frac{1}{\alpha_2} + \Bigg(\underbrace{\frac{x_{21}^{(1-\lambda_4)}}{a_{14}^{(1)\lambda_1} a_{14}^{(2)\lambda_2} a_{14}^{(3)\lambda_3}}} \Bigg) + \frac{1}{\alpha_3} + \\ & \Bigg(\underbrace{\frac{x_{22}^{(1-\lambda_4)}}{a_{23}^{(1)\lambda_1} a_{23}^{(2)\lambda_2} a_{23}^{(3)\lambda_3}}} \Bigg) + \frac{1}{\alpha_4} + \Bigg(\underbrace{\frac{x_{23}^{(1-\lambda_4)}}{a_{24}^{(1)\lambda_1} a_{24}^{(2)\lambda_2} a_{24}^{(3)\lambda_3}}} \Bigg) + \frac{1}{\alpha_5} + \Bigg(\underbrace{\frac{x_{24}^{(1-\lambda_4)}}{a_{34}^{(1)\lambda_1} a_{34}^{(2)\lambda_2} a_{34}^{(3)\lambda_3}}} \Bigg) + \frac{1}{\alpha_6} \Bigg] + \frac{1}{4} \Bigg], \end{aligned} \tag{66}$$

where α_1 equals the term over the first brace, α_2 equals the term over the second brace, and so on. In addition, $sign = 1$ is defined to obtain the minimization and $sign = -1$ for the maximization.

Let us then apply the SQP algorithm for $A_4^4 = (\overline{a}_{ij}^{(4)}\ \overset{+}{a}_{ij}^{(4)})$ with $\overline{GCI} = 1.1$ and $\epsilon = 0.001$.

As soon as the optimization algorithm converges, the new Group Consensus Indices from Equation (23) is read as:

$$GCI_{\check{A}_1} = 1.0132,\;\; GCI_{\check{A}_2} = 1.0921,\;\; GCI_{\check{A}_3} = 1.0508,\;\; GCI_{\check{A}_4} = 1.0697. \tag{67}$$

In addition, from Equation (24):

$$GCI_{\overline{A}_1} = 1.0008,\;\; GCI_{\overline{A}_2} = 1.0722,\;\; GCI_{\overline{A}_3} = 1.0495,\;\; GCI_{\overline{A}_4} = 1.0777. \tag{68}$$

The priority vector is $w_1 = [0.268889\ 0.459874]$, $w_2 = [0.284454\ 0.400261]$, $w_3 = [0.09285\ \ 0.186909]$, $w_4 = [0.123878\ 0.2380]$, and the final ranking of priorities is given by $w_1 \overset{.6153}{\succ} w_2 \overset{1}{\succ} w_4 \overset{0.8149}{\succ} w_3$, or $x_1 > x_2 > x_4 > x_3$.

Second case: For an equal experts' weighting (For example, $\lambda = [1/4, 1/4, 1/4, 1/4]$), one has that the Group Consensus Indices for each decision maker (by using Equation (23)) are:

$$GCI_{\check{A}_1} = 1.0902,\;\; GCI_{\check{A}_2} = 1.0597,\;\; GCI_{\check{A}_3} = 1.1458,\;\; GCI_{\check{A}_4} = 1.4162, \tag{69}$$

where the third and fourth DM have provided their I-MPRs not in consensus.

*Then, apply again **Algorithm GC-I-MPR** for A_3^3, A_4^4, $\overline{GCI} = 1.1$ and $\epsilon = 0.001$, but this time with the new $\lambda = [1/4, 1/4, 1/4, 1/4]$.*

As soon as the optimization algorithm converges, the new Group Consensus Indices from Equation (23) is read as:

$$GCI_{\check{A}_1} = 1.0436,\;\; GCI_{\check{A}_2} = 1.0528,\;\; GCI_{\check{A}_3} = 1.0983,\;\; GCI_{\check{A}_4} = 1.0463. \tag{70}$$

In addition, from Equation (24):

$$GCI_{\overline{A}_1} = 1.0058,\;\; GCI_{\overline{A}_2} = 1.0348,\;\; GCI_{\overline{A}_3} = 1.0365,\;\; GCI_{\overline{A}_4} = 1.0590, \tag{71}$$

where the set of I-MPRs are Individually Consistent and in an acceptable Group Consensus. They are read as:

$$
\begin{aligned}
A_1 &= \begin{pmatrix} 1 & [0.7507\ 1.0] & [1.6237\ 2.0] & [1.8408\ 2.0] \\ * & 1 & [1.1212\ 3.0] & [1.1860\ 2.0] \\ * & * & 1 & [0.5505\ 1.0] \\ * & * & * & 1 \end{pmatrix}, \\
A_2 &= \begin{pmatrix} 1 & [2.0\ 2.1349] & [3\ 3.1074] & [2.0\ 4.3848] \\ * & 1 & [3/2\ 2.5193] & [3.0\ 3.1426] \\ * & * & 1 & [9/10\ 0.9112] \\ * & * & * & 1 \end{pmatrix}, \\
A_3 &= \begin{pmatrix} 1 & [0.9092\ 1.0] & [2.0\ 2.5686] & [1.7908\ 2.9962] \\ * & 1 & [2.9035\ 3.0] & [2.0\ 3.1838] \\ * & * & 1 & [0.388\ 1/2] \\ * & * & * & 1 \end{pmatrix}, \\
A_4 &= \begin{pmatrix} 1 & [1.0\ 1.4055] & [1.4248\ 2.0] & [2.2640\ 2.4575] \\ * & 1 & [1.3641\ 1.3654] & [2.0\ 2.9146] \\ * & * & 1 & [1.7183\ 1.8753] \\ * & * & * & 1 \end{pmatrix}.
\end{aligned}
\tag{72}
$$

Their priority vector is $w_1 = [0.308967\ 0.488215]$ $w_2 = [0.25554\ 0.397948]$, $w_3 = [0.104711\ 0.235793]$, $w_4 = [0.113979\ 0.198821]$, *and the final ranking of priorities is given by* $w_1 \overset{0.8449}{\succ} w_2 \overset{1}{\succ} w_3 \overset{0.6057}{\succ} w_4$, *or* $x_1 > x_2 > x_3 > x_4$.

We note that, through Definitions, Theorem and Algorithms introduced here, we obtain the same results of [51,52] when the experts' weight is the same. However, when the weighting of the experts is different, some small differences are found. The above is due to the variation in the weighting provided here and in those articles.

Case Study Discussions and Managerial Implications

In the first example, the second and third DMs have provided inconsistent I-MPRs, and it was sufficient to improve their Individual Consistency to have the three I-MPRs in Group Consensus. Note that the ranking order of each DM was initially slightly different among them; nevertheless, at the end of the process, the Group Consensus points to an acceptable and analytic decision.

As we have have seen along the first example, when a group of DMs needs to state a point of agreement, they can define their assessments through I-MPRs which do not necessarily need to be exactly in the same judgment direction. Sometimes, they only need to be confident in the quality of their decisions based on an MCDM system which analyzes them through a well-known decision model.

For the second example, only two DMs evaluate in the same direction the set of criteria (alternatives). The other two experts are even in clear contradiction, since the first DM states that the third criteria is lesser in importance and the second DM states that this criteria is the most important. Furthermore, by assigning to the third expert a highest evaluation confidence (weight), the methodology produced a very interesting result, by selecting a global ranking order that none of them had chosen.

On the other hand, when every DM has the same weight, the methodology produced also a different global ranking order that none of them had chosen. The DMs A_1, A_2 and A_3, have preserved their individual ranking order at the end of the process, and only the fourth DM has had his individual ranking order changed. This result can be used in his/her operating department as an internal feedback to reconsider their position with respect to the other departments and the organization objectives.

As soon as a DM obtains the results, s/he can use them so that, with this information, he can state new ways of organizing her/his operating department. For example, this method can also be carried out within the operating departments since each one of their business operations have criteria and alternatives that can be better emphasized.

In summary, the MCDM methodology can provide results which could reinforce the position of a set of DMs or point out to a new direction.

6. Concluding Remarks and Future Work

In this paper, we provide a methodology based on a couple of algorithms and a nonlinear optimization approach to be used when a heterogeneous managers group needs to solve an MCDM problem.

Our approach can use Interval Multiplicative Preference Relations or Multiplicative Preference Relations, and demonstrates the utilization of the methodology to synthesize reliable intervals where consistency and consensus constraints hold. Once decision makers have proposed their I-MPRs, our method can solve for these I-MPRs from well-known decision support models. One advantage of our algorithm is that DM can re-express their preferences within an interval where, usually, they have to observe some constraints based on decision targets, framework rules and advice. When the DM is confident on the pair of criteria (o alternatives) under evaluation, s/he can utilize a crisp value. On other other hand, when s/he is hesitant or uncertain about the assessment, s/he could use an interval. Our algorithm can solve independently of the used approach.

In this work, reliable I-MPRs provide a distinct advantage in interpretation of hesitancy and uncertainty about the final consistency and consensus.

Main advantages of the present approach:

- It is provided through a couple of algorithms and a nonlinear optimization approach (Sequential Quadratic Programming) concurrently applied.
- Through the Hadamard's operator and some easy algebraic manipulations, objective functionals were synthesized to be used in the optimization algorithm.
- When the I-MPRs improved by the methodology are reduced into an MPR (defined in the I-MPR), our approach can still give reliable results. For example, for this MPR, we can verify the results of IC or GC with an alternative method.
- The IC or the GC accepted indices (threshold values) have been previously investigated and fixed. Nevertheless, the project designer could assign a different value depending on the project requirements.
- Obtained results are independent of the method of prioritization utilized in the consensus operation.

Main drawbacks of the present approach:

- The computational cost increases as the I-MPRs dimension and the number of DMs involved in the evaluation process are increased.
- For a real project where a high number of criteria and experts participate, it can be necessary to program this method through an exhaustive parallel computation system.
- For a real project where a high number of criteria and experts participate, the notation can be cumbersome.

Future works aim to make an implementation on:

- The application of our approach to various study cases where heterogenous groups of DMs with different weights participate in a collaborative manner.
- The integration of the complete methodology in a benchmark to compare the results of a diverse set of MCDM tools.
- The definition or employment of this methodology on different frameworks, v.gr. fuzzy or hesitant MCDM.

Funding: This research was funded by PFCE-PRODEP grant number DIP/DI/CI/2015-1160.

Acknowledgments: I would like to thank the anonymous reviewers and the Journal's staff for their constructive advice and support for improving this paper.

Conflicts of Interest: The author declares no conflict of interest.

References

1. Saaty, T. *The Analytic Hierarchy Process*; McGraw-Hill: New York, NY, USA, 1980.
2. Yntema, D.B.; Torgerson, W. Man-computer cooperation in decisions requiring common sense. *IRE Trans. Hum. Factors Electron.* **1961**, *2*, 20–26.
3. Xia, W.; Wu, Z. Supplier selection with multiple criteria in volume discount environments. *Omega* **2007**, *35*, 494–504.
4. Zadeh, L. Fuzzy sets. *Inf. Control* **1965**, *8*, 338–353.
5. Seiford, M. A DEA Bibliography (1978 1992). In *Data Envelopment Analysis: Theory, Methodology and Applications*; Charnes, A., Cooper, W.W., Lewin, A.Y., Seiford, L.M., Eds.; Springer: Dordrecht, The Netherlands, 1994.
6. Aamodt, A.; Plaza, E. Case-based reasoning: Foundational issues. methodological variations and system approaches. *Artif. Intell. Commun.* **1994**, *7*, 39–59.
7. Kusumadewi, S.; Hartati, S.; H, A.; Wardoyo, R. *Fuzzy multi-attribute decision-making (FUZZY MADM)*; GrahaIlmu Publisher: Dresden, Germany, 2010.
8. Benayoun, R.; Roy, B.; Sussman, B. ELECTRE: Une méthode pour guider le choix en présence de points de vue multiples. In *SEMA*; METRA International: Paris, France; 1996.
9. Gwo-Hshiung, T.; Tzeng, G.; Huang, J.J. *Multiple Attribute Decision Making: Methods and Applications*; CRC Press: Boca Raton, FL, USA, 2011.
10. Vincke, J.; Brans, P. A preference ranking organization method (the PROMETHEE method for MCDM). *Manag. Sci.* **1985**, *31*, 641–656.
11. Wang, P.; Zhu, Z.; Wang, Y. A novel hybrid MCDM model combining the SAW, TOPSIS and GRA methods based on experimental design. *Inf. Sci.* **2016**, *345*, 27–45.
12. Vicenc, T. Hesitant fuzzy sets. *Int. J. Intell. Syst.* **2010**, *25*, 529–539.
13. Liu, J.; Zhaoa, H.K.; Li, Z.B.; Liu, S.F. Decision process in MCDM with large number of criteria and heterogeneous risk preferences. *Oper. Res. Perspect.* **2017**, *4*, 106–112.
14. Zhang, X.; Xu, Z. The Extended TOPSIS Method for Multi-criteria Decision Making Based on Hesitant Heterogeneous Information. In Proceedings of the 2nd International Conference on Software Engineering, Knowledge Engineering and Information Engineering (SEKEIE 2014), Singapore, 5–6 August 2014; pp. 81–86.
15. Faizi, S.; Salabun, W.; Rashid, T.; Watróbski, J.; Zafar, S. Group Decision-Making for Hesitant Fuzzy Sets Based on Characteristic Objects Method. *Symmetry* **2017**, *9*, 136.
16. Cheng, J.; Zhang, Y.; Feng, Y.; Liu, Z.; Tan, J. Structural Optimization of a High-Speed Press Considering Multi-Source Uncertainties Based on a New Heterogeneous TOPSIS. *Appl. Sci.* **2018**, *8*, 126.
17. Faizi, S.; Rashid, T.; Salabun, W.; Zafar, S.; Watróbski, J. Decision Making with Uncertainty Using Hesitant Fuzzy Sets. *Int. J. Fuzzy Syst.* **2018**, *20*, 93–103.
18. Watróbski, J.; Jankowski, J.; Ziemba, P.; Karczmarczyk, A.; Ziolo, M. Generalised framework for multi-criteria method selection. *Omega* **2018**, in press.
19. Linkov, I.; Bates, M.E.; Canis, L.J.; Seager, T.P.; Keisler, J.M. A decision-directed approach for prioritizing research into the impact of nanomaterials on the environment and human health. *Nat. Nanotechnol.* **2011**, *6*, 784–787.
20. Karlson, M.; Karlsson, C.S.J.; Mortberg, U.; Olofsson, B.; Balfors, B. Design and evaluation of railway corridors based on spatial ecological and geological criteria. *Transp. Res. Part D Transp. Environ.* **2016**, *46*, 207–228.
21. Liu, F.; Aiwu, G.; Lukovac, V.; Vukic, M. A multicriteria model for the selection of the transport service provider: A single valued neutrosophic DEMATEL multicriteria model. *Decis. Mak. Appl. Manag. Eng.* **2018**, *1*, 121–130.
22. Veskovic, S.; Stevic, e.; Stojic, G.; Vasiljevic, M.; Milinkovic, S. Evaluation of the railway management model by using a new integrated model DELPHI-SWARA-MABAC. *Decis. Mak. Appl. Manag. Eng.* **2018**, *1*, 34–50.
23. Petrovic, I.B.; Kankaraš, M. DEMATEL-AHP multi-criteria decision-making model for the determination and evaluation of criteria for selecting an air traffic protection aircraft. *Decis. Mak. Appl. Manag. Eng.* **2018**, *1*, 93–110.
24. Pape, T. Prioritising data items for business analytics: Framework and application to human resources. *Eur. J. Oper. Res.* **2016**, *252*, 687–698.

25. Cid-López, A.; Hornos, M.J.; Carrasco, R.A.; Herrera-Viedma, E. Applying a linguistic multi-criteria decision-making model to the analysis of ICT suppliers' offers. *Expert Syst. Appl.* **2016**, *57*, 127–138.
26. Shen, J.; Lu, H.; Zhang, Y.; Song, X.; He, L. Vulnerability assessment of urban ecosystems driven by water resources, human health and atmospheric environment. *J. Hydrol.* **2016**, *536*, 457–470.
27. Mansour, F.; Al-Hindi, M.; Saad, W.; Salam, D. Environmental risk analysis and prioritization of pharmaceuticals in a developing world context. *Sci. Total Environ.* **2016**, *557*, 31–43.
28. Azarnivand, A.; Malekian, A. Analysis of flood risk management strategies based on a group decision-making process via interval-valued intuitionistic fuzzy numbers. *Water Resour. Manag.* **2016**, *30*, 1903–1921.
29. Ruiz-Padillo, A.; Torija, A.J.; Ramos-Ridao, A.; Ruiz, D.P. Application of the fuzzy analytic hierarchy process in multi-criteria decision in noise action plans: Prioritizing road stretches. *Environ. Model. Softw.* **2016**, *81*, 45–55.
30. Maimoun, M.; Madani, K.; Reinhart, D. Multi-level multi-criteria analysis of alternative fuels for waste collection vehicles in the United States. *Sci. Total Environ.* **2016**, *550*, 349–361.
31. Kim, Y.; Chung, E.S. Assessing climate change vulnerability with group multi-criteria decision-making approaches. *Clim. Chang.* **2013**, *121*, 301–315.
32. Potić, I.; Golić, R.; Joksimović, T. Analysis of insolation potential of Knjazevac Municipality (Serbia) using multi-criteria approach. *Renew. Sustain. Energy Rev.* **2016**, *56*, 235–245.
33. Franco, C.; Bojesen, M.; Hougaard, J.L.; Nielsen, K. A fuzzy approach to a multiple criteria and Geographical Information System for decision support on suitable locations for biogas plants. *Appl. Energy* **2015**, *140*, 304–315.
34. Schmitz, S.; McCullagh, L.; Adams, R.; Barry, M.; Walsh, C. Identifying and revealing the importance of decision-making criteria for health technology assessment: A retrospective analysis of reimbursement recommendations in Ireland. *PharmacoEconomics* **2016**, *34*, 925–937.
35. Linkov, I.; Trump, B.; Jin, D.; Mazurczak, M.; Schreurs, M. A decision-analytic approach to predict state regulation of hydraulic fracturing. *Environ. Sci. Eur.* **2014**, *26*, 20.
36. Cabrerizo, F.J.; Herrera-Viedma, E.; Pedrycz, W. A method based on PSO and granular computing of linguistic information to solve group decision-making problems defined in heterogeneous contexts. *Eur. J. Oper. Res.* **2013**, *230*, 624–633. [CrossRef]
37. Herrera-Viedma, E.; Cabrerizo, F.J.; Kacprzyk, J.; Pedrycz, W. A review of soft consensus models in a fuzzy environment. *Inf. Fusion* **2014**, *17*, 4–13. [CrossRef]
38. Sahin, R.; Yigider, M. A Multi-criteria neutrosophic group decision-making method based TOPSIS for supplier selection. *arXiv* **2014**, arXiv:1412.5077.
39. Wu, Z.; Xu, J. A consistency and consensus based decision support model for group decision-making with multiplicative preference relations. *Decis. Support Syst.* **2012**, *52*, 757–767.
40. Saaty, T.L.; Vargas, L.G. Uncertainty and rank order in the analytic hierarchy process. *Eur. J. Oper. Res.* **1987**, *32*, 107–117.
41. Saaty, T. A ratio scale metric and the compatibility of ratio scales: The possibility of arrow's impossibility theorem. *Appl. Math. Lett.* **1994**, *7*, 45–49.
42. Wang, L. Compatibility and group decision-making. *Syst. Eng. Theory Pract.* **2002**, *20*, 92–96.
43. López-Morales, V. A Reliable Method for Consistency Improving of Interval Multiplicative Preference Relations Expressed under Uncertainty. *Int. J. Inf. Technol. Decis. Mak.* **2018**. [CrossRef]
44. López-Morales, V. Reliable Group Decision-Making under Uncertain Judgments. *Int. Tech. Rep.* **2018**, *1*, 1–30.
45. Beale, E.M.L. *Numerical Methods in: Nonlinear Programming*; Abadie, J., Ed.; North-Holland: Amsterdam, The Netherlands, 1967.
46. Kou, G.; Lin, C. A cosine maximization method for the priority vector derivation in AHP. *Eur. J. Oper. Res.* **2014**, *235*, 225–232.
47. Saaty, T. Decision-making with the AHP: Why is the principal eigenvector necessary? *Eur. J. Oper. Res.* **2003**, *145*, 85–91.
48. Crawford, G.; Williams, C. A note on the analysis of subjective judgement matrices. *J. Math. Psychol.* **1985**, *29*, 387–405.

49. Meng, F.; Zeng, X.; Li, Z. Research the priority methods of interval numbers complementary judgment matrix. In Proceedings of the 2007 IEEE International Conference on Grey Systems and Intelligent Services, Nanjing, China, 18–20 November 2007.
50. Meng, F.; Chen, X.; Zhu, M.; Lin, J. Two new methods for deriving the priority vector from interval multiplicative preference relations. *Inf. Fusion* **2015**, *26*, 122–135.
51. Ye, F. An extended TOPSIS method with interval-valued intuitionistic fuzzy numbers for virtual enterprise partner selection. *Expert Syst. Appl.* **2010**, *37*, 7050–7055.
52. Liu, F.; Zhang, W.G.; Shang, Y.F. A group decision-making model with interval multiplicative reciprocal matrices based on the geometric consistency index. *J. Comput. Ind. Eng.* **2016**, *101*, 184–193.

information

MDPI

Article

A Quick Algorithm for Binary Discernibility Matrix Simplification using Deterministic Finite Automata

Nan Zhang [1,*], Baizhen Li [1,2], Zhongxi Zhang [1] and Yanyan Guo [1]

1 School of Computer and Control Engineering, Yantai University, 264005 Yantai, China; 1833009@tongji.edu.cn (B.L.); zhangzhongxi89@gmail.com (Z.Z.); smallgyy@ytu.edu.cn (Y.G.)
2 Department of Computer Science and Technology, Tongji University, 201804 Shanghai, China
* Correspondence: zhangnan@ytu.edu.cn; Tel.: +86-186-1535-1131

Received: 30 October 2018; Accepted: 6 December 2018; Published: 7 December 2018

Abstract: The binary discernibility matrix, originally introduced by Felix and Ushio, is a binary matrix representation for storing discernible attributes that can distinguish different objects in decision systems. It is an effective approach for feature selection, knowledge representation and uncertainty reasoning. An original binary discernibility matrix usually contains redundant objects and attributes. These redundant objects and attributes may deteriorate the performance of feature selection and knowledge acquisition. To overcome this shortcoming, row relations and column relations in a binary discernibility matrix are defined in this paper. To compare the relationships of different rows (columns) quickly, we construct deterministic finite automata for a binary discernibility matrix. On this basis, a quick algorithm for binary discernibility matrix simplification using deterministic finite automata (BDMSDFA) is proposed. We make a comparison of BDMR (an algorithm of binary discernibility matrix reduction), IBDMR (an improved algorithm of binary discernibility matrix reduction) and BDMSDFA. Finally, theoretical analyses and experimental results indicate that the algorithm of BDMSDFA is effective and efficient.

Keywords: rough sets; binary discernibility matrices; deterministic finite automata

1. Introduction

Decision making can be considered as the process of choosing the best alternative from the feasible alternatives. With the development of research, decision making is extended from one attribute to multiple attributes. To solve problems in multiple attribute decision making, various theories such as fuzzy sets, rough sets and utility theory, etc. have been used. Many of significant results [1–8] have been achieved in multiple attribute decision making. Researchers in rough set theory [9] are usually concerned with attribute reduction (or feature selection) problems of multiple attribute decision making. The binary discernibility matrix, proposed by Felix and Ushio [10], is a useful tool for attribute reduction and knowledge acquisition. Recently, many algorithms of attribute reduction based on binary discernibility matrices have been developed [11–13]. In 2014, Zhang et al. [14] proposed a binary discernibility matrix for an incomplete information system, and designed a novel algorithm of attribute reduction based on the proposed binary discernibility matrix. In the paper [15], Li et al. developed an attribute reduction algorithm in terms of the improved binary discernibility matrix, and applied the algorithm in customer relationship management. Tiwari et al. [16] developed hardware for a binary discernibility matrix which can be used for attribute reduction and rule acquisition in an information system. Considering mathematical properties of a binary discernibility matrix, Zhi and Miao [17] introduced the so-called binary discernibility matrix reduction (BDMR), which was actually an algorithm for binary discernibility matrix simplification. On the basis of BDMR, two algorithms for attribute reduction and reduction judgement were presented. A binary discernibility matrix with a vertical partition [18] was proposed to deal with big data in attribute reduction. Ren et al. [19] constructed an improved binary discernibility matrix which can be used in an inconsistent information system. Ding et al. [20] discussed several problems about a binary discernibility matrix in an incomplete system. Combining the binary discernibility matrix in an incomplete system, an algorithm of incremental attribute reduction was proposed. In the paper [20], a novel method for calculation of incremental core attribute was introduced firstly. On this basis, an algorithm of attribute reduction was proposed. As is well known that core attributes play a crucial role in heuristic attribute reduction algorithms. Core attributes are computationally expensive in attribute reduction. Hu et al. [21] gave a quick algorithm of the core attribute calculation using a binary discernibility matrix. The computational complexity of the algorithm is $O(|C||U|)$, where $|C|$ is the number of condition attributes and $|U|$ is the number of objects in the universe.

An original binary discernibility matrix usually contains redundant objects and attributes. These redundant objects and attributes may deteriorate the performance of feature selection (attribute reduction) and knowledge acquisition based on binary discernibility matrices. In other words, storing or processing all objects and attributes in an original binary discernibility matrix could be computationally expensive, especially in dealing with large scale data sets with high dimensions. So far, however, few works about the binary discernibility matrix simplification have been investigated. The existing algorithms regarding binary discernibility matrix simplification are time-consuming. To tackle this problem, our works in this paper concern on how to improve the time efficiency of algorithms of binary discernibility matrix simplification. On this purpose, we construct deterministic finite automata in a binary discernibility matrix to compare the relationships of different rows (or columns) quickly. By using deterministic finite automata, we develop a quick algorithm of binary discernibility matrix simplification. Experimental results show that the proposed algorithm is effective and efficient. The contributions of this paper are summarized as follows: First, we define row and column relations which can be used for constructing deterministic finite automata in a binary discernibility matrix. Second, deterministic finite automata in a binary discernibility matrix are proposed to compare the relationships of different rows (or columns) quickly. Third, based on this method, a quick algorithm for binary discernibility matrix simplification (BDMSDFA) is proposed. The proposed method in this paper is meaningful in practical applications. First, by using BDMSDFA, we obtain the simplified binary discernibility matrices quickly. These simplified binary discernibility matrices can significantly improve the efficiency of attribute reduction (feature selection) in decision systems. Second, a binary discernibility matrix without redundant objects and attributes will have the high performance of learning algorithms, and need less space for data storage.

The rest of this paper is structured as follows. We review basic notions about rough set theory in the next section. In Section 3, we propose a general binary discernibility matrix, and define row relations and column relations in a binary discernibility matrix. In Section 4, we develop a quick algorithm for binary discernibility matrix simplification which is called BDMSDFA. Experimental results in Section 5 show that the algorithm of BDMSDFA is effective and efficient, it can be applicable to simplification of large-scale binary discernibility matrices. Finally, the whole paper is summarized in Section 6.

2. Preliminaries

Basic notions about rough set theory are briefly reviewed in this section. Some further details about rough set theory can be found in the paper [9]. A Pawlak decision system can be regarded as an original information system with decision attributes which give decision classes for objects.

A Pawlak decision system [9] can be denoted by 4-tuple $DS = (U, AT, V, f)$, where universe $U = \{x_1, x_2, ..., x_{|U|}\}$ is a finite non-empty set of objects; attribute set $AT = C \cup D$, $C \cap D = \emptyset$, where $C = \{a_1, a_2, ..., a_{|C|}\}$ is called a condition attribute set and $D = \{d\}$ is called a decision attribute set in a decision system; V_{a_m} is the domain of a condition attribute $a_m \in AT$, $V = \cup_{a_m \in AT} V_{a_m}$ and $f : U \times AT \to V$ is a function such that $f(x_i, a_m) = a_m(x_i) \in V_{a_m \in AT}$, $f(x_i, d) = d(x_i) \in V_{d \in AT}$, where $x_i \in U$.

Given a Pawlak decision system $DS = (U, C \cup D, V, f)$, for $\forall x_i, x_j \in U$, an indiscernibility relation regarding attribute set $B \subseteq C$ is defined as $IND(B) = \{(x_i, x_j) : \forall b \in B, f(x_i, b) = f(x_j, b)\}$. Therefore, the discernibility relation regarding attribute set $B \subseteq C$ is given by $DIS(B) = \{(x_i, x_j) : \exists b \in B, f(x_i, b) \neq f(x_j, b)\}$. The indiscernibility relation regarding $B \subseteq C$ is reflexive, symmetric and transitive. Meanwhile, the discernibility relation is irreflexive, symmetric, but not transitive. A partition of U derived from $IND(B)$ is denoted by $U/IND(B)$. The equivalence class in $U/IND(B)$ containing object x_i is defined as $[x_i]_{IND(B)} = [x_i]_B = \{x_j \in U : (x_i, x_j) \in IND(B)\}$.

For $\forall B \subseteq C$, the relative indiscernibility relation and discernibility relation with respect to decision attribute set [9] are defined by:

$$IND(B|D) = \{(x_i, x_j) : x_i, x_j \in U, (\forall b \in B \rightarrow (f(x_i, b) = f(x_j, b)) \vee (f(x_i, d) = f(x_j, d))\},$$

$$DIS(B|D) = \{(x_i, x_j) : x_i, x_j \in U, (\exists b \in B \rightarrow (f(x_i, b) \neq f(x_j, b)) \wedge (f(x_i, d) \neq f(x_j, d))\}.$$

A relative indiscernibility relation $IND(B|D)$ with respect to $B \subseteq C$ is reflexive, symmetric, but not transitive. A relative discernibility relation $DIS(B|D)$ with respect to $B \subseteq C$ is irreflexive, symmetric, but not transitive.

A discernibility matrix, proposed by Skowron and Rauszer [22], suggests a matrix representation for storing condition attribute sets which can discern objects in the universe. Discernibility matrix is an effective method in reduct construction, data representation and rough logic reasoning, and it is also useful mathematical tool in data mining, machine learning, etc. Many extended models of dicernibility matrices have been studied in recent years [23–30]. Considering the classification property Δ, Miao et al. [31] constructed a general discernibility matrix $M_\Delta = (m_\Delta(x_i, x_j))$, where $m_\Delta(x_i, x_j)$ is denoted by:

$$m_\Delta(x_i, x_j) = \begin{cases} \{a \in C : f(x_i, a) \neq f(x_j, a)\}, & (x_i, x_j) \in DIS_\Delta(C|D) \\ \emptyset & \text{otherwise} \end{cases},$$

where $(x_i, x_j) \in DIS_\Delta(C|D)$ denotes objects x_i and x_j are discernible with respect to the classification property Δ in a decision system DS. It should be noted that Δ is a general definition on classification property. A general discernibility matrix provides a common solution to attribute reduction algorithms based on discernibility matrices. By constructing different discernibility matrices, the relative attribute reducts with different reduction targets can be obtained. Based on the relative discernibility relation $DIS(C|D)$, Miao et al. [31] introduced a relationship preservation discernibility matrix which can be denoted as follows:

Definition 1. *[31] Let $DS = (U, C \cup D, V, f)$ be a decision system, for $\forall x_i, x_j \in U$, $\forall a \in C$, $1 \leq i < j \leq |U|$, $M_{relationship} = (m_{relationship}(x_i, x_j))$ is a relationship preservation discernibility matrix, where $m_{relationship}(x_i, x_j)$ is defined by:*

$$m_{relationship}(x_i, x_j) = \begin{cases} \{a \in C : f(x_i, a) \neq f(x_j, a)\} & (x_i, x_j) \in DIS(C|D) \\ \emptyset & \textit{otherwise} \end{cases}.$$

3. Binary Discernibility Matrices and Their Simplifications

The binary discernibility matrix, initiated by Felix and Ushio [10], is a binary presentation of original discernibility matrix. In this section, we suggest a general binary discernibility matrix. Relations of row pairs and column pairs are discussed respectively. Formally, a binary discernibility matrix [10] is introduced as follows:

Definition 2. [10] *Given a decision system* $DS = (U, C \cup D, V, f)$, *for* $\forall x_i, x_j \in U$ *and* $\forall a_m \in C$. $M_{BDM} = (m_{BDM}(x_i, x_j))$ *is a binary discernibility matrix, where the element* $m_{BDM}(x_i, x_j)$ *is denoted by:*

$$m_{BDM}(x_i, x_j) = \begin{cases} 1 & f(x_i, a_m) \neq f(x_j, a_m) \wedge d(x_i) \neq d(x_j) \\ 0 & otherwise \end{cases}.$$

Based on a binary discernibility matrix, discernible attributes about x_i and x_j can be easily obtained. A binary discernibility matrix brings us an understandable approach for representations of discernible attributes, and can be used for designing reduction algorithms. To satisfy more application requirements, we extend original binary discernibility matrix to general binary discernibility matrix as follows:

Definition 3. *Given a decision system* $DS = (U, C \cup D, V, f)$, *for* $\forall x_i, x_j \in U$, $\forall a_m \in C$, $M^{\Delta}_{BDM} = (m^{\Delta}_{BDM}(x_i, x_j))$ *regarding* Δ *is a general binary discernibility matrix, in which* $m^{\Delta}_{BDM}(x_i, x_j)$ *is defined by:*

$$m^{\Delta}_{BDM}(x_i, x_j) = \begin{cases} 1 & (x_i, x_j) \in DIS_{\Delta}(C|D) \\ 0 & otherwise \end{cases}.$$

$DIS_{\Delta}(C|D)$ *is the discernibility relation regarding classification property* Δ. *The set of rows in* M^{Δ}_{BDM} *is presented by* $R = \{r_1, r_2, ..., r_{|R|}\}$, *where* $|R| = (|U| \times (|U| - 1))/2$. *The set of columns in* M^{Δ}_{BDM} *is presented by* $C = \{a_1, a_2, ..., a_{|C|}\}$, *where* $|C|$ *is the cardinality of attribute sets in a decision system. For convenience, a general binary discernibility matrix* $M^{\Delta}_{BDM} = (m^{\Delta}_{BDM}(x_i, x_j))$ *can be also denoted by* $M^{\Delta}_{BDM} = (m^{\Delta}_{BDM}(e^m_p, e^m_q))$. *For* $\forall a_m \in C$, $\forall r_p, r_q \in R$, e^m_p *is the matrix element at row* r_p *and column* a_m *in* M^{Δ}_{BDM}, *and* e^m_q *is the matrix element at row* r_q *and column* a_m *in* M^{Δ}_{BDM}, *where* $1 \leq p < q \leq (|U| \times (|U| - 1))/2, 1 \leq m \leq |C|$.

Since a general binary discernibility matrix provides a common structure of binary discernibility matrices in rough set theory, one can construct a binary discernibility matrix according to a given classification property. Any binary discernibility matrix can be also regarded as the special case of the general binary discernibility matrix. Therefore, a general definition of binary discernibility matrix is necessary and important. Based on the relative discernibility relation with respect to D, Definition 2 can be also rewritten as follows:

Definition 4. *Given a decision system* $DS = (U, C \cup D, V, f)$, *for* $\forall x_i, x_j \in U$, $\forall a_m \in C$, $DIS(C|D)$ *is the relative discernibility relation with respect to a condition attribute set C.* $M_{BDM} = (m_{BDM}(x_i, x_j))$ *is a binary discernibility matrix, in which the element* $m_{BDM}(x_i, x_j)$ *is denoted by:*

$$m_{BDM}(x_i, x_j) = \begin{cases} 1 & (x_i, x_j) \in DIS(C|D) \\ 0 & otherwise \end{cases}.$$

This definition is equivalent to Definition 2 [10]. It is noted that we calculate binary discernibility matrix in this paper by using the relationship preservation discernibility matrix.

Definition 5. *For* $\forall a_m \in C$, $\forall r_p, r_q \in R$, e^m_p *and* e^m_q *are elements in a binary discernibility matrix* $M_{BDM} = (m_{BDM}(x_i, x_j))$, *a row pair with respect to attribute* a_m *is denoted by* $< e^m_p, e^m_q > \in \{<0,0>, <0,1>, <1,0>, <1,1>\}$, *a binary relation between row* r_p *and* r_q *is defined as* $R_{row} = \{< e^m_p, e^m_q >: 1 \leq p < q \leq |R|\}$.

Similar to Definition 5, we define a column pair and a binary relation with respect to columns as follows.

Definition 6. *For $\forall a_m, a_n \in C$, $\forall r_p \in R$, elements e_p^m and e_p^n in a binary discernibility matrix $M_{BDM} = (m_{BDM}(x_i, x_j))$, a column pair with respect to row r_p is denoted by $< e_p^m, e_p^n > \in \{ < 0,0 >, < 0,1 >, < 1,0 >, < 1,1 > \}$, a binary relation between column a_m and a_n is defined as $R_{col} = \{ < e_p^m, e_p^n >: 1 \leq m < n \leq |C| \}$.*

For the matrix element e_p^m and e_q^m in the same column a_m, we define three row relations in a binary discernibility matrix as follows.

Definition 7. *Given a binary discernibility matrix $M_{BDM} = (m_{BDM}(x_i, x_j))$, $\forall r_p, r_q \in R$,*

(1) *for $\forall a_m \in C$, $\exists a_n \in C$, $r_p \supset r_q$ if and only if $e_p^m + e_q^m = e_p^m$ and $e_p^n \neq e_q^n$; for $\forall a_m \in C$, $\exists a_n \in C$, $r_q \supset r_p$ if and only if $e_q^m + e_p^m = e_q^m$ and $e_q^n \neq e_p^n$;*
(2) *for $\forall a_m \in C$, $r_p = r_q$ if and only if $e_p^m = e_q^m$;*
(3) *for $\exists a_m, a_n \in C$, $r_p \neq r_q$ if and only if $e_p^m + e_q^m = e_q^m (e_p^m \neq e_q^m)$ and $e_p^n + e_q^n = e_p^n (e_p^n \neq e_q^n)$.*

Analogous to Definition 7, for matrix elements e_p^m and e_p^n in the same row r_p, we define column relations in a binary discernibility matrix as follows.

Definition 8. *Given a binary discernibility matrix $M_{BDM} = (m_{BDM}(x_i, x_j))$, $\forall a_m, a_n \in C$,*

(1) *for $\forall r_p \in R$, $\exists r_q \in R$, $a_m \supset a_n$ if and only if $e_p^m + e_p^n = e_p^m$ and $e_q^m \neq e_q^n$; for $\forall r_p \in R$, $\exists r_q \in R$, $a_n \supset a_m$ if and only if $e_p^n + e_p^m = e_p^n$ and $e_q^n \neq e_q^m$;*
(2) *for $\forall r_p \in R$, $a_m = a_n$ if and only if $e_p^m = e_p^n$;*
(3) *for $\exists r_p, r_q \in R$, $a_m \neq a_n$ if and only if $e_p^m + e_p^n = e_p^m (e_p^m \neq e_p^n)$ and $e_q^m + e_q^n = e_q^n (e_q^m \neq e_q^n)$.*

Let A_p be the elements' set of a prime implicant in a disjunctive normal form with row r_p and A_q be the elements' set of a prime implicant in a disjunctive normal form with row r_q, then $r_p \supset r_q$ means that A_p is the superset of A_q. For $\forall a_m, a_n \in C$, $a_m \supset a_n$ indicates attribute a_m can distinguish more objects in the universe. In a binary discernibility matrix, the row in which all elements are 0s indicates there are no attribute can discern the related objects, and the column in which all elements are 0s indicates that this attribute cannot discern objects in the universe.

In [17], Zhi and Miao first proposed an algorithm of a binary discernibility matrix simplification shown in Algorithm 1. To improve the efficiency of BDMR, Wang et al. [32] introduced an improved algorithm of binary discernibility matrix reduction shown in Algorithm 2.

Algorithm 1 : An algorithm of binary discernibility matrix reduction, BDMR.

Input: Original binary discernibility matrix M_{BDM};
Output: Simplified binary discernibility matrix M_{BDM}'
  1: delete the row in which all elements are 0 s;
  2: **for** $p = 1$ **to** $|R|$ **do**
  3:      **for** $q = 1$ **to** $|R|$ **do**
  4:          **if** $r_p \supset r_q$ **then**
  5:              delete row r_p
  6:              break
  7:          **end if**
  8:      **end for**
  9: **end for**
10: delete the column in which all elements are 0 s;
11: **for** $m = 1$ **to** $|C|$ **do**
12:      **for** $n = 1$ **to** $|C|$ **do**
13:          **if** $a_n \supset a_m$ **then**
14:              delete column a_m
15:              break
16:          **end if**
17:      **end for**
18: **end for**
19: output a simplified binary discernibility matrix M_{BDM}';

Algorithm 2 : An improved algorithm of binary discernibility matrix reduction, IBDMR.

Input: Original binary discernibility matrix M_{BDM};
Output: Simplified binary discernibility matrix M'_{BDM}

1: delete the row in which all elements are 0 s;
2: sort rows in ascending order by the quantity of the number '1' in each row
3: **for** $p = 1$ to $|R|$ **do**
4:      **for** $q = 1$ to $|R|$ **do**
5:          **if** $r_p \supset r_q$ **then**
6:              delete row r_p
7:              break
8:          **end if**
9:      **end for**
10: **end for**
11: delete the column in which all elements are 0 s;
12: **for** $m = 1$ to $|C|$ **do**
13:      **for** $n = 1$ to $|C|$ **do**
14:          **if** $a_n \supset a_m$ **then**
15:              delete column a_m
16:              break
17:          **end if**
18:      **end for**
19: **end for**
20: output a simplified binary discernibility matrix M'_{BDM};

---

4. A Quick Algorithm for Binary Discernibility Matrix Simplification

In this section, we investigate two theorems related to row relations and column relations respectively. Based on the two theorems, deterministic finite automata for row and column relations are introduced. Deterministic finite automata can be carried out to obtain the row relations and column relations quickly. By using deterministic finite automata, we propose an algorithm of binary discernibility matrix simplification using deterministic finite automata (BDMSDFA).

Theorem 1. *Let $M_{BDM} = (m_{BDM}(x_i, x_j))$ be a binary discernibility matrix, for $\forall a_m \in C$, $\forall r_p, r_q \in R$, $|R| = (|U| \times (|U| - 1))/2$, we have:*

(1)   *if $r_p \supset r_q$, then there exists $< 1, 0 > \in \{< e_p^m, e_q^m >: 1 \leq p < q \leq |R|\}$;*
(2)   *if $r_q \supset r_p$, then there exists $< 0, 1 > \in \{< e_p^m, e_q^m >: 1 \leq p < q \leq |R|\}$;*
(3)   *if $r_p = r_q$, then there exists $< 0, 0 > \in \{< e_p^m, e_q^m >: 1 \leq p < q \leq |R|\}$ or*
$$< 1, 1 > \in \{< e_p^m, e_q^m >: 1 \leq p < q \leq |R|\};$$
(4)   *if $r_p \neq r_q$, then there exists $< 1, 0 > \in \{< e_p^m, e_q^m >: 1 \leq p < q \leq |R|\}$ and*
$$< 0, 1 > \in \{< e_p^m, e_q^m >: 1 \leq p < q \leq |R|\}.$$

Proof.

(1)   If there does not exist $< 1, 0 > \in \{< e_p^m, e_q^m >: 1 \leq p < q \leq |R|\}$, then $< 0, 0 > \in \{< e_p^m, e_q^m >: 1 \leq p < q \leq |R|\}$ or $< 0, 1 > \in \{< e_p^m, e_q^m >: 1 \leq p < q \leq |R|\}$ or $< 1, 1 > \in \{< e_p^m, e_q^m >: 1 \leq p < q \leq |R|\}$. We have seven binary relations as follows: $\{< 0, 0 >\}$, $\{< 0, 1 >\}$, $\{< 1, 1 >\}$, $\{< 0, 0 >, < 0, 1 >\}$, $\{< 0, 0 >, < 1, 1 >\}$, $\{< 0, 1 >, < 1, 1 >\}$ and $\{< 0, 0 >, < 0, 1 >, < 1, 1 >\}$. From seven binary relations above, if $\forall a_m \in C$, $\exists a_n \in C$, one cannot get $e_p^m + e_q^m = e_p^m (e_p^m \neq e_q^n)$. Thus, there exists $< 1, 0 > \in \{< e_p^m, e_q^m >: 1 \leq p < q \leq |R|\}$ in M_{BDM}.

(2)   If there does not exist $< 0, 1 > \in \{< e_p^m, e_q^m >: 1 \leq p < q \leq |R|\}$, then $< 0, 0 > \in \{< e_p^m, e_q^m >: 1 \leq p < q \leq |R|\}$ or $< 1, 0 > \in \{< e_p^m, e_q^m >: 1 \leq p < q \leq |R|\}$ or $< 1, 1 > \in \{< e_p^m, e_q^m >:$

$1 \leq p < q \leq |R|\}$. Thus, we can also have seven binary relations as follows: $\{<0,0>\}$, $\{<1,0>\}$, $\{<1,1>\}$, $\{<0,0>,<1,0>\}$, $\{<0,0>,<1,1>\}$, $\{<1,0>,<1,1>\}$ and $\{<0,0>,<1,0>,<1,1>\}$. From seven binary relations above, if $\forall a_m \in C$ and $\exists a_n \in C$, one cannot get $e_p^m + e_q^m = e_q^m (e_q^n \neq e_p^n)$. Thus, there exists $<0,1> \in \{<e_p^m, e_q^m>: 1 \leq p < q \leq |R|\}$ in M_{BDM}.

(3) If there does not exist $<0,0> \in \{<e_p^m, e_q^m>: 1 \leq p < q \leq |R|\}$ or $<1,1> \in \{<e_p^m, e_q^m>: 1 \leq p < q \leq |R|\}$. There must have seven binary relations as follows: $\{<0,1>\}$, $\{<1,0>\}$, $\{<0,1>,<1,0>\}$, $\{<0,0>,<0,1>\}$, $\{<0,0>,<1,0>\}$, $\{<1,1>,<0,1>\}$, $\{<1,1>,<1,0>\}$. From seven binary relations above, we cannot have $e_p^m = e_q^m$. Thus, there exists $<0,0> \in \{<e_p^m, e_q^m>: 1 \leq p < q \leq |R|\}\}$ or $<1,1> \in \{<e_p^m, e_q^m>: 1 \leq p < q \leq |R|\}\}$ in M_{BDM}.

(4) If there does not exist $<1,0> \in \{<e_p^m, e_q^m>: 1 \leq p < q \leq |R|\}$ and $<0,1> \in \{<e_p^m, e_q^m>: 1 \leq p < q \leq |R|\}$. We may obtain eleven binary relations as follows: $\{<0,0>\}$, $\{<1,1>\}$, $\{<0,1>\}$, $\{<1,0>\}$, $\{<0,0>,<0,1>\}$, $\{<0,0>,<1,0>\}$, $\{<1,1>,<0,1>\}$, $\{<1,1>,<1,0>\}$, $\{<0,0>,<1,1>\}$, $\{<0,0>,<1,1>,<0,1>\}$, $\{<0,0>,<1,1>,<1,0>\}$. From eleven binary relations, for $\forall a_m \in C$, we cannot have $e_p^m \neq e_q^m$. Thus, there exists $<1,0> \in \{<e_p^m, e_q^m>: 1 \leq p < q \leq |R|\}$ and $<0,1> \in \{<e_p^m, e_q^m>: 1 \leq p < q \leq |R|\}$ in M_{BDM}.

This completes the proof. □

Analogous to Theorem 1, we can easily obtain the following theorem as:

Theorem 2. *Let $M_{BDM} = (m_{BDM}(x_i, x_j))$ be a binary discernibility matrix, for $\forall a_m, a_n \in C$, $\forall r_p \in R$, we can have:*

(1) if $a_m \supset a_n$, then there exists $<1,0> \in \{<e_p^m, e_p^n>: 1 \leq m < n \leq |C|\}$;
(2) if $a_n \supset a_m$, then there exists $<0,1> \in \{<e_p^m, e_p^n>: 1 \leq m < n \leq |C|\}$;
(3) if $a_m = a_n$, then there exists $<0,0> \in \{<e_p^m, e_p^n>: 1 \leq m < n \leq |C|\}$ or $<1,1> \in \{<e_p^m, e_p^n>: 1 \leq m < n \leq |C|\}$;
(4) if $a_m \neq a_n$, then there exists $<1,0> \in \{<e_p^m, e_p^n>: 1 \leq m < n \leq |C|\}$ and $<0,1> \in \{<e_p^m, e_p^n>: 1 \leq m < n \leq |C|\}$.

Proof.

(1) If there does not exist $<1,0> \in \{<e_p^m, e_p^n>: 1 \leq m < n \leq |C|\}$, then $<0,0> \in \{<e_p^m, e_p^n>: 1 \leq m < n \leq |C|\}$ or $<0,1> \in \{<e_p^m, e_p^n>: 1 \leq m < n \leq |C|\}$ or $<1,1> \in \{<e_p^m, e_p^n>: 1 \leq m < n \leq |C|\}$. We have seven binary relations as follows: $\{<0,0>\}$, $\{<0,1>\}$, $\{<1,1>\}$, $\{<0,0>,<0,1>\}$, $\{<0,0>,<1,1>\}$, $\{<0,1>,<1,1>\}$ and $\{<0,0>,<0,1>,<1,1>\}$. From seven binary relations above, if $\forall r_p \in R$, $\exists r_q \in R$, one cannot get $e_p^m + e_p^n = e_p^m (e_q^m \neq e_q^n)$. Thus, there exists $<1,0> \in \{<e_p^m, e_p^n>: 1 \leq m < n \leq |C|\}$ in M_{BDM}.

(2) If there does not exist $<0,1> \in \{<e_p^m, e_p^n>: 1 \leq m < n \leq |C|\}$, then $<0,0> \in \{<e_p^m, e_p^n>: 1 \leq m < n \leq |C|\}$ or $<1,0> \in \{<e_p^m, e_p^n>: 1 \leq m < n \leq |C|\}$ or $<1,1> \in \{<e_p^m, e_p^n>: 1 \leq m < n \leq |C|\}$. Thus, we can also have seven binary relations as follows: $\{<0,0>\}$, $\{<1,0>\}$, $\{<1,1>\}$, $\{<0,0>,<1,0>\}$, $\{<0,0>,<1,1>\}$, $\{<1,0>,<1,1>\}$ and $\{<0,0>,<1,0>,<1,1>\}$. From seven binary relations above, if $\forall r_p \in R$ and $\exists r_q \in R$, one cannot get $e_p^m + e_p^n = e_p^n (e_q^m \neq e_q^n)$. Thus, there exists $<0,1> \in \{<e_p^m, e_p^n>: 1 \leq m < n \leq |C|\}$ in M_{BDM}.

(3) If there does not exist $<0,0> \in \{<e_p^m, e_p^n>: 1 \leq m < n \leq |C|\}$ or $<1,1> \in \{<e_p^m, e_p^n>: 1 \leq m < n \leq |C|\}$. There must have seven binary relations as follows: $\{<0,1>\}$, $\{<1,0>\}$, $\{<0,1>,<1,0>\}$, $\{<0,0>,<0,1>\}$, $\{<0,0>,<1,0>\}$, $\{<1,1>,<0,1>\}$, $\{<1,1>,<1,0>\}$. From seven binary relations above, we cannot have $e_p^m = e_p^n$. Thus,

there must exists $< 0,0 >\in \{< e_p^m, e_p^n >: 1 \leq m < n \leq |C|\}$ or $< 1,1 >\in \{< e_p^m, e_p^n >: 1 \leq m < n \leq |C|\}$ in M_{BDM}.

(4) If there does not exist $< 1,0 >\in \{< e_p^m, e_p^n >: 1 \leq m < n \leq |C|\}$ and $< 0,1 >\in \{< e_p^m, e_p^n >: 1 \leq m < n \leq |C|\}$. We may obtain eleven binary relations as follows: $\{< 0,0 >\}$, $\{< 1,1 >\}$, $\{< 0,1 >\}$, $\{< 1,0 >\}$, $\{< 0,0 >, < 0,1 >\}$, $\{< 0,0 >, < 1,0 >\}$, $\{< 1,1 >, < 0,1 >\}$, $\{< 1,1 >, < 1,0 >\}$, $\{< 0,0 >, < 1,1 >\}$, $\{< 0,0 >, < 1,1 >, < 0,1 >\}$, $\{< 0,0 >, < 1,1 >, < 1,0 >\}$. From eleven binary relations, for $\forall r_p \in R$, we cannot have $e_p^m \neq e_p^n$. Thus, there exists $< 1,0 >\in \{< e_p^m, e_p^n >: 1 \leq m < n \leq |C|\}$ and $< 0,1 >\in \{< e_p^m, e_p^n >: 1 \leq m < n \leq |C|\}$ in M_{BDM}.

This completes the proof. □

Deterministic finite automaton, also called deterministic finite acceptor, is an important concept in theory of computation. A deterministic finite automaton constructs a finite-state machine which can accept or reject symbol strings, and produce a computation of automation for each input string. In what follows, we adopt deterministic finite automata to obtain row relations and column relations in a binary discernibility matrix. Here, we first review the definition of deterministic finite automaton as follows.

Definition 9. *A deterministic finite automaton is a 5-tuple $(Q, \sum, \delta, S_0, F)$, where Q is a finite nonempty set of states, $\sum$ is a finite set of input symbols, δ is a transition function, $S_0 \in Q$ is a start state, F is a set of accept states.*

Regarding object pair '$e_p^m e_q^m$' as the basic granule in input symbols, a deterministic finite automaton for row relations in a binary discernibility matrix is illustrated by the following theorem:

Theorem 3. *A deterministic finite automaton for row relations, denoted by DFA_{row}, is a 5-tuple $(Q, \sum, \delta, S_0, F)$, where $Q = \{S_0, S_1, S_2, S_3, S_4\}$ is a finite set of states, $\sum = \{e_p^0 e_q^0 e_p^1 e_q^1 \ldots e_p^m e_q^m e_p^{m+1} e_q^{m+1} \ldots e_p^{|C|} e_q^{|C|}\} (1 < m < |C|, r_p, r_q \in R)$ is an input binary character string, δ is a transition function, $S_0 \in Q$ is a start state, $F = \{S_1, S_2, S_3, S_4\}$ is a set of accept states. A deterministic finite automaton for row relations can be illustrated in Figure 1 as follows.*

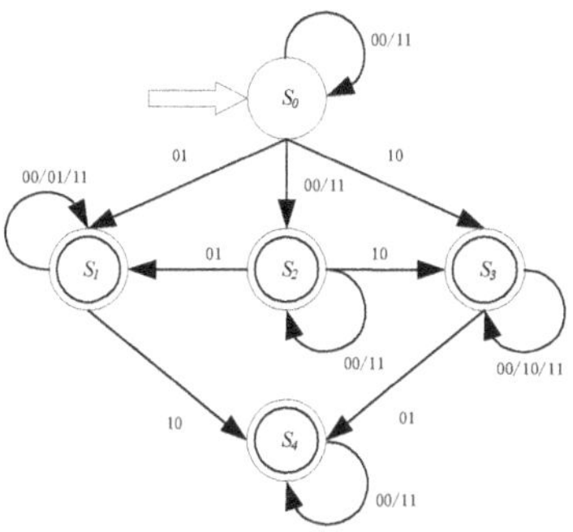

Figure 1. A Deterministic Finite Automaton for Row Relations.

Proof. In a binary discernibility matrix, relations between r_p and r_q can be concluded as $r_q \supset r_p$, $r_p = r_q$, $r_p \supset r_q$ and $r_p \neq r_q$.

We discuss a deterministic finite automaton for row relations from four parts separately, as follows.

(1) According to Definition 5 and Theorem 1, for $\forall a_m \in C$, $\forall r_p, r_q \in R$, there must be $< 0,1 >\in \{< e_p^m, e_q^m >: 1 \leq p < q \leq |R|\}$. Thus, the regular expression for $r_q \supset r_p$ can be defined as

$[(00/11)^*(01)^+(00/11)^*]^+$. We can easily have the corresponding deterministic finite automaton in Figure 2 as:

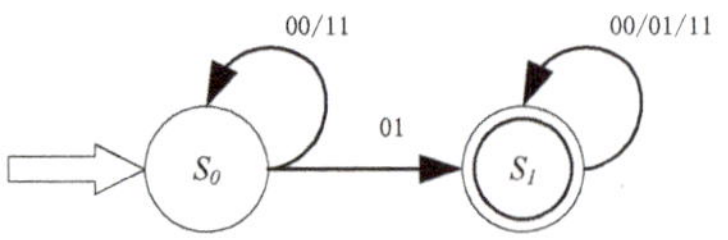

Figure 2. A Deterministic Finite Automaton for $r_q \supset r_p$.

(2) For $r_p = r_q$, there must be $< 0,0 > \in \{< e_p^m, e_q^m >: 1 \leq p < q \leq |R|\}$ or $< 1,1 > \in \{< e_p^m, e_q^m >: 1 \leq p < q \leq |R|\}$. The regular expression for $r_p = r_q$ is denoted by $(00/11)^+$. So, the corresponding deterministic finite automaton can be illustrated in Figure 3 as:

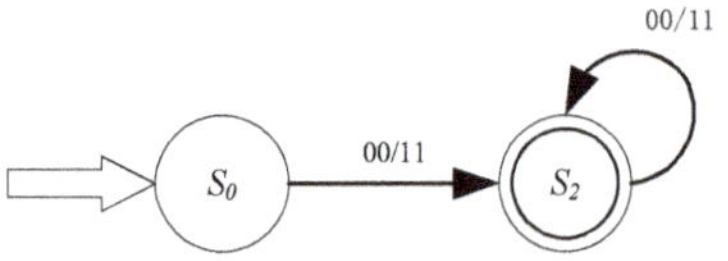

Figure 3. A Deterministic Finite Automaton for $r_p = r_q$.

(3) Analogous to $r_p \supset r_q$, for $r_p, r_q \in R$, there must be $< 1,0 > \in \{< e_p^m, e_q^m >: 1 \leq p < q \leq |R|\}$. Therefore, the regular expression for $r_p \supset r_q$ can be obtained as $[(00/11)^*(10)^+(00/11)^*]^+$. We can easily have the corresponding deterministic finite automaton in Figure 4 as:

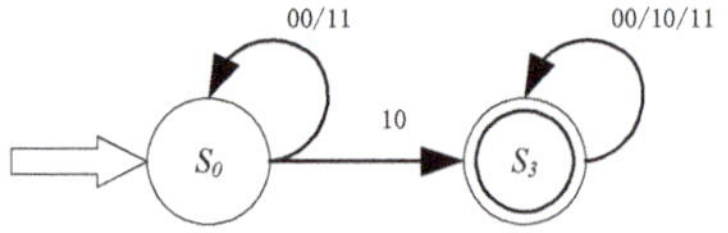

Figure 4. A Deterministic Finite Automaton for $r_p \supset r_q$.

(4) For $r_p \neq r_q$, there must be $< 1,0 > \in \{< e_p^m, e_q^m >: 1 \leq p < q \leq |R|\}$ and $< 0,1 > \in \{< e_p^m, e_q^m >: 1 \leq p < q \leq |R|\}$. The regular expression for $r_p \neq r_q$ is denoted by $[(00/11)^*(01)^+(00/11)^*(10)^+(00/11)^*]^+/[(00/11)^*(10)^+(00/11)^*(01)^+(00/11)^*]^+$. Hence, the corresponding deterministic finite automaton can be illustrated in Figure 5 as:

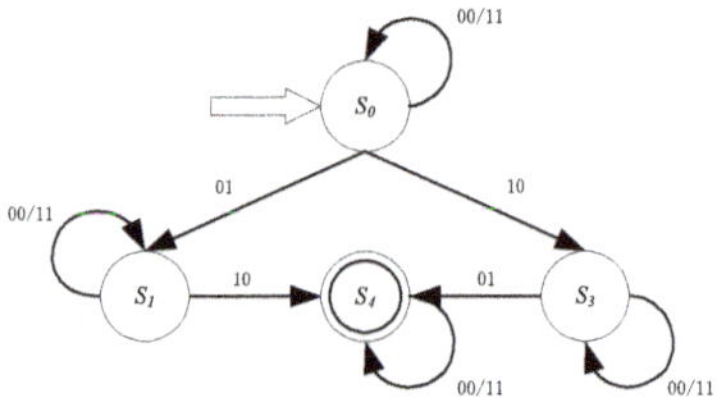

Figure 5. A Deterministic Finite Automaton for $r_p \neq r_q$.

One can construct a deterministic finite automaton for row relations by four deterministic finite automata shown in Figure 1.

This completes the proof. □

Similar to the deterministic finite automaton for row relations, we present the deterministic finite automaton for column relations in a binary discernibility matrix as follows.

Theorem 4. *A deterministic finite automaton for column relations DFA_{col} is a 5-tuple $(Q, \Sigma, \delta, S_0, F)$, where $Q = \{S_0, S_1, S_2, S_3, S_4\}$ is a finite set of states, $\Sigma = \{e_0^m e_0^n e_1^m e_1^n \ldots e_p^m e_p^n e_{p+1}^m e_{p+1}^n \ldots e_{|R|}^m e_{|R|}^n\}$ ($1 < p < |R|, \forall a_m, a_n \in C$) is an input binary character string, δ is a transition function, $S_0 \in Q$ is a start state. $F = \{S_1, S_2, S_3, S_4\}$ is a set of accept states. A deterministic finite automaton for column relations can be illustrated in Figure 6 as follows:*

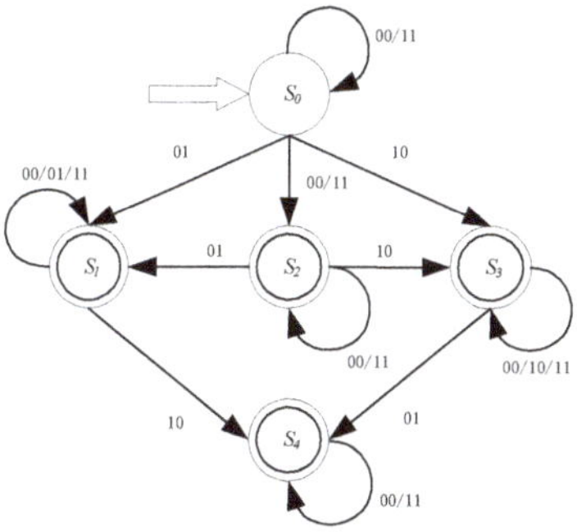

Figure 6. A Deterministic Finite Automaton for Column Relations.

Proof. This proof is similar to the proof of Theorem 3. □

By means of the proposed deterministic finite automata for row and column relations, we propose a quick algorithm for binary discernibility matrix simplification using deterministic finite automata (BDMSDFA) as follows:

We present the following example to explain Algorithm 3 as follows.

Algorithm 3 : A quick algorithm for binary discernibility matrix simplification using deterministic finite automata, BDMSDFA.

Input: Original binary discernibility matrix M_{BDM};
Output: Simplified binary discernibility matrix M'_{BDM}

1: delete the row in which all elements are 0 s;
2: compare the row relation between r_p and r_q by DFA_{row}
3: **for** $p = 1$ to $|R|$ **do**
4:     **for** $q = 1$ to $|R|$ **do**
5:         **if** $r_p \supset r_q$ **then**
6:             delete row r_p from M_{BDM}
7:             break
8:         **end if**
9:     **end for**
10: **end for**
11: delete the column in which all elements are 0 s;
12: compare the column relation between a_m and a_n by DFA_{col}
13: **for** $m = 1$ to $|C|$ **do**
14:     **for** $n = 1$ to $|C|$ **do**
15:         **if** $a_n \supset a_m$ **then**
16:             delete column a_m from M_{BDM}
17:             break.
18:         **end if**
19:     **end for**
20: **end for**
21: output a simplified binary discernibility matrix M'_{BDM};

Example 1. *Let $DS = (U, C \cup D, V, f)$ be a decision system shown Table 1, where the universe $U = \{x_1, x_2, x_3, x_4, x_5\}$, the condition attribute set $C = \{a_1, a_2, a_3, a_4\}$, the decision attribute set $D = \{d\}$.*

Table 1. A decision system.

	a_1	a_2	a_3	a_4	d
x_1	1	2	0	0	1
x_2	1	2	0	0	1
x_3	1	2	1	1	2
x_4	1	3	1	1	3
x_5	2	4	1	1	4

For the decision system above, we have the corresponding binary discernibility matrix as follows:

$$M^1_{BDM} = \begin{pmatrix} 0 & 0 & 0 & 0 \\ 0 & 0 & 1 & 1 \\ 0 & 1 & 1 & 1 \\ 1 & 1 & 1 & 1 \\ 0 & 0 & 1 & 1 \\ 0 & 1 & 1 & 1 \\ 1 & 1 & 1 & 1 \\ 0 & 1 & 0 & 0 \\ 1 & 1 & 0 & 0 \\ 1 & 1 & 0 & 0 \end{pmatrix}.$$

We delete the row in which all elements are 0 s in M^1_{BDM}, and obtain the binary discernibility matrix M^2_{BDM} as follows.

$$M^2_{BDM} = \begin{pmatrix} 0 & 0 & 1 & 1 \\ 0 & 1 & 1 & 1 \\ 1 & 1 & 1 & 1 \\ 0 & 0 & 1 & 1 \\ 0 & 1 & 1 & 1 \\ 1 & 1 & 1 & 1 \\ 0 & 1 & 0 & 0 \\ 1 & 1 & 0 & 0 \\ 1 & 1 & 0 & 0 \end{pmatrix}.$$

In the binary discernibility matrix M^2_{BDM}, $r_1 : 0011$, $r_2 : 0111$, $r_3 : 1111$, $r_4 : 0011$, $r_5 : 0111$, $r_6 : 1111$, $r_7 : 0100$, $r_8 : 1100$, $r_9 : 1100$. According to the definition of the deterministic finite automaton for row relations, we have $\sum_{12} = 00011111$, $\sum_{13} = 01011111$, $\sum_{14} = 00001111$, $\sum_{15} = 00011111$, $\sum_{16} = 01011111$, $\sum_{17} = 00011010$, $\sum_{18} = 01011010$, $\sum_{19} = 01011010$. By using the deterministic finite automaton for row relations shown in Figure 1, we can get the row relations as follows. $r_2 \supset r_1$, $r_3 \supset r_1$, $r_4 = r_1$, $r_5 \supset r_1$, $r_6 \supset r_1$, $r_7 \neq r_1$, $r_8 \neq r_1$, $r_9 \neq r_1$. Therefore, we delete r_2, r_3, r_5 and r_6. Similarly, we get $r_8 \supset r_7$ and $r_9 \supset r_7$, $r_4 \neq r_7$, and then delete r_8 and r_9. Therefore, we have the following binary discernibility matrix:

$$M^3_{BDM} = \begin{pmatrix} 0 & 0 & 1 & 1 \\ 0 & 0 & 1 & 1 \\ 0 & 1 & 0 & 0 \end{pmatrix}.$$

We delete the column in which all elements are 0 s in M^3_{BDM}, and have

$$M^4_{BDM} = \begin{pmatrix} 0 & 1 & 1 \\ 0 & 1 & 1 \\ 1 & 0 & 0 \end{pmatrix}.$$

In the binary discernibility matrix M^4_{BDM}, $a_1 : 001$, $a_2 : 110$, $a_3 : 110$. According to the definition of deterministic finite automaton for column relations, we have $\sum_{12} = 010110$, $\sum_{13} = 010110$, $\sum_{23} = 111100$. By using the deterministic finite automaton for column relations shown in Figure 6, we have $a_1 \neq a_2$, $a_1 \neq a_3$, $a_2 = a_3$. Thus, we cannot delete any column in M^4_{BDM}, and get the following binary discernibility matrix.

$$M^5_{BDM} = \begin{pmatrix} 0 & 1 & 1 \\ 0 & 1 & 1 \\ 1 & 0 & 0 \end{pmatrix}.$$

A 10×4 matrix M^1_{BDM} is compressed to a 3×3 matrix M^5_{BDM}. The simplified binary discernibility matrix with fewer objects or columns will be help in improving the efficiency of attribute reduction.

Assume that $t = |R|$ and $s = |C|$, the upper bound of time complexity of BDMR is $3ts(t + s - 2)$, the lower bound of time complexity of BDMR is $2ts(t + s - 2)$. The upper bound of time complexity of IBDMR is $3ts(t + s - 2)$ and the lower bound of the worst-case time complexity of IBDMR is $2ts(t + s - 2)$. By employing deterministic finite automata, the algorithm complexity of BDMSDFA is $ts(t + s - 2)$. Obviously, the time complexity of BDMSDFA is lower than that of BDMR and IBDMR. Therefore, it is concluded that the proposed algorithm BDMSDFA reduces the computational time for binary discernibility matrix simplification in general.

The advantages of the proposed method are expressed as follows. (1) Deterministic finite automata in a binary discernibility matrix are constructed, it can provide an understandable approach to comparing the relationships of different rows (columns) quickly. (2) Based on deterministic finite automata, a high efficiency algorithm of binary discernibility matrix simplification is developed. Theoretical analyses and experimental results indicate that the proposed algorithm is effective and efficient. It should be noted that the proposed method is based on Pawlak decision systems, but not suitable for generalized decision systems, such as incomplete decision systems, interval-valued decision systems and fuzzy decision systems. Deterministic finite automata in generalized decision systems will be investigated in the future.

5. Experimental Results and Analyses

The objective of the following experiments in this section is to demonstrate the high efficiency of the algorithm BDMSDFA. The experiments are divided into two aspects. In one aspect, we employ 10 datasets in Table 2 to verify the performance of time consumption of BDMR, IBDMR and BDMSDFA. In the other aspect, the computational times of algorithms BDMR, IBDMR and BDMSDFA with the increase of the size of attributes (or objects) are calculated respectively. We carry out three algorithms on a personal computer with Windows 8.1 (64 bit) and Inter(R) Core(TM) i5-4200U, 1.6 GHz and 4 GB memory. The software is Microsoft Visual Studio 2017 version 15.9 and C++. Data sets used in the experiments are all downloaded from UCI repository of machine learning data sets (http://archive.ics.uci.edu/ml/datasets.html).

Table 2 indicates the computational time of BDMR, IBDMR and BDMSDFA on the 10 data sets. We can see that the algorithm BDMSDFA is much faster than the algorithms BDMR and IBDMR. The computational times of three algorithms follows this order: BDMR $\geq$ IBDMR $>$ BDMSDFA. The computational time of BDMSDFA is the minimum among the three algorithms. For the data set Auto in Table 2, the computational times of BDMR and IBDMR are 75 ms and 68 ms, while that of BDMSDFA is 36 ms. For the data set Credit_a, the computational times of BDMR and IBDMR are

113 ms and 105 ms, while that of BDMSDFA is 55 ms. For some data sets in Table 2, the computational time of BDMSDFA can reduce over half the computational time of BDMR or IBDMR. In Table 2, for the data set Breast_w, the computational times of BDMR and IBDMR are 75 ms and 73 ms, while that of BDMSDFA is 29 ms. For the data set Promoters, the computational times of BDMR and IBDMR are 1517 ms and 936 ms, while that of BDMSDFA is only 398 ms. For the date sets such as Lung-cancer, Credit_a, Breast_w, Anneal, the computational time of BDMR is close to that of IBDMR. For the data set Labor_neg, the computational time of BDMR is equivalent to that of IBDMR. For each data set in Table 2, difference between BDMR and IBDMR is relatively smaller than difference between BDMR (IBDMR) and BDMSDFA.

Table 2. Time consumption of BDMR, IBDMR and BDMSDFA.

Data Sets	Num. of Objects	Num. of Attributes	Num. of Rows	Num. of Columns	Time of BDMR (ms)	Time of IBDMR (ms)	Time of BDMSDFA (ms)
Labor_neg	40	15	52	14	2	2	1
Lung-cancer	32	57	206	53	11	9	3
Heart_statlog	270	14	96	13	32	27	14
Autos	250	26	36	19	75	68	36
Credit_a	690	16	27	12	113	105	55
Breast_w	699	10	20	9	75	73	29
Anneal	898	39	9	9	206	194	134
Promoters	106	58	2761	57	1517	936	398
Dermatology	366	35	1347	31	3239	2638	1318
Connect_4	67,557	43	697	42	2,444,328	2,433,863	1,759,917

We compare the computational times of BDMR, IBDMR and BDMSDFA with the increase of the size of objects. In Figure 7a–f, the x-coordinate pertains to the size of objects in the universe, while the y-coordinate concerns the time consumption of algorithms. We employ 6 data sets (Dermatlogy, Credit_a, Controceptive_Method_Choice, Letter, Flag and Mushroom) to verify the performance of time consumption of BDMR, IBDMR and BDMSDFA. When dealing with the same UCI data sets, the computational time of BDMSDFA is less than that of BDMR and IBDMR, in other words, BDMSDFA is more efficient than BDMR and IBDMR. Figure 7 shows more detailed change trends of each algorithm with the number of objects increasing. The computational times of three algorithms increase with the increase of the number of objects simultaneously. It is obvious to see that the slope of the curve of BDMSDFA is smaller than the curve of BDMR or IBDMR, and the computational time of BDMSDFA increases slowly. The differences between BDMR (IBDMR) and BDMSDFA become distinctly larger when the size of the objects increases. In Figure 7c, the difference of BDMR (IBDMR) and BDMSDFA is not obviously different at the beginning. The computational time of DBMR (IBDMR) increases distinctly when the number of objects is over 450. The computational time of algorithm BDMR increases by 479 ms when the number of objects rises from 450 to 1473, whereas the computational time of algorithm BDMSDFA increases by only 141 ms. In Figure 7e, the computational time of the algorithm IBDMR increases by 104 ms when the number of objects rises from 20 to 160, whereas the time consumption of algorithm BDMSDFA increases by only 49.

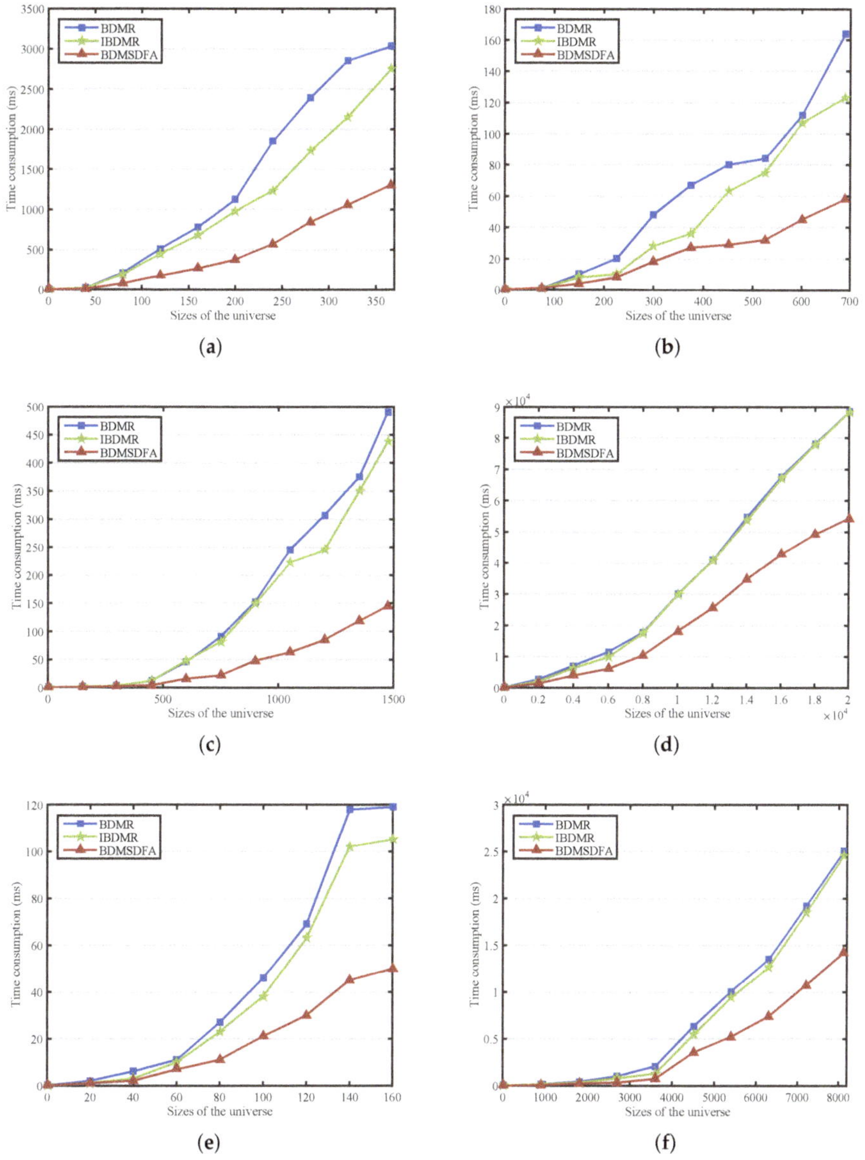

Figure 7. (**a**) Dermatlogy; (**b**) Credit_a; (**c**) Controceptive_Method_Choice; (**d**) Letter; (**e**) Flag; (**f**) Mushroom.

In Figure 8a–f, the x-coordinate pertains to the size of attributes, while the y-coordinate concerns the time consumption of algorithms. We also take 6 data sets (Dermatlogy, Credit_a, Controceptive_Method_Choice, Letter, Flag and Mushroom) to verify the performance of the computational times of BDMR, IBDMR and BDMSDFA. The curve of BDMR is similar to that of IBDMR. The curve of BDMSDFA is under the curves of BDMR and IBDMR. Then, the computational time of BDMSDFA is less than that of BDMR or IBDMR. In Figure 8b, the computational time of algorithms BDMR and IBDMR increase by 164 ms and 123 ms respectively, while the computational time of algorithm BDMSDFA increases by 58 ms. In Figure 8c, the curves of BDMR and IBDMR raise profoundly when the size of the attributes increases. In Figure 8e, the computational time of algorithm IBDMR increases from 4 ms to 105 ms when the number of objects rises from 3 to 24, while the computational time of algorithm BDMSDFA increasedly from 2 ms to 50 ms. For Figure 8a–f, it is concluded that the efficiency of BDMSDFA is higher than that of BDMR or IBDMR with the increase of the number of attributes. Difference between BDMR and IBDMR is relatively smaller than difference between BDMR (IBDMR) and BDMSDFA. The computational times of three algorithms increase with the increase of the number of attributes monotonously. When dealing with the same situation, the computational time of BDMSDFA is the minimum among the three algorithms.

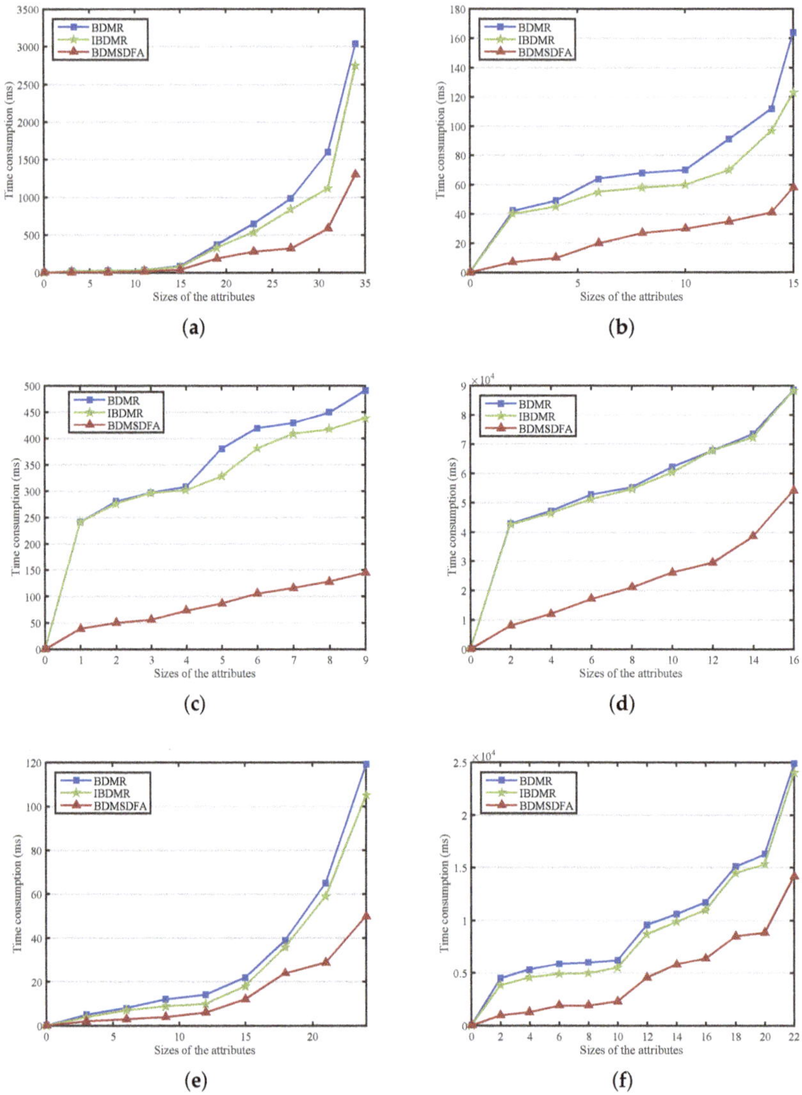

Figure 8. (**a**) Dermatlogy; (**b**) Credit_a; (**c**) Controceptive_Method_Choice; (**d**) Letter; (**e**) Flag; (**f**) Mushroom.

Experimental analyses and results show a high efficiency of the algorithm BDMSDFA. The proposed simplification algorithm using deterministic finite automata can be applied as a preprocessing technique for data compression and attribute reduction in large-scale data sets.

6. Conclusions

Original binary discernibility matrices which are not simplified usually have irrelative objects and attributes. These irrelative objects and attributes may lead to inefficiency in attribute reduction, knowledge acquisition, etc. To tackle this problem, a quick method of comparing the relationships of different rows (columns) are introduced in binary discernibility matrices. By using deterministic finite automata, a quick algorithm for binary discernibility matrix simplification (BDMSDFA) is developed. The experiment results indicate that DBMSDFA can get higher performance in the efficiency of binary discernibility matrix simplification. The contributions of this paper can be summarized as follows.

(1) We define row (or column) relations which are used for constructing deterministic finite automata.
(2) Deterministic finite automata are firstly used for comparing the relationships of different rows (columns) in a binary discernibility matrix.
(3) Based on deterministic finite automata, a quick algorithm for binary discernibility matrix simplification is developed. Experimental results indicate that the relationship between the time consumption of BDMSDFA and the number of objects (attributes) is strictly monotonic. With the increase of the size of objects (attributes), the algorithm BDMSDFA is more efficient than BDMR and IBDMR.

It is noted that the proposed quick simplification algorithm for discernibility matrix is only suitable for completed decision systems. However, in practical applications, there exists many generalized decision systems, such as incomplete decision systems, interval-valued decision systems, etc. Researches on quick simplification algorithms in generalized decision systems will be investigated. Combing the researches on fuzzy sets [33–36], we will propose the fuzzy binary discernibility matrix. Some applications of the (fuzzy) binary discernibility matrix simplification will also be studied in the future.

Author Contributions: N.Z. and B.L. designed the research. N.Z gave theoretical analysis and wrote the whole paper. N.Z. and B.L. designed and conducted the experiments. Z.Z. and Y.G. provided some suggestions on the paper.

Funding: This research was funded by the National Natural Science Foundation of China (No. 61403329, No. 61572418, No. 61502410, No. 61572419).

Acknowledgments: The authors thank the anonymous referees for the constructive comments and suggestions.

Conflicts of Interest: The authors declare no conflict of interest.

References

1. Arora, R.; Garg, H. A robust correlation coefficient measure of dual hesitant fuzzy soft sets and their application in decision making. *Eng. Appl. Artif. Intell.* **2018**, *72*, 80–92. [CrossRef]
2. Garg, H.; Nancy. New Logarithmic operational laws and their applications to multiattribute decision making for single-valued neutrosophic numbers. *Cogn. Syst. Res.* **2018**, *52*, 931–946. [CrossRef]
3. Garg, H.; Nancy. Linguistic single-valued neutrosophic prioritized aggregation operators and their applications to multiple-attribute group decision-making. *J. Ambient Intell. Hum. Comput.* **2018**, *9*, 1975–1997. [CrossRef]
4. Garg, H.; Arora, R. Dual hesitant fuzzy soft aggregation operators and their application in decision making. *Cogn. Comput.* **2018**, *10*, 769–789. [CrossRef]
5. Garg, H.; Kumar, K. An advanced study on the similarity measures of intuitionistic fuzzy sets based on the set pair analysis theory and their application in decision making. *Soft Comput.* **2018**, *22*, 4959–4970. [CrossRef]

6. Garg, H.; Kumar, K. Distance measures for connection number sets based on set pair analysis and its applications to decision-making process. *Appl. Intell.* **2018**, *48*, 3346–3359. [CrossRef]
7. Badi, I.; Ballem, M. Supplier selection using the rough BWM-MAIRCA model: a case study in pharmaceutical supplying in Libya. *Decis. Mak. Appl. Manag. Eng.* **2018**, *1*, 16–33. [CrossRef]
8. Chatterjee, K.; Pamucar, D.; Zavadskas, E.K. Evaluating the performance of suppliers based on using the R'AMATEL-MAIRCA method for green supply chain implementation in electronics industry. *J. Clean. Prod.* **2018**, *184*, 101–129. [CrossRef]
9. Pawlak, Z. Rough sets. *Int. J. Inf. Comput. Sci.* **1982**, *11*, 341–356. [CrossRef]
10. Felix, R.; Ushio, T. Rough sets-based machine learning using a binary discernibility matrix. In Proceedings of the Second International Conference on Intelligent Processing and Manufacturing of Materials, Honolulu, HI, USA, 10–15 July 1999; pp. 299–305.
11. Ding, M.W.; Zhang, T.F.; Ma, F.M. Incremental attribute reduction algorithm based on binary discernibility matrix. *Comput. Eng.* **2017**, *43*, 201–206.
12. Qian, W.B.; Xu, Z.Y.; Yang, B.R.; Huang, L.Y. Efficient incremental updating algorithm for computing core of decision table. *J. Chin. Comput. Syst.* **2010**, *31*, 739–743.
13. Wang, Y.Q.; Fan, N.B. Improved algorithms for attribute reduction based on simple binary discernibility matrix. *Comput. Sci.* **2015**, *42*, 210–215.
14. Zhang, T.F.; Yang, X.X.; Ma, F.M. Algorithm for attribute relative reduction based on generalized binary discernibility matrix. In Proceedings of the 26th Chinese Control and Decision Conference, Changsha, China, 30 May–2 June 2014; pp. 2626–2631.
15. Li, J.; Wang, X.; Fan, X.W. Improved binary discernibility matrix attribute reduction algorithm in customer relationship management. *Procedia Eng.* **2010**, *7*, 473–476. [CrossRef]
16. Tiwari, K.S.; Kothari, A.G.; Keskar, A.G. Reduct generation from binary discernibility matrix: an hardware approach. *Int. J. Future Comput. Commun.* **2012**, *1*, 270–272. [CrossRef]
17. Zhi, T.Y.; Miao, D.Q. The binary discernibility matrix's transformation and high Efficiency attributes reduction algorithm's conformation. *Comput. Sci.* **2002**, *29*, 140–142.
18. Yang, C.J.; Ge, H.; Li, L.S. Attribute reduction of vertically partitioned binary discernibility matrix. *Control Decis.* **2013**, *28*, 563–568.
19. Ren, Q.; Luo, Y.T.; Yao, G.S. An new method for modifying binary discernibility matrix and computation of core. *J. Chin. Comput. Syst.* **2013**, *34*, 1437–1440.
20. Ding, M.W.; Zhang, T.F.; Ma, F.M. Incremental attribute reduction algorithm based on binary discernibility matrix in incomplete information System. *Comput. Sci.* **2017**, *44*, 244–250.
21. Hu, S.P.; Zhang, Q.H.; Yao, L.Y. Effective algorithm for computing attribute core based on binary representation. *Comput. Sci.* **2016**, *43*, 79–83.
22. Skowron, A.; Rauszer, C. The discernibility matrices and functions in information systems. In *Intelligent Decision Support: Handbook of Applications and Advances of the Rough Sets Theory*; Słowiński, R., Ed.; Kluwer Academic Publishers: Dordrecht, The Netherlands, 1992; pp. 331–362.
23. Guan, Y.Y.; Wang, H.K.; Wang, Y.; Yang, F. Attribute reduction and optimal decision rules acquisition for continuous valued information systems. *Inf. Sci.* **2009**, *179*, 2974–2984. [CrossRef]
24. Sun, B.Z.; Ma, W.M.; Gong, Z.T. Dominance-based rough set theory over interval-valued information systems. *Expert Syst.* **2014**, *31*, 185–197. [CrossRef]
25. Qian, Y.H.; Liang, J.Y.; Dang, C.Y. Interval ordered information systems. *Comput. Math. Appl.* **2008**, *56*, 1994–2009. [CrossRef]
26. Qian, Y.H.; Dang, C.Y.; Liang, J.Y.; Tang, D.W. Set-valued ordered information systems. *Inf. Sci.* **2009**, *179*, 2809–2832. [CrossRef]
27. Zhang, H.Y.; Leung, Y.; Zhou, L. Variable-precision-dominance-based rough set approach to interval-valued information systems. *Inf. Sci.* **2013**, *244*, 75–91. [CrossRef]
28. Kryszkiewicz, M. Rough set approach to incomplete information systems. *Inf. Sci.* **1998**, *112*, 39–49. [CrossRef]
29. Leung, Y.; Li, D.Y. Maximal consistent block technique for rule acquisition in incomplete information systems. *Inf. Sci.* **2003**, *153*, 85–106. [CrossRef]
30. Guan, Y.Y.; Wang, H.K. Set-valued information systems. *Inf. Sci.* **2006**, *176*, 2507–2525. [CrossRef]

31. Miao, D.Q.; Zhao, Y.; Yao, Y.Y.; Li, H.; Xu, F. Relative reducts in consistent and inconsistent decision tables of the Pawlak rough set model. *Inf. Sci.* **2009**, *179*, 4140–4150. [CrossRef]
32. Wang, X.; Ma, Y.; Wang, L. Research on space complexity of binary discernibility matrix. *J. Tianjin Univ. Sci. Technol.* **2006**, *21*, 50–53. [CrossRef]
33. Kaur, G.; Garg, H. Generalized cubic intuitionistic fuzzy aggregation operators using t-norm operations and their applications to group decision-making process. *Arab. J. Sci. Eng.* **2018**. [CrossRef]
34. Garg, H. New logarithmic operational laws and their aggregation operators for Pythagorean fuzzy set and their applications. *Int. J. Intell. Syst.* **2019**, *34*, 82–106. [CrossRef]
35. Garg, H.; Singh, S. A novel triangular interval type-2 intuitionistic fuzzy sets and their aggregation operators. *Iran. J. Fuzzy Syst.* **2018**, *15*, 69–93.
36. Garg, H.; Kumar, K. Improved possibility degree method for ranking intuitionistic fuzzy numbers and their application in multiattribute decision making. *Granul. Comput.* **2018**. [CrossRef]

MDPI
St. Alban-Anlage 66
4052 Basel
Switzerland
Tel. +41 61 683 77 34
Fax +41 61 302 89 18
www.mdpi.com

Information Editorial Office
E-mail: information@mdpi.com
www.mdpi.com/journal/information

www.ingramcontent.com/pod-product-compliance
Lightning Source LLC
Chambersburg PA
CBHW051715210326
41597CB00032B/5485
9783038976424